Knut Stamnes and Jakob J. Stamnes

Radiative Transfer in Coupled Environmental Systems

Related Titles

Salam, A.

Molecular Quantum Electrodynamics

Long-Range Intermolecular Interactions

2010
Print ISBN: 978-0-470-25930-6; also available in electronic formats

Ambaum, M.M.

Thermal Physics of the Atmosphere

2010
Print ISBN: 978-0-470-74515-1; also available in electronic formats

Bohren, C.F., Huffman, D.R., Clothiaux, E.E.

Absorption and Scattering of Light by Small Particles

2 Edition

2015
Print ISBN: 978-3-527-40664-7; also available in electronic formats

Wendisch, M., Yang, P.

Theory of Atmospheric Radiative Transfer

A Comprehensive Introduction

2012
Print ISBN: 978-3-527-40836-8; also available in electronic formats

Wendisch, M., Brenguier, J. (eds.)

Airborne Measurements for Environmental Research

Methods and Instruments

2013
Print ISBN: 978-3-527-40996-9; also available in electronic formats

North, G.R., Kim, K.

Energy Balance Climate Models

2013
Print ISBN: 978-3-527-41132-0; also available in electronic formats

Knut Stamnes and Jakob J. Stamnes

Radiative Transfer in Coupled Environmental Systems

An Introduction to Forward and Inverse Modeling

WILEY-VCH
Verlag GmbH & Co. KGaA

Authors

Prof. Knut Stamnes
Stevens Inst. of Technology
Dept. Physics and Eng. Physics
1 Castle Point on Hudson Hoboken
New Jersey 07030
United States

Prof. Jakob J. Stamnes
University of Bergen
Dept. Physics and Technology
P.O. Box 7803
5020 Bergen
Norway

Library of Congress Card No.: applied for

British Library Cataloguing-in-Publication Data
A catalogue record for this book is available from the British Library.

Bibliographic information published by the Deutsche Nationalbibliothek
The Deutsche Nationalbibliothek lists this publication in the Deutsche Nationalbibliografie; detailed bibliographic data are available on the Internet at <http://dnb.d-nb.de>.

Print ISBN: 978-3-527-41138-2
ePDF ISBN: 978-3-527-69662-8
ePub ISBN: 978-3-527-69659-8
Mobi ISBN: 978-3-527-69661-1
oBook ISBN: 978-3-527-69660-4

Cover Design Schulz Grafik-Design, Fußgönheim, Germany
Typesetting SPi Global, Chennai, India
Printing and Binding Markono Print Media Pte Ltd, Singapore

Printed on acid-free paper

Contents

Preface

In a *transparent medium*, it follows from energy conservation that *light attenuation* is proportional to the inverse square of the distance traveled. In a translucent or turbid medium, light suffers attenuation in addition to that given by the *inverse-square law* caused by absorption and scattering, which give an exponential decay in accordance with the *Beer–Lambert law* to a first-order approximation. Application of calculus to interpret experimental results in light attenuation led to the foundation of *radiometry* and the development of appropriate physical theories and mathematical models.

The most important quantity of classical radiative transfer (RT) theory is the *specific intensity* (or *radiance*), defined as the radiant power transported through a surface element in directions confined to a solid angle around the direction of propagation. To treat the *polarization* properties of radiation, Stokes introduced four parameters that describe the state of polarization, which were used by Chandrasekhar to replace the specific intensity with a four-element column vector to describe polarized radiation.

In many applications, an accurate description is required of light propagation in two adjacent slabs of *turbid media* which are separated by an interface, across which the *refractive index* changes. Three important examples of such *coupled media* are atmosphere–water systems, atmosphere–sea ice systems, and air–tissue systems, in which the change in the refractive index across the interface between the two media plays an important role for the transport of light throughout the coupled system. For *imaging* of biological tissues or satellite *remote sensing* of water bodies, an accurate RT model for a coupled system is an indispensable tool. In both cases, an accurate RT tool is essential for obtaining satisfactory solutions of retrieval problems through iterative forward/inverse modeling.

In optical remote sensing of the Earth from space, an important goal is to retrieve atmospheric and surface parameters from measurements of the reflected solar radiation emerging at the top of the atmosphere at a number of wavelengths. These *retrieval parameters*, such as *aerosol type* as well as *loading* and concentrations of aquatic constituents in an open ocean or coastal water area, depend on the *inherent optical properties*, that is, the *scattering* and *absorption* coefficients of the atmosphere and the water. By having a model that provides a link between retrieval parameters and inherent optical properties, one can use a *forward RT model* to compute how the radiation measured by an instrument deployed on a satellite will respond to changes in the retrieval parameters, and then formulate and solve an *inverse RT problem* to derive information about the retrieval parameters. A forward RT model employing inherent optical properties that describes how atmospheric and aquatic constituents absorb and scatter light can be used to compute the *multiply scattered light field* in any particular direction at any particular depth level in a *vertically stratified medium*, such as a coupled atmosphere–water system. In order to solve an inverse RT problem, it is important to have an accurate and efficient forward RT model. Accuracy is important in order to obtain reliable and robust retrievals, and efficiency is an issue because standard iterative solutions of *nonlinear inverse RT problems* require executing a forward RT model repeatedly to compute the radiation field as well as the partial derivatives with respect to the *retrieval parameters* (the *Jacobians*).

This book is aimed at students with a good undergraduate background in mathematics and physics, but it is kept at a fairly fundamental level. It will teach the reader how to formulate and solve forward and inverse RT problems related to coupled media and gives examples of how to solve concrete problems in remote sensing of coupled atmosphere–surface systems. Thus, it is suitable as a teaching tool for the next generation of environmental scientists and engineers and remote-sensing specialists. It will also be useful for researchers (in academia, industry, and government institutions) in atmospheric and planetary sciences as well as in remote sensing of the environment.

This book discusses RT in coupled media such as the atmosphere–ocean system with *Lambertian* and non-Lambertian reflecting ground surfaces for *polarized* as well as *unpolarized radiation*. The *polarized reflectance* from natural ground surfaces such as plant canopies and wind-roughened water surfaces is discussed, and emphasis is placed on the mathematical description of the *inherent optical properties* of natural media including atmospheric gases and particles, water bodies with embedded impurities (particles), and snow/ice bodies. The spectral range from the *ultraviolet* to the *microwave* region of the electromagnetic spectrum is considered, and *multispectral* and *hyperspectral* remote sensing are discussed, as well as solutions of forward problems for unpolarized and polarized radiation in *coupled media*. A unique feature of this book is that it contains a basic description of inverse methods together with a comprehensive and systematic coverage of formulations and solutions of inverse problems related to coupled media.

Knut Stamnes
Hoboken, New Jersey, USA
June 30, 2015
Jakob J. Stamnes
Bergen, Norway
June 30, 2015

Acknowledgments

Many people have helped us with the creation of this book by contributing to the research behind it, providing feedback on various parts of the text, and helping to create graphic illustrations. We are indebted to several students and colleagues for the contributions, comments, and suggestions, including Nan Chen, Dennis Cohen, Lingling Fan, Yongzhen Fan, Øyvind Frette, Karl I. Gjerstad, Børge Hamre, Min He, Jeff Koskulics, Wei Li, Zhenyi Lin, Jon K. Lotsberg, Matteo Ottaviani, Monika Sikand, Endre R. Sommersten, Snorre Stamnes, and Qiang Tang. We are also grateful to Hilding Lindquist, who helped us in getting the permissions required to use material (illustrations) from published sources. The writing started while KS was on a sabbatical leave from Stevens Institute of Technology. During the period of writing, KS was partially supported by the National Aeronautics and Space Administration (NASA) through a grant from NASA's Remote Sensing Theory Program, a contract from the Japan Aerospace Exploration Agency (JAXA), and a grant from the Air Force Research Laboratory. JJS was partially supported by the Norwegian Research Council.

1
Introduction

1.1
Brief History

The idea or notion that *light attenuation* is proportional to the inverse square of the distance traveled can be traced to Kepler [1]. Its experimental verification was provided by Bouguer [2], who used the inverse square dependence to establish the exponential *extinction law* by studying the attenuation of light passing through *translucent media*. A mathematical foundation of *radiometry* was provided by Lambert [3], who used calculus to interpret experimental results and thereby develop appropriate mathematical models and physical theories. As pointed out by Mishchenko [4], although the first introduction of the *radiative transfer equation (RTE)* has traditionally been attributed to Schuster [5], the credit should go instead to Lommel [6], who derived an integral form of the RTE by considering the directional flow of radiant energy crossing a surface element; almost identical results were obtained independently by Chwolson [7].

The *specific intensity* (or *radiance*) $I(\mathbf{r}, \hat{\mathbf{\Omega}})$ is the most important quantity of classical *radiative transfer theory (RTT)*. Planck [8] defined it by stating that the amount of radiant energy dE transported through a surface element dA in directions confined to a *solid angle* $d\omega$ around the direction of propagation $\hat{\mathbf{\Omega}}$ in a time interval dt is given by $dE = I(\mathbf{r}, \hat{\mathbf{\Omega}}) \cos\theta dA dt d\omega$, where $\mathbf{r}$ is the position vector of the surface element dA, and θ is the angle between $\hat{\mathbf{\Omega}}$ and the normal to dA. This definition was adopted in the works of Milne [9], Hopf [10], and Chandrasekhar [11], and has since been used in many monographs [12–16] and textbooks [17–21] on RTT. To treat the polarization properties of radiation Stokes [22] introduced four parameters to describe the state of polarization. These so-called *Stokes parameters* were used by Chandrasekhar [11, 23] to replace the specific intensity with the four-element column vector $\mathbf{I}(\mathbf{r}, \hat{\mathbf{\Omega}})$ to describe polarized radiation.

The heuristic derivation of the RTE adopted in Chapter 3 of this book for *unpolarized* as well as *polarized radiation* is based on classical RTT invoking the specific intensity and simple energy conservation arguments. Such a derivation is easy to understand and sufficient for our purpose. Mandel and Wolf [24] noted that a more fundamental derivation that can be traced to the *Maxwell equations* was

Radiative Transfer in Coupled Environmental Systems: An Introduction to Forward and Inverse Modeling,
First Edition. Knut Stamnes and Jakob J. Stamnes.

desirable, and stated "In spite of the extensive use of the theory of radiative energy transfer, no satisfactory derivation of its basic equation... from electromagnetic theory... has been obtained up to now." Recently, however, much progress toward such a derivation has been made, as reported by Mishchenko [25].

1.2 What is Meant by a Coupled System?

In many applications, an accurate description is required of light propagation in two adjacent slabs of *turbid media* that are separated by an interface, across which the refractive index changes. Such a two-slab configuration will be referred to as a *coupled system*. Three important examples are atmosphere–water systems [26, 27], atmosphere–sea ice systems [28, 29], and air–tissue systems [30]. In each of these three examples, the change in the refractive index across the interface between the two media must be accounted for in order to model the transport of light throughout the respective coupled system correctly. In the second example, the refractive-index change, together with *multiple scattering*, leads to a significant trapping of light inside the strongly scattering, optically thick sea-ice medium [28, 29]. For *imaging* of biological tissues or satellite remote sensing of water bodies, an accurate radiative transfer (RT) model for a coupled system is an indispensable tool [31, 32]. In both cases, an accurate RT tool is essential for obtaining satisfactory solutions of retrieval problems through iterative forward/inverse modeling [33, 34].

In *remote sensing* of the Earth from space, one goal is to retrieve atmospheric and surface parameters from measurements of the radiation emerging at the top of the atmosphere (TOA) at a number of wavelengths [35, 36]. These *retrieval parameters* (RPs), such as aerosol type and loading and concentrations of aquatic constituents in an open ocean or coastal water area, depend on the *inherent optical properties* (IOPs) of the atmosphere and the water. If there is a model providing a link between the RPs and the IOPs, a forward RT model can be used to compute how the measured TOA radiation field should respond to changes in the RPs, and an inverse RT problem can be formulated and solved to derive information about the RPs [37, 38]. A *forward RT model*, employing IOPs that describe how atmospheric and aquatic constituents absorb and scatter light can be used to compute the *multiply scattered light field* in any particular direction (with specified polar and azimuth angles) at any particular depth level (including the TOA) in a vertically *stratified medium*, such as a coupled atmosphere–water system [34, 39]. In order to solve the *inverse RT problem*, it is important to have an accurate and efficient forward RT model. Accuracy is important in order to obtain reliable and robust retrievals, and efficiency is an issue because standard iterative solutions of the *nonlinear inverse RT problem* require executing the forward RT model repeatedly to compute the radiation field and its partial derivatives with respect to the RPs (the *Jacobians*) [37, 38].

1.3 Scope

While solutions to the *scalar RTE*, which involve only the first component of the *Stokes vector* (the radiance or intensity), are well developed, modern RT models that solve the *vector RTE* are capable of also accounting for polarization effects described by the second, third, and fourth components of the Stokes vector. Even if one's interest lies primarily in the radiance, it is important to realize that solutions of the *scalar RTE*, which ignores *polarization effects*, introduce errors in the computed radiances [40–42].

In this book, we will consider the theory and applications based on both scalar and vector RT models, which include polarization effects. There are numerous RT models available that include polarization effects (see Zhai *et al.* [43] and references therein for a list of papers), and the interest in applications based on *polarized radiation* is growing. There is also a growing interest in applications based on vector RT models that apply to coupled systems. Examples of vector RT modeling pertinent to a coupled atmosphere–water system include applications based on the *doubling-adding method* (e.g., Chowdhary [44], Chowdhary *et al.*, [45–47]), the *successive order of scattering method* (e.g., Chami *et al.*, [48], Min and Duan [49], Zhai *et al.*, [43]), the *matrix operator method* (e.g., Fisher and Grassl, [50], Ota *et al.*, [51]), and *Monte Carlo methods* (e.g., Kattawar and Adams [40], Lotsberg and Stamnes [52]).

Chapter 2 provides definitions of IOPs including *absorption* and *scattering coefficients* as well as the *normalized angular scattering cross section*, commonly referred to as the *scattering phase function*, and the corresponding *scattering phase matrix* needed for vector RT modeling and applications. In several subsections basic scattering theory with emphasis on spherical particles (*Mie–Lorenz theory*) is reviewed, and IOPs for atmospheric gases and aerosols as well those for surface materials including snow/ice, liquid water, and land surfaces are discussed. The impact of a rough interface between the two adjacent slabs is also discussed.

In Chapter 3, an overview is given of the *scalar RTE* as well as the *vector RTE* applicable to a coupled system consisting of two adjacent slabs with different refractive indices. Several methods of solution are discussed: the *successive order of scattering method*, the *discrete-ordinate method*, the *doubling-adding method*, and the *Monte Carlo method*. In Chapter 4, we discuss forward RT modeling in coupled environmental systems based on the discrete-ordinate method, while Chapter 5 is devoted to a discussion of the inverse problem. Finally, in Chapter 6, a few typical applications are discussed including (i) how spectral redundancy can be exploited to reduce the computational burden in atmospheric RT problems, (ii) simultaneous retrieval of total ozone column amount and cloud effects from ground-based irradiance measurements, (iii) retrieval of aerosol and snow-ice properties in coupled atmosphere–cryosphere systems from space, (iv) retrieval of aerosol and aquatic parameters in coupled atmosphere–water systems from space, (v) vector RT in coupled systems, and (vi) how polarization measurements can be used to improve retrievals of atmospheric and surface parameters in coupled atmosphere–surface systems.

1.4
Limitations of Scope

We restrict our attention to scattering by molecules and small particles such as aerosols and cloud particles in an atmosphere, hydrosols in water bodies such as oceans, lakes, and rivers, and inclusions (air bubbles and brine pockets) embedded in ice. To explain the meaning of *independent scattering*, let us consider an infinitesimal volume element filled with small particles that are assumed to be randomly distributed within the volume element. Such infinitesimal volume elements are assumed to constitute the elementary scattering agents. *Independent scattering* implies that each particle in each of the infinitesimal volume elements is assumed to scatter radiation independently of all other volume elements.

Although there are many applications that require a three-dimensional (3-D) RT treatment, in this book we limit our discussion to *plane-parallel* systems with an emphasis on the coupling between the atmosphere and the underlying surface consisting of a water body, a snow/ice surface, or a vegetation canopy. For a clear (cloud- and aerosol-free) atmosphere, 3-D effects are related to the impact of the Earth's curvature on the radiation field. To include such effects, a *pseudo-spherical* treatment (see Dahlback and Stamnes [53]) may be sufficient, in which the direct solar beam illumination is treated using spherical geometry, whereas *multiple scattering* is done using a plane-parallel geometry. This pseudo-spherical approach has been implemented in many RT codes [54, 55]. There is a large body of literature on *3-D RT modeling* with applications to broken clouds. Readers interested in RT in cloudy atmospheres may want to consult books like that of Marshak and Davis [12] or visit the Web site http://i3rc.gsfc.nasa.gov/.

3-D RT modeling may also be important for analysis and interpretation of *lidar* data. In this context, the classical "searchlight problem" [56], which considers the propagation of a *laser beam* through a *turbid medium*, is relevant. Long-range propagation of a lidar beam has been studied both theoretically and experimentally [57]. Monte Carlo simulations are well suited for such studies [58], and use of deterministic models such the *discrete-ordinate method*, discussed in Chapters 3 and 4 of this book, have also been reported [59, 60].

Most RT studies in the ocean have been concerned with understanding the propagation of sunlight, as discussed by Mobley *et al.* [26]. For these applications, the transient or *time-dependent* term in the RTE can be ignored, because changes in the incident illumination are much slower than the changes imposed by the propagation of the light field through the medium. While this assumption is satisfied for solar illumination, lidar systems can use pulses that are shorter than the attenuation distance of seawater divided by the speed of light in water. Also, as pointed out by Mitra and Churnside [61], due to multiple light scattering, understanding the lidar signal requires a solution of the *time-dependent RTE*. Although such studies are beyond the scope of this book, the transient RT problem can be reduced to solving a series of *time-independent* RT problems, as discussed by Stamnes *et al.* [62].

We restrict our attention to *elastic scattering*, although *inelastic scattering* processes (Raman and Brillouin) certainly can be very important and indeed essential in some atmospheric [63–65] and aquatic [66, 67] applications. Although most particles encountered in nature have nonspherical shapes – cloud droplets being the notable exception – we will not consider nonspherical particles in this book. Although the general introduction to the scattering problem provided in Chapter 2 is generic in nature and thus applies to particles of arbitrary shape, our more detailed review is limited to spherical particles (Mie–Lorenz theory). The reader is referred to the books by Bohren and Huffman [68] and Zdunkowski *et al.* [20] for a more comprehensive discussion of the Mie-Lorenz theory and to the recent book by Wendisch and Yang [21] for an excellent introduction to scattering by nonspherical particles.

2
Inherent Optical Properties (IOPs)

2.1
General Definitions

2.1.1
Absorption Coefficient and Volume Scattering Function

The optical properties of a medium can be categorized as inherent or apparent. An *inherent optical property* (IOP) depends only on the medium itself, and not on the ambient light field within the medium [69]. An *apparent optical property* (AOP) depends also on the illumination, and hence on light propagating in particular directions inside and outside the medium[1].

The *absorption coefficient* α and the *scattering coefficient* β are important IOPs, defined as [18]

$$\alpha(s) = \frac{1}{I^i}\left(\frac{dI^\alpha}{ds}\right) \tag{1}$$

$$\beta(s) = \frac{1}{I^i}\left(\frac{dI^\beta}{ds}\right). \tag{2}$$

Here, I^i is the *radiance* of the incident light beam entering a volume element $dV = dA\,ds$ of the medium of cross-sectional area dA and length ds, and $dI^\alpha > 0$ and $dI^\beta > 0$ are, respectively, the radiances that are absorbed and scattered in all directions as the light beam propagates the distance ds, which is the thickness of the volume element dV along the direction of the incident light beam. If the distance ds is measured in meters, the unit for the absorption or scattering coefficient defined in Eq. (1) or Eq. (2) becomes [m^{-1}]. The *extinction coefficient* γ is the sum of the absorption and scattering coefficients

$$\gamma(s) = \alpha(s) + \beta(s) \tag{3}$$

1) Apparent optical properties (i) depend both on the medium (the IOPs) and on the geometric (directional) structure of the radiance distribution, and (ii) display enough regular features and stability to be useful descriptors of a water body [69]. Hence, a radiance or an irradiance would satisfy only the first part of the definition, while a radiance or irradiance reflectance, obtained by division of the radiance or the upward irradiance by the downward irradiance, would satisfy also the second part of the definition.

Radiative Transfer in Coupled Environmental Systems: An Introduction to Forward and Inverse Modeling,
First Edition. Knut Stamnes and Jakob J. Stamnes.

and the *single-scattering albedo* ϖ is defined as the ratio of β to γ

$$\varpi(s) = \frac{\beta(s)}{\gamma(s)}. \tag{4}$$

Thus, given an interaction between an incident light beam and the medium, the single-scattering albedo, which varies between 0 and 1, gives the probability that the light beam will be scattered rather than absorbed.

The angular distribution of the scattered light is given in terms of the *volume scattering function* (vsf), which is defined as

$$\text{vsf}(s, \hat{\mathbf{\Omega}}', \hat{\mathbf{\Omega}}) = \frac{1}{I^i} \frac{d^2 I^\beta}{ds\, d\omega} = \frac{1}{I^i} \frac{d}{ds} \left(\frac{dI^\beta}{d\omega} \right) \quad [\text{m}^{-1}\text{sr}^{-1}]. \tag{5}$$

Here, $d^2 I^\beta$ is the *radiance* scattered from an incident direction $\hat{\mathbf{\Omega}}'$ into a cone of solid angle $d\omega$ around the direction $\hat{\mathbf{\Omega}}$ as the light propagates the distance ds along $\hat{\mathbf{\Omega}}'$. The plane spanned by the unit vectors $\hat{\mathbf{\Omega}}'$ and $\hat{\mathbf{\Omega}}$ is called the *scattering plane*, and the *scattering angle* Θ is given by $\cos\Theta = \hat{\mathbf{\Omega}}' \cdot \hat{\mathbf{\Omega}}$. Integration on the far right side of Eq. (5) over all scattering directions yields, using Eq. (2)

$$\begin{aligned} \beta(s) &= \frac{1}{I^i} \frac{d}{ds} \int_{4\pi} \left(\frac{dI^\beta}{d\omega} \right) d\omega = \frac{1}{I^i} \left(\frac{dI^\beta}{ds} \right) \\ &= \int_{4\pi} \text{vsf}(s, \hat{\mathbf{\Omega}}', \hat{\mathbf{\Omega}}) d\omega = \int_0^{2\pi} \int_0^{\pi} \text{vsf}(s, \cos\Theta, \phi) \sin\Theta d\Theta d\phi \end{aligned} \tag{6}$$

where Θ and ϕ are, respectively, the polar angle and the azimuth angle in a spherical coordinate system, in which the polar axis is along $\hat{\mathbf{\Omega}}'$. As indicated in Eq. (6), the vsf is generally a function of both Θ and ϕ, but for randomly oriented scatterers one may assume that the scattering potential is spherically symmetric, implying that there is no azimuthal dependence, so that vsf $=$ vsf$(s, \cos\Theta)$. Then one finds, with $x = \cos\Theta$

$$\beta(s) = 2\pi \int_0^{\pi} \text{vsf}(s, \cos\Theta) \sin\Theta d\Theta = 2\pi \int_{-1}^{1} \text{vsf}(s, x) dx. \tag{7}$$

2.1.2 Scattering Phase Function

A normalized *vsf*, denoted by $p(s, \cos\Theta)$ and referred to hereafter as the *scattering phase function*, may be defined as follows:

$$p(s, \cos\Theta) = 4\pi \frac{\text{vsf}(s, \cos\Theta)}{\int_{4\pi} \text{vsf}(s, \cos\Theta) d\omega} = \frac{\text{vsf}(s, \cos\Theta)}{\frac{1}{2} \int_{-1}^{1} \text{vsf}(s, x) dx} \tag{8}$$

so that

$$\frac{1}{4\pi} \int_{4\pi} p(s, \cos\Theta) d\omega = \frac{1}{2} \int_{-1}^{1} p(s, x) dx = 1. \tag{9}$$

The scattering phase function has the following physical interpretation: Given that a scattering event has occurred, $p(s, \cos\Theta) d\omega / 4\pi$ is the probability that a light beam traveling in the direction $\hat{\mathbf{\Omega}}'$ is scattered into a cone of solid angle $d\omega$ around the direction $\hat{\mathbf{\Omega}}$ within the volume element dV with thickness ds along $\hat{\mathbf{\Omega}}'$.

The scattering phase function $p(s, \cos\Theta)$ describes the angular distribution of the scattering, while the scattering coefficient $\beta(s)$ describes its magnitude. A convenient measure of the "shape" of the scattering phase function is the average over all scattering directions (weighted by $p(s, \cos\Theta)$) of the cosine of the scattering angle Θ, that is,

$$g(s) = \langle \cos\Theta \rangle = \frac{1}{4\pi} \int_{4\pi} p(s, \cos\Theta)\ \cos\Theta\, d\omega$$

$$= \frac{1}{2} \int_0^{\pi} p(s, \cos\Theta)\ \cos\Theta\ \sin\Theta\, d\Theta = \frac{1}{2} \int_{-1}^{1} p(s, x)\, x\, dx \tag{10}$$

where $x = \cos\Theta$. The average cosine $g(s)$ is called the *asymmetry factor* of the scattering phase function. Equation (10) yields complete forward scattering if $g = 1$ and complete backward scattering if $g = -1$, and $g = 0$ if $p(s, \cos\Theta)$ is symmetric about $\Theta = 90°$. Thus, *isotropic scattering* also gives $g = 0$. The *scattering phase function* $p(s, \cos\Theta)$ depends on the *refractive index* as well as the size of the scattering particles, and will thus depend on the physical situation and the practical application of interest. The probability of scattering into the backward hemisphere is given by the *backscattering ratio* (or *backscatter fraction*) b, defined as

$$b(s) = \frac{1}{2} \int_{\pi/2}^{\pi} p(s, \cos\Theta)\ \sin\Theta\, d\Theta = \frac{1}{2} \int_0^{1} p(s, -x)\, dx. \tag{11}$$

The scattering phase function may be approximated by a finite sum of $(M + 1)$ Legendre polynomials (dropping for simplicity the dependence on the position s)

$$p(\cos\Theta) \approx \sum_{\ell=0}^{M} (2\ell + 1)\chi_\ell P_\ell(\cos\Theta) \tag{12}$$

where P_ℓ is the ℓth *Legendre polynomial*, and the *expansion coefficient* is given by

$$\chi_\ell = \frac{1}{2} \int_{-1}^{1} P_\ell(x) p(x) dx. \tag{13}$$

The Legendre polynomials satisfy an orthogonality relation

$$\frac{1}{2} \int_{-1}^{+1} P_\ell(x) P_k(x) dx = \frac{1}{2\ell + 1} \delta_{\ell k} \tag{14}$$

as well as an *Addition Theorem*:

$$P_\ell(\cos\Theta) = P_\ell(u') P_\ell(u) + 2 \sum_{m=1}^{\ell} \Lambda_\ell^m(u') \Lambda_\ell^m(u) \cos m(\phi' - \phi) \tag{15}$$

where $u = \cos\theta$, $u' = \cos\theta'$,

$$\Lambda_\ell^m(u) = \sqrt{\frac{(\ell - m)!}{(\ell + m)!}} P_\ell^m(u)$$

and $P_\ell^m(u)$ is the *associated Legendre polynomial*. For $m = 0$, we have $\Lambda_\ell^0(u) = P_\ell^0(u) = P_\ell(u)$. The scattering angle Θ is related to the polar and azimuthal angles by

$$\cos\Theta = uu' + \sqrt{1-u^2}\sqrt{1-u'^2}\cos(\phi' - \phi) \tag{16}$$

where (θ', ϕ') and (θ, ϕ) are the polar and azimuthal angles before and after scattering, respectively. Substituting Eq. (15) into Eq. (12), we have

$$\begin{aligned} p(\cos\Theta) = p(u', \phi'; u, \phi) &\approx \sum_{\ell=0}^{M}(2\ell+1)\chi_\ell \\ &\times \left\{P_\ell(u')P_\ell(u) + 2\sum_{m=1}^{\ell}\Lambda_\ell^m(u')\Lambda_\ell^m(u)\cos m(\phi' - \phi)\right\} \end{aligned} \tag{17}$$

which can be rewritten as a *Fourier cosine series*

$$p(u', \phi'; u, \phi) \approx \sum_{m=0}^{M}(2 - \delta_{m0})p^m(u', u)\cos m(\phi' - \phi) \tag{18}$$

where δ_{0m} is the Kronecker delta, that is, $\delta_{0m} = 1$ for $m = 0$ and $\delta_{0m} = 0$ for $m \neq 0$, and

$$p^m(u', u) \approx \sum_{\ell=m}^{M}(2\ell+1)\chi_\ell\Lambda_\ell^m(u')\Lambda_\ell^m(u). \tag{19}$$

In a *plane-parallel* or *slab geometry*, *irradiances* and the *scalar radiance* (*mean intensity*) depend on the azimuthally averaged phase function. Application of azimuthal averaging, that is, $\frac{1}{2\pi}\int_0^{2\pi} d\phi \cdots$, to both sides of Eq. (12), combined with Eq. (15) or Eq. (18) gives

$$\begin{aligned} p(u', u) \equiv p^0(u', u) &= \frac{1}{2\pi}\int_0^{2\pi} p(u', \phi'; u, \phi)d\phi \\ &\approx \sum_{\ell=0}^{M}(2\ell+1)\chi_\ell P_\ell(u)P_\ell(u'). \end{aligned} \tag{20}$$

From Eq. (20), it follows that

$$\frac{1}{2}\int_{-1}^{1} p(u', u)P_k(u')du' \approx \sum_{\ell=0}^{M}(2\ell+1)\chi_\ell P_\ell(u)\frac{1}{2}\int_{-1}^{1} P_\ell(u')P_k(u')du' \tag{21}$$

which by the use of orthogonality [Eq. (14)] leads to

$$\chi_\ell = \frac{1}{P_\ell(u)}\frac{1}{2}\int_{-1}^{1} p(u', u)P_\ell(u')du'. \tag{22}$$

Thus, to calculate the *expansion coefficients* or *moments* χ_ℓ, we can use the azimuthally averaged phase function $p(u', u)$ given by Eq. (20).

Four different scattering phase functions, which are useful in practical applications, are discussed below.

2.2 Examples of Scattering Phase Functions

2.2.1 Rayleigh Scattering Phase Function

When the size d of a scatterer is small compared to the wavelength of light ($d < \frac{1}{10}\lambda$), the *Rayleigh scattering phase function* gives a good description of the angular distribution of the scattered light. It is given by [see Eq. (59)]

$$p_{\text{Ray}}(\cos\Theta) = \frac{3}{3+f}(1 + f\cos^2\Theta) \tag{23}$$

where the parameter $f = \frac{1-\rho}{1+\rho}$, and ρ is the *depolarization factor* defined in Eq. (63), attributed to the anisotropy of the scatterer (the molecule) [70–73]. Originally, this *scattering phase function* was derived for light scattering by an electric dipole [74]. Since the Rayleigh scattering phase function is symmetric about $\Theta = 90°$, the *asymmetry factor* is $g = \chi_1 = 0$. If the Rayleigh scattering phase function is expanded in *Legendre polynomials*, the *expansion coefficients* χ_ℓ [see Eqs. (13) and (22)] are simply given by $\chi_0 = 1$, $\chi_1 = 0$, $\chi_2 = \frac{2f}{5(3+f)}$, and $\chi_\ell = 0$ for $\ell > 2$ (see Problem 2.1). For Rayleigh scattering, a value of $\rho = 0.04$ for air gives $f_{\text{air}} = 0.923$, while for water the numerical value $\rho = 0.09$ is commonly used, implying $f_{\text{water}} = 0.835$. For inelastic *Raman scattering*, the scattering phase function is the same as for Rayleigh scattering except that $\rho = 0.29$, implying $f_{\text{water}}^{\text{Raman}} = 0.55$.

2.2.2 Henyey–Greenstein Scattering Phase Function

In 1941, Henyey and Greenstein [75] proposed a one-parameter scattering phase function given by (suppressing the dependence on the position s)

$$p_{\text{HG}}(\cos\Theta) = \frac{1-g^2}{(1+g^2-2g\cos\Theta)^{3/2}} \tag{24}$$

where the parameter g is the asymmetry factor defined in Eq. (10). The *Henyey–Greenstein (HG) scattering phase function* has no physical basis, but is very useful for describing a highly scattering medium, such as turbid water or sea ice, for which the actual scattering phase function is unknown. The HG scattering phase function is convenient for Monte Carlo simulations and other numerical calculations because of its analytical form. In deterministic plane-parallel RT models, it is also very convenient because the *addition theorem* of spherical harmonics can be used to expand the scattering phase function in a series of Legendre polynomials [18], as reviewed in the previous section. For the HG scattering phase function, the expansion coefficients χ_ℓ in this series [see Eqs. (13) and (22)] are simply given by $\chi_\ell = g^\ell$, implying that the HG scattering phase function can be approximated by a finite sum of $(M + 1)$ Legendre polynomials [see Eq. (12) and Problem 2.2)]

$$p_{HG}(\cos\Theta) = \frac{1-g^2}{(1+g^2-2g\cos\Theta)^{3/2}} \approx \sum_{\ell=0}^{M}(2\ell+1)g^{\ell}P_{\ell}(\cos\Theta). \qquad (25)$$

The HG scattering phase function is useful for scatterers with sizes comparable to or larger than the wavelength of light.

The probability of scattering into the backward hemisphere, the *backscattering ratio* (or backscatter fraction) becomes [see Eq. (11)]:

$$\begin{aligned} b_{HG} &= \frac{1}{2}\int_0^1 p(s,-x)\,dx = \frac{1-g^2}{2}\int_0^1 \frac{dx}{(1+g^2+2g\,x)^{3/2}} \\ &= \frac{1-g}{2g}\left[\frac{(1+g)}{\sqrt{(1+g^2)}}-1\right]. \end{aligned} \qquad (26)$$

Figure 1 shows the scattering phase functions computed for a collection of particles with a log-normal size distribution (see Section 2.4.3 and Eq. (102)). The left panel pertains to nonabsorbing aerosol particles with *refractive index* $n = 1.385$, mode radius $r_n = 0.3$ μm, and standard deviation $\sigma_n = 0.92$, and the smallest and largest radii are selected to be $r_1 = 0.005$ μm and $r_2 = 30$ μm. The right panel pertains to nonabsorbing cloud droplets with refractive index $n = 1.339$, mode radius $r_n = 5$ μm, and standard deviation $\sigma_n = 0.4$, and the smallest and largest radii are selected to be $r_1 = 0.005$ μm and $r_2 = 100$ μm.

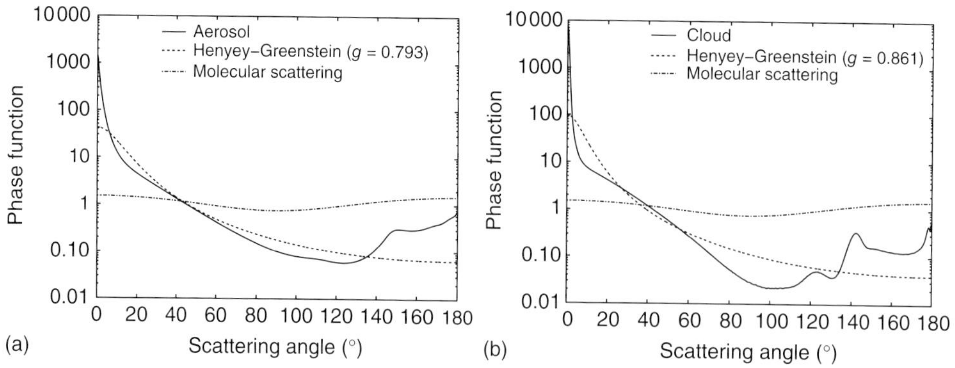

Figure 1 Scattering phase functions calculated using a Mie code. (a) For aerosols with asymmetry factor 0.79275. (b) For clouds with asymmetry factor 0.86114. HG scattering phase functions [see Eq. (24)] with asymmetry factors equal to those for the cloud and aerosol particles are shown for comparison. The Rayleigh scattering phase function [Eq (23)] describing molecular scattering is also shown for comparison.

2.2.3 Fournier–Forand Scattering Phase Function

Measurements have shown that the *particle size distribution* (PSD) in oceanic water can be accurately described by a power law (*Junge distribution*) $n(r) = C(\xi, r_1, r_2)/r^{\xi}$, where $n(r)$ is the number of particles per unit volume per unit bin width, r [μm] is the radius of the assumed spherical particles, and r_1 and r_2 denote the smallest and largest particle size, respectively. The normalization constant $C(\xi, r_1, r_2)$ $[\mathrm{cm}^{-3} \cdot \mu\mathrm{m}^{\xi-1}]$ is called the *Junge coefficient*, and $\xi > 0$ is the PSD slope, which typically varies between 3.0 and 5.0 (Diehl and Haardt [76]; McCave [77]). The power-law PSD is further described in Section 2.4.3. By assuming a power law for the PSD, and letting each particle scatter in accordance with the anomalous diffraction approximation, Fournier and Forand [78] derived an analytic expression for the scattering phase function of oceanic water (hereafter referred to as the FF scattering phase function). A commonly used version of the *FF scattering phase function* is given by (Mobley *et al.*, [79])

$$p_{\mathrm{FF}}(\Theta) = \frac{1}{4\pi(1-\delta)^2\delta^{\nu}}\left\{\nu(1-\delta) - (1-\delta^{\nu}) + \frac{4}{\tilde{u}^2}[\delta(1-\delta^{\nu}) - \nu(1-\delta)]\right\}$$
$$+ \frac{1-\delta_{180}^{\nu}}{16\pi(\delta_{180}-1)\delta_{180}^{\nu}}[3\cos^2\Theta - 1] \tag{27}$$

where $\nu = 0.5(3-\xi)$,

$$\delta \equiv \delta(\Theta) = \frac{\tilde{u}^2(\Theta)}{3(n-1)^2}$$

$\tilde{u}(\Theta) = 2\sin(\Theta/2)$, $\delta_{180} = \delta(\Theta = 180°) = \frac{4}{3(n-1)^2}$, Θ is the *scattering angle*, and n is the real part of the *refractive index*. Note that, in addition to the scattering angle Θ, the *FF scattering phase function* depends also on the real part of the refractive index of the particle relative to water and the *slope parameter* ξ characterizing the PSD.

Setting $x = -\cos\Theta$, and integrating the FF scattering phase function over the backward hemisphere, one obtains the *backscattering ratio* or *backscatter fraction* defined in Eq. (11), that is, [79]

$$b_{\mathrm{FF}} = \frac{1}{2}\int_{\pi/2}^{\pi} p_{\mathrm{FF}}(\cos\Theta)\sin\Theta d\Theta = \frac{1}{2}\int_0^1 p_{\mathrm{FF}}(-x)\,dx$$
$$= 1 - \frac{1-\delta_{90}^{\nu+1} - 0.5(1-\delta_{90}^{\nu})}{(1-\delta_{90})\delta_{90}^{\nu}} \tag{28}$$

where $\delta_{90} = \delta(\Theta = 90°) = \frac{4}{3(n-1)^2}\sin^2(45°) = \frac{2}{3(n-1)^2}$. Equation (28) can be solved for ν in terms of b_{FF} and δ_{90}, implying that ν and thus ξ can be determined if the real part of the refractive index n and the *backscattering ratio* b_{FF} are specified. As a consequence, the FF scattering phase function can be evaluated from the measured value of b_{FF} if the real part of the refractive index n is known.

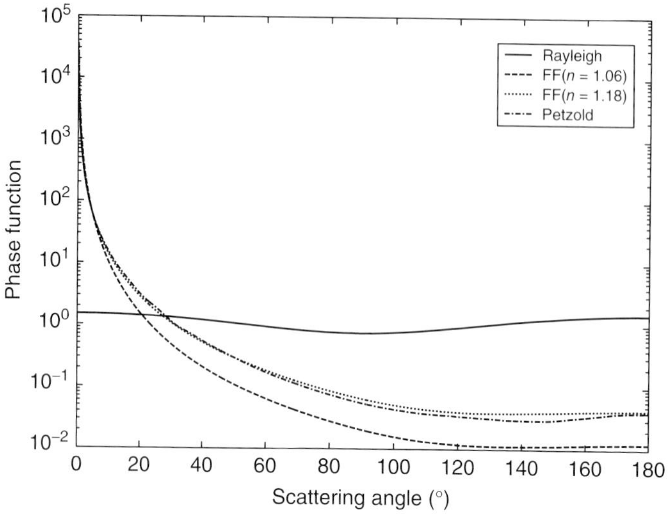

Figure 2 Rayleigh [Eq. (23)], FF [Eq. (27)], and Petzold scattering phase functions.

2.2.4 The Petzold Scattering Phase Function

The vsfs measured by Petzold [80] have been widely used by ocean optics researchers. These *scattering phase functions* are discussed by Mobley [69], who tabulated the vsfs for clear ocean, coastal ocean, and turbid harbor waters. Based on these three vsfs, an average scattering phase function for supposedly "typical" ocean waters was created (Mobley *et al.* [26], Table 2). This average *Petzold scattering phase function*, which has an asymmetry factor $g = 0.9223$ and a backscattering ratio $b_{FF} = 0.019$, is shown in Figure 2 together with the *Rayleigh scattering phase function* and the *FF scattering phase function*. For the FF phase function, the power-law slope was set to $\xi = 3.38$, but results for two different values of the real part of the refractive index are shown: $n = 1.06$ and $n = 1.18$. These values yield an asymmetry factor $g = 0.9693$ and a backscattering ratio $b_{FF} = 0.0067$ for $n = 1.06$ and $g = 0.9160$ and $b_{FF} = 0.022$ for $n = 1.18$. Note the similarity between the FF scattering phase function for $n = 1.18$ and the average Petzold scattering phase function, which indicates that the average Petzold scattering phase function is more suitable for mineral-dominated waters than for pigment-dominated waters.

2.3 Scattering Phase Matrix

The theoretical development of *vector radiative transfer theory* starts with the *Stokes vector* representation $\mathbf{I} = [I_{\parallel}, I_{\perp}, U, V]^T$, where the superscript T denotes the transpose. In terms of the complex transverse electric field components of the radiation field $E_{\|} = |E_{\|}|e^{-i\epsilon_1}$ and $E_{\perp} = |E_{\perp}|e^{-i\epsilon_2}$, these Stokes vector components are given by

$$\begin{aligned} I_{\|} &= E_{\|}E_{\|}^{*} \\ I_{\perp} &= E_{\perp}E_{\perp}^{*} \\ U &= 2|E_{\|}||E_{\perp}|\cos\delta \\ V &= 2|E_{\|}||E_{\perp}|\sin\delta \end{aligned} \tag{29}$$

where $\delta = \epsilon_1 - \epsilon_2$. The connection between this Stokes vector representation, $\mathbf{I} = [I_{\|}, I_{\perp}, U, V]^T$, and the more commonly used representation $\mathbf{I}_S = [I, Q, U, V]^T$, where $I = I_{\|} + I_{\perp}$ and $Q = I_{\|} - I_{\perp}$, is given by

$$\mathbf{I}_S = \mathbf{DI} \tag{30}$$

where

$$\mathbf{D} = \begin{pmatrix} 1 & 1 & 0 & 0 \\ 1 & -1 & 0 & 0 \\ 0 & 0 & 1 & 0 \\ 0 & 0 & 0 & 1 \end{pmatrix}, \qquad \mathbf{D}^{-1} = \frac{1}{2}\begin{pmatrix} 1 & 1 & 0 & 0 \\ 1 & -1 & 0 & 0 \\ 0 & 0 & 2 & 0 \\ 0 & 0 & 0 & 2 \end{pmatrix}. \tag{31}$$

The *degree of polarization* is defined as

$$p = \frac{[Q^2 + U^2 + V^2]^{1/2}}{I} \tag{32}$$

so that $0 \leq p \leq 1$, where $p = 1$ corresponds to *completely polarized light* and $p = 0$ to *natural (unpolarized) light*. The *degree of circular polarization* is defined as

$$p_c = V/I \tag{33}$$

the *degree of linear polarization* as

$$p_l = \frac{[Q^2 + U^2]^{1/2}}{I} \tag{34}$$

and, alternatively, when $U = 0$, as

$$p_l = -\frac{Q}{I} = \frac{I_{\perp} - I_{\|}}{I_{\perp} + I_{\|}}. \tag{35}$$

The *transverse electric field vector* $[E_{\|}, E_{\perp}]^T$ of the scattered field can be obtained in terms of the transverse electric field vector $[E_{\|0}, E_{\perp 0}]^T$ of the incident field by a linear transformation:

$$\begin{pmatrix} E_{\|} \\ E_{\perp} \end{pmatrix} = \mathbf{A}\begin{pmatrix} E_{\|0} \\ E_{\perp 0} \end{pmatrix}$$

where $\mathbf{A}$ is a 2×2 matrix, referred to as the *amplitude scattering matrix*. The corresponding linear transformation connecting the *Stokes vectors* of the incident and scattered fields in the *scattering plane* is called the *Mueller matrix* (in the case of a single scattering event). For scattering by a small volume containing an ensemble of particles, the ensemble-averaged Mueller matrix is referred to as the *Stokes scattering matrix* **F**. Finally, when transforming from the scattering plane to a fixed laboratory frame, the corresponding matrix is referred to as the *scattering phase matrix* **P**.

2.3.1
Stokes Vector Representation $\mathbf{I}_S = [I, Q, U, V]^T$

The scattering geometry is illustrated in Figure 3. The plane **AOB**, defined as the *scattering plane*, is spanned by the directions of propagation of the incident parallel beam with Stokes vector $\mathbf{I}_S^{\text{inc}}$ and the scattered parallel beam with Stokes vector $\mathbf{I}_S^{\text{sca}}$. Here the subscript S pertains to the Stokes vector representation $\mathbf{I}_S = [I, Q, U, V]^T$. The scattered radiation, represented by the Stokes vector $\mathbf{I}_S^{\text{sca}}$, is related to the incident radiation, represented by the Stokes vector $\mathbf{I}_S^{\text{inc}}$, by a 4×4 scattering matrix [see Eqs. (36) and (37) below] and two rotations are required to properly connect the two Stokes vectors as explained below. We describe the Stokes vector of the incident beam in terms of two unit vectors $\hat{\ell}'$ and $\hat{\mathbf{r}}'$, which are normal to each other and to the unit vector $\hat{\mathbf{\Omega}}' = \hat{\mathbf{r}}' \times \hat{\ell}'$ along the propagation direction of the incident beam. Similarly, we describe the Stokes vector of the scattered beam in terms of two unit vectors $\hat{\ell}$ and $\hat{\mathbf{r}}$, which are normal to each other and to the unit vector $\hat{\mathbf{\Omega}} = \hat{\mathbf{r}} \times \hat{\ell}$ along the propagation direction of the scattered beam. The unit vector $\hat{\ell}'$ is along the direction of $\mathbf{E}'_{\|}$ of the incident beam and lies in the *meridian plane* of that beam, which is defined as the plane **OAC** in Figure 3. Similarly, the unit vector $\hat{\ell}$ is along the direction of $\mathbf{E}_{\|}$ of the scattered beam and lies in the meridian plane

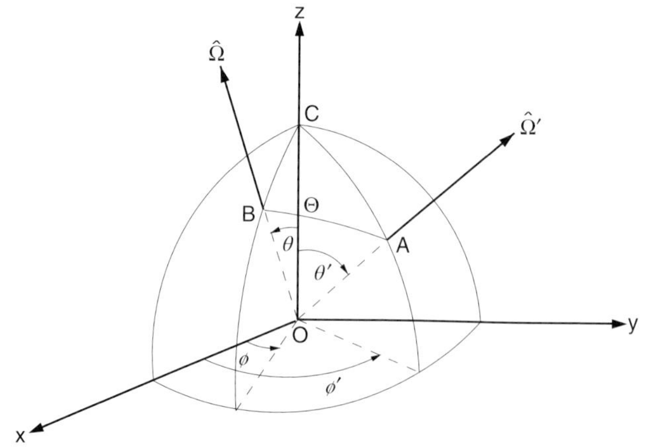

Figure 3 Coordinate system for scattering by a volume element at **O**. The points **C**, **A**, and **B** are located on the unit sphere. The incident light beam with Stokes vector $\mathbf{I}_S^{\text{inc}}$ is in direction $\mathbf{OA}(\theta', \phi')$ with unit vector $\hat{\mathbf{\Omega}}'$, and the scattered beam with Stokes vector $\mathbf{I}_S^{\text{sca}}$ is in direction $\mathbf{OB}(\theta, \phi)$ with unit vector $\hat{\mathbf{\Omega}}$ [16].

of that beam, which is defined as the plane **OBC** in Figure 3. For the incident beam, the unit vector $\hat{\ell}'$ may be defined as the tangent at the point **A** to the unit circle passing through the points **A** and **C** in Figure 3. For the scattered beam, the unit vector $\hat{\ell}$ may be defined as tangent at the point **B** to the unit circle passing through the points **B** and **C** in Figure 3. For either beam, its meridian plane acts as a plane of reference for the Stokes vector, so that the point **A** in Figure 3 is the starting point for the unit vector $\hat{\mathbf{\Omega}}' = \hat{\mathbf{r}}' \times \hat{\ell}'$ along the direction of propagation of the incident beam, and the point **B** in Figure 3 is the starting point for the unit vector $\hat{\mathbf{\Omega}} = \hat{\mathbf{r}} \times \hat{\ell}$ along the direction of propagation of the scattered beam.

As explained above, the *Mueller matrix* describes scattering by a single particle, and for scattering by a small volume of particles, the ensemble-averaged Mueller matrix is referred to as the *Stokes scattering matrix* $\mathbf{F}_S$. If any of the following conditions are fulfilled [81], that is

1) each particle in the volume element has a plane of symmetry, and the particles are randomly oriented,
2) each volume element contains an equal number of particles and their mirror particles in random orientation,
3) the particles are much smaller than the wavelength of the incident light,

then the Stokes scattering matrix pertaining to the representation $\mathbf{I}_S = [I, Q, U, V]^T$ has the following form:

$$\mathbf{F}_S(\Theta) = \begin{bmatrix} a_1(\Theta) & b_1(\Theta) & 0 & 0 \\ b_1(\Theta) & a_2(\Theta) & 0 & 0 \\ 0 & 0 & a_3(\Theta) & b_2(\Theta) \\ 0 & 0 & -b_2(\Theta) & a_4(\Theta) \end{bmatrix}. \tag{36}$$

Each of the six independent matrix elements in Eq. (36) depends on the *scattering angle* Θ, and will in general also depend on the position in the medium. For spherical particles, the matrix in Eq. (36) simplifies, since $a_1 = a_2$ and $a_3 = a_4$, so that only four independent elements remain.

As already mentioned, two rotations are required to connect the Stokes vector of the scattered radiation to that of the incident radiation. As illustrated in Figure 3, the first rotation is from the *meridian plane* **OAC**, associated with the Stokes vector $\mathbf{I}_S^{\text{inc}}$, into the scattering plane **OAB**, whereas the second rotation is from the *scattering plane* **OAB** into the meridian plane **OBC**, associated with the Stokes vector $\mathbf{I}_S^{\text{sca}}$. Hence, the Stokes vector for the scattered radiation is given by (Chandrasekhar [11])

$$\mathbf{I}_S^{\text{sca}} = \mathbf{R}_S(\pi - i_2)\mathbf{F}_S(\Theta)\mathbf{R}_S(-i_1)\mathbf{I}_S^{\text{inc}} \equiv \mathbf{P}_S(\Theta)\mathbf{I}_S^{\text{inc}}. \tag{37}$$

The matrix $\mathbf{R}_S$ is called the *Stokes rotation matrix*. It represents a rotation in the clockwise direction with respect to an observer looking into the direction of propagation, and can be written as $(0 \le \omega \le 2\pi)$

$$\mathbf{R}_S(\omega) = \begin{bmatrix} 1 & 0 & 0 & 0 \\ 0 & \cos(2\omega) & -\sin(2\omega) & 0 \\ 0 & \sin(2\omega) & \cos(2\omega) & 0 \\ 0 & 0 & 0 & 1 \end{bmatrix}. \tag{38}$$

Hence, according to Eq. (37), the *scattering phase matrix*, which connects the Stokes vector of the scattered radiation to that of the incident radiation, is obtained from the *Stokes scattering matrix* $\mathbf{F}_S(\Theta)$ in Eq. (36) by

$$\mathbf{P}_S(\theta', \phi'; \theta, \phi) = \mathbf{R}_S(\pi - i_2)\mathbf{F}_S(\Theta)\mathbf{R}_S(-i_1) = \mathbf{R}_S(-i_2)\mathbf{F}_S(\Theta)\mathbf{R}_S(-i_1) \tag{39}$$

where $\mathbf{R}_S$ is the rotation matrix described in Eq. (38) [11], and $\mathbf{R}_S(\pi - i_2) = \mathbf{R}_S(-i_2)$ since the rotation matrix is periodic with a period π.

According to Eq. (37) (see also Figure 3), the Stokes vector $\mathbf{I}_S^{\mathrm{inc}}$ of the incident parallel beam must be multiplied by the rotation matrix $\mathbf{R}_S(-i_1)$ before it is multiplied by the Stokes scattering matrix $\mathbf{F}_S(\Theta)$, whereafter it must be multiplied by the rotation matrix $\mathbf{R}_S(\pi - i_2)$. These matrix multiplications are carried out explicitly in some radiative transfer (RT) models including *Monte Carlo* simulations, while they are implicitly taken care of in other RT models such as the *adding-doubling method* [82] and the *discrete-ordinate method* [83, 84] which use the expansion of the phase matrix in *generalized spherical functions* [85, 86].

On carrying out the matrix multiplications in Eq. (39), one finds (see Problem 2.3):

$$\mathbf{P}_S(\Theta) = \begin{bmatrix} a_1 & b_1 C_1 & -b_1 S_1 & 0 \\ b_1 C_2 & C_2 a_2 C_1 - S_2 a_3 S_1 & -C_2 a_2 S_1 - S_2 a_3 C_1 & -b_2 S_2 \\ b_1 S_2 & S_2 a_2 C_1 + C_2 a_3 S_1 & -S_2 a_2 S_1 + C_2 a_3 C_1 & -b_2 C_2 \\ 0 & -b_2 S_1 & -b_2 C_1 & a_4 \end{bmatrix} \tag{40}$$

where $a_i = a_i(\Theta), i = 1, \ldots, 4, b_i = b_i(\Theta), i = 1, 2$, and

$$C_1 = \cos 2i_1, \qquad C_2 = \cos 2i_2 \tag{41}$$

$$S_1 = \sin 2i_1, \qquad S_2 = \sin 2i_2. \tag{42}$$

A comparison of Eqs. (36) and (40) shows that only the corner elements of $\mathbf{F}_S(\Theta)$ remain unchanged by the rotations of the reference planes. The (1,1)-element of the *scattering phase matrix* $\mathbf{P}_S(\Theta)$ (and of the Stokes scattering matrix $\mathbf{F}_S(\Theta)$) is the *scattering phase function*. Since also the (4,4)-element of the scattering phase matrix remains unchanged by the rotations, the state of circular polarization of the incident light does not affect the intensity of the scattered radiation after one scattering event.

To compute $\mathbf{P}_S(\theta', \phi'; \theta, \phi)$ given by Eq. (39), we must relate the angles θ', ϕ', θ, and ϕ on the left side with the angles i_1, i_2, and Θ on the right side. Using spherical geometry, we may apply the cosine rule for Θ, θ, and θ' successively, in Figure 3, to obtain ($u = \cos\theta, u' = \cos\theta'$) [16] (see Problem 2.4)

$$\cos\Theta = uu' + (1-u^2)^{1/2}1-u'^2)^{1/2}\cos(\phi'-\phi) \tag{43}$$

$$\cos i_1 = \frac{-u + u'\cos\Theta}{(1-u'^2)^{1/2}(1-\cos^2\Theta)^{1/2}} \tag{44}$$

$$\cos i_2 = \frac{-u' + u\cos\Theta}{(1-u^2)^{1/2}(1-\cos^2\Theta)^{1/2}}. \tag{45}$$

Note that Eq. (43) is the same as Eq. (16). The trigonometric functions for the double angles can be obtained by using

$$\cos 2i = 2\cos^2 i - 1 \tag{46}$$

and

$$\sin 2i = 2\sin i\cos i \tag{47}$$

or

$$\sin 2i = \begin{cases} 2(1-\cos^2 i)^{1/2}\cos i & \text{if} \quad 0 < \phi' - \phi < \pi \\ -2(1-\cos^2 i)^{1/2}\cos i & \text{if} \quad \pi < \phi' - \phi < 2\pi \end{cases} \tag{48}$$

where i is i_1 or i_2. We now have all the information needed to compute the scattering phase matrix [see Eq. (39)] as a function of the three variables u, u', and $\phi' - \phi$:

$$\mathbf{P}_S(u', u, \phi' - \phi) = \mathbf{R}_S(-i_2)\mathbf{F}_S(\Theta)\mathbf{R}_S(-i_1).$$

If there is no difference in the azimuth (i.e., $\phi' - \phi = 0$), then the *meridian planes* of the incident and scattered beams in Figure 3 coincide with the scattering plane. Hence there is no need to rotate the reference planes ($\mathbf{R}(-i_2)$ and $\mathbf{R}(-i_1)$ both reduce to the identity matrix), so that

$$\mathbf{P}_S(u', u, 0) = \mathbf{P}_S(u', u, \pi) = \mathbf{F}_S(\Theta). \tag{49}$$

Symmetry Relations

The Stokes scattering matrix satisfies the *reciprocity relation*

$$\mathbf{F}_S(\Theta) = \mathbf{\Delta}_3\mathbf{F}_S^T(\Theta)\mathbf{\Delta}_3 \tag{50}$$

where $\mathbf{\Delta}_3 = \text{diag}(1, 1, -1, 1)$ and the *mirror symmetry relation*

$$\mathbf{F}_S(\Theta) = \mathbf{\Delta}_{3,4}\mathbf{F}_S^T(\Theta)\mathbf{\Delta}_{3,4} \tag{51}$$

where $\mathbf{\Delta}_{3,4} = \text{diag}(1, 1, -1, -1)$ and $\mathbf{F}_S^T(\Theta)$ denotes the transpose of $\mathbf{F}_S(\Theta)$.

The scattering phase matrix has similar symmetry relations. Thus, it can be seen from Eq. (43) that the scattering phase matrix is invariant to three basic changes in the polar angles u' and u and azimuthal angles ϕ' and ϕ, which leave the *scattering angle* unaltered. These are

1) changing the signs of u and u' simultaneously: $\mathbf{P}_S(-u', -u, \phi' - \phi) = \mathbf{P}_S(u', u, \phi' - \phi)$,

2) interchange of u and u': $\mathbf{P}_S(u', u, \phi' - \phi) = \mathbf{P}_S(u, u', \phi' - \phi)$,
3) interchange of ϕ and ϕ': $\mathbf{P}_S(u', u, \phi' - \phi) = \mathbf{P}_S(u', u, \phi - \phi')$.

Also, if the b_2-element in Eq. (40) is zero, the circular polarization component decouples from the other three. Then, the Stokes parameter V is scattered independently of the others according to the phase function $a_4(\Theta)$, and the remaining part of the scattering phase matrix referring to I, Q, and U becomes a 3×3 matrix:

$$\mathbf{P}_S(\Theta) = \begin{bmatrix} a_1 & b_1 C_1 & -b_1 S_1 \\ b_1 C_2 & C_2 a_2 C_1 - S_2 a_3 S_1 & -C_2 a_2 S_1 - S_2 a_3 C_1 \\ b_1 S_2 & S_2 a_2 C_1 + C_2 a_3 S_1 & -S_2 a_2 S_1 + C_2 a_3 C_1 \end{bmatrix}. \quad (52)$$

Finally, in a plane-parallel or slab geometry, there is no azimuth dependence for light beams traveling in directions perpendicular to the slab (either up or down). Thus, if either the incident or the scattered beam travels in a perpendicular direction, we may use the *meridian plane* of the other beam as a reference plane for both beams. Since this plane coincides with the scattering plane, Eq. (49) applies in this situation also.

2.3.2 Stokes Vector Representation $\mathbf{I} = [I_{\|}, I_{\perp}, U, V]^T$

The Stokes vector $\mathbf{I} = [I_{\|}, I_{\perp}, U, V]^T$ is related to $\mathbf{I}_S = [I, Q, U, V]^T$ by

$$\mathbf{I}_S = \mathbf{D}\mathbf{I} \quad (53)$$

where $\mathbf{D}$ is given by Eq. (31), so that $I = I_{\|} + I_{\perp}$, and $Q = I_{\|} - I_{\perp}$. Denoting the *Stokes vector* obtained after a rotation by

$$\mathbf{I}'_S = \mathbf{R}_S(\omega)\mathbf{I}_S \quad (54)$$

we find

$$\mathbf{I}' = \mathbf{D}^{-1}\mathbf{I}'_S = \mathbf{D}^{-1}\mathbf{R}_S(\omega)\mathbf{I}_S = \mathbf{D}^{-1}\mathbf{R}_S(\omega)\mathbf{D}\,\mathbf{I} = \mathbf{R}(\omega)\mathbf{I}. \quad (55)$$

Hence, the rotation matrix for the Stokes vector in the representation $\mathbf{I} = [I_{\|}, I_{\perp}, U, V]^T$ becomes

$$\mathbf{R}(\omega) = \mathbf{D}^{-1}\mathbf{R}_S(\omega)\mathbf{D} = \begin{bmatrix} \cos^2\omega & \sin^2\omega & -\frac{1}{2}\sin(2\omega) & 0 \\ \sin^2\omega & \cos^2\omega & \frac{1}{2}\sin(2\omega) & 0 \\ \sin(2\omega) & -\sin(2\omega) & \cos(2\omega) & 0 \\ 0 & 0 & 0 & 1 \end{bmatrix}. \quad (56)$$

The scattering phase matrix $\mathbf{P}(\Theta)$ in the Stokes vector representation $\mathbf{I} = [I_{\|}, I_{\perp}, U, V]^T$ is related to the scattering phase matrix $\mathbf{P}_S(\Theta)$ in the Stokes vector representation $\mathbf{I}_S = [I, Q, U, V]^T$ by

$$\mathbf{P}(\Theta) = \mathbf{D}^{-1}\mathbf{P}_S(\Theta)\mathbf{D}. \quad (57)$$

Similarly, the *Stokes scattering matrix* $\mathbf{F}(\Theta)$ associated with the Stokes vector representation $\mathbf{I} = [I_{\|}, I_{\perp}, U, V]^T$ is related to the Stokes scattering matrix $\mathbf{F}_S(\Theta)$ in Eq. (36) by

$$\mathbf{F}(\Theta) = \mathbf{D}^{-1}\mathbf{F}_S(\Theta)\mathbf{D} = \begin{pmatrix} \frac{1}{2}(a_1 + a_2 + 2b_1) & \frac{1}{2}(a_1 - a_2) & 0 & 0 \\ \frac{1}{2}(a_1 - a_2) & \frac{1}{2}(a_1 + a_2 - 2b_1) & 0 & 0 \\ 0 & 0 & a_3 & b_2 \\ 0 & 0 & -b_2 & a_4 \end{pmatrix}. \quad (58)$$

For *Rayleigh scattering* with parameter $f = \frac{1-\rho}{1+\rho}$, where ρ is the *depolarization factor* defined in Eq. (63), the Stokes scattering matrix in the Stokes vector representation $\mathbf{I}_S = [I, Q, U, V]^T$ is given by [11, 87]

$$\mathbf{F}_S(\Theta) = \frac{3}{3+f}\begin{bmatrix} 1 + f\cos^2\Theta & -f\sin^2\Theta & 0 & 0 \\ -f\sin^2\Theta & f(1+\cos^2\Theta) & 0 & 0 \\ 0 & 0 & 2f\cos\Theta & 0 \\ 0 & 0 & 0 & (3f-1)\cos\Theta \end{bmatrix}. \quad (59)$$

For the first scattering event of unpolarized light, only the (1,1)-element of Eq. (59) matters, and leads to the *scattering phase function* given by Eq. (23).

In the Stokes vector representation $\mathbf{I} = [I_{\|}, I_{\perp}, U, V]^T$, the corresponding *Stokes scattering matrix* for Rayleigh scattering becomes (using Eqs. (58) and (59) [11] see Problem 2.5)

$$\mathbf{F}(\Theta) = \frac{3}{2(1+2\zeta)}\begin{pmatrix} \cos^2\Theta + \zeta\sin^2\Theta & \zeta & 0 & 0 \\ \zeta & 1 & 0 & 0 \\ 0 & 0 & (1-\zeta)\cos\Theta & 0 \\ 0 & 0 & 0 & (1-3\zeta)\cos\Theta \end{pmatrix} \quad (60)$$

where $\zeta = \rho/(2-\rho) = \frac{1-f}{1+3f}$.

From Eq. (60) we see that for an incident beam of *natural unpolarized light* given by $\mathbf{I}^{\mathrm{inc}} = [I_{\|}^{\mathrm{inc}}, I_{\perp}^{\mathrm{inc}}, U^{\mathrm{inc}}, V^{\mathrm{inc}}]^T = [\frac{1}{2}I^{\mathrm{inc}}, \frac{1}{2}I^{\mathrm{inc}}, 0, 0]^T$, the scattered intensities in the plane parallel and perpendicular to the *scattering plane* are (obtained by carrying out the multiplication $\mathbf{I}^{\mathrm{sca}} = [I_{\|}^{\mathrm{sca}}, I_{\perp}^{\mathrm{sca}}, U^{\mathrm{sca}}, V^{\mathrm{sca}}]^T = \mathbf{F}(\Theta)\mathbf{I}^{\mathrm{inc}} = \mathbf{F}(\Theta)[I_{\|}^{\mathrm{inc}}, I_{\perp}^{\mathrm{inc}}, U^{\mathrm{inc}}, V^{\mathrm{inc}}]^T$):

$$I_{\|}^{\mathrm{sca}} \propto \frac{3}{4(1+2\zeta)}[2\zeta + (1-\zeta)\cos^2\Theta]I^{\mathrm{inc}} \quad (61)$$

$$I_{\perp}^{\mathrm{sca}} \propto \frac{3}{4(1+2\zeta)}[(1+\zeta)]I^{\mathrm{inc}}. \quad (62)$$

Thus, for unpolarized incident light, the scattered light at right angles ($\Theta = 90°$) to the direction of incidence defines the *depolarization factor*:

$$\rho \equiv \left(\frac{I_{\|}^{\mathrm{sca}}}{I_{\perp}^{\mathrm{sca}}}\right)_{\Theta = 90°} = \frac{2\zeta}{1+\zeta} \quad (63)$$

whereas the degree of linear polarization becomes [Eq. (35)]

$$p_l = \frac{I_{\perp} - I_{\|}}{I_{\perp} + I_{\|}} = \frac{(1-\zeta)(1-\cos^2\Theta)}{1 + 3\zeta + (1-\zeta)\cos^2\Theta} \to \frac{1-\zeta}{1+3\zeta} = \frac{1-\rho}{1+\rho} = f \quad \text{as} \quad \Theta \to 90°.$$

2.3.3
Generalized Spherical Functions

For unpolarized radiation, only the $a_1(\Theta)$ element of the Stokes scattering matrix Eq. (36) is relevant, and this element is the scattering phase function given by Eq. (8) in general, and by Eq. (23) for Rayleigh scattering. As discussed above, the scattering phase function can be expanded in Legendre polynomials [see Eq. (12)], enabling one to express it as a Fourier cosine series [see Eq. (18)].

In a similar manner, the scattering phase matrix can be expanded in *generalized spherical functions.* In the Stokes vector representation $\mathbf{I}_S = [I, Q, U, V]^T$, the scattering phase matrix is $\mathbf{P}_S(\Theta) = \mathbf{P}_S(u', u; \phi' - \phi)$, with $u = \cos\theta$, θ being the polar angle after scattering and $u' = \cos\theta'$, θ' the polar angle prior to scattering. Similarly, ϕ and ϕ' are the azimuth angles after and before scattering, respectively. To accomplish the expansion in generalized spherical functions, the scattering phase matrix is first expanded in an $(M+1)$-term *Fourier series* in the azimuth angle difference ($\Delta\phi = \phi' - \phi$):

$$\mathbf{P}_S(u', u; \Delta\phi) = \sum_{m=0}^{M} \{\mathbf{P}_c^m(u', u) \cos m(\Delta\phi) + \mathbf{P}_s^m(u', u) \sin m(\Delta\phi)\} \tag{64}$$

where $\mathbf{P}_c^m(u', u)$ and $\mathbf{P}_s^m(u', u)$ are the coefficient matrices of the cosine and sine terms, respectively, of the Fourier series.

Symmetry Relations

As discussed by Hovenier *et al.* [16], because of *mirror symmetry* the cosine terms of $\mathbf{P}_S(u', u; \phi' - \phi)$ occur in 2×2 submatrices in the upper left corner and the lower right corner, while the sine terms occur in the remaining submatrices. As a consequence, each of the coefficient matrices $\mathbf{P}_c^m(u', u)$ for the cosine terms in Eq. (64) has two 2×2 zero submatrices, one in the upper right corner and another in the lower left corner. Similarly, the coefficient matrices $\mathbf{P}_s^m(u', u)$ for the sine terms have two 2×2 zero submatrices, one in the upper left corner and another in the lower right corner. Because of the mirror symmetry, the following relations apply [16]:

$$\boldsymbol{\Delta}_{3,4} \mathbf{P}_c^m(u', u) \boldsymbol{\Delta}_{3,4} = \mathbf{P}_c^m(u', u) \tag{65}$$

and

$$\boldsymbol{\Delta}_{3,4} \mathbf{P}_s^m(u', u) \boldsymbol{\Delta}_{3,4} = -\mathbf{P}_s^m(u', u) \tag{66}$$

where $\boldsymbol{\Delta}_{3,4} = \text{diag}(1, 1, -1, -1)$. In addition, the following symmetry relations due to reciprocity apply:

$$\mathbf{P}_c^m(-u', -u) = \boldsymbol{\Delta}_3 (\mathbf{P}_c^m(u, u'))^T \boldsymbol{\Delta}_3 = \mathbf{P}_c^m(u', u) \tag{67}$$

$$\mathbf{P}_s^m(-u', -u) = -\boldsymbol{\Delta}_3 (\mathbf{P}_s^m(u, u'))^T \boldsymbol{\Delta}_3 = -\mathbf{P}_s^m(u', u) \tag{68}$$

where $\boldsymbol{\Delta}_3 = \text{diag}(1, 1, -1, 1)$.

An *addition theorem* for the generalized spherical functions can be used to express the Fourier expansion coefficient matrices directly in terms of the expansion coefficients of the *Stokes scattering matrix* $\mathbf{F}_S(\Theta)$ [see Eq. (36)] as follows [85, 86, 88]:

$$\mathbf{P}_c^m(u', u) = \mathbf{A}^m(u', u) + \mathbf{\Delta}_{3,4}\mathbf{A}^m(u', u)\mathbf{\Delta}_{3,4} \tag{69}$$

$$\mathbf{P}_s^m(u', u) = \mathbf{A}^m(u', u)\mathbf{\Delta}_{3,4} - \mathbf{\Delta}_{3,4}\mathbf{A}^m(u', u). \tag{70}$$

The matrix $\mathbf{A}^m(u', u)$ is given by

$$\mathbf{A}^m(u', u) = \sum_{\ell=m}^{M} \mathbf{P}_\ell^m(u)\mathbf{\Lambda}_\ell \mathbf{P}_\ell^m(u') \tag{71}$$

where

$$\mathbf{\Lambda}_\ell = \begin{pmatrix} \alpha_{1,\ell} & \beta_{1,\ell} & 0 & 0 \\ \beta_{1,\ell} & \alpha_{2,\ell} & 0 & 0 \\ 0 & 0 & \alpha_{3,\ell} & \beta_{2,\ell} \\ 0 & 0 & -\beta_{2,\ell} & \alpha_{4,\ell} \end{pmatrix} \tag{72}$$

and

$$a_1(\Theta) = \sum_{\ell=0}^{M} \alpha_{1,\ell} P_\ell^{0,0}(\cos\Theta) \tag{73}$$

$$a_2(\Theta) + a_3(\Theta) = \sum_{\ell=2}^{M} (\alpha_{2,\ell} + \alpha_{3,\ell}) P_\ell^{2,2}(\cos\Theta) \tag{74}$$

$$a_2(\Theta) - a_3(\Theta) = \sum_{\ell=2}^{M} (\alpha_{2,\ell} - \alpha_{3,\ell}) P_\ell^{2,-2}(\cos\Theta) \tag{75}$$

$$a_4(\Theta) = \sum_{\ell=0}^{M} \alpha_{4,\ell} P_\ell^{0,0}(\cos\Theta) \tag{76}$$

$$b_1(\Theta) = \sum_{\ell=2}^{M} \beta_{1,\ell} P_\ell^{0,2}(\cos\Theta) \tag{77}$$

$$b_2(\Theta) = \sum_{\ell=2}^{M} \beta_{2,\ell} P_\ell^{0,2}(\cos\Theta). \tag{78}$$

Here, $\alpha_{j,\ell}$ and $\beta_{j,\ell}$ are *expansion coefficients,* and $a_j(\Theta)$ and $b_j(\Theta)$ are the elements of the *Stokes scattering matrix* $\mathbf{F}_S(\Theta)$ in Eq. (36).

The matrix $\mathbf{P}_\ell^m(u)$ occurring in Eq. (71) is defined as follows:

$$\mathbf{P}_\ell^m(u) = \begin{pmatrix} P_\ell^{m,0}(u) & 0 & 0 & 0 \\ 0 & P_\ell^{m,+}(u) & P_\ell^{m,-}(u) & 0 \\ 0 & P_\ell^{m,-}(u) & P_\ell^{m,+}(u) & 0 \\ 0 & 0 & 0 & P_\ell^{m,0}(u) \end{pmatrix} \tag{79}$$

where

$$P_\ell^{m,\pm}(u) = \frac{1}{2}[P_\ell^{m,-2}(u) \pm P_\ell^{m,2}(u)] \tag{80}$$

and the functions $P_\ell^{m,0}(u)$ and $P_\ell^{m,\pm 2}(u)$ are the *generalized spherical functions*. Readers interested in more details about the generalized spherical functions and how they are computed are advised to consult Section 2.8 and Appendix B of the book by Hovenier *et al.* [16].

Before proceeding, we note that in the scalar (unpolarized) case all the components of the Stokes scattering matrix $\mathbf{F}_S(\Theta)$ [see Eq. (36)] are zero except for $a_1(\Theta)$, and

$$a_1(\Theta) = \sum_{\ell=0}^{M} \alpha_{1,\ell}(\tau) P_\ell^{0,0}(\cos\Theta) \equiv p(\tau, \cos\Theta) \approx \sum_{\ell=0}^{M} (2\ell+1)\chi_\ell(\tau) P_\ell(\cos\Theta) \quad (81)$$

since $P_\ell^{0,0}(\cos\Theta) \equiv P_\ell(\cos\Theta)$, where $P_\ell(\cos\Theta)$ is the *Legendre polynomial* of order ℓ, and $\alpha_{1,\ell}(\tau) \equiv (2\ell+1)\chi_\ell(\tau)$.

Note that the expansion coefficients given above are for the scattering phase matrix $\mathbf{P}_S(\Theta)$, which relates the incident and scattered Stokes vectors in the representation $\mathbf{I}_S = [I, Q, U, V]^T$. It is related to the representation $\mathbf{I} = [I_{||}, I_\perp, U, V]^T$ through $\mathbf{I}_S = \mathbf{DI}$ [Eq. (53)], where $\mathbf{D}$ is given by Eq. (31), and where the scattering phase matrix $\mathbf{P}(\Theta)$ is related to the scattering phase matrix $\mathbf{P}_S(\Theta)$ as follows: $\mathbf{P}(\Theta) = \mathbf{D}^{-1}\mathbf{P}_S(\Theta)\mathbf{D}$ [see Eq. (57)].

2.4
IOPs of a Polydispersion of Particles – Integration over the Size Distribution

Particles encountered in nature consist of a variety of chemical compositions, sizes, and shapes. The chemical composition determines the *refractive index* of the particle, and unless the composition is the same throughout the particle, the refractive index will depend on the location inside the particle. The computation of *IOPs* for such a collection of particles requires solutions of *Maxwell's equations* for electromagnetic radiation interacting with an inhomogeneous, nonspherical particle of a given size. Then one needs to integrate over the size and shape for particles of a given chemical composition, and, finally, average over the particle composition. To avoid having to deal with this complexity and simplify the problem, it is frequently assumed that the particles are homogeneous with a constant refractive index, and that the shapes can be taken to be spherical. Even with these assumptions, one still needs to deal with the variety of sizes of particles encountered in nature.

For a spherical particle with given radius and refractive index, one may use the *Mie–Lorenz theory* (to be reviewed in the next section) and the corresponding computational Mie code to generate IOPs for a single particle. Thus, assuming that the particles are spherical and that we have computed the IOPs for a single particle with specified refractive index and a given size, we may compute the *absorption* and *scattering coefficients* and the *scattering phase function* for a *polydispersion* of particles by integrating over the PSD:

$$\alpha_p(\lambda) = \int_{r_{\min}}^{r_{\max}} \alpha_n(\lambda, r) n(r) dr = \int_{r_{\min}}^{r_{\max}} \pi r^2 Q'_\alpha(\lambda, r) n(r) dr \tag{82}$$

$$\beta_p(\lambda) = \int_{r_{\min}}^{r_{\max}} \beta_n(\lambda, r) n(r) dr = \int_{r_{\min}}^{r_{\max}} \pi r^2 Q'_\beta(\lambda, r) n(r) dr \tag{83}$$

$$p_p(\lambda, \Theta) = \frac{\int_{r_{\min}}^{r_{\max}} p(\lambda, \Theta, r) n(r) dr}{\int_{r_{\min}}^{r_{\max}} n(r) dr} \tag{84}$$

where $n(r)$ is the PSD, and $\alpha_n(\lambda, r)$, $\beta_n(\lambda, r)$, and $p(\lambda, \Theta, r)$ are the *absorption cross section*, the *scattering cross section*, and the *scattering phase function* per particle of radius r. The absorption or scattering "efficiency," $Q'_\alpha(r)$ or $Q'_\beta(r)$, is defined as the ratio of the absorption or scattering cross section for a spherical particle of radius r to the geometrical cross section πr^2. The *scattering phase function* $p(\lambda, \Theta, r)$ in Eq. (84) is the $a_1(\Theta)$ element of the *Stokes scattering matrix* [Eq. (36)]. Since a Mie code can be used to compute all elements of the Stokes scattering matrix in Eq. (36), we may use an expression analogous to Eq. (84) to carry out the integration over the PSD for each of the matrix elements.

2.4.1 IOPs for a Mixture of Different Particle Types

Consider a particle mixture consisting of a total of N particles per unit volume in a layer of thickness Δz, and let $N = \sum_i n_i$ and $f_i = n_i/N$, where n_i is the concentration and f_i the fraction of homogeneous particles (with fixed chemical composition or refractive index) of type labeled i. To compute IOPs for the mixture of particles, we define $\beta_{n,i}$ = *scattering cross section*, $\alpha_{n,i}$ = *absorption cross section*, $\gamma_{n,i} = \beta_{n,i} + \alpha_{n,i}$ = *extinction cross section*, and $\varpi_i = \beta_{n,i} n_i / \gamma_{n,i} n_i = \beta_i / \gamma_i$ = *single-scattering albedo*, where the subscript i stands for the particle type. Weighting by number concentration may be used to create IOPs for the particle mixture. Thus, by combining the absorption and scattering cross sections and the moments of the scattering phase matrix elements, one obtains the following *IOPs* of the mixture (subscript m stands for mixture):

$$\Delta\tau_m = \Delta z \sum_i n_i \gamma_{n,i} = \Delta z \sum_i \gamma_i = \gamma_m \Delta z \tag{85}$$

$$\varpi_m = \frac{\beta_m}{\gamma_m} = \frac{\sum_i \beta_{n,i} n_i}{\sum_i \gamma_{n,i} n_i} = \frac{\sum_i \beta_i}{\sum_i \gamma_i} = \frac{\sum_i \varpi_i \gamma_{n,i} f_i}{\sum_i \gamma_{n,i} f_i} \tag{86}$$

$$\chi_{m,\ell} = \frac{\sum_i \beta_{n,i} n_i \chi_{i,\ell}}{\sum_i \beta_{n,i} n_i} = N \frac{\sum_i \beta_{n,i} f_i \chi_{i,\ell}}{\beta_m} = \frac{\sum_i f_i \varpi_i \gamma_{n,i} \chi_{i,\ell}}{\sum_i f_i \beta_{n,i}} \tag{87}$$

where $\Delta\tau_m$ = layer optical depth; β_m = total scattering coefficient; γ_m = total extinction coefficient; ϖ_m = single-scattering albedo; and $\chi_{m,\ell}$ = scattering phase function *expansion coefficient* for the particle mixture [see Eqs. (13) and (22)].

A mixing rule similar to Eq. (87) should be used for each element of the scattering phase matrix.

2.4.2 Treatment of Strongly Forward-Peaked Scattering

In scalar RT models, strongly forward-peaked scattering phase functions such as those illustrated in Figures 1–2 may be approximated by a combination of a delta function to represent the forward-scattering peak and a smoother residual scattering phase function as follows [89]:

$$p'(\cos\Theta) = 2f_{\delta M}\delta(1 - \cos\Theta) + (1 - f_{\delta M})\hat{p}(\cos\Theta) \tag{88}$$

where

$$\hat{p}(\cos\Theta) = \sum_{\ell=0}^{2N-1}(2\ell + 1)\hat{\chi}_{\ell}P_{\ell}(\cos\Theta). \tag{89}$$

In the *delta-M method* [89], the expansion coefficients $\hat{\chi}_{\ell}$ are obtained from the *Legendre polynomial* expansion coefficients $\chi_{\ell} = \frac{1}{2}\int_{-1}^{1}P_{\ell}(\cos\Theta)p_{\mathrm{ac}}(\cos\Theta)d\cos\Theta$ of the actual phase function $p_{\mathrm{ac}}(\cos\Theta)$ [see Eqs. (13) and (22)]

$$\hat{\chi}_{\ell} = \frac{\chi_{\ell} - f_{\delta M}}{1 - f_{\delta M}} \qquad \ell = 0, \ldots, 2N - 1 \tag{90}$$

and the fraction $f_{\delta M}$ is set to

$$f_{\delta M} = \chi_{2N}. \tag{91}$$

Note that the (1,1) element of the *Stokes scattering matrix* [see Eq. (36)], expanded in the *generalized spherical functions* in Eq. (72), is given by [see Eq. (81)]

$$\alpha_{1,\ell} = (2\ell + 1)\chi_{\ell}. \tag{92}$$

To extend the *delta-M method* to apply to the vector RT case, the diagonal elements of the matrix $\mathbf{\Lambda}_{\ell}$ given by Eq. (72) are assumed to scale like $\alpha_{1,\ell}$:

$$\hat{\alpha}_{i,\ell} = \frac{\alpha_{i,\ell} - f_{\alpha}(2\ell + 1)}{1 - f_{\alpha}}; \qquad i = 1, \ldots, 4 \tag{93}$$

where $f_{\alpha} \equiv f_{\delta M}/(2N + 1)$, while the off-diagonal elements of Eq. (72) are assumed to scale like [48, 49, 90]

$$\hat{\beta}_{i,\ell} = \frac{\beta_{i,\ell}}{1 - f_{\alpha}}; \qquad i = 1, 2. \tag{94}$$

When $\hat{\alpha}_{1,\ell} = [\alpha_{1,\ell} - f_{\alpha}(2\ell + 1)]/(1 - f_{\alpha})$ is substituted into the *RTE* applicable to the scalar case [see Eq. (350) or Eq. (360)], an identical form of the RTE is obtained, but with $\hat{p}$, $\hat{\tau}$, $\hat{\varpi}$ replacing p, τ, ϖ, respectively, where $\hat{p}$ is given by Eq. (89), and

$$d\hat{\tau} = (1 - \varpi f_{\alpha})d\tau; \qquad \hat{\varpi} = \frac{\varpi(1 - f_{\alpha})}{1 - \varpi f_{\alpha}}. \tag{95}$$

Likewise, when Eqs. (93)–(94) are substituted into the RTE applicable to the vector case [see Eq. (353) or Eq. (382)], an identical form of the vector RTE is obtained with scaled optical depth and *single-scattering albedo* given by Eq. (95).

The *delta-M* method yields accurate irradiance results, but may lead to significant errors in computed radiances. To improve the performance of this approach for radiance computations, the $\delta - $*fit method* [91] was introduced, in which the expansion coefficients were selected so as to minimize the least squares residual ϵ between the approximate phase function $\hat{p}'(\cos\Theta)$ and the actual *scattering phase function*, where

$$\epsilon = \sum_k w_k \left(\frac{\hat{p}'(\cos\Theta_k)}{p_{\rm ac}(\cos\Theta_k)} - 1\right)^2 \tag{96}$$

where Θ_k is the *scattering angle*, w_k is the associated weight, and

$$\hat{p}(\cos\Theta_k) \rightarrow \hat{p}'(\cos\Theta_k) = \sum_{\ell=0}^{2N-1} (2\ell+1) c_\ell P_\ell(\cos\Theta_k). \tag{97}$$

The new *expansion coefficients* c_ℓ are determined by solving the least-squares fitting problem defined by Eq. (96). By setting the weights w_k for the forward-scattering angle to zero, the forward peak will automatically be separated. The separation constant is set to $f_{\delta_{\rm fit}} = 1 - c_0$, and the normalization of the *scattering phase function* is obtained by dividing by $c_0 = 1 - f_{\delta_{\rm fit}}$:

$$p_{\delta-{\rm fit}}(\cos\Theta) = \frac{1}{1 - f_{\delta_{\rm fit}}} \hat{p}'(\cos\Theta). \tag{98}$$

To extend the δ–*fit method* to polarized RT, the elements of the scaled *scattering matrix* are obtained from (see Zhai *et al.* [92])

$$\hat{F}_{ij}(\cos\Theta_k) = \frac{\hat{p}'(\cos\Theta_k)}{p_{\rm ac}(\cos\Theta_k)} F_{ij}(\Theta_k) \tag{99}$$

where $F_{ij}(\cos\Theta_k)$ (other than $F_{11}(\cos\Theta_k) = a_1(\cos\Theta_k) = p_{\rm ac}(\cos\Theta_k)$) are the elements of the Stokes scattering matrix given by Eq. (36). The expansion coefficients $\hat{c}_{ij,\ell}$ for the elements $\hat{F}_{ij}$ are found by minimizing the residual

$$\epsilon_{ij} = \sum_k \left[\sum_{\ell=0}^{2N-1} \hat{c}_{ij,\ell} d_\ell^{mn}(\cos\Theta_k) - \hat{F}_{ij}(\cos\Theta_k) \right]^2 \tag{100}$$

where $d_\ell^{mn}(\cos\Theta)$ are the *Wigner d functions*, which are related to the *generalized spherical functions* by (Zhai *et al.* [92])

$$P_\ell^{mn}(\cos\Theta) = i^{m-n} d_\ell^{mn}(\cos\Theta). \tag{101}$$

For $mn = 00$ and $(i,j) = (4,4)$, $\hat{c}_{ij,\ell}$ corresponds to the element $\alpha_{4,\ell}$ of the matrix $\mathbf{\Lambda}_\ell$ in Eq. (72). For $mn = 20$ and $(i,j) = (1,2)$ and $(3,4)$, $\hat{c}_{ij,\ell}$ correspond to $\beta_{1,\ell}$ and $\beta_{2,\ell}$ in Eq. (72). For $mn = 22$ and $(i,j) = (2,2)$ and $(3,3)$, $\hat{c}_{ij,\ell}$ correspond to $\alpha_{2,\ell}$ and $\alpha_{3,\ell}$ in Eq. (72). Finally, to apply the δ–fit method, the optical depth and *single-scattering albedo* must be scaled according to the expressions given in Eq. (95) with f_α replaced by $f_{\delta_{\rm fit}}$.

2.4.3
Particle Size Distributions (PSDs)

Several types of size distributions have appeared in the literature, including the *gamma distribution* [93], the *modified gamma distribution* [94], the *power-law distribution* [95], and the *log-normal distribution* [95]. For the sake of discussion, let us assume that the *(*PSD) can be described by a log-normal distribution given by

$$n(r) = \frac{dN(r)}{dr} = \frac{1}{r}\frac{dN(r)}{d\ln r} = \frac{N}{\sqrt{2\pi}\sigma}\frac{1}{r}\exp\left[-\left(\frac{\ln r - \ln r_n}{\sqrt{2}\sigma}\right)^2\right] \tag{102}$$

where r_n is the mode radius, $n(r)$ is the number density or PSD in units of $[\mathrm{m}^{-3} \cdot \mathrm{m}^{-1}]$. Per definition, $n(r)dr$ is the fraction of spheres per unit volume with radii between r and $r + dr$, and $N = \int_0^\infty n(r)dr$ $[\mathrm{m}^{-3}]$ is the total number of particles per unit volume since (see Problem 2.6)

$$\int_0^\infty \frac{dr}{\sqrt{2\pi}\sigma}\frac{1}{r}\exp\left[-\left(\frac{\ln r - \ln r_n}{\sqrt{2}\sigma}\right)^2\right] = 1. \tag{103}$$

Note that, since the numerator in the exponential of Eq. (102), $\ln(r/r_n)$, is dimensionless, so is σ. In analogy to the *liquid water content* defined in Eq. (221) below, we may introduce the *particle mass content (PMC)* given by

$$\mathrm{PMC} = \rho_p \int_{r_{\min}}^{r_{\max}} \left(\frac{4\pi}{3}\right) r^3 n(r)dr \equiv \rho_p f_V \qquad [\mathrm{kg}\cdot\mathrm{m}^{-3}] \tag{104}$$

where ρ_p is the bulk particle mass density $[\mathrm{kg} \cdot \mathrm{m}^{-3}]$. Hence, we may define a particle *volume fraction* as

$$f_V \equiv \int_{r_{\min}}^{r_{\max}} \left(\frac{4\pi}{3}\right) r^3 n(r)dr = \mathrm{PMC}/\rho_p \qquad \text{(dimensionless)}. \tag{105}$$

Relationship Between Effective Radius and Mode Radius

The PSD may also be characterized by an *effective radius*

$$r_{\mathrm{eff}} = \frac{\int_{r_{\min}}^{r_{\max}} n(r)r^3dr}{\int_{r_{\min}}^{r_{\max}} n(r)r^2dr} \tag{106}$$

and an *effective variance*

$$v_{\mathrm{eff}} = \frac{\int_{r_{\min}}^{r_{\max}} (r - r_{\mathrm{eff}})^2 n(r)r^2dr}{r_{\mathrm{eff}}^2 \int_{r_{\min}}^{r_{\max}} n(r)r^2dr} \tag{107}$$

where r_{eff}^2 is included in the denominator to make v_{eff} dimensionless (Hansen and Travis [95]). The effective radius r_{eff} can be used to describe the IOPs in an approximate manner as will be discussed below for cloud as well as snow/ice materials. To determine how the effective radius r_{eff} is related to the mode radius r_n,

we make the change of integration variable $x = \frac{1}{\sqrt{2}\sigma}\ln(r/r_n)$ in Eq. (106), so that $dx = \frac{1}{\sqrt{2}\sigma}\frac{dr}{r}$ and $\exp[\sqrt{2}\sigma x] = \frac{r}{r_n}$. Further, we have
$x_{\max} = \frac{1}{\sqrt{2}\sigma}\ln(r_{\max}/r_n) \to +\infty$, when $r_{\max} \to +\infty$
and $x_{\min} = \frac{1}{\sqrt{2}\sigma}\ln(r_{\min}/r_n) \to -\infty$, when $r_{\min} \to 0$.
Thus, Eq. (106) becomes

$$r_{\text{eff}} = r_n \frac{\int_{-\infty}^{+\infty} \exp[-(x^2 - 3\sqrt{2}\sigma x)]dx}{\int_{-\infty}^{+\infty} \exp[-(x^2 - 2\sqrt{2}\sigma x)]dx} \tag{108}$$

which, on completing the square in each of the exponents, leads to (see Problem 2.7)

$$r_{\text{eff}} = r_n \exp[2.5\sigma^2]. \tag{109}$$

Proceeding in a similar manner, one finds that the *effective variance* v_{eff} is related to the variance σ^2 as follows [95]:

$$v_{\text{eff}} = \exp[\sigma^2] - 1. \tag{110}$$

Finally, we note that the sizes of nonspherical particles often are described in terms of their volume-equivalent spheres or projected-surface-area equivalent spheres. The radii of these spheres may then be described by a log-normal or another size distribution, as mentioned above.

2.5
Scattering of an Electromagnetic Wave by Particles

Mie–Lorenz theory [96] refers to scattering of electromagnetic radiation by a homogeneous (nonmagnetic) dielectric particle of spherical shape. Thus, assuming that the particles encountered in nature can be represented by spheres, we may obtain their *IOPs* from Mie computations, which require the *refractive index* and PSD as inputs. Then, the IOPs, that is, the absorption and scattering coefficients and the scattering phase function, $\alpha_p(\lambda)$, $\beta_p(\lambda)$, and $p_p(\lambda, \Theta)$, can be obtained from Eqs. (82)–(84), and all the other elements of the Stokes scattering matrix in Eq. (36) can be obtained using an equation similar to Eq. (84). This approach frequently leads to results that agree surprisingly well with available observations. The following reasons why one does not make large errors by assuming spherical particles have been put forward by Craig Bohren as quoted by Grenfell *et al.* [97]: *The orientationally averaged extinction cross section of a convex particle that is large compared with the wavelength is one-half its surface area. The absorption cross section of a large, nearly transparent particle is proportional to its volume, almost independently of its shape. The closer the real part of the particle's refractive index is to 1, the more irrelevant becomes the particle shape. The asymmetry parameter of a large particle is dominated by near-forward scattering, which does not depend greatly on the particle shape.*

In view of the frequent use of Mie–Lorenz theory in a variety of applications including some reviewed in this book, we provide a brief description of this theory below with details to be provided in Appendix A.

2.5.1
Summary of Electromagnetic Scattering

In this section, we review the scattering of an electromagnetic wave by a finite-sized particle with particular emphasis on scattering of a linearly polarized plane electromagnetic wave by a *spherical* particle, commonly referred to as the *Mie–Lorenz theory*. Details of the theoretical development is given in Appendix A. The main sources used in the preparation of this review are the books by Bohren and Huffman [68] and Born and Wolf [98].

Considering a time-harmonic electromagnetic field $(\mathbf{E}^i, \mathbf{H}^i)$ that is incident upon a particle, and that generates an internal field $(\mathbf{E}^{\text{int}}, \mathbf{H}^{\text{int}})$ inside the particle as well as a scattered field $(\mathbf{E}^{\text{sc}}, \mathbf{H}^{\text{sc}})$ outside the particle, one may formulate the scattering problem as follows:

1) The incident field $(\mathbf{E}^i, \mathbf{H}^i)$ is considered to be known and the internal field $(\mathbf{E}^{\text{int}}, \mathbf{H}^{\text{int}})$ and the scattered field $(\mathbf{E}^{\text{sc}}, \mathbf{H}^{\text{sc}})$ are to be determined.
2) The scattered field $(\mathbf{E}^{\text{sc}}, \mathbf{H}^{\text{sc}})$ must satisfy the *vector wave equations* (Helmholtz equations), given by

$$(\nabla^2 + k^2)\mathbf{E} = (\nabla^2 + k^2)\mathbf{H} = \mathbf{0} \tag{111}$$

   with $k^2 = k_2^2 = \frac{\omega^2}{c^2}\mu_2\varepsilon_2$, and subscript 2 referring to the surrounding medium.
3) The internal field $(\mathbf{E}^{\text{int}}, \mathbf{H}^{\text{int}})$ must satisfy the vector wave equations [Eqs. (111)] with $k^2 = k_1^2 = \frac{\omega^2}{c^2}\mu_1\left(\varepsilon_1 + 4\pi i\frac{\sigma_1}{\omega}\right)$, and subscript 1 referring to the medium of the particle.
4) The scattered field $(\mathbf{E}^{\text{sc}}, \mathbf{H}^{\text{sc}})$ must satisfy the radiation condition at infinity, that is, it must behave as an outgoing spherical wave.
5) The internal field $(\mathbf{E}^{\text{int}}, \mathbf{H}^{\text{int}})$ must be finite at the origin.
6) The electromagnetic field must satisfy certain *boundary conditions* at the surface of the particle:
   a. If $\sigma_1 = \infty$, the material of the particle is a perfect conductor, which is also a perfect reflector, and the tangential component of $\mathbf{E}$ must vanish at the surface of the particle;
   b. If σ_1 is finite, the tangential components of $\mathbf{E}$ and $\mathbf{H}$ must be continuous across the interface between the particle and the surrounding medium, that is,

$$\left(\mathbf{E}^i + \mathbf{E}^{\text{sc}} \quad \mathbf{E}^{\text{int}}\right) \times \hat{\mathbf{n}} - \left(\mathbf{II}^i + \mathbf{H}^{\text{sc}} - \mathbf{H}^{\text{int}}\right) \times \hat{\mathbf{n}} = 0 \tag{112}$$

   where $\hat{\mathbf{n}}$ is a unit vector along the outward surface normal, and the electric field in the medium surrounding the particle is the sum of the incident electric field $\mathbf{E}^i$ and the scattered electric field $\mathbf{E}^{\text{sc}}$.

2.5.2
Amplitude Scattering Matrix

Consider a particle with material properties $\epsilon_1,\ \mu_1,\ \sigma_1$ that is embedded in a medium with material properties $\epsilon_2,\ \mu_2,\ \sigma_2 = 0$, and let a time-harmonic electromagnetic *plane wave* given by

$$\mathbf{E}^i = \Re\left\{\mathbf{E}_0^i e^{ik_2 z} e^{-i\omega t)}\right\}\ ;\ \mathbf{H}^i = \Re\left\{\mathbf{H}_0^i e^{ik_2 z} e^{-i\omega t)}\right\}\ ;\ k_2^2 = \frac{\omega_2^2}{c^2}\epsilon_2\mu_2 \tag{113}$$

be incident upon the particle and give rise to a scattered field ($\mathbf{E}^{\mathrm{sc}}, \mathbf{H}^{\mathrm{sc}}$) outside the particle and a field ($\mathbf{E}^{\mathrm{int}}, \mathbf{H}^{\mathrm{int}}$) inside the particle. Here, $k_2 = \frac{2\pi}{\lambda} n_2$, with λ and $n_2 = \sqrt{\varepsilon_2\mu_2}$ being, respectively, the wavelength in vacuum and the refractive index of the surrounding medium.

One may decompose the electric field into components parallel and perpendicular to the *scattering plane*, which is spanned by the propagation direction of the incident plane wave (along $\hat{\mathbf{e}}_z$) and the observation direction (along $\hat{\mathbf{e}}_r$). Thus, for linear polarization the amplitude part of the incident plane wave becomes

$$\mathbf{E}_0^i = \left(E_{0\|}\hat{\mathbf{e}}_{\|i} + E_{0\perp}\hat{\mathbf{e}}_{\perp i}\right) = E_{\|}^i\hat{\mathbf{e}}_{\|i} + E_{\perp}^i\hat{\mathbf{e}}_{\perp i}. \tag{114}$$

Let the projection of $\hat{\mathbf{e}}_r$ onto the plane $z = 0$ make an angle ϕ with the x-axis. Since this projection is along $\hat{\mathbf{e}}_{\|i}$, it follows that

$$E_{\|}^i = E_x^i \cos\phi + E_y^i \sin\phi \tag{115}$$

$$E_{\perp}^i = E_x^i \sin\phi - E_y^i \cos\phi. \tag{116}$$

In the *far zone*, where $kr \gg 1$, the expression for the scattered electric field becomes

$$\mathbf{E}^{\mathrm{sc}} \sim i\frac{e^{ik_2 r}}{k_2 r}\mathbf{X}\ ;\ \mathbf{X}\cdot\hat{\mathbf{e}}_r = 0\ ;\ \mathbf{H}^{\mathrm{sc}} = \frac{c}{\omega\mu_2}\mathbf{k}^{\mathrm{sc}}\times\mathbf{E}^{\mathrm{sc}} \sim i\frac{ck_2}{\omega\mu_2}\frac{e^{ik_2 r}}{k_2 r}\hat{\mathbf{e}}_r\times\mathbf{X} \tag{117}$$

where

$$\mathbf{X} = E_{\|}^{\mathrm{sc}}\hat{\mathbf{e}}_{\|s} + E_{\perp}^{\mathrm{sc}}\hat{\mathbf{e}}_{\perp s}\ ;\ \hat{\mathbf{e}}_{\|s} = \hat{\mathbf{e}}_\theta\ ;\ \hat{\mathbf{e}}_{\perp s} = -\hat{\mathbf{e}}_\phi\ ;\ \hat{\mathbf{e}}_{\perp s}\times\hat{\mathbf{e}}_{\|s} = \hat{\mathbf{e}}_r. \tag{118}$$

Because *Maxwell's equations* and the *boundary conditions* in Eq. (112) are linear, the scattered electric field components $E_{\|}^{\mathrm{sc}}$ and $E_{\perp}^{\mathrm{sc}}$ must be linearly related to the incident electric field components $E_{\|}^i$ and $E_{\perp}^i$, that is,

$$\begin{pmatrix} E_{\|}^{\mathrm{sc}} \\ E_{\perp}^{\mathrm{sc}} \end{pmatrix} = i\frac{e^{ik_2 r}}{k_2 r}\begin{pmatrix} S_2 & S_3 \\ S_4 & S_1 \end{pmatrix}\begin{pmatrix} E_{\|}^i \\ E_{\perp}^i \end{pmatrix} \tag{119}$$

where the elements S_j ($j = 1, 2, 3, 4$) of the *amplitude scattering matrix* depend on the *scattering angle* Θ and the *azimuth angle* ϕ. From Eqs. (118) and (119), it follows that

$$\mathbf{X} = (S_2 E_{\|}^i + S_3 E_{\perp}^i)\hat{\mathbf{e}}_{\|s} + (S_4 E_{\|}^i + S_1 E_{\perp}^i)\hat{\mathbf{e}}_{\perp s}. \tag{120}$$

2.5.3 Scattering Matrix

The time-averaged *Poynting vector* of the field in the medium surrounding the particle is

$$\mathbf{S}_2 = \frac{c}{8\pi}\Re\{\mathbf{E}_2 \times \mathbf{H}_2^*\} = \mathbf{S}^i + \mathbf{S}^{sc} + \mathbf{S}^{ext} \tag{121}$$

where

$$\mathbf{E}_2 = \mathbf{E}^i + \mathbf{E}^{sc} \quad ; \quad \mathbf{H}_2 = \mathbf{H}^i + \mathbf{H}^{sc} \tag{122}$$

so that

$$\mathbf{S}^i = \frac{c}{8\pi}\Re\{\mathbf{E}^i \times (\mathbf{H}^i)^*\} \tag{123}$$

$$\mathbf{S}^{sc} = \frac{c}{8\pi}\Re\{\mathbf{E}^{sc} \times (\mathbf{H}^{sc})^*\} \tag{124}$$

$$\mathbf{S}^{ext} = \frac{c}{8\pi}\Re\{\mathbf{E}^i \times (\mathbf{H}^{sc})^* + \mathbf{E}^{sc} \times (\mathbf{H}^i)^*\} \tag{125}$$

where $\mathbf{S}^i$ and $\mathbf{S}^{sc}$ are the Poynting vectors of the incident and scattered fields, respectively, whereas $\mathbf{S}^{ext}$ is the *Poynting vector* of the field that is due to the interaction between the incident and scattered fields.

For a scattering arrangement in which a detector "sees" only the scattered field, the recorded signal is proportional to

$$\mathbf{S}^{sc} \cdot \hat{\mathbf{e}}_r \Delta A \tag{126}$$

where ΔA is the detector area. From Eqs. (117) and (124), it follows that

$$\begin{aligned} \mathbf{S}^{sc} \cdot \hat{\mathbf{e}}_r \Delta A &= \frac{c}{8\pi}\Re\,(\mathbf{E}^{sc} \times (\mathbf{H}^{sc})^*) \cdot \hat{\mathbf{e}}_r \Delta A \\ &= \frac{c}{8\pi}\Re\left(i\frac{e^{ik_2 r}}{k_2 r}\mathbf{X} \times (-i)\frac{ck_2}{\omega\mu_2}\frac{e^{-ik_2 r}}{k_2 r}\hat{\mathbf{e}}_r \times \mathbf{X}^* \right) \cdot \hat{\mathbf{e}}_r \Delta A \\ &= \frac{c}{8\pi}\frac{ck_2}{\omega\mu_2}|C|^2|\mathbf{X}|^2\Delta\Omega \quad ; \quad C = i\frac{e^{ik_2 r}}{k_2 r} \end{aligned} \tag{127}$$

where $\Delta\Omega = \frac{\Delta A}{r^2}$ is the solid angle subtended by the detector.

For a time-harmonic incident electromagnetic *plane wave* with $\mathbf{H}^i = \frac{c}{\omega\mu_2}\mathbf{k}_2 \times \mathbf{E}^i$, the Poynting vector becomes

$$\mathbf{S}^i = \frac{c}{8\pi}\Re\left(\mathbf{E}^i \times (\mathbf{H}^i)^*\right) = \frac{c}{8\pi}\Re\left(\mathbf{E}^i \times \frac{c}{\omega\mu_2}\mathbf{k}_2 \times (\mathbf{E}^i)^*\right) = S^i\hat{\mathbf{s}} \tag{128}$$

where $\hat{\mathbf{s}}$ is a unit vector in the direction of the Poynting vector and S^i is its magnitude given by

$$S^i = \frac{c}{8\pi}\frac{ck_2}{\omega\mu_2}|\mathbf{E}^i|^2 = \frac{c}{8\pi}\frac{ck_2}{\omega\mu_2}|E|^2. \tag{129}$$

Letting $\mathbf{X} = E\mathbf{X}'$, where E is the electric field amplitude of the incident plane wave, one obtains from Eq. (127)

$$\mathbf{S}^{\mathrm{sc}} \cdot \hat{\mathbf{e}}_r = S^i |C|^2 |\mathbf{X}'|^2 \quad ; \quad C = i\frac{e^{ik_2 r}}{k_2 r}. \tag{130}$$

Thus, by recording the detector response at various positions on a hemisphere surrounding the particle, one can determine the angular variation of $|\mathbf{X}|^2$. The *Stokes parameters* of the light scattered by the particle are given by

$$I^{\mathrm{sc}} = E_{\|}^{\mathrm{sc}}(E_{\|}^{\mathrm{sc}})^* + E_{\perp}^{\mathrm{sc}}(E_{\perp}^{\mathrm{sc}})^* \tag{131}$$
$$Q^{\mathrm{sc}} = E_{\|}^{\mathrm{sc}}(E_{\|}^{\mathrm{sc}})^* - E_{\perp}^{\mathrm{sc}}(E_{\perp}^{\mathrm{sc}})^* \tag{132}$$
$$U^{\mathrm{sc}} = E_{\|}^{\mathrm{sc}}(E_{\perp}^{\mathrm{sc}})^* + E_{\perp}^{\mathrm{sc}}(E_{\|}^{\mathrm{sc}})^* \tag{133}$$
$$V^{\mathrm{sc}} = i[E_{\|}^{\mathrm{sc}}(E_{\perp}^{\mathrm{sc}})^* - E_{\perp}^{\mathrm{sc}}(E_{\|}^{\mathrm{sc}})^*] \tag{134}$$

consistent with the representation $I_S = [I, Q, U, V]^T$, with $I = I_{\|} + I_{\perp}$ and $Q = I_{\|} - I_{\perp}$, where $I_{\|}$, $I_{\perp}$, U, and V are given in Eq. (29). The *scattering matrix*, which by definition is the *Mueller matrix* for scattering by a single particle, follows from Eqs. (119) and (131)–(134):

$$\begin{pmatrix} I^{\mathrm{sc}} \\ Q^{\mathrm{sc}} \\ U^{\mathrm{sc}} \\ V^{\mathrm{sc}} \end{pmatrix} = \begin{pmatrix} S_{11} & S_{12} & S_{13} & S_{14} \\ S_{21} & S_{22} & S_{23} & S_{24} \\ S_{31} & S_{32} & S_{33} & S_{34} \\ S_{41} & S_{42} & S_{43} & S_{44} \end{pmatrix} \begin{pmatrix} I^i \\ Q^i \\ U^i \\ V^i \end{pmatrix} \tag{135}$$

where

$$S_{11} = \frac{1}{2}\left(|S_1|^2 + |S_2|^2 + |S_3|^2 + |S_4|^2\right) \tag{136}$$
$$S_{12} = \frac{1}{2}\left(|S_2|^2 - |S_1|^2 + |S_4|^2 - |S_3|^2\right) \tag{137}$$
$$S_{13} = \Re\left(S_2 S_3^* + S_1 S_4^*\right) \tag{138}$$
$$S_{14} = \Im\left(S_2 S_3^* - S_1 S_4^*\right) \tag{139}$$
$$S_{21} = \frac{1}{2}\left(|S_2|^2 - |S_1|^2 - |S_4|^2 + |S_3|^2\right) \tag{140}$$
$$S_{22} = \frac{1}{2}\left(|S_2|^2 + |S_1|^2 - |S_4|^2 - |S_3|^2\right) \tag{141}$$
$$S_{23} = \Re\left(S_2 S_3^* - S_1 S_4^*\right) \tag{142}$$
$$S_{24} = \Im\left(S_2 S_3^* + S_1 S_4^*\right) \tag{143}$$
$$S_{31} = \Re\left(S_2 S_4^* + S_1 S_3^*\right) \tag{144}$$
$$S_{32} = \Re\left(S_2 S_4^* - S_1 S_3^*\right) \tag{145}$$
$$S_{33} = \Re\left(S_1 S_2^* + S_3 S_4^*\right) \tag{146}$$
$$S_{34} = \Im\left(S_2 S_1^* + S_4 S_3^*\right) \tag{147}$$
$$S_{41} = \Im\left(S_2^* S_4 + S_3^* S_1\right) \tag{148}$$
$$S_{42} = \Im\left(S_2^* S_4 - S_3^* S_1\right) \tag{149}$$
$$S_{43} = \Im\left(S_1 S_2^* - S_3 S_4^*\right) \tag{150}$$
$$S_{44} = \Re\left(S_1 S_2^* - S_3 S_4^*\right). \tag{151}$$

These relations are derived in Appendix A (see Problem 2.8).

2.5.4 Extinction, Scattering, and Absorption

When one or more particles are placed in a beam of electromagnetic radiation, the rate at which electromagnetic energy is received by a detector placed in the forward direction is reduced compared to that received in the absence of the particles. Thus, the presence of the particles results in *extinction* of the incident beam. If the medium in which the particles are placed is nonabsorbing, the extinction must be due to *absorption* in the particles and *scattering* by the particles.

Considering extinction by a single particle that is embedded in a nonabsorbing medium and illuminated by a plane wave, one finds that the net rate at which electromagnetic energy crosses the surface A of a sphere of radius r around the particle is

$$W^{\rm ab} = -\int_A \mathbf{S}_2 \cdot \hat{\mathbf{e}}_r dA \tag{152}$$

where the *Poynting vector* $\mathbf{S}_2$ of the fields in the medium surrounding the particle is given by Eq. (121). If $W^{\rm ab} > 0$, energy is absorbed within the sphere of radius r, and if the medium surrounding the particle is nonabsorbing, the energy must be absorbed by the particle.

Using Eq. (121), one may may express $W^{\rm ab}$ as a sum of three terms, that is,

$$W^{\rm ab} = W^i - W^{\rm sc} + W^{\rm ext} \tag{153}$$

where

$$W^i = -\int_A \mathbf{S}^i \cdot \hat{\mathbf{e}}_r dA \quad ; \quad W^{\rm sc} = \int_A \mathbf{S}^{\rm sc} \cdot \hat{\mathbf{e}}_r dA \quad ; \quad W^{\rm ext} = -\int_A \mathbf{S}^{\rm ext} \cdot \hat{\mathbf{e}}_r dA \tag{154}$$

but since $W^i = 0$ for a nonabsorbing medium, Eq. (153) yields

$$W^{\rm ext} = W^{\rm ab} + W^{\rm sc}. \tag{155}$$

For an x polarized incident plane wave with electric field given by Eq. (113), the scattered electric field in the *far zone* is [cf. Eq. (117)]

$$\mathbf{E}^{\rm sc} \sim i\frac{e^{ik_2 r}}{k_2 r}\mathbf{X} \quad ; \quad \mathbf{X} \cdot \hat{\mathbf{e}}_r = 0 \quad ; \quad \mathbf{H}^{\rm sc} = \frac{c}{\omega\mu_2}\mathbf{k}^{\rm sc} \times \mathbf{E}_s \sim i\frac{c}{\omega\mu_2}\frac{e^{ik_2 r}}{r}\hat{\mathbf{e}}_r \times \mathbf{X} \tag{156}$$

where $\mathbf{X}$ is the *vector scattering amplitude* given in Eq. (120) with $E_y^i = 0$ and $E_x^i = E$:

$$\mathbf{X} = E\left[\left(S_2 \cos\phi + S_3 \sin\phi\right)\hat{\mathbf{e}}_{\parallel s} + \left(S_4 \cos\phi + S_1 \sin\phi\right)\hat{\mathbf{e}}_{\perp s}\right] = \mathbf{X}'E. \tag{157}$$

For the time-harmonic incident electromagnetic plane wave given by Eq. (113), the Poynting vector is given by [see Eqs. (128) and (129) of Appendix A]

$$\mathbf{S}^i = S^i\hat{\mathbf{s}} \quad ; \quad S^i = \frac{c}{8\pi}\frac{ck_2}{\omega\mu_2}E^2 \tag{158}$$

where $\hat{\mathbf{s}}$ is a unit vector in the direction of the Poynting vector, S^i is the magnitude of the Poynting vector, and E is the electric field amplitude of the incident plane wave.

In Appendix A, it is shown that W^{ext} given in Eq. (154) can be expressed as

$$W^{\text{ext}} = \frac{4\pi}{k_2^2} S^i \Re\left(\hat{\mathbf{e}}_x \cdot \mathbf{X}'\right)_{\Theta=0} \tag{159}$$

where Θ is the scattering angle. The ratio of W^{ext} to S^i, denoted by C^{ext}, has the dimension of area and is called the *extinction cross section*:

$$C^{\text{ext}} = \frac{W^{\text{ext}}}{S^i} = \frac{4\pi}{k_2^2} \Re\left(\hat{\mathbf{e}}_x \cdot \mathbf{X}'\right)_{\Theta=0}. \tag{160}$$

This result is known as the *optical theorem* and is common to all kinds of scattering phenomena involving, for example, acoustic waves, electromagnetic waves, and elementary particles.

2.6 Absorption and Scattering by Spherical Particles – Mie–Lorenz Theory

Solutions of the *vector wave equations* [Eqs. (111)] may be constructed by considering a scalar function ψ and a vector $\mathbf{c}$, which either may be a constant vector or have zero curl, that is, $\nabla \times \mathbf{c} = 0$, and fulfill $\nabla^2\mathbf{c} = 0$. From ψ and $\mathbf{c}$, one may construct a divergence-free vector function $\mathbf{M}$, given by

$$\mathbf{M} = \nabla \times (\mathbf{c}\psi) = \psi\nabla \times \mathbf{c} - \mathbf{c} \times \nabla\psi = -\mathbf{c} \times \nabla\psi \tag{161}$$

which fulfills $\nabla \cdot \mathbf{M} = 0$. Also, $\mathbf{M}$ satisfies the vector wave equation (see Appendix A) provided ψ is a solution of the *scalar wave equation*

$$(\nabla^2 + k^2)\psi = 0. \tag{162}$$

From $\mathbf{M} = -\mathbf{c} \times \nabla\psi$ [cf. Eq. (161)] it follows that $\mathbf{c} \cdot \mathbf{M} = 0$. Next, one may construct another vector function $\mathbf{N}$ given by

$$\mathbf{N} = \frac{1}{k}\nabla \times \mathbf{M} \tag{163}$$

which clearly has zero divergence ($\nabla \cdot \mathbf{N} = \frac{1}{k}\nabla \cdot (\nabla \times \mathbf{M}) = 0$), and which satisfies

$$\nabla \times \mathbf{N} = \frac{1}{k}\nabla \times (\nabla \times \mathbf{M}) = -\frac{1}{k}\nabla^2\mathbf{M} = k\mathbf{M} \tag{164}$$

where we have made use of $(\nabla^2 + k^2)\mathbf{M} = 0$. Therefore, the vector functions $\mathbf{M}$ and $\mathbf{N}$ have all the required properties of an electromagnetic field: that is

- they satisfy the vector wave equation,
- they have zero divergence, and
- as required by Maxwell's curl equations [see Eq. (940) and (941) of Appendix A], the curl of $\mathbf{M}$ is proportional to $\mathbf{N}$, and the curl of $\mathbf{N}$ is proportional to $\mathbf{M}$.

Thus, one may start by finding solutions ψ of the scalar wave equation. The scalar function ψ is called a *generating function* for the *vector harmonics* **M** and **N**.

Considering scattering by a spherical particle, one may choose the functions ψ that are solutions of the *scalar wave equation* in spherical polar coordinates r, θ, and ϕ. By choosing $\mathbf{c} = \mathbf{r} = x\hat{\mathbf{e}}_x + y\hat{\mathbf{e}}_y + z\hat{\mathbf{e}}_z = r\hat{\mathbf{e}}_r$, one obtains

$$\mathbf{M} = \nabla \times (\mathbf{r}\psi) = -\mathbf{r} \times \nabla\psi \quad ; \quad \mathbf{r} \cdot \mathbf{M} = 0 \tag{165}$$

where $\mathbf{r}$ is the radius vector and **M** is a solution of the *vector wave equation* in spherical polar coordinates. Since $\mathbf{r} \cdot \mathbf{M} = 0$, **M** is tangential to any sphere $|\mathbf{r}| =$ constant.

The scalar wave equation in spherical polar coordinates is

$$\frac{1}{r^2}\frac{\partial}{\partial r}\left(r^2\frac{\partial\psi}{\partial r}\right) + \frac{1}{r^2\sin\theta}\frac{\partial}{\partial\theta}\left(\sin\theta\frac{\partial\psi}{\partial\theta}\right) + \frac{1}{r^2\sin^2\theta}\frac{\partial^2\psi}{\partial\phi^2} + k^2\psi = 0. \tag{166}$$

As shown in Appendix A, from the solutions of the scalar wave equation in spherical polar coordinates, one may construct even and odd *generating functions*:

$$\psi_{emn} = \cos m\phi P_n^{(m)}(\cos\theta) z_n(kr) \quad ; \quad \psi_{omn} = \sin m\phi P_n^{(m)}(\cos\theta) z_n(kr) \tag{167}$$

where $P_n^{(m)}(\cos\theta)$ are *associated Legendre functions* of the first kind of degree n and order m, $z_n(kr)$ are spherical Bessel functions of the first kind $[j_n(kr)]$, second kind $[y_n(kr)]$, or third kind $[h_n^{(1)}(kr) = j_n(kr) + iy_n(kr)$ or $h_n^{(2)}(kr) = j_n(kr) - iy_n(kr)]$.

The *vector spherical harmonics* generated by ψ_{emn} and ψ_{omn} are [see Eqs. (163) and (165)]

$$\mathbf{M}_{emn} = \nabla \times (\mathbf{r}\psi_{emn}) \quad ; \quad \mathbf{M}_{omn} = \nabla \times (\mathbf{r}\psi_{omn}) \tag{168}$$

$$\mathbf{N}_{emn} = \frac{1}{k}\nabla \times \mathbf{M}_{emn} \quad ; \quad \mathbf{N}_{omn} = \frac{1}{k}\nabla \times \mathbf{M}_{omn} \tag{169}$$

which in component form can be written as

$$\mathbf{M}_{emn} = -\frac{m}{\sin\theta}\sin m\phi P_n^{(m)}(\cos\theta) z_n(z)\hat{\mathbf{e}}_\theta - \cos m\phi \frac{dP_n^{(m)}(\cos\theta)}{d\theta} z_n(z)\hat{\mathbf{e}}_\phi \tag{170}$$

$$\mathbf{M}_{omn} = \frac{m}{\sin\theta}\cos m\phi P_n^{(m)}(\cos\theta) z_n(z)\hat{\mathbf{e}}_\theta - \sin m\phi \frac{dP_n^{(m)}(\cos\theta)}{d\theta} z_n(z)\hat{\mathbf{e}}_\phi \tag{171}$$

$$\begin{aligned}\mathbf{N}_{emn} = {} & \frac{z_n(z)}{z}\cos m\phi \, n(n+1) P_n^{(m)}(\cos\theta)\hat{\mathbf{e}}_r \\ & + \cos m\phi \frac{dP_n^{(m)}(\cos\theta)}{d\theta}\frac{1}{z}\frac{d}{dz}\left[z z_n(z)\right]\hat{\mathbf{e}}_\theta \\ & - m \sin m\phi \frac{P_n^{(m)}(\cos\theta)}{\sin\theta}\frac{1}{z}\frac{d}{dz}\left[z z_n(z)\right]\hat{\mathbf{e}}_\phi \end{aligned} \tag{172}$$

$$\mathbf{N}_{omn} = \frac{z_n(z)}{z} \sin m\phi \, n(n+1) P_n^{(m)}(\cos\theta)\hat{\mathbf{e}}_r + \sin m\phi \frac{dP_n^{(m)}(\cos\theta)}{d\theta} \frac{1}{z}\frac{d}{dz}\left[z z_n(z)\right]\hat{\mathbf{e}}_\theta + m\cos m\phi \frac{P_n^{(m)}(\cos\theta)}{\sin\theta}\frac{1}{z}\frac{d}{dz}\left[z z_n(z)\right]\hat{\mathbf{e}}_\phi. \quad (173)$$

Expansion of Incident Plane Wave in Vector Spherical Harmonics

Let the incident field be an x-polarized *plane wave* propagating in the z direction, so that in spherical polar coordinates [omitting the time dependence $\exp(-i\omega t)$]

$$\mathbf{E}^i = E e^{ikr\cos\theta}\hat{\mathbf{e}}_x \quad (174)$$

where

$$\hat{\mathbf{e}}_x = \sin\theta\cos\phi\hat{\mathbf{e}}_r + \cos\theta\cos\phi\hat{\mathbf{e}}_\theta - \sin\phi\hat{\mathbf{e}}_\phi. \quad (175)$$

Note that, since the propagation direction of the incident plane wave is along the z-axis, the *scattering angle* Θ is equal to the polar angle θ. It follows from Eqs. (115) and (116) with $E_y^i = 0$ and $E_x^i = E$ that

$$E_{\|}^i = E\cos\phi \;\; ; \;\; E_{\perp}^i = E\sin\phi. \quad (176)$$

Expanding the incident field in *vector spherical harmonics*, one obtains

$$\mathbf{E}^i = \sum_{m=0}^{\infty}\sum_{n=m}^{\infty}\{B_{emn}\mathbf{M}_{emn} + B_{omn}\mathbf{M}_{omn} + A_{emn}\mathbf{N}_{emn} + A_{omn}\mathbf{N}_{omn}\} \quad (177)$$

where the coefficients B_{emn}, B_{omn}, A_{emn}, and A_{omn} are to be determined. As shown in Appendix A, $B_{emn} = A_{omn} = 0$, and all remaining coefficients vanish unless $m = 1$. Further, the appropriate radial function to be used in the vector spherical harmonics is $j_n(k_1 r)$ since it is finite at the origin. Labeling such functions $\mathbf{M}_{pmn}$ and $\mathbf{N}_{pmn}$ $(p = o, e)$ by the superscript (1), one obtains from Eq. (177)

$$\mathbf{E}^i = \sum_{n=1}^{\infty}\{B_{o1n}\mathbf{M}_{o1n}^{(1)} + A_{e1n}\mathbf{N}_{e1n}^{(1)}\}. \quad (178)$$

Further, it can be shown that

$$B_{o1n} = -A_{e1n} = i^n E\frac{2n+1}{n(n+1)} \quad (179)$$

so that Eq. (178) gives

$$\mathbf{E}^i = \sum_{n=1}^{\infty} E_n(\mathbf{M}_{o1n}^{(1)} - i\mathbf{N}_{e1n}^{(1)}) \;\; ; \;\; E_n = i^n E\frac{2n+1}{n(n+1)} \quad (180)$$

and the corresponding magnetic field becomes [cf. Eqs. (163) and (164)]

$$\mathbf{H}^i = -i\frac{c}{\omega\mu_2}\nabla\times\mathbf{E}^i = -\frac{ck_2}{\omega\mu_2}\sum_{n=1}^{\infty}E_n(\mathbf{M}^{(1)}_{e1n} + i\mathbf{N}^{(1)}_{o1n}). \tag{181}$$

Expansions of Internal and Scattered Fields

The *boundary conditions* to be satisfied at the surface of the sphere, where $r = a$, are [cf. Eqs. (112)]

$$(\mathbf{E}^i + \mathbf{E}^{\mathrm{sc}} - \mathbf{E}^{\mathrm{int}})\times\hat{\mathbf{e}}_r = 0 \quad ; \quad (\mathbf{H}^i + \mathbf{H}^{\mathrm{sc}} - \mathbf{H}^{\mathrm{int}})\times\hat{\mathbf{e}}_r = 0 \tag{182}$$

where $(\mathbf{E}^{\mathrm{sc}}, \mathbf{H}^{\mathrm{sc}})$ is the scattered field, and $(\mathbf{E}^{\mathrm{int}}, \mathbf{H}^{\mathrm{int}})$ is the field inside the sphere. The *boundary conditions*, the orthogonality of the *vector spherical harmonics*, and the form of the expansion for the incident field dictate the form of the expansions for the scattered field and the field inside the sphere:

$$\mathbf{E}^{\mathrm{int}} = \sum_{n=1}^{\infty}E_n(c_n\mathbf{M}^{(1)}_{o1n} - id_n\mathbf{N}^{(1)}_{e1n}) \tag{183}$$

$$\mathbf{H}^{\mathrm{int}} = -\frac{ck_1}{\omega\mu_1}\sum_{n=1}^{\infty}E_n(d_n\mathbf{M}^{(1)}_{e1n} + ic_n\mathbf{N}^{(1)}_{o1n}) \tag{184}$$

$$\mathbf{E}^{\mathrm{sc}} = \sum_{n=1}^{\infty}E_n(ia_n\mathbf{N}^{(3)}_{e1n} - b_n\mathbf{M}^{(3)}_{o1n}) \tag{185}$$

$$\mathbf{H}^{\mathrm{sc}} = \frac{ck_2}{\omega\mu_2}\sum_{n=1}^{\infty}E_n(ib_n\mathbf{N}^{(3)}_{o1n} + a_n\mathbf{M}^{(3)}_{e1n}) \tag{186}$$

where the superscript (3) has been used to denote vector spherical harmonics with radial dependence given by $h^{(1)}_n(k_2r)$, which for each order n behaves as an outgoing spherical wave when $k_2r \gg 1$. Here the coefficients c_n and d_n of the field inside the sphere and the scattering coefficients a_n and b_n are to be determined.

Angle-Dependent Functions

In terms of the angular functions

$$\pi_n = \frac{P^{(1)}_n}{\sin\theta} \quad ; \quad \tau_n = \frac{dP^{(1)}_n}{d\theta} \tag{187}$$

which have the property that their sum and difference are orthogonal functions, that is

$$\int_0^{\pi}(\tau_n + \pi_n)(\tau_m + \pi_m)\sin\theta d\theta = \int_0^{\pi}(\tau_n - \pi_n)(\tau_m - \pi_m)\sin\theta d\theta = 0 \quad (m \neq n) \tag{188}$$

the vector sperical harmonics in Eqs. (170)–(173) with $m = 1$ become

$$\mathbf{M}_{e1n} = -\sin\phi\ \pi_n(\cos\theta)z_n(z)\hat{\mathbf{e}}_\theta - \cos\phi\ \tau_n(\cos\theta)z_n(z)\hat{\mathbf{e}}_\phi \tag{189}$$

$$\mathbf{M}_{o1n} = \cos\phi\, \pi_n(\cos\theta) z_n(z)\hat{\mathbf{e}}_\theta - \sin\phi \tau_n(\cos\theta) z_n(z)\hat{\mathbf{e}}_\phi \tag{190}$$

$$\begin{aligned}\mathbf{N}_{e1n} &= \cos\phi\, n(n+1)\sin\theta\, \pi_n(\cos\theta)\frac{z_n(z)}{z}\hat{\mathbf{e}}_r \\ &+ \cos\phi\, \tau_n(\cos\theta)\frac{\left[zz_n(z)\right]'}{z}\hat{\mathbf{e}}_\theta \\ &- \sin\phi\, \pi_n(\cos\theta)\frac{\left[zz_n(z)\right]'}{z}\hat{\mathbf{e}}_\phi\end{aligned} \tag{191}$$

$$\begin{aligned}\mathbf{N}_{o1n} &= \sin\phi\, n(n+1)\sin\theta\, \pi_n(\cos\theta)\frac{z_n(z)}{z}\hat{\mathbf{e}}_r \\ &+ \sin\phi\, \tau_n(\cos\theta)\frac{\left[zz_n(z)\right]'}{z}\hat{\mathbf{e}}_\theta \\ &+ \cos\phi\, \pi_n(\cos\theta)\frac{\left[zz_n(z)\right]'}{z}\hat{\mathbf{e}}_\phi.\end{aligned} \tag{192}$$

To determine the elements S_j ($j = 1, 2, 3, 4$) of the *amplitude scattering matrix* [see Eq. (119)], one may use Eqs. (190) and (191) to express the scattered electric field in Eq. (185) in terms of its components parallel and perpendicular to the *scattering plane* and in terms of the corresponding components given in Eq. (176) for the incident electric field. As shown in Appendix A, one finds $S_3 = S_4 = 0$, and when one uses also asymptotic forms of the expressions for $h_n^{(1)}(k_2 r)$, then

$$S_2 = \sum_{n=1}^{\infty} \frac{2n+1}{n(n+1)}[a_n\tau_n(\cos\Theta) + b_n\pi_n(\cos\Theta)] \tag{193}$$

$$S_1 = -\sum_{n=1}^{\infty} \frac{2n+1}{n(n+1)}[a_n\pi_n(\cos\Theta) + b_n\tau_n(\cos\Theta)] \tag{194}$$

where Θ is the *scattering angle*. To determine the *extinction cross section* given by [cf. Eq. (160)]

$$C^{\text{ext}} = \frac{4\pi}{k_2^2}\Re\left(\hat{\mathbf{e}}_x \cdot \mathbf{X}'\right)_{\Theta=0} \tag{195}$$

one may use Eq. (157) with $S_3 = S_4 = 0$, from which it follows that (see Appendix A)

$$C^{\text{ext}} = \frac{4\pi}{k_2^2}\Re\left(\hat{\mathbf{e}}_x \cdot \mathbf{X}\right)_{\Theta=0,\phi=0} = \frac{4\pi}{k_2^2}\Re[S_2(\Theta = 0, \phi = 0)]. \tag{196}$$

As shown in Appendix A, from the definitions of the angular functions in Eq. (187), it follows that

$$\pi_n(\Theta = 0) = \frac{1}{2}n(n+1) \;\; ; \;\; \tau_n(\Theta = 0) = \frac{1}{2}n(n+1) \tag{197}$$

and hence Eq. (193) gives (on noting that $\Theta = \theta$)

$$S_2(\Theta = 0, \phi = 0) = \frac{1}{2}\sum_{n=1}^{\infty}(2n+1)[a_n + b_n] \tag{198}$$

so that Eq. (196) leads to

$$C^{\text{ext}} = \frac{4\pi}{k_2^2}\Re[S_2(\Theta = 0, \phi = 0)] = \frac{2\pi}{k_2^2}\Re\sum_{n=1}^{\infty}(2n+1)[a_n + b_n]. \tag{199}$$

Also, as shown in Appendix A, the *scattering cross section* is given by

$$C^{\text{sc}} = \frac{W^{\text{sc}}}{S^i} = \frac{2\pi}{k_2^2}\sum_{n=1}^{\infty}(2n+1)(|a_n|^2 + |b_n|^2). \tag{200}$$

Scattering Coefficients

In component form, the boundary conditions in Eq. (182) at $r = a$ are given by

$$E_\theta^i + E_\theta^s - E_\theta^{\text{int}} = 0 \quad ; \quad E_\phi^i + E_\phi^s - E_\phi^{\text{int}} = 0 \tag{201}$$

$$H_\theta^i + H_\theta^s - H_\theta^{\text{int}} = 0 \quad ; \quad H_\phi^i + H_\phi^s - H_\phi^{\text{int}} = 0. \tag{202}$$

For a given value of n in the expansions in Eqs. (182)–(186), there are four unknown coefficients a_n, b_n, c_n, and d_n. Thus, one needs four independent equations, which are obtained by substituting the field expansions in Eqs. (180)–(181) and (182)–(186) into the four boundary conditions in Eqs. (201) and (202). Then one obtains, as shown in Appendix A, the following results for the scattering coefficients a_n and b_n (valid when $\mu_2 = \mu_1$):

$$a_n = \frac{m\psi_n(mx)\psi_n'(x) - \psi_n(x)\psi_n'(mx)}{m\psi_n(mx)\xi_n'(x) - \xi_n(x)\psi_n'(mx)} \tag{203}$$

$$b_n = \frac{\psi_n(mx)\psi_n'(x) - m\psi_n(x)\psi_n'(mx)}{\psi_n(mx)\xi_n'(x) - m\xi_n(x)\psi_n'(mx)} \tag{204}$$

where $\psi_n(z)$ and $\xi_n(z)$ are the *Ricatti–Bessel functions* given by

$$\psi_n(z) = zj_n(z) \quad ; \quad \xi_n(z) = zh_n^{(1)}(z) \tag{205}$$

and the *size parameter* x and the *relative refractive index* m are given by

$$x = k_2 a = \frac{2\pi n_2 a}{\lambda} \quad ; \quad m = \frac{k_1}{k_2} = \frac{n_1}{n_2} \tag{206}$$

with $n_1 = \sqrt{(\varepsilon_1 + i\frac{4\pi\sigma_1}{\omega})\mu_2}$ and $n_2 = \sqrt{\varepsilon_2\mu_2}$ being the refractive indices of the medium inside the particle and the surrounding medium, respectively. Note that both a_n and b_n vanish as $m \to 1$, as expected.

Rayleigh Scattering

When the size of the sphere becomes very small compared to the wavelength of the incident light, so that $a \ll \lambda$, the results will reduce to those known as Rayleigh scattering. In this limit, the size parameter $x = k_2 a = \frac{2\pi n_2 a}{\lambda} \ll 1$, and, as shown in Appendix A, the scattering coefficients in Eqs. (203)–(204) simplify considerably to give the follwing result for the amplitude scattering matrix and the Mueller matrix for Rayleigh scattering:

$$\begin{pmatrix} E_{\parallel}^{sc} \\ E_{\perp}^{sc} \end{pmatrix} = \frac{m^2-1}{m^2+2}(k_2 a)^3 \frac{e^{ik_2 r}}{k_2 r} \begin{pmatrix} \cos\Theta & 0 \\ 0 & 1 \end{pmatrix} \begin{pmatrix} E_{\parallel}^{i} \\ E_{\perp}^{i} \end{pmatrix} \tag{207}$$

$$\begin{pmatrix} I^{sc} \\ Q^{sc} \\ U^{sc} \\ V^{sc} \end{pmatrix} = C \begin{pmatrix} 1+\cos^2\Theta & -\sin^2\Theta & 0 & 0 \\ -\sin^2\Theta & 1+\cos^2\Theta & 0 & 0 \\ 0 & 0 & 2\cos\Theta & 0 \\ 0 & 0 & 0 & 2\cos\Theta \end{pmatrix} \begin{pmatrix} I^{i} \\ Q^{i} \\ U^{i} \\ V^{i} \end{pmatrix} \tag{208}$$

where

$$C = \frac{1}{2}(k_2 a)^6 \left| \frac{m^2-1}{m^2+2} \right|^2 . \tag{209}$$

As expected, the result in Eq. (208) has the same form as Eq. (59) with $f = \frac{1-\rho}{1+\rho} = 1$, since the *depolarization factor* $\rho = 0$ for scattering by a homogeneous spherical particle.

2.7 Atmosphere IOPs

2.7.1 Vertical Structure

The stratified vertical structure of the bulk properties of an atmosphere is a consequence of *hydrostatic balance*. For an atmosphere in a state of rest, the pressure $p(z)$ must support the weight of the fluid above it. By equating pressure forces and gravitational forces, one finds that $dp(z) = -g\,\rho(z)\,dz$, where g is the acceleration due to gravity (assumed to be constant), $\rho(z)$ is the air density, and dp is the differential change in pressure over the small height interval dz. Combining this equation with the ideal gas law, $pV = NkT$ for a volume V of gas at temperature T containing N molecules, which can be written as $(n = N/V, \rho = n\overline{M})$ $\rho(z) = \overline{M}\,p(z)/RT(z) = \overline{M}\,n(z)$, one finds upon integration

$$p(z) = p(z_0)\exp[-\int_{z_0}^{z} dz'/H(z')] \tag{210}$$

where k is the Boltzmann constant, $\overline{M}$ is the mean molecular weight, $R = kN_a$ is the gas constant, N_a is the Avogadro number, and $H(z) = RT(z)/\overline{M}g$ is the *atmospheric scale height*. The ideal gas law allows one to write similar expressions for the density $\rho(z)$ and the concentration $n(z)$. Clearly, from a knowledge of the surface pressure $p(z_0)$ and the variation of the scale height $H(z)$ with height z, the hydrostatic equation (210) allows one to determine the bulk gas properties at any height. Equation (210) applies to well-mixed gases but not to short-lived species such as ozone, which is chemically created and destroyed, or water, which undergoes phase changes on short time scales. Assuming that g, T, and $\overline{M}$ (and hence H) are independent of height z, we may integrate Eq. (210) to obtain

$$\frac{p(z)}{p(z_0)} \approx e^{-(z-z_0)/H}; \qquad \frac{n(z)}{n(z_0)} \approx e^{-(z-z_0)/H}; \qquad \frac{\rho(z)}{\rho(z_0)} \approx e^{-(z-z_0)/H}. \tag{211}$$

Thus, the atmospheric scale height H is an e-fold height for density.

Going back to the ideal gas law $pV = NkT \Leftrightarrow pDA = NkT$, where A is the area of a vertical column and D its height, we may define an *equivalent depth* as

$$D \equiv \frac{\mathcal{N}kT}{p} \tag{212}$$

where $\mathcal{N} = N/A$ is the column amount [molecules m^{-2}].

2.7.2 Gases in the Earth's Atmosphere

The total number of air molecules in a 1 m^2 wide vertical column extending from the sea level to the top of the atmosphere is about 2.15×10^{29}. In comparison, the total column amount of ozone in the same vertical column is about 1.0×10^{23}. The *Dobson unit* (DU) is used to quantify the height in millicentimeters (10^{-5} m) that the ozone gas in the atmosphere would occupy if it were compressed to standard pressure (1 013 [hP]; 1 hP (hectopascal) = 1 N m^{-2}) at standard temperature (0 °C). Thus, one DU refers to a layer of ozone that would be 10 μm = 10^{-5} m thick under standard temperature and pressure. The conversion is 1DU = 2.69×10^{20} mol m^{-2}. The 1976 *US Standard Atmosphere* contains about 348 DU of ozone gas [99].

To represent typical atmospheric conditions, six model atmospheres were described by Anderson *et al.* [99]. These model atmospheres are included in a band model based on LowTran/ModTran [100], and they are tabulated in Appendix C in Thomas and Stamnes [18] as follows:

Table C1: AFGL atmospheric constituent profiles, US Standard 1976
Table C2: AFGL atmospheric constituent profiles, tropical
Table C3: AFGL atmospheric constituent profiles, midlatitude summer
Table C4: AFGL atmospheric constituent profiles, midlatitude winter
Table C5: AFGL atmospheric constituent profiles, subarctic summer
Table C6: AFGL atmospheric constituent profiles, subarctic winter.

These atmospheric models were based on the best data available at the time of publication [99], and they contain profiles of temperature, pressure, and gaseous concentrations of constituents including H_2O, CO_2, O_3, CH_4, and NO_2. These five species are the most important infrared-active gases in the Earth's atmosphere.

The clear atmosphere (no clouds or aerosols) gaseous absorption coefficients as a function of wavelength and height in the atmosphere may be generated as discussed below, and scattering by molecules is described by the *Rayleigh scattering phase function* [Eq. (23)] in the scalar case and by the *Rayleigh scattering phase matrix* [see Eqs. (39), (57), (59), and (60)] when polarization is taken into account. Computation of *scattering* and *absorption coefficients* as well as the scattering phase function (in the scalar case) and scattering phase matrix (in the case of polarization) for particles embedded in an atmosphere containing aerosols and cloud particles is also discussed below. Expansion of the scattering phase function in Legendre polynomials (scalar case) as well as of the scattering phase matrix in generalized spherical functions (in the case of polarization) was discussed in Section 2.3.3. The atmosphere should be divided into an adequate number of layers to resolve the vertical variation in the IOPs.

2.7.3 Molecular IOPs

A suitable band model can be used to provide specification of *absorption coefficients* $\alpha(\lambda)$ (or cross sections $\alpha_n(\lambda)$) for the radiatively active atmospheric gases including ozone and water vapor. A specification with moderate spectral resolution based on "fixed-wavenumber" sampling of 1 cm^{-1} (and a nominal resolution of 2 cm^{-1}) is sufficient for many purposes, because the RT process is assumed to be *quasi-monochromatic* within this spectral band [18]. For wavenumbers larger than 17,905 cm^{-1} ($\lambda < 558.5$ nm), one may use a lower spectral resolution of 20 cm^{-1}, since the RT process is assumed to be "quasi-monochromatic" for $\lambda < 558.5$ nm [18].

Spectral Absorption by Atmospheric Gases

Absorption coefficients for gases can be obtained from the *LOWTRAN/MODTRAN band model* [100], which also provides computation of molecular (Rayleigh) scattering coefficients. Alternatively, for wavelengths in the ultraviolet and visible spectral ranges, one may use number density profiles and gas absorption cross sections to compute *absorption coefficients*, and the total amount of molecules to compute Rayleigh (molecular) scattering coefficients. This approach may be adequate if only ozone absorption in addition to molecular scattering is considered to be important. The RT in the atmosphere is strongly affected by gaseous absorption. For example, in the visible part of the spectrum, the oxygen A band centered at 760 nm is a prominent absorption feature. At wavelengths longer than those of visible light, there is strong absorption by trace gases (water vapor, H_2O; carbon dioxide, CO_2; methane, CH_4, and others) with absorption coefficients that vary rapidly and erratically with wavelength. Therefore, we provide in Appendix B a review of some methods, including band models, *k-distributions*, *exponential-sum-fitting of transmissions (ESFT)*, *principal component analysis (PCA)*, *optimal spectral sampling (OSS)*, available to deal with this problem. A brief summary of these methods is given here.

- The *random band model* is based on the assumption that the transmission of the band can be expressed in terms of the products of the individual line transmittances so that the beam transmittance of randomly placed lines is equal to the exponential attenuation due to the single-line beam absorptance. The *MODTRAN* band model [100], based on this assumption, covers the spectral range between 0 and 50,000 cm^{-1}, which spans ultraviolet (UV) through far-infrared (IR) wavelengths (from 0.2 to >40 μm). *MODTRAN* has the capability for rapid calculations of atmospheric absorption, extinction, and emission using molecular band model techniques.
- The *k-distribution method* and its associated *correlated k-distribution* method [18, 101] provide adequate accuracy for many applications, such as computation of warming/cooling rates, and require much less computer time than *line-by-line (LBL) methods*. By use of this method or the ESFT approximation [102], a non-gray problem is reduced to a finite number of gray or monochromatic problems. The correlated-k method is based on the assumption that there is a perfect *spectral correlation* at different pressure levels.
- Like the correlated-k distribution method, spectral mapping methods identify spectral intervals with similar IOPs without making any assumptions about the spectral correlation along the optical path [103, 104]. Compared to the correlated-k method, this approach is less efficient because it requires fine spectral binning.
- The *principal component (PC) method* [105] attempts to overcome the limitations of the correlated-k and spectral mapping methods by making use of empirical orthogonal functions.
- Finally, using the *optimal spectral sampling method (OSS)*, one avoids the problems with the correlated-k approach by relying on monochromatic computations at a small set of spectral points called OSS "nodes".

In all these methods, LBL data and models play central roles in development and verification (see Appendix B for details).

Spectral Averaging: The Chandrasekhar Mean

The TOA solar irradiance decreases rapidly with wavelength for $\lambda < 350$ nm, whereas the ozone *absorption cross section* increases rapidly between 350 and 250 nm. These steep gradients in the solar irradiance and ozone absorption cross sections suggest that it may be useful to define a mean absorption cross section by weighting it with the solar irradiance $F_0(\lambda)$ as follows:

$$\langle \alpha_n \rangle \equiv \frac{\int_{\lambda_1}^{\lambda_2} d\lambda \alpha_n(\lambda) F_0(\lambda)}{\int_{\lambda_1}^{\lambda_2} d\lambda F_0(\lambda)}. \tag{213}$$

Here, $\Delta\lambda = \lambda_2 - \lambda_1$ defines the spectral resolution of the "quasi-monochromatic" computation. Analogously, one may define a mean scattering cross section $\langle \beta_n \rangle$. For rapid energy budget calculations, the spectral resolution adopted in k-distribution band models may be a useful option (see Appendix B).

2.7.4 IOPs of Suspended Particles in the Atmosphere

A variety of particles present in the atmosphere will, for simplicity, be categorized into two broad types: *aerosol and cloud particles.* The first category consists of particles with a variety of shapes, sizes, and chemical compositions that scatter and absorb radiation and thereby impact visibility and the radiative energy budget. The second category consists of water droplets in "warm" clouds and ice particles with a variety of shapes and sizes in "cold" clouds. For the sake of argument, let us assume that each particle is spherical with radius r and homogeneous with a known *refractive index*. Also, let us further assume that the *PSD* can be described by a log-normal distribution given by Eq. (102).

2.7.5 Aerosol IOPs

Assuming that aerosol particles are spherical, we may use a Mie code to generate aerosol IOPs based on available aerosol models. Thus, if such a model specifies the refractive index and the PSD, we may use Eqs. (82)–(84) to compute $\alpha_p(\lambda)$, $\beta_p(\lambda)$, and $p_p(\lambda, \Theta)$, as well as all the other elements of the Stokes scattering matrix in Eq. (36) as discussed above. One option is to use the aerosol models employed in the Sea-viewing Wide Field-of-view Sensor (SeaWiFS) Database Analysis System (SeaDAS) and described by Ahmad *et al.* [106]. Another option is to use the OPAC models described by Hess *et al.* [107]. For atmospheric correction of ocean color imagery, Gordon and co-workers [35, 108] selected 16 candidate aerosol models consisting of several types of particles, each having its own characteristic chemical composition, size distribution, and hygroscopicity.

It is customary to assume a log-normal distribution of aerosol sizes as proposed by Davies [109]. Based on AERONET data (Holben *et al.* [110, 111]), Ahmad *et al.* [106] proposed a bimodal log-normal volume size distribution

$$v(r) = \frac{dV(r)}{dr} = \frac{1}{r}\frac{dV(\ln r)}{d\ln r} = \sum_{i=1}^{2} \frac{V_i}{\sqrt{2\pi}\sigma_i}\frac{1}{r}\exp\left[-\left(\frac{\ln r - \ln r_{vi}}{\sqrt{2}\sigma_i}\right)^2\right] \tag{214}$$

where the subscript i represents the mode, V_i is the total volume of particles belonging to mode i, r_{vi} is the mode radius, also called the volume geometric mean radius, and σ_i is the geometric standard deviation. Integration over all sizes for both modes yields

$$\int_0^\infty v(r)dr = V_1 + V_2 = V$$

due to the normalization [see Eq. (103)]. In terms of the number density, Eq. (214) becomes [see Eq. (102)]

$$n(r) = \frac{dN(r)}{dr} = \frac{1}{r}\frac{dN(r)}{d\ln r} = \sum_{i=1}^{2} \frac{N_i}{\sqrt{2\pi}\sigma_i}\frac{1}{r}\exp\left[-\left(\frac{\ln r - \ln r_{ni}}{\sqrt{2}\sigma_i}\right)^2\right] \tag{215}$$

where the mean geometric (or mode) radius r_{ni} and the number of particles N_i are related to r_{vi} and V_i as follows (see Problem 2.9):

$$\ln r_{ni} = \ln r_{vi} - 3\sigma_i^2 \tag{216}$$

$$N_i = \frac{V_i}{\frac{4}{3}\pi r_{ni}^3}\exp(-4.5\sigma_i^2) \tag{217}$$

and integration over all sizes for both modes yields

$$\int_0^\infty n(r)dr = N_1 + N_2 = N.$$

If instead of using $i = 1, 2$, as above, we use the subscript $i = f$ to denote the fine mode, and the subscript $i = c$ to denote the coarse mode, we have $V = V_f + V_c$, and the volume fraction f_v [not to be confused with the particle volume fraction f_V defined in Eq. (105)] of fine mode particles becomes $f_v = V_f/V$.

Impact of Relative Humidity/Hygroscospicity

A change in the *relative humidity (RH)* will affect the size of the particle as well as the refractive index. The particle radius can be determined as a function of RH from the wet-to-dry mass ratio:

$$r(a_w) = r_0\left[1 + \rho\frac{m_w(a_w)}{m_0}\right]^{1/3} \tag{218}$$

where the *water activity* a_w of a soluble aerosol at radius r [μm] can be expressed as

$$a_w = \text{RH}\exp\left[\frac{-2\sigma_{\text{st}}V_m}{R_wT}\frac{1}{r(a_w)}\right]. \tag{219}$$

Here, r_0 is the dry particle radius (RH = 0), ρ is the particle density relative to that of water, $m_w(a_w)$ is the mass of condensed water, m_0 is the dry particle mass (RH = 0), σ_{st} is the surface tension on the wet surface, V_m is the specific volume of water, R_w is the gas constant for water vapor, and T is the absolute temperature [K] (Hänel [112]). Similarly, the change in *refractive index* with RH can be determined from (Hänel [112]):

$$n = n_w + (n_0 - n_w)\left[\frac{r_0}{r_{RH}}\right]^3 \tag{220}$$

where n_w and n_0 are the complex refractive indices of water and dry aerosols, respectively, and r_0 and r_{RH} are the radii of the aerosols in the dry state and at the given RH, respectively. From these formulas, it follows that the magnitude of the particle growth and the change of refractive index with increasing RH depend (i) on the size r_0 of the dry aerosol and (ii) on the aerosol type because the water uptake [the ratio $m_w(a_w)/m_0$ in Eq. (218)] depends on the aerosol type (Hänel [112], Shettle and Fenn [113], Yan *et al.* [114]). Since the particle radius r depends on a_w and vice versa, Eqs. (218)–(220) must be solved in an iterative manner. However, retrieval of the relative humidity would provide information about the *aerosol particle size* and the complex index of refraction.

From Eq. (105), it follows that the aerosol volume fraction becomes $f_V = \text{AMC}/\rho_a$, where AMC is the *aerosol mass content*, and ρ_a its mass density. Typical values of atmospheric aerosol densities are $\rho_a \approx 1$ g·cm^{-3} = 1×10^6 g·m^{-3}. Hence, an AMC value of 10^{-6} g·m^{-3} would yield $f_V = 10^{-12}$.

2.7.6
Cloud IOPs

Clouds consist of liquid water droplets or ice (frozen water) particles. The liquid water droplets making up warm clouds can be assumed to be spherical, whereas ice crystals have a variety of nonspherical shapes. If we assume for simplicity that all cloud particles consist of spherical water droplets or spherical ice particles, we can use a *Mie code* to compute their IOPs because their *refractive index* is known. Hence, we may use Eqs. (82)–(84) to compute $\alpha_p(\lambda)$, $\beta_p(\lambda)$, and $p_p(\lambda, \Theta)$, as well as all the other elements of the Stokes scattering matrix in Eq. (36) in much the same manner as we did for aerosols.

The real part of the refractive index of pure water needed in the Mie computations may be adopted from Segelstein [115], while the imaginary part can be calculated from the absorption coefficient $\alpha_w(\lambda)$ obtained from data published by Smith and Baker [116], by Sogandares and Fry [117], by Pope and Fry [118] for wavelengths between 340 and 700 nm, and by Kou *et al.* [119] for wavelengths between 720 and 900 nm.

It is customary to introduce the liquid water content (LWC) defined as

$$\text{LWC} \equiv \rho_w \int_{r_{\min}}^{r_{\max}} \left(\frac{4\pi}{3}\right) r^3 n(r) dr \equiv \rho_w f_V \qquad [\text{kg} \cdot \text{m}^{-3}] \tag{221}$$

where $n(r)$ is the cloud droplet size distribution [m^{-3} · m^{-1}] and ρ_w is the liquid water mass density [kg · m^{-3}]. Further, f_V stands for the dimensionless liquid (cloud) *particle volume fraction* defined in a similar way as in Eq. (105), that is, $f_V = \text{LWC}/\rho_w$. For a liquid water cloud, a typical value of the LWC is about 0.5 g ·m^{-3}, implying that $f_V = 5 \times 10^{-7}$, since the density of water is $\rho_w = 10^6$ g ·m^{-3}.

In the expression for the *effective radius*, given by [see Eq. (106)]

$$r_{\text{eff}} = \frac{\int_{r_{\min}}^{r_{\max}} n(r) r^3 dr}{\int_{r_{\min}}^{r_{\max}} n(r) r^2 dr}$$

the numerator is proportional to the *LWC*, while the denominator is related to the scattering coefficient:

$$\beta_c = \int_0^\infty dr(\pi r^2) Q_\beta(r)\, n(r) dr \qquad [\text{m}^{-1}].$$

If the size r of the droplet is large compared to the wavelength λ, then $Q_\beta(r) \rightarrow 2$. Therefore, in the visible spectral range, where $2\pi r/\lambda >> 1$, we find

$$\beta_c \approx \frac{3}{2}\frac{1}{\rho_w}\frac{\text{LWC}}{r_{\text{eff}}} = \frac{3}{2}\frac{f_V}{r_{\text{eff}}} \qquad [\text{m}^{-1}]. \tag{222}$$

For ice *cloud particles* that are assumed to be spherical, a similar expression for the scattering coefficient is obtained with f_V being the ice particle volume fraction. For a liquid water cloud with $f_V = 5 \times 10^{-7}$ and $r_{\text{eff}} = 5 \times 10^{-6}$ m, we get $\beta_c = \frac{3}{2}\frac{f_V}{r_{\text{eff}}} = 0.15\ \text{m}^{-1}$, and hence an optical thickness of 15 for a 100-m-thick cloud layer.

Parameterized Cloud IOPs

Equation (222) suggests that cloud IOPs can be parameterized in terms of the *effective radius* and the volume fraction f_V. In fact, the cloud IOPs can be computed from the following simple algebraic expressions [120]:

$$\gamma_c/f_V = a_1 r_{\text{eff}}^{b_1} + c_1 \tag{223}$$

$$1 - \varpi_c = a_2 r_{\text{eff}}^{b_2} + c_2 \tag{224}$$

$$g_c = a_3 r_{\text{eff}}^{b_3} + c_3. \tag{225}$$

Here, γ_c is the cloud extinction coefficient, $\varpi_c = \beta_c/\gamma_c$ is the *single-scattering albedo*, and g_c is the *asymmetry factor*, which can be used in conjunction with the HG scattering phase function [Eq. (24)]. The coefficients $a_1, \ldots, a_3$ and $c_1, \ldots, c_3$ were obtained by comparisons with detailed Mie computations [120].

2.8 Snow and Ice IOPs

2.8.1 General Approach

Assuming that *snow grains* and *sea ice inclusions* consist of spherical particles, we may obtain their IOPs from Mie computations, which require the refractive index and the PSD as inputs. Then, the IOPs, that is, the absorption and scattering coefficients and the scattering phase function, $\alpha_p(\lambda)$, $\beta_p(\lambda)$, and $p_p(\lambda, \Theta)$, can be obtained from Eqs. (82)–(84), and all the other elements of the *Stokes scattering matrix* in Eq. (36) can be obtained using an equation similar to Eq. (84). This approach leads to computed snow albedo values that agree surprisingly well with available observations. In Section 2.5, we briefly discussed when the use of Mie theory may be acceptable. Based on that discussion, two options are available:

1) Direct Mie calculations based on specified information about the particle refractive index and size for a given (e.g., log-normal) PSD;
2) A fast, yet accurate parameterization based on Mie calculations.

Direct Mie Calculations

For this option, we assume that snow grains and ice inclusions, (*air bubbles*, and *brine pockets*) consist of homogeneous spheres with a single-mode log-normal volume size distribution [see Eq. (214)], and we may use the refractive index data base for ice compiled by Warren and Brandt [121].

If one specifies the effective radius $r_{\rm eff}$ and the width σ of the PSD, the geometrical mean radius r_n can be obtained from Eq. (109). The parameters r_n and σ constitute the only inputs required to specify the PSD. Then the absorption and scattering coefficients, as well the scattering phase function ($\alpha_p(\lambda)$, $\beta_p(\lambda)$, and $p_p(\lambda, \Theta)$), can be computed by using Eqs. (82)–(84), and all the other elements of the Stokes scattering matrix in Eq. (36) can be obtained using an equation similar to Eq. (84). An option to use only the first moment of the phase function in conjunction with the HG scattering phase function [Eq. (24)] is useful because the Mie scattering phase function is unrealistic for nonspherical *snow grains* and *ice inclusions*.

Parameterization Based on Mie Calculations

Sea ice optical properties were described by Jin *et al.* [122], and refined by Hamre *et al.* [28] and Jiang *et al.* [29]. Here we follow the most recent development described by Stamnes *et al.* [123], who created a generic tool [*ISIOP*] for computing ice/snow *IOPs* (τ, ϖ, and g). The *ISIOP tool* can be used to generate snow and sea ice IOPs for any desired wavelength from sea ice physical parameters: real and imaginary parts of the sea ice refractive index, brine pocket concentration and effective size, air bubble concentration and effective size, volume fraction and absorption coefficient of sea ice *impurities*, asymmetry factors for scattering by *brine pockets* and *air bubbles*, and sea ice thickness.

For a specific value of r, we can compute $Q'_\alpha(r)$, $Q'_\beta(r)$, and $p_p(\lambda, \Theta, r)$ using a *Mie code*, but evaluation of Eqs. (82)–(84) requires knowledge of the *PSD* $n(r)$, which is usually unknown. Equations (82)–(84) can be considerably simplified by making the following assumptions [123]:

- The particle distribution is characterized by an *effective radius* given by Eq. (106), which obviates the need for an integration over r.
- The particles are weakly absorbing, so that

$$Q'_\alpha(r) \equiv Q'_\alpha \approx \frac{16\pi\, r_{\rm eff}\, n'_p}{3\lambda} \frac{1}{m_{\rm rel}} [m^3_{\rm rel} - (m^2_{\rm rel} - 1)^{3/2}] \tag{226}$$

  where n'_p is the imaginary part of the refractive index of the particle, λ is the wavelength in vacuum, and $m_{\rm rel} = n_p/n_{\rm med}$ is the ratio of the real part of the refractive index of the particle (n_p) to that of the surrounding medium ($n_{\rm med}$).
- The particles are large compared to the wavelength ($2\pi r/\lambda >> 1$), which implies

$$Q'_\beta(r) \equiv Q'_\beta = 2. \tag{227}$$

- The scattering phase function may be represented by the one-parameter HG scattering phase function [Eq. (24)], which depends only on the asymmetry factor g [Eq. (10)].

With these assumptions, Eqs. (82)–(83) become

$$\alpha_p(\lambda) = \alpha(\lambda)\frac{1}{m_{\rm rel}}\left[1 - (m_{\rm rel}^2 - 1)^{3/2}\right]f_V \qquad (228)$$

$$\beta_p(\lambda) = \frac{3}{2}\frac{f_V}{r_{\rm eff}} \qquad (229)$$

and in Eq. (84) we can use the HG scattering phase function [Eq. (24)]. Here, $\alpha(\lambda) = 4\pi n'_p/\lambda$ is the *absorption coefficient* of the material of which the particle is composed, and $f_V \equiv \frac{4\pi}{3}\int n(r)r^3dr \approx \frac{4}{3}\pi r_{\rm eff}^3 n_e$, where n_e is the number of particles per unit volume with radius $r_{\rm eff}$. Note that Eq. (229) is identical to Eq. (222). Thus, it is clear that f_V represents the *volume fraction* of the particles as defined in Eq. (105). Typical values of f_V for air bubbles and brine pockets in sea ice are shown in Figure 4.

2.8.2
Extension of Particle IOP Parameterization to Longer Wavelengths

For wavelengths $\lambda \leq 1.2$ µm, the absorption and scattering efficiency for *snow grains*, brine inclusions in sea ice, and *air bubbles* in ice may be parameterized by Eqs. (226) and (227), and the *asymmetry factor* g can be held constant with

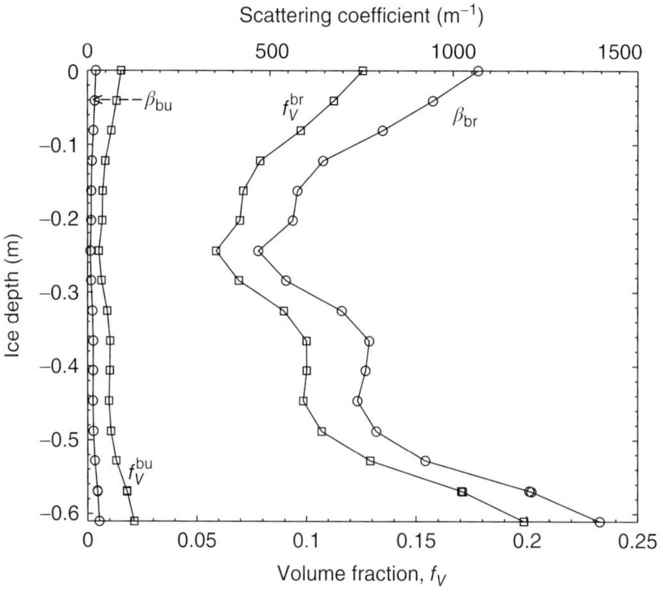

Figure 4 Volume fractions of brine pockets ($f_V^{\rm br}$) and air bubbles ($f_V^{\rm bu}$) (squares), and scattering coefficients of brine pockets ($\beta_{\rm br}$) and air bubbles ($\beta_{\rm bu}$) (circles). The two curves to the left represent air bubbles, and the two curves to the right represent brine pockets (after Hamre *et al.* [28] with permission).

wavelength and set equal to 0.85, 0.89, and 0.997 for air bubbles, snow grains, and brine pockets, respectively, and we may use the HG scattering phase function [Eq. (24)]. To extend the validity to near-infrared wavelengths, we may use the following modified parameterizations, which are based in part on fits to results from Mie calculations [123]:

$$Q_\alpha = 0.94[1 - \exp(Q'_\alpha/0.94)]; \qquad Q_\beta = 2 - Q_\alpha; \qquad g = g_0^{(1-Q_\alpha)^{0.6}} \tag{230}$$

where Q'_α is given by Eq. (226). Here, g is the asymmetry factor of the scattering phase function, and g_0 is the asymmetry factor for nonabsorbing particles ($n'_p = 0$). For large particles ($r > \sim 50$ μm), g_0 depends only on the real part of the *refractive index*. For a medium consisting of several absorbing and scattering constituents, the total *absorption* and *scattering efficiencies* are just the sum of those due to the separate constituents. The optical thickness τ and *single-scattering albedo* ϖ for a slab of thickness h become [123]

$$\tau = \pi r_{\text{eff}}^2 Nh(Q_\alpha + Q_\beta); \qquad \varpi = \frac{Q_\beta}{Q_\alpha + Q_\beta} \tag{231}$$

where N is the total number of particles per unit volume, and Q_α and Q_β are the total absorption and scattering efficiencies. In highly scattering media such as snow and sea ice, we may use the HG scattering phase function [Eq. (24)] to describe the angular scattering behavior. The modified parameterizations, which are represented by the dash-dot curves in Figure 5 (Parameterizion 2), work well for all wavelengths for Q_α, while for Q_β and g they work well for wavelengths shorter than about 2.8 μm, but deviate significantly from predictions by Mie computations for longer wavelengths. Thus, for wavelengths longer than 2.8 μm, it may be preferable to use results from a *Mie code*. Note that for wavelengths shorter than 2.8 μm, where the parameterizations work well, the variations in n_p and n'_p are large. Thus, one would expect these parameterizations to be representative for most types of large particles [123].

2.8.3
Impurities, Air Bubbles, Brine Pockets, and Snow

If the volume fraction of *impurities* within a *snow grain* or *brine pocket* is not too large, which is the case for typical situations occurring in nature, scattering by impurities can be ignored, so that their effects can be included by simply adding the imaginary part n'_{imp} of the refractive index for impurities to n'_p in Eq. (226). For typical impurities in snow and ice, the wavelength dependence of n'_{imp} can be parameterized as

$$n'_{\text{imp}} = n'_{\text{imp}}(\lambda_0)\left(\lambda_0/\lambda\right)^\eta \tag{232}$$

where η would be close to zero for black carbon, but larger for other impurities, and n'_{imp}(440 nm) has values that depend on the type of impurity. Equation (232) is based on the observation that non-algal impurities tend to have a smooth increase

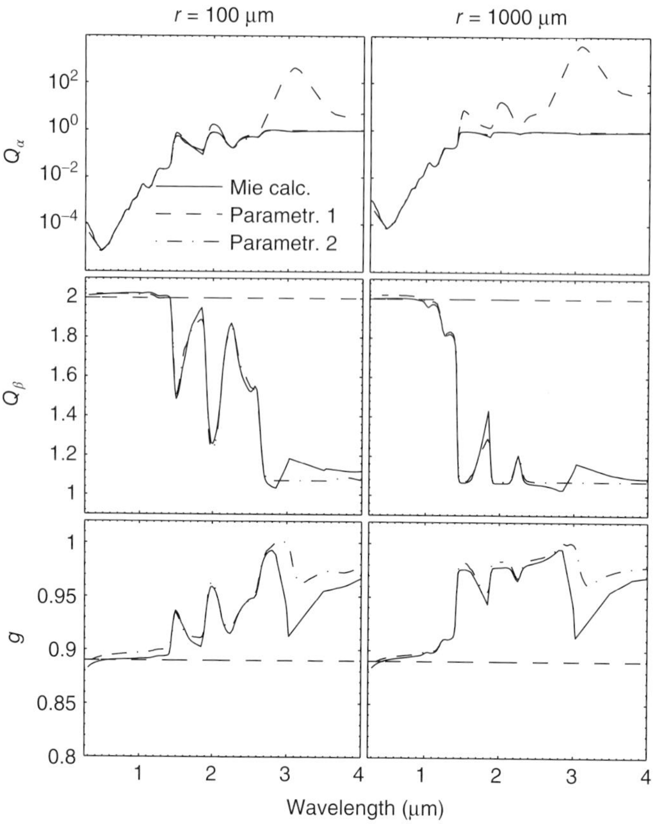

Figure 5 Comparisons of IOPs calculated using Mie computations with those obtained using Eq. (226) (Parameterization 1), which is valid for wavelengths shorter than about 1.2 μm [28], and Eq. (230) (Parameterization 2), which is valid also in the near-infrared region (after Stamnes *et al.* [123] with permission).

in the absorption coefficient toward shorter wavelengths [124–127], and it is connected to the absorption coefficient through $\alpha = 4\pi\, n'_{\text{imp}}/\lambda$. For snow, the number of snow grain particles per unit volume is $N = \frac{1}{\frac{4}{3}\pi r_{\text{eff}}^3}\frac{\rho_s}{\rho_i}$, where r_{eff} is the effective particle radius, while ρ_s and ρ_i are the mass densities of snow and pure ice, respectively. The optical thickness and the *single-scattering albedo* can be calculated from Eqs. (230) and (231), using the refractive indices of pure ice [121] and impurities [Eq. (232)]. Sea ice is assumed to consist of pure ice with embedded brine pockets, air bubbles, and impurities. To include the effects of the embedded components, one may express the *absorption coefficient* α for sea ice as follows:

$$\alpha = \pi r_{\text{br}}^2 N_{\text{br}} Q_{\alpha,br} + \left[1 - \frac{4}{3}\pi r_{\text{br}}^3 N_{\text{br}} - \frac{4}{3}\pi r_{\text{bu}}^3 N_{\text{bu}}\right] \frac{4\pi(n'_p + V_{\text{imp}} n'_{\text{imp}})}{\lambda} \tag{233}$$

where V_{imp} is the volume fraction of impurities, N_{br} and N_{bu} are the number concentrations of brine pockets and air bubbles, respectively, r_{br} and r_{bu} are the corresponding effective radii, and $Q_{\alpha,\text{br}}$ is the *absorption efficiency* for *brine pockets*. The two terms on the right side of Eq. (233) represent the absorption coefficients of brine pockets and surrounding ice (including *impurities*), respectively. In Eq. (233), $\alpha = 4\pi\, n_p'/\lambda$, where λ is the wavelength in vacuum, and the expression inside the square brackets is the *volume fraction* of the ice surrounding all brine pockets and bubbles.

The air bubbles were assumed to be nonabsorbing ($Q_{\alpha,\text{bu}} = 0$), and the impurities were assumed to be uniformly distributed in the ice with n_p' and n_{imp}' being the imaginary parts of the refractive indices for pure ice and impurities, respectively. For brine pockets that are in the liquid phase, the *refractive index* of sea water was used. The *volume fraction* V_{imp} of *impurities* typically lies in the range 1×10^{-7}–1×10^{-5}. The *scattering coefficient* β of sea ice is given by

$$\beta = \beta_{\text{br}} + \beta_{\text{bu}}; \quad \beta_{\text{br}} = \pi r_{\text{br}}^2 N_{\text{br}} Q_{\beta,\text{br}}; \quad \beta_{\text{bu}} = \pi r_{\text{bu}}^2 N_{\text{bu}} Q_{\beta,\text{bu}} \tag{234}$$

where β_{br} and β_{bu} are the scattering coefficients for *brine pockets* and *air bubbles*, respectively, and $Q_{\beta,\text{br}}$ and $Q_{\beta,\text{bu}}$ are the corresponding scattering efficiencies. Here, the scattering coefficient for pure sea ice has been ignored because it is very small compared to either β_{br} or β_{bu}. The optical thickness τ, the *single-scattering albedo* ϖ, and the *asymmetry factor* g for sea ice now become

$$\tau = (\alpha + \beta)h; \qquad \varpi = \frac{\beta}{\alpha + \beta}; \qquad g = \frac{\beta_{\text{br}} g_{\text{br}} + \beta_{\text{bu}} g_{\text{bu}}}{\beta_{\text{br}} + \beta_{\text{bu}}} \tag{235}$$

where h is the sea ice thickness.

2.9 Water IOPs

In open ocean water, it is customary to assume that the IOPs of dissolved and particulate matter can be parameterized in terms of the chlorophyll concentration. In turbid waters such as coastal waters, lakes, and rivers, the IOPs will depend on the presence of several types of particulate matter including *inorganic (mineral) particles* and *organic (algae) particles*, as well as *colored dissolved organic matter (CDOM)* in addition to pure water. The IOPs produced in this manner are said to result from a bio-optical model. For illustration purposes, we describe one such *bio-optical model* below [128], which henceforth will be referred to as the *CCRR bio-optical model*.

2.9.1 Absorption and Scattering by Pure Water

For *pure water*, we use the absorption coefficient $\alpha_w(\lambda)$ based on the data published by Pope and Fry [118] for wavelengths between 400 and 700 nm, and by

Kou *et al.* [119] for wavelengths between 720 and 900 nm. Pure water scattering coefficients $\beta_w(\lambda)$ are based on data published by Morel [72], and the Rayleigh scattering phase function is given by Eq. (23) with *depolarization ratio* $\rho = 0.09$, and thus $f = (1 - \rho)/(1 + \rho) = 0.835$.

2.9.2
Absorption and Scattering by Water Impurities

The CCRR bio-optical model [128] has three input parameters that are allowed to vary: the *chlorophyll concentration (CHL)*, the concentration of mineral particles (MIN), and the absorption coefficient due to CDOM ($\alpha_{\mathrm{CDOM}}(443)$). The CCRR model may be used for wavelengths between 280 and 900 nm if the absorption by algae particles is extrapolated below 400 nm and above 700 nm where data are currently missing as indicated in Figure 6. It should be noted that, according to this decomposition into three basic components, the "mineral particle" component can include also non-algae particles which are not associated with, that is, whose absorption does not covary with that of, the *algae particles*. [128]

Mineral Particles

The absorption coefficient for *mineral particles* at 443 nm is given by (Babin *et al.* [129]):

$$\alpha_{\mathrm{MIN}}(443) = 0.041 \times 0.75 \times \mathrm{MIN}$$

and its spectral variation is described by (Babin *et al.* [129]):

$$\alpha_{\mathrm{MIN}}(\lambda) = \alpha_{\mathrm{MIN}}(443)\exp[-0.0123(\lambda - 443)]. \tag{236}$$

Note on units: $\alpha_{\mathrm{MIN}}(\lambda)/\mathrm{MIN} = 0.041$ has units $[\mathrm{m}^2\mathrm{g}^{-1}]$, so that if MIN has units of $[\mathrm{g\ m}^{-3}]$, then the units of $\alpha_{\mathrm{MIN}}(\lambda)$ will be $[\mathrm{m}^{-1}]$.

The scattering coefficient at 555 nm is given by (Babin *et al.* [130]):

$$\beta_{\mathrm{MIN}}(555) = 0.51 \times \mathrm{MIN}$$

and the spectral variation of the *attenuation coefficient* is

$$\gamma_{\mathrm{MIN}}(\lambda) = \gamma_{\mathrm{MIN}}(555) \times (\lambda/\lambda_0)^{-c}, \qquad c = 0.3749, \qquad \lambda_0 = 555\ \mathrm{nm} \tag{237}$$

where

$$\begin{aligned}\gamma_{\mathrm{MIN}}(555) &= \alpha_{\mathrm{MIN}}(555) + \beta_{\mathrm{MIN}}(555)\\ &= [0.041 \times \exp(-0.0123(555 - 443)) + 0.51] \times \mathrm{MIN} = 0.52 \times \mathrm{MIN}.\end{aligned}$$

The spectral variation of the *scattering coefficient* for mineral particles follows from

$$\beta_{\mathrm{MIN}}(\lambda) = \gamma_{\mathrm{MIN}}(\lambda) - \alpha_{\mathrm{MIN}}(\lambda). \tag{238}$$

The average *Petzold scattering phase function* with a *backscattering ratio* of 0.019 (see Section 2.2 and Figure 2) is used to describe the *scattering phase function* for mineral particles.

Algae Particles

The absorption coefficient for *pigmented particles* (*algae particles* or phytoplankton) can be written as (Bricaud *et al.* [131]):

$$\alpha_{\mathrm{pig}}(\lambda) = A_{\phi}(\lambda) \times [\mathrm{CHL}]^{E_{\phi}(\lambda)} \tag{239}$$

where $A_{\phi}(\lambda)$ and $E_{\phi}(\lambda)$ are given by Bricaud *et al.* [131], and CHL is the chlorophyll concentration, which represents the concentration of pigmented particles (algae particles or phytoplankton). The functions $A_{\phi}(\lambda)$ and $E_{\phi}(\lambda)$ are shown in Figure 6.

The attenuation coefficient for pigmented particles at 660 nm is given by Loisel and Morel [132]:

$$\gamma_{\mathrm{pig}}(660) = \gamma_0 \times [\mathrm{CHL}]^{\eta}; \qquad \gamma_0 = 0.407; \qquad \eta = 0.795$$

and its spectral variation is taken to be (Morel *et al.* [133]):

$$\gamma_{\mathrm{pig}}(\lambda) = \gamma_{\mathrm{pig}}(660) \times (\lambda/660)^{\nu} \tag{240}$$

where

$$\nu = \begin{cases} 0.5 \times [\log_{10}\mathrm{CHL} - 0.3] & 0.02 < \mathrm{CHL} < 2.0 \\ 0 & \mathrm{CHL} > 2.0. \end{cases}$$

The spectral variation of the scattering coefficient for *pigmented particles* follows from

$$\beta_{\mathrm{pig}}(\lambda) = \gamma_{\mathrm{pig}}(\lambda) - \alpha_{\mathrm{pig}}(\lambda). \tag{241}$$

The scattering phase function for pigmented particles is assumed to be described by the *FF scattering phase function* [see Eq. 27] with a *backscattering ratio* [see Eq. (28)] equal to $b_{\mathrm{FF}} = 0.006$ (Mobley *et al.* [79], Morel *et al.* [133]).

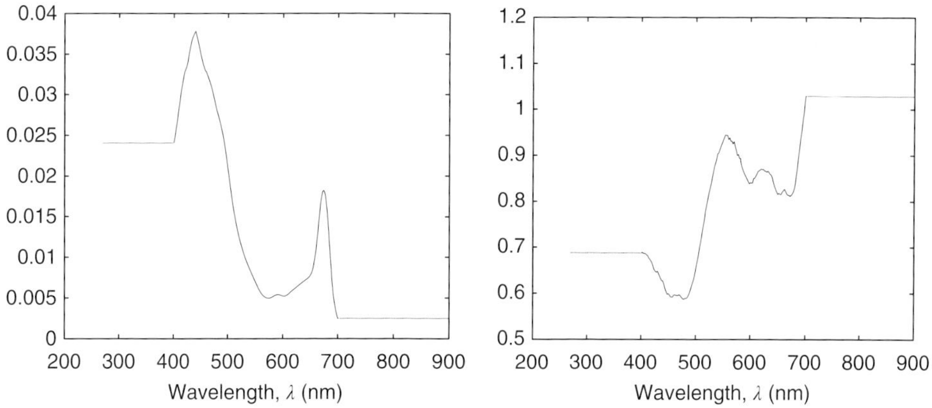

Figure 6 Spectral variation of the coefficients $A_{\phi}(\lambda)$ (left) and $E_{\phi}(\lambda)$ (right) in Eq. (239). The solid curves indicate original data provided by Bricaud *et al.* [131]. The dotted horizontal lines indicate extrapolated values.

Colored Dissolved Organic Matter

The absorption by *CDOM* is given by Babin *et al.* [129] as

$$\alpha_{\mathrm{CDOM}}(\lambda) = \alpha_{\mathrm{CDOM}}(443) \times \exp[-S(\lambda - 443)] \tag{242}$$

where the slope parameter of $S = 0.0176$ represents an average value for different types of water.

Total Absorption and Scattering Coefficients

The total *absorption* and *scattering coefficients* due to water *impurities* for the CCRR bio-optical model are given by

$$\alpha_{\mathrm{tot}}(\lambda) = \alpha_{\mathrm{MIN}}(\lambda) + \alpha_{\mathrm{pig}}(\lambda) + \alpha_{\mathrm{CDOM}}(\lambda) \tag{243}$$

$$\beta_{\mathrm{tot}}(\lambda) \equiv b_p(\lambda) = \beta_{\mathrm{MIN}}(\lambda) + \beta_{\mathrm{pig}}(\lambda). \tag{244}$$

Scattering Phase Function

To describe the angular variation of the scattering, we may use the *FF scattering phase function* [see Eq. (27) and Figure 2] for *pigmented (phytoplankton) particles* and the *Rayleigh scattering phase function* [see Eq. (23) and Figure 2] for scattering by pure water. It was shown by Mobley *et al.* [79] that with proper choice of the real part of the *refractive index* n and the *PSD slope* ξ, the FF scattering phase function is an excellent proxy for the well-known Petzold [80] measurements. In one particular study, Li *et al.* [38] used $n = 1.069$ and $\xi = 3.38$, which correspond to a *backscattering fraction* of $b_{\mathrm{FF}} = 0.0067$. As noted by Mobley *et al.* [79], this choice of $\{n, \xi\}$ values is consistent with a certain mixture of living microbes and resuspended sediments. We may use Eq. (13) or Eq. (22) to compute the Legendre *expansion coefficients* (or *moments* χ_ℓ) of the *scattering phase function*. For strongly forward-peaked scattering typical of the Petzold and FF scattering phase functions, the moment-fitting methods of Wiscombe [89] and Hu *et al.* [91] are very useful for computing $\chi_{\ell,\mathrm{PETZ}}$ and $\chi_{\ell,\mathrm{FF}}$. Thus, the total scattering phase function Legendre expansion coefficients are given by

$$\chi_\ell(\lambda) = \frac{\beta_{\mathrm{MIN}}(\lambda)\chi_{\ell,\mathrm{PETZ}} + \beta_{\mathrm{pig}}(\lambda)\chi_{\ell,\mathrm{FF}} + \beta_w(\lambda)\chi_{\ell,\mathrm{water}}}{\beta_{\mathrm{MIN}}(\lambda) + \beta_{\mathrm{pig}}(\lambda) + \beta_w(\lambda)}. \tag{245}$$

2.9.3 Bio-Optical Model Based on the Particle Size Distribution (PSD)

Current *bio-optical models (BOMs)* are based on field measurements of bulk *absorption, extinction, and backscattering coefficients.* An exception is the approach advocated by Stramski *et al.* (SBM01) [134], who provided detailed information about absorption cross sections of 18 planktonic species present in sea water. Each of the 18 species makes its own contribution to the total *chlorophyll concentration*. Using Table 3 in SBM01 [134], we may compute the number concentration for each species, and multiply it by the absorption cross section to get the corresponding *absorption coefficient*. The total *absorption coefficient* of the 18 planktonic particle types is obtained by summing over all 18 species.

Absorption coefficients estimated from the MERIS CoastColour processing were provided by the CCRR project [128] for 12 sites worldwide. A comparison of the absorption coefficient of *pigmented particles* at 443 nm estimated from the MERIS CoastColour processing with both the CCRR model and the SBM01 model shows that, if one multiplies the SBM01 model results by a factor $1.6948 \times \text{CHL}^{-0.03846}$, where CHL is the chlorophyll concentration, one obtains an excellent fit for all sites with absorption coefficients estimated from the MERIS CoastColour processing [128].

The SMB01 approach indicates that it should be possible to construct a *BOM* from the size distribution and the complex *refractive index* of the *pigmented particles* in the water. If, for simplicity, one assumes that the particles are homogeneous and spherical, one may use a standard *Mie code* to compute their absorption and scattering properties. This approach was adopted by Babin *et al.* [130] and by Kostadinov *et al.* [135–137], who noted that the PSD of most particles in sea water (including *pigmented particles*) follow a power law (Junge-type)

$$n(r) = C(\xi, r_1, r_2)\, r^{-\xi} \qquad r_1 \le r \le r_2 \tag{246}$$

where r is the particle radius, and ξ is the slope of the distribution. The radius r typically varies between $r_1 = 0.23$ μm and $r_2 = 32$ μm. Requiring the distribution to have a total of N_0 particles, so that $\int_{r_1}^{r_2} dr\, n(r) = N_0$, one finds

$$C(\xi, r_1, r_2) = N_0 \left[\int_{r_1}^{r_2} dr r^{-\xi} \right]^{-1} = \begin{cases} \frac{N_0(1-\xi)}{r_2^{1-\xi} - r_1^{1-\xi}}, & \xi \neq 1, \\ N_0 / \ln(r_2/r_1), & \xi = 1. \end{cases} \tag{247}$$

The *effective radius* [Eq. (106)] for this PSD is [16]

$$r_{\text{eff}} = \frac{3-\xi}{4-\xi} \frac{r_2^{4-\xi} - r_1^{4-\xi}}{r_2^{3-\xi} - r_1^{3-\xi}} \tag{248}$$

and the *effective variance* [Eq. (107)] is given by (Hovenier *et al.* [16])

$$v_{\text{eff}} = \frac{(4-\xi)^2}{(5-\xi)(3-\xi)} \frac{(r_2^{3-\xi} - r_1^{3-\xi})(r_2^{5-\xi} - r_1^{5-\xi})}{(r_2^{4-\xi} - r_1^{4-\xi})^2} - 1. \tag{249}$$

From this distribution, one may divide the pigmented particles into phytoplankton functional types (PFTs) based on size as follows [135, 136, 138]:

- *pico-phytoplankton*: $r < 1.0$ μm;
- *nano-phytoplankton*: $1.0 < r < 10.0$ μm;
- *micro-phytoplankton*: $r > 10.0$ μm

as illustrated schematically in Figure 7.

In this approach, different water types characterized in terms of PFTs are determined by the value of the Junge exponent ξ, which enables partitioning between

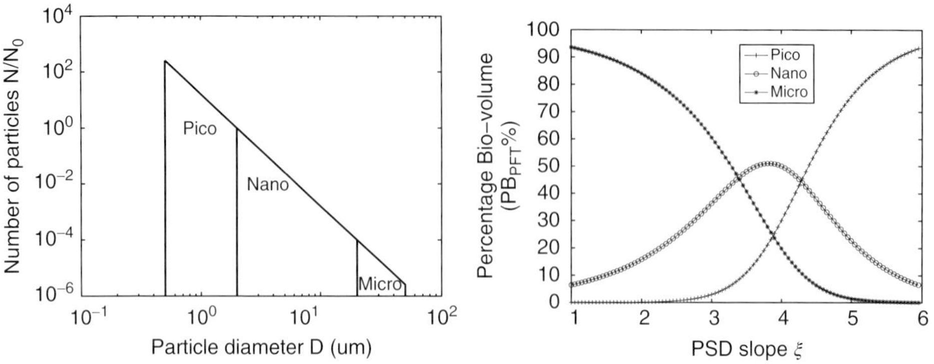

Figure 7 Schematic illustration of the connection between size distribution and phytoplankton functional types (PFTs). The right panel shows the percentage bio-volume versus the PSD slope ξ for each of the three different phytoplankton groups. The particle size range for each phytoplankton group is provided in the text.

pico-, nano-, and micro-sized phytoplankton particles, whereas the *chlorophyll concentration* will be proportional to N_0. To determine the absorption and scattering properties of the *phytoplankton particles*, one needs the complex *refractive index*. Let us assume, for simplicity, that the real part of the refractive index can be approximated by 1.05 (independent of wavelength), while the imaginary part is adopted from Babin *et al.* [130] as illustrated in Figure 8.

To quantify such absorption effects, Babin *et al.* [130] considered two idealized categories of particles: a purely mineral component, and a purely algal component. The *mineral particles* were assumed to have a real part of refractive index equal to 1.18 and an imaginary part that increases with decreasing wavelength (see Figure 8) in such a way that the resulting spectral absorption conforms to an exponential law similar to that given for CDOM in Eq. (242). The algal component was assumed to have a real part of the refractive index of 1.05, while the imaginary part n' was assumed to vary with wavelength in such a way that a typical absorption spectrum of large phytoplanktonic organisms can be reproduced. The spectral dependence of n' shown in Figure 8 was adopted from Ahn *et al.* [139]. To complete the *bio-optical model*, one needs absorption coefficients for CDOM [see Eq. (242] and scattering coefficients for particulate matter.

A frequently used bio-optical model is the *GSM model* (Garver and Siegel [140]; Maritorena *et al.* [141]), which is included in NASA's SeaDAS software package. In the *GSM model* the *absorption coefficient* due to *pigmented particles* is assumed to be directly proportional to the *chlorophyll concentration* (CHL), and the CDOM absorption is assumed to depend exponentially on the wavelength [see Eq. (242)]. The GSM scattering coefficient is parameterized as

$$\beta(\lambda) \propto \beta(\lambda_0)(\lambda/\lambda_0)^{-\eta(\xi)} \tag{250}$$

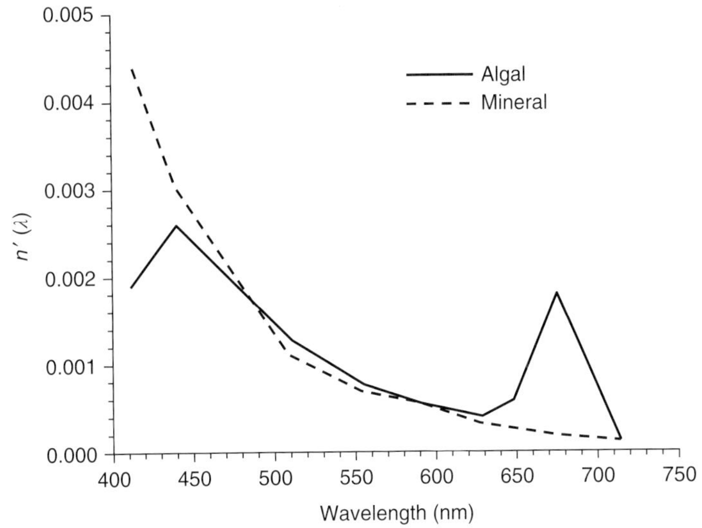

Figure 8 Imaginary part of the refractive index of phytoplankton and mineral particles (redrawn after Babin *et al.* [130]) with permission.

where λ_0 is a reference wavelength (assumed to be 443 nm). Based on a fit to results obtained from detailed Mie computations, one finds

$$\eta(\xi) = 4.4762 - 1.8168\,\xi + 0.11744\,\xi^2 \tag{251}$$

which yields the numerical value $\eta(\xi) = 0.9048$ for $\xi = 4$. In comparison, the GSM model employs a constant value of $\eta = 1.0337$.

Figure 9 shows the absorption coefficients obtained from Mie computations for several values of the slope parameter ξ. The values obtained from the GSM model are shown for comparison. We note that a value of ξ close to 4 gives good general agreement with the standard GSM model with small discrepancies at short and long wavelengths. A similar comparison of the scattering coefficients (Figure 10) shows that a value of ξ close to 4 gives good general agreement with the GSM model. Zhang *et al.* [142, 143] provided the following empirical relation between chlorophyll content per cell $[\text{CHL}]_{\text{cell}}$ [g per cell] and mean cell size (particle radius r [μm]):

$$[\text{CHL}]_{\text{cell}}(r) = (0.0030 \pm 0.006) \times 10^{-12} \times r^{2.876 \pm 0.115}. \tag{252}$$

Using this relation, we have [see Eq. (247)]

$$\text{CHL} = C(\xi, r_1, r_2) \int_{r_1}^{r_2} [\text{CHL}]_{\text{cell}}(r)\, r^{-\xi}\, dr \tag{253}$$

which was used in Figures 9 and 10 to relate CHL and

$$N_0 = \begin{cases} \frac{C(r_2^{1-\xi} - r_1^{1-\xi})}{1-\xi}, & \xi \neq 1, \\ C(\xi, r_1, r_2)/\ln(r_1/r_2), & \xi = 1. \end{cases} \tag{254}$$

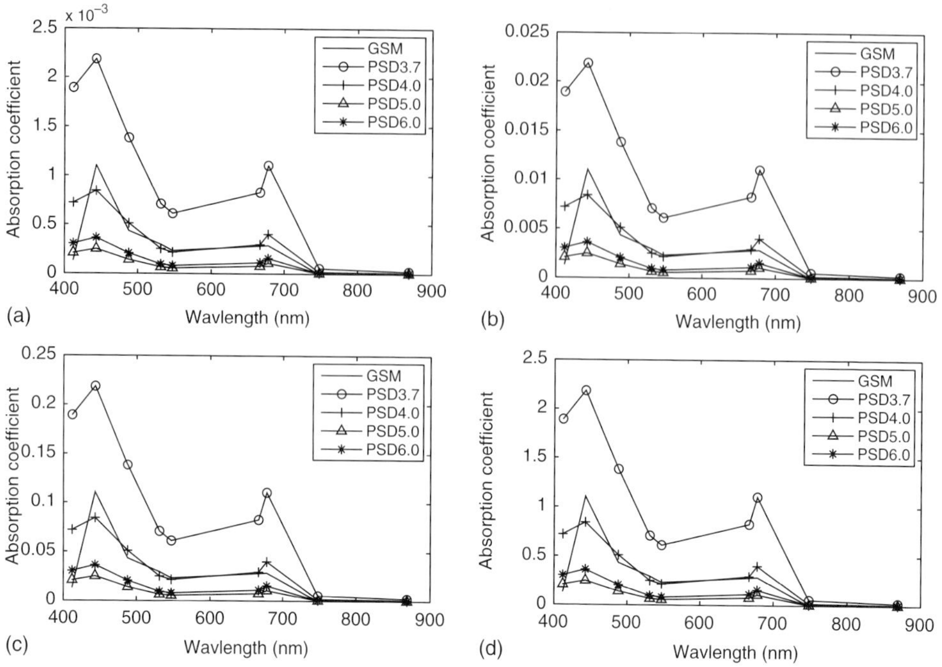

Figure 9 Absorption coefficients [m^{-1}] derived from Mie computations for slope parameters $\xi = 3.7$, 4.0, 5.0, and 6.0 of the PSD. The absorption coefficient used in the GSM model is also shown for comparison. (a) CHL = 0.01976 mg·m^{-3}. (b) CHL = 0.1976 mg·m^{-3}. (c) CHL = 1.976 mg·m^{-3}. (d) CHL = 19.76 mg·m^{-3}.

The lower and upper limits of the integration were taken to be $r_1 = 0.25$ μm and $r_2 = 25.0$ μm.

In order to compute the *apparent optical properties*, we need a *RT model*, which also requires the vsf (or *scattering phase function*) as input. In our CRTM for the coupled atmosphere–ocean system, we used the analytic Fournier–Forand (FF) *scattering phase function* [78] to model the angular scattering. As discussed by Mobley *et al.* [79], the FF scattering phase function [Eq. (27)] depends on the *Junge slope parameter* as well as on the real part of the *refractive index*, and can be used to model the angular scattering by high-index mineral particles as well as low-index *organic particles*. Figure 11 shows the remote sensing reflectances (apparent optical properties (AOPs)) computed with our CRTM (to be discussed in Chapter 3) for the absorption and scattering coefficients shown in Figures 9 and 10 using the FF scattering phase function [Eq. (27)]. Results using the IOPs (absorption and scattering coefficients) provided by the *GSM model* in the CRTM are also shown. To make the comparison more meaningful, the CDOM absorption coefficients were taken from the GSM model in both simulations. The close match between the results obtained using our PSD-based bio-optical model for absorption by pigmented phytoplankton particles and the parameterization used in the GSM bio-optical model, where the absorption is assumed to be proportional to

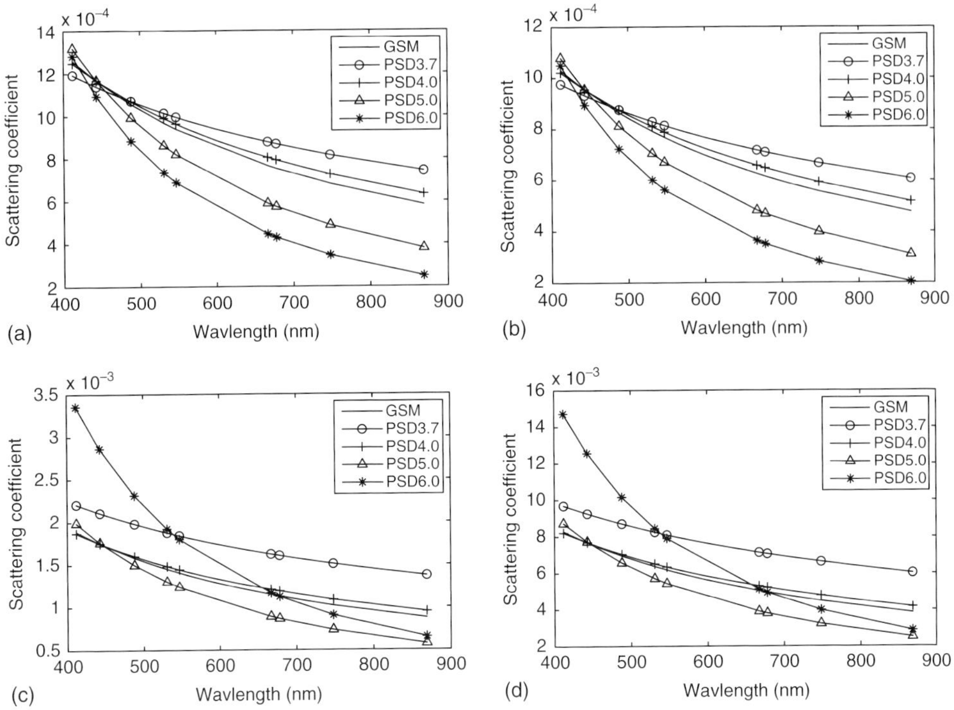

Figure 10 Scattering coefficients [m^{-1}] derived from Mie computations for slope parameters $\xi = 3.7$, 4.0, 5.0, and 6.0 of the PSD. The scattering coefficient used in the GSM model is also shown for comparison. (a) CHL = 0.004323 mg·m^{-3}. (b) CHL = 0.04323 mg·m^{-3}. (c) CHL = 0.4323 mg·m^{-3}. (d) CHL = 4.323 mg·m^{-3}.

the chlorophyll concentration, indicates that our methodology for deriving AOPs from the IOPs is basically sound.

To compute the IOPs of *detritus* and *mineral particles*, Stramski *et al.* [134] proposed to estimate the background contributions of detritus and minerals as follows: Detritus was modeled as an assemblage of particles with a wavelength-independent real part of the refractive index, given by $n = 1.04$ (relative to water), and with diameters ranging from 0.05 to 500 μm having a *Junge size distribution* [see Eq. (246)] with the *slope parameter* $\xi = 4$. The spectral dependence of the imaginary part of the refractive index $n'(\lambda)$ for detritus was taken to be $n'(\lambda) = 0.010658 \exp\{-0.0071868\lambda\}$ where the unit of λ is [nm]. Mineral particles were assumed to have the same size distribution and $n'(\lambda)$-values as organic detritus, while the real part of the refractive index was set to $n = 1.18$ for all wavelengths. Thus, while *mineral particles* represent high-index *inorganic particles*, detritus can be considered to represent low-index organic particles (primarily nonliving particles but also possibly heterotrophic organisms not included in the 18 species description of the planktonic community).

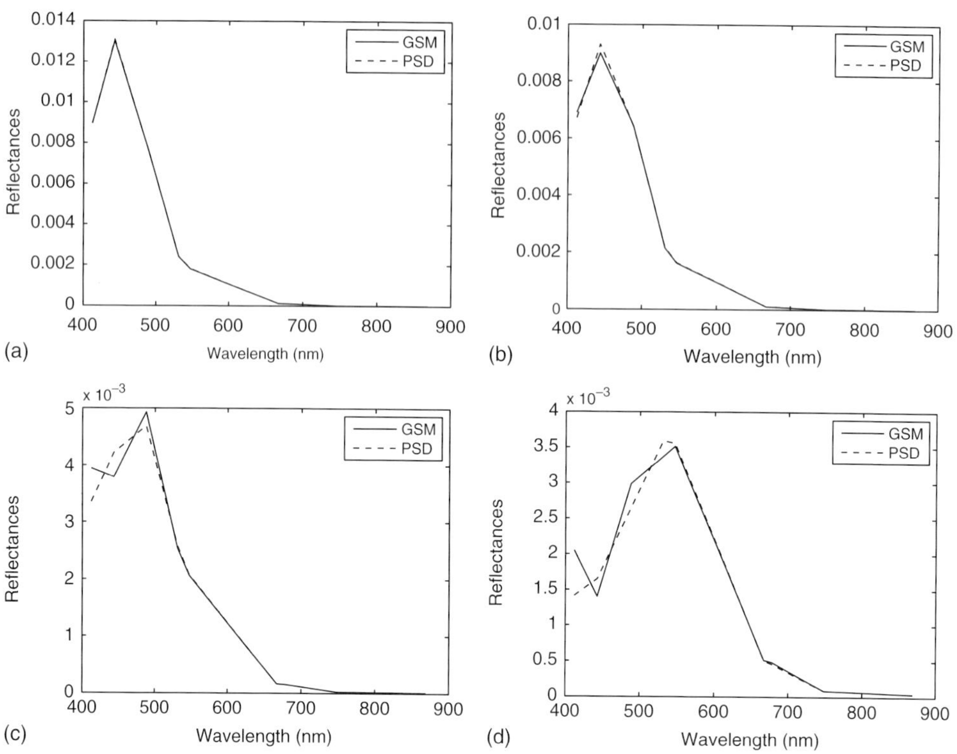

Figure 11 Remote sensing reflectances R_{rs} derived from our CRTM (coupled atmosphere–ocean radiative transfer model) using our PSD bio-optical model with $\xi = 4$ and the GSM bio-optical model. (a): CHL = 0.001976 mg·m^{-3}. (b): CHL = 0.1976 mg·m^{-3}. (c): CHL = 1.976 mg·m^{-3}. (d): CHL = 19.76 mg·m^{-3}.

A similar rationale and type of approximation can be used to model scattering and absorption by *air bubbles* in the water. Thus, Stramski *et al.* [134] assumed that the size of air bubbles ranged in diameter from 20 to 500 μm, that the real part of the refractive index relative to water was 1/1.34, and that the imaginary part was zero (no absorption).

In essence, the approach discussed above can be summarized as follows:

1) Adopt two different groups of particle populations; one group to mimic *pigmented particles* and another group to mimic *inorganic particles*, each characterized by its own size distribution and refractive index.
2) Use Mie theory to calculate IOPs for each of these populations, and then mix them properly according to concentrations.
3) Use *optimal estimation* to determine the combination of size distributions, refractive indices, and mixing proportions that gives the optimal match between the modeled and measured IOPs.

This approach can easily be extended to include several subgroups within each of these populations. In fact, as discussed above, Stramski *et al.* [134] and Zhang *et al.*

[142, 143] have already demonstrated that such an approach is viable. A significant advantage is that this approach lends itself readily to an extension of the PSD-based BOM methodology to complex turbid waters. Another advantage is that we may use Eqs. (82)–(84) to compute $\alpha_p(\lambda)$, $\beta_p(\lambda)$, and $p_p(\lambda, \Theta)$, as well as all the other elements of the *Stokes scattering matrix* in Eq. (36) in much the same manner as we did for aerosol, cloud, snow, and ice particles.

From Eq. (105), it follows that the *hydrosol volume fraction* is given by $f_V = \mathrm{HMC}/\rho_a$, where HMC is the *hydrosol mass content*, and ρ_a its mass density. Typical values of hydrosol mass densities are $\rho_a \approx (0.2 - 0.5) \times 10^6$ g·m^{-3} for living organic matter, and $\rho_a \approx (2 - 3) \times 10^6$ g·m^{-3} for inorganic matter [130]. Hence, an HMC value of 1.0 g·m^{-3} would yield f_V-values of about $(2 - 5) \times 10^{-6}$ for *organic particles* and about $(3 - 5) \times 10^{-7}$ for *inorganic particles.*

2.10 Fresnel Reflectance and Transmittance at a Plane Interface Between Two Coupled Media

Any method for solving the *scalar RTE* or the *vector RTE* (a review of some methods are provided in Chapter 3) must deal with a lower boundary at which incident radiation will be partly reflected, transmitted, or absorbed. Surfaces also emit thermal radiation.

Thus, special consideration must be given to what happens at the interface between two coupled media such as air and water (and ice), where there is a change in refractive index between the two media. In this section we discuss a *plane interface* (such as a calm water surface or a smooth ice surface). The extension needed to take *surface roughness* into account will be discussed in the following section.

Figure 12 shows a parallel beam of radiation that is incident upon a plane interface between two media with different refractive indices at an angle of incidence θ_i $(0 \le \theta_i \le \pi/2)$ with the interface normal. It gives rise to a reflected parallel

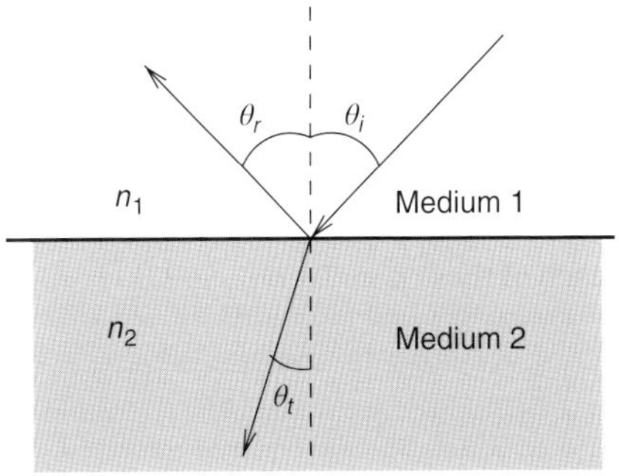

Figure 12 Angles relevant for reflection and refraction of a plane wave at a plane interface between two dielectric media with different refractive indices.

beam and a refracted (transmitted) parallel beam at angles θ_r $(0 \le \theta_r \le \pi/2)$ and θ_t $(0 \le \theta_t \le \pi/2)$, respectively, with the interface normal. According to the reflection law, $\theta_r = \theta_i$, and the reflected beam must lie in the *plane of incidence* defined by the incident beam direction and the interface normal. According to *Snell's law*, $n_1 \sin\theta_i = n_2 \sin\theta_t$, and the transmitted beam must lie in the plane of incidence. Here, n_1 and n_2 are the real parts of the refractive indices in medium 1 (containing the incident beam) and medium 2 (containing the transmitted beam), respectively. Both for reflection and transmission, the *scattering plane* is identical to the plane of incidence.

We decompose the electric field vector (polarization direction) of a plane electromagnetic wave into components parallel and perpendicular to the scattering plane (plane of incidence). For reflection and transmission of a *plane wave* at a plane interface, there is no coupling between these two polarization components, implying that the reflected and transmitted plane waves will have the same polarization as the incident plane wave. The reflection and transmission coefficients are given by the *Fresnel formulas* [68, 98]. For *parallel polarization*, we have

$$R_\parallel = \frac{E_{\parallel r}}{E_{\parallel i}} = \frac{m_2\cos\theta_i - m_1\cos\theta_t}{m_1\cos\theta_t + m_2\cos\theta_i} = \frac{\tan(\theta_i - \theta_t)}{\tan(\theta_i + \theta_t)} = \frac{\sqrt{1-(1-\mu_i^2)/m_{\rm rel}^2} - m_r\mu_i}{\sqrt{1-(1-\mu_i^2)/m_{\rm rel}^2} + m_r\mu_i} \tag{255}$$

$$T_\parallel = \frac{E_{\parallel t}}{E_{\parallel i}} = \frac{2m_1\cos\theta_i}{m_1\cos\theta_t + m_2\cos\theta_i} = \frac{2\mu_i}{\sqrt{1-(1-\mu_i^2)/m_{\rm rel}^2} + m_r\mu_i} \tag{256}$$

and for *perpendicular polarization*, we have

$$R_\perp = \frac{E_{\perp r}}{E_{\perp i}} = \frac{m_1\cos\theta_i - m_2\cos\theta_t}{m_1\cos\theta_i + m_2\cos\theta_t} = -\frac{\sin(\theta_i - \theta_t)}{\sin(\theta_i + \theta_t)} = \frac{\mu_i - m_r\sqrt{1-(1-\mu_i^2)/m_{\rm rel}^2}}{\mu_i + m_r\sqrt{1-(1-\mu_i^2)/m_{\rm rel}^2}} \tag{257}$$

$$T_\perp = \frac{E_{\perp t}}{E_{\perp i}} = \frac{2m_1\cos\theta_i}{m_1\cos\theta_i + m_2\cos\theta_t} = \frac{2\mu_i}{\mu_i + m_r\sqrt{1-(1-\mu_i^2)/m_{\rm rel}^2}} \tag{258}$$

where $\mu_i = \cos\theta_i$, $m_r = m_2/m_1$, and we have used Snell's law $\sin\theta_t = \sin\theta_i/m_{\rm rel}$ $(m_{\rm rel} = n_2/n_1)$. Here, $m_j = n_j + in'_j$ $(j = 1, 2)$ is the *complex refractive index* with real part n_j and imaginary part n'_j.

The following relations apply:

$$R_\perp + 1 = T_\perp \qquad R_\parallel + 1 = \frac{m_1}{m_2} T_\parallel \tag{259}$$

and for *unpolarized light* we have

$$R_{\text{unpol}} = \frac{1}{2}(R_\perp^2 + R_\parallel^2) = \frac{1}{2}\left[\frac{\sin^2(\theta_i - \theta_t)}{\sin^2(\theta_i + \theta_t)} + \frac{\tan^2(\theta_i - \theta_t)}{\tan^2(\theta_i + \theta_t)}\right] \tag{260}$$

$$\begin{aligned} T_{\text{unpol}} &= \frac{1}{2}(T_\perp^2 + T_\parallel^2) \\ &= \frac{1}{2}\left[\left(\frac{2m_1\cos\theta_i}{m_1\cos\theta_i + m_2\cos\theta_t}\right)^2 + \left(\frac{2m_1\cos\theta_i}{m_1\cos\theta_t + m_2\cos\theta_i}\right)^2\right]. \end{aligned} \tag{261}$$

Finally, in models describing polarized reflectance from land surfaces, the *polarized reflectance* is assumed to be given by (see Section 2.12.2)

$$R_{\text{pol}} = \frac{1}{2}(R_\perp^2 - R_\parallel^2) = \frac{1}{2}\left[\frac{\sin^2(\theta_i - \theta_t)}{\sin^2(\theta_i + \theta_t)} - \frac{\tan^2(\theta_i - \theta_t)}{\tan^2(\theta_i + \theta_t)}\right]. \tag{262}$$

2.10.1 Stokes Vector of Reflected Radiation

For reflection and transmission of a parallel beam that is incident upon a *plane interface*, the *Stokes vector* of the reflected radiation is related to the Stokes vector of the incident radiation by the following 4×4 matrix [68]:

$$\begin{pmatrix} I_{\parallel r} \\ I_{\perp r} \\ U_r \\ V_r \end{pmatrix} = \begin{pmatrix} S_{11}^r & 0 & 0 & 0 \\ 0 & S_{22}^r & 0 & 0 \\ 0 & 0 & S_{33}^r & S_{34}^r \\ 0 & 0 & S_{43}^r & S_{44}^r \end{pmatrix} \begin{pmatrix} I_{\parallel i} \\ I_{\perp i} \\ U_i \\ V_i \end{pmatrix} = \mathbf{R}_F(u_i, m_r) \begin{pmatrix} I_{\parallel i} \\ I_{\perp i} \\ U_i \\ V_i \end{pmatrix} \tag{263}$$

where $\mathbf{R}_F(\mu_i, m_r)$ is the reflection matrix for a parallel beam of light incident at an angle θ_i ($\mu_i = \cos\theta_i$) with elements S_{ij}^r given by

$$\begin{aligned} S_{11}^r &= R_\parallel R_\parallel^* = |R_\parallel|^2 \\ S_{22}^r &= R_\perp R_\perp^* = |R_\perp|^2 \\ S_{33}^r &= S_{44}^r = \text{Re}(R_\parallel R_\perp^*) \\ S_{34}^r &= \text{Im}(R_\parallel R_\perp^*) \\ S_{43}^r &= -S_{34}^r = \text{Im}(R_\parallel^* R_\perp). \end{aligned}$$

If both media are nonabsorbing ($n_1' = n_2' = 0$), and $n_2 > n_1$, as illustrated in Figure 12, then both $R_\parallel$ and $R_\perp$ are real, implying that $S_{34}^r = S_{43}^r = 0$, and that the reflection matrix is diagonal.

2.10.2 Total Reflection

Let the real part n_1 of the *refractive index* in medium 1 (e.g., air) be smaller than the real part n_2 of the refractive index in medium 2 (e.g., water), so that $n_2 > n_1$, as depicted in Figure 12. Now consider a parallel beam incident *in medium 2 (water)* upon the interface at an angle θ_i and transmitted at an angle θ_t into medium 1 (air). Then, if θ_i is greater than or equal to the critical angle, given by $\theta_c = \arcsin(1/m_{\rm rel})$ ($m_{\rm rel} = n_2/n_1 > 1$), the beam will be totally reflected at the interface. Thus, from Snell's law, $n_2 \sin\theta_i = n_1 \sin\theta_t$ or $\sin\theta_t = m_{\rm rel} \sin\theta_i$, and it follows that when $\sin\theta_i = \sin\theta_c = 1/m_{\rm rel}$, then $\theta_t = \pi/2$. Also, when $\theta_i > \theta_c$, then $\sin\theta_t > 1$, implying that $\cos\theta_t$ is purely imaginary (real part is zero):

$$\cos\theta_t = \pm i\sqrt{m_{\rm rel}^2 \sin^2\theta_i - 1} = \pm i\sqrt{m_{\rm rel}^2(1-\mu_i^2) - 1}. \tag{264}$$

Here, the lower sign must be discarded because it would give an exponentially growing field in medium 1. Substituting this expression for $\cos\theta_t$ into Eqs. (255) and (257), we obtain the reflection coefficients

$$R_\parallel \approx -\frac{m_{\rm rel}\mu_i - i\sqrt{m_{\rm rel}^2(1-\mu_i^2)-1}}{m_{\rm rel}\mu_i + i\sqrt{m_{\rm rel}^2(1-\mu_i^2)-1}} = -\frac{z_\parallel}{z_\parallel^*} = -e^{i\phi_\parallel} \tag{265}$$

$$R_\perp \approx \frac{\mu_i - i m_{\rm rel}\sqrt{m_{\rm rel}^2(1-\mu_i^2)-1}}{\mu_i + i m_{\rm rel}\sqrt{m_{\rm rel}^2(1-\mu_i^2)-1}} = \frac{z_\perp}{z_\perp^*} = e^{i\phi_\perp} \tag{266}$$

where

$$z_\parallel = m_{\rm rel}\mu_i - i\sqrt{m_{\rm rel}^2(1-\mu_i^2)-1}$$
$$z_\perp = \mu_i - i m_{\rm rel}\sqrt{m_{\rm rel}^2(1-\mu_i^2)-1} \tag{267}$$

$$\phi_\parallel = -2\arctan\left(\frac{\sqrt{m_{\rm rel}^2(1-\mu_i^2)-1}}{m_{\rm rel}\mu_i}\right)$$
$$\phi_\perp = -2\arctan\left(\frac{m_{\rm rel}\sqrt{m_{\rm rel}^2(1-\mu_i^2)-1}}{\mu_i}\right). \tag{268}$$

The approximation made in the first term on the right side of Eq. (265) or Eq. (266) is obtained by assuming that both media are nonabsorbing, so that $m_1 \to n_1$ and $m_2 \to n_2$. For $\theta_i \geq \theta_c$, we use Eqs. (265) and (266) to obtain the elements of the *reflection matrix* in Eq. (263). Clearly, $|R_\parallel| = |R_\perp| = 1$, as expected for *total reflection*, implying that $S_{11}^r = S_{22}^r = 1$. Thus, from Eq. (263), it follows that $I_{\parallel r} = I_{\parallel i}$ and $I_{\perp r} = I_{\perp i}$, implying that total reflection does not change the radiation intensity parallel with or perpendicular to the *plane of incidence*.

The off-diagonal elements $S_{34}^r = \mathrm{Im}(R_\parallel R_\perp^*)$ and $S_{43}^r = -S_{34}^r$ are nonvanishing, implying that total reflection gives a coupling between the U and V elements of the *Stokes vector*.

2.10.3 Stokes Vector of Transmitted Radiation

For reflection and transmission of a parallel beam that is incident upon a plane interface, the Stokes vector of the transmitted radiation is related to the Stokes vector of the incident radiation by the following 4×4 matrix [68]:

$$\begin{pmatrix} I_{\|t} \\ I_{\perp t} \\ U_t \\ V_t \end{pmatrix} = \begin{pmatrix} S^t_{11} & 0 & 0 & 0 \\ 0 & S^t_{22} & 0 & 0 \\ 0 & 0 & S^t_{33} & S^t_{34} \\ 0 & 0 & S^t_{43} & S^t_{44} \end{pmatrix} \begin{pmatrix} I_{\|i} \\ I_{\perp i} \\ U_i \\ V_i \end{pmatrix} = \mathbf{T}_F(\mu_i, m_r) \begin{pmatrix} I_{\|i} \\ I_{\perp i} \\ U_i \\ V_i \end{pmatrix} \tag{269}$$

where $\mathbf{T}_F(\mu_i, m_r)$ is the transmission matrix for a parallel beam of light incident at an angle θ_i ($\mu_i = |\cos\theta_i|$) with elements S^t_{ij} given by

$$\begin{aligned} S^t_{11} &= |K_t|^2 T_\| T^*_\| = |K_t|^2 |T_\||^2 \\ S^t_{22} &= |K_t|^2 T_\perp T^*_\perp = |K_t|^2 |T_\perp|^2 \\ S^t_{33} &= S^t_{44} = |K_t|^2 \mathrm{Re}(T_\| T^*_\perp) \\ S^t_{34} &= |K_t|^2 \mathrm{Im}(T_\| T^*_\perp) \\ S^t_{43} &= -S^t_{34} = |K_t|^2 \mathrm{Im}(T^*_\| T_\perp). \end{aligned}$$

The factor $|K_t|^2$ accounts for the different directions of propagation of the incident and transmitted beams relative to the interface normal and for the difference in *refractive index* between the two media. For a parallel beam that is incident upon a surface at an angle of incidence θ_i and transmitted through it at an angle θ_t, the radiative energy of each of the two beams must be projected on to the interface, giving a factor $\cos\theta_i$ for the incident beam and a factor $\cos\theta_t$ for the transmitted beam, as illustrated in Figure 13. Also, the radiative energy, given by the *Poynting vector*, is proportional to the refractive index. Therefore, the transmittance of a parallel beam is given by (Born and Wolf [98])

$$\mathcal{T}_p = \frac{n_2 \cos\theta_t}{n_1 \cos\theta_i} |T_p|^2 = m_{\mathrm{rel}} \frac{\mu_t}{\mu_i} |T_p|^2 \;\; ; \;\; (p = \|, \perp) \tag{270}$$

where $T_\|$ and $T_\perp$ are given by Eqs. (256) and (258).

In addition, one must account for ray bending at the interface. To that end, we consider an angular beam of solid angle $d\omega_i$ that is incident at an angle θ_i upon an interface and transmitted through it at an angle θ_t into a *solid angle* $d\omega_t$. If one neglects losses due to reflection and transmission, the energy of the incident angular beam, given by $I_{pi} \cos\theta_i d\omega_i$ ($p = \|, \perp$), must be equal to the energy of the transmitted beam, given by $I_{pt} \cos\theta_t d\omega_t$. By using *Snell's law*, one can readily show that this requirement leads to conservation of the *basic intensity* I_{pt}/n^2, and hence to a transmittance factor m^2_{rel}, which in combination with Eq. (270) gives

$$T_{F,p}(\mu_i, m_r) = \frac{I_{pt}}{I_{pi}} \mathcal{T}_p = m^2_{\mathrm{rel}} \mathcal{T}_p = m^3_{\mathrm{rel}} \frac{\mu_t}{\mu_i} |T_p|^2$$

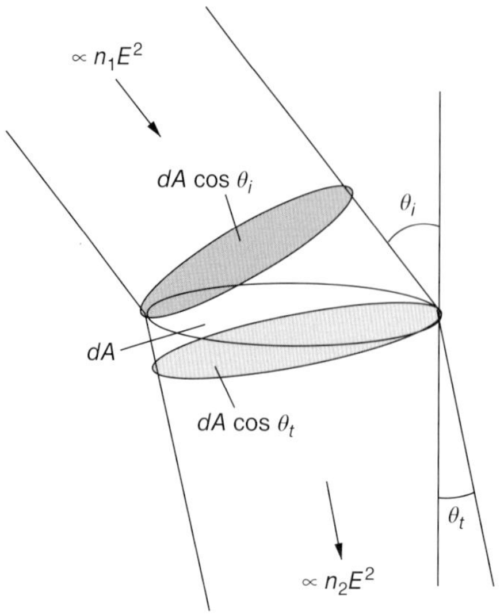

Figure 13 A beam of light with electric field amplitude E and intensity proportional to n_1E^2 is incident in a medium with refractive index n_1 upon an interface with a second medium with refractive index n_2 at an angle θ_i with the interface normal. The beam is transmitted through an area dA of the interface at an angle θ_t with the interface normal into the second medium with intensity proportional to n_2E^2 [from Sommersten *et al.* [87] with permission].

so that the factor $|K_t|^2$ becomes

$$|K_t|^2 = m_{\mathrm{rel}}^3 \frac{\mu_t}{\mu_i}. \tag{271}$$

The theory described above applies to a plane (smooth) interface between two adjacent slabs such as a calm air–water interface. The extensions required to deal with realistic, nonplanar interfaces are discussed below and in Appendix C.

2.11 Surface Roughness Treatment

2.11.1 Basic Definitions

Consider a parallel beam of radiation with electric field $\mathbf{E}_i$ that is incident on a target area A in Figure 14 and gives rise to a scattered electric field $\mathbf{E}_s$, which is the difference between the total field $\mathbf{E}$ and the incident field, that is,

$$\mathbf{E}_s = \mathbf{E} - \mathbf{E}_i. \tag{272}$$

If the power of the incident beam is proportional to $|E_{ai}|^2 A \cos\theta_i d\omega_s$, where the subscript a denotes the electric field component in the direction with unit vector

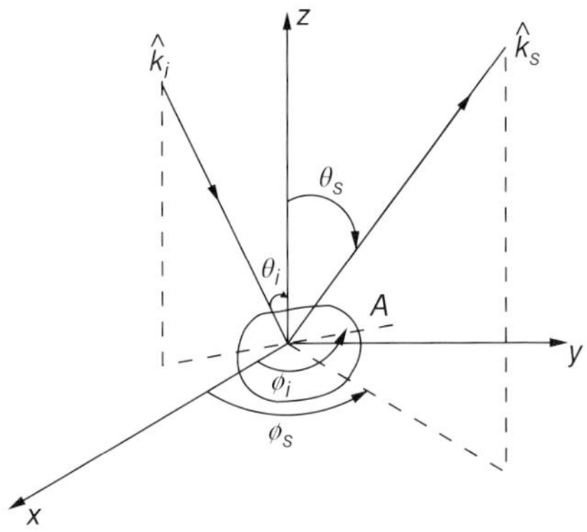

Figure 14 Illustration of the scattering geometry used to calculate bidirectional reflection coefficients.

$\hat{a}$, then the fractional power scattered into the direction (θ_s, ϕ_s) within the *solid angle* $d\omega_s$ with polarization along the unit vector $\hat{b}$ may be expressed in terms of the *bidirectional reflection coefficient*:

$$\rho_{ba}(\theta_i, \phi_i; \theta_s, \phi_s)\, d\omega_s = \lim_{r\to\infty} \frac{4\pi r^2 |E_{bs}|^2}{|E_{ai}|^2 A \cos\theta_i} d\omega_s. \tag{273}$$

Here, r is the distance between the target and the point of measurement. From Figure 14, it follows that the incident and scattered directions can be expressed in terms of the unit vectors $\hat{k}_i$ and $\hat{k}_s$ given by

$$\begin{aligned} \hat{k}_i &= \sin\theta_i \cos\phi_i\, \hat{i}_1 + \sin\theta_i \sin\phi_i\, \hat{i}_2 - \cos\theta_i\, \hat{i}_3 \\ \hat{k}_s &= \sin\theta_s \cos\phi_s\, \hat{i}_1 + \sin\theta_s \sin\phi_s\, \hat{i}_2 + \cos\theta_s\, \hat{i}_3 \end{aligned} \tag{274}$$

where $(\hat{i}_1, \hat{i}_2, \hat{i}_3) = (\hat{x}, \hat{y}, \hat{z})$ are unit vectors in the Cartesian coordinate system. In the particular backscattering direction $\theta_s = \theta_i$ and $\phi_s = \pi + \phi_i$, the *backscattering coefficient* is given by

$$\beta_{ba}(\theta_i, \phi_i) = \cos\theta_i \rho_{ba}(\theta_i, \phi_i; \theta_s = \theta_i, \phi_s = \pi + \phi_i). \tag{275}$$

The *spectral intensity* or *radiance* I_ν is defined as the ratio

$$I_\nu = \frac{d^4E}{\cos\theta dAdtd\omega d\nu} \qquad \left[\mathrm{W\cdot m^{-2}\cdot sr^{-1}\cdot Hz^{-1}}\right] \tag{276}$$

where d^4E is the energy flow at frequency ν within the *solid angle* $d\omega$ in time interval dt. The radiation in a *blackbody cavity* is *isotropic* and *unpolarized*. When this radiation escapes from a small hole in the cavity into a medium with *permeability* μ and *permittivity* ϵ, the radiance at frequency ν is given by the *Planck function* multiplied by $\mu\epsilon$, that is,

$$I_\nu = \mu\epsilon \frac{h\nu^3}{e^{h\nu/k_BT} - 1}. \tag{277}$$

In the *Rayleigh–Jeans limit* $h\nu << k_B T$, Eq. (277) becomes $[(e^{h\nu/k_B T} - 1)^{-1} \approx k_B T/h\nu]$

$$I_\lambda = \frac{k_B T}{\lambda^2} \frac{\mu\epsilon}{\mu_0\epsilon_0} \tag{278}$$

where
$\lambda = c/\nu$ is the wavelength, and $c = 1/\sqrt{\epsilon_0\mu_0}$ is the speed of light in vacuum. Thus, in free space

$$I_\lambda = \frac{k_B T}{\lambda^2}. \tag{279}$$

Real materials emit less than an *ideal blackbody*, and the *emitted radiance depends* on the *direction* (θ, ϕ) and the *polarization* direction $\hat{b}$. When $h\nu << k_B T$, the radiance can be converted into an equivalent *brightness temperature* [Eq. (279)]

$$T_{Bb}(\theta, \phi) = I_{b\lambda}(\theta, \phi)\frac{\lambda^2}{k_B}. \tag{280}$$

The *emissivity* for a *body* with *uniform temperature* T is defined as

$$\mathcal{E}_b(\theta, \phi) = \frac{T_{Bb}(\theta, \phi)}{T}. \tag{281}$$

For a blackbody, $\mathcal{E}_b(\theta, \phi) \to 1$. The radiative power $dP_b(\theta, \phi)$ emitted in a particular direction with polarization $\hat{b}$ per unit frequency interval is

$$dP_b(\theta, \phi) \equiv \frac{d^4 E_b}{dt} = I_{b\nu}(\theta, \phi) \cos\theta dA d\omega d\nu. \tag{282}$$

2.11.2
Reciprocity Relation and Kirchhoff's Law

The following *reciprocity relation* holds for bidirectional scattering coefficients:

$$\cos\theta_i \rho_{ab}(\theta_r, \phi_r; \theta_i, \phi_i) = \cos\theta_r \rho_{ab}(\theta_i, \phi_i; \theta_r, \phi_r). \tag{283}$$

Kirchhoff's law describes the relation between *absorptivity* and *emissivity* of a body. For a *plane wave* with electric field E_{ai} incident on a body with surface area A, the power intercepted by A is proportional to $|E_{ai}|^2 A \cos\theta_i$ and the power scattered into the upper hemisphere (uh) is proportional to $\sum_{b=\perp,\parallel} \int_{\text{uh}} r^2 |E_{bs}|^2 d\omega_s$. The *fraction of the power* of the incident radiation with polarization direction $\hat{a}$ that is *absorbed by the surface* is $[\rho_{ba}(\theta_i, \phi_i; \theta_s, \phi_s) = \lim_{r\to\infty} \frac{4\pi r^2 |E_{bs}|^2}{|E_{ai}|^2 A \cos\theta_i}$, see Eq. (273)]

$$\begin{aligned} \mathcal{A}_a(\theta_i, \phi_i) &= 1 - \frac{\sum_{b=\perp,\parallel} \int_0^{2\pi} d\phi_s \int_0^{\pi/2} d\theta_s \sin\theta_s\, r^2 |E_{bs}|^2}{|E_{ai}|^2 A \cos\theta_i} \\ &= 1 - \frac{1}{4\pi} \sum_{b=\perp,\parallel} \int_0^{2\pi} d\phi_s \int_0^{\pi/2} d\theta_s \sin\theta_s \rho_{ba}(\theta_i, \phi_i; \theta_s, \phi_s). \end{aligned} \tag{284}$$

Here, $b = \perp$ denotes the polarization (direction of the electric field) perpendicular to and $b = \parallel$ polarization parallel to the *plane of incidence*, which is defined by the unit vectors $\hat{n}$ and $\hat{k}_i$, that is, the normal to the surface and the direction of the incident wave [see Eq. (1200) in Appendix C].

If the body with surface area A is assumed to be in thermal equilibrium with blackbody radiation in the half-space above it, then according to the *principle of detailed balance* (see, e.g., Reif [144]), the radiation with a given polarization *leaving* the surface in a specified direction must equal the radiation *received* from the same direction with the same polarization. The power of thermal radiation with polarization direction $\hat{b}$ *incident* on the surface is [using Eq. (282)]

$$dP_{\mathrm{bi}}(\theta_i, \phi_i) = I_{b\nu}(\theta_i, \phi_i) d\omega_i A \cos\theta_i d\nu. \tag{285}$$

The power of the thermal radiation *leaving* the surface in the same direction consists of *two parts*. One is due to *thermal emission*

$$dP_{\mathrm{be}}(\theta_i, \phi_i) = \mathcal{E}_b(\theta_i, \phi_i) I_{b\nu}(\theta_i, \phi_i) d\omega_i A \cos\theta_i d\nu \tag{286}$$

where $\mathcal{E}_b(\theta_i, \phi_i)$ is the emissivity of the surface. The other part originates from incoming *thermal radiation* from all directions (θ, ϕ) that is *scattered* into direction (θ_i, ϕ_i) with polarization direction $\hat{b}$

$$dP_{\mathrm{br}}(\theta_i, \phi_i) = \int_{\mathrm{uh}} d\omega \cos\theta A \, I_{b\nu}(\theta, \phi) \sum_{a=\perp,\parallel} \rho_{ba}(\theta_i, \phi_i; \theta, \phi) \frac{d\omega_i}{4\pi} d\nu. \tag{287}$$

Since the body is assumed to be in *thermodynamic equilibrium* with the half-space above it, the incident power must equal the emitted power:

$$dP_{\mathrm{bi}}(\theta_i, \phi_i) = dP_{\mathrm{be}}(\theta_i, \phi_i) + dP_{\mathrm{br}}(\theta_i, \phi_i). \tag{288}$$

Substituting in Eq. (288) $dP_{\mathrm{bi}}(\theta_i, \phi_i)$ from Eq. (285), $dP_{\mathrm{be}}(\theta_i, \phi_i)$ from Eq. (286), and $dP_{\mathrm{br}}(\theta_i, \phi_i)$ from Eq. (287), we obtain (see Problem 2.10)

$$\mathcal{E}_b(\theta_i, \phi_i) = 1 - \frac{1}{4\pi} \sum_{a=\perp,\parallel} \int_0^{2\pi} d\phi \int_0^{\pi/2} d\theta \sin\theta \rho_{ab}(\theta, \phi; \theta_i, \phi_i) = \mathcal{A}_b(\theta_i, \phi_i) \tag{289}$$

where we have used the *reciprocity relation* [Eq. (283)]. Using Eq. (289), we may compute the emissivity $\mathcal{E}_b(\theta_i, \phi_i)$ from the bidirectional scattering coefficient $\rho_{ab}(\theta, \phi; \theta_i, \phi_i)$.

2.11.3
Specular Versus Lambertian and Non-Lambertian Reflection at the Lower Boundary

For simplicity, it is often assumed that the reflection by the underlying surface is independent of the direction as well as the state of polarization of the incident radiation. As a consequence, the reflected radiation will be isotropic and unpolarized, and the bidirectional reflectance of such a *Lambertian surface* is given by a single number $\rho_L(\lambda)$, which in general will depend on the frequency or wavelength.

Plane Dielectric Interface

The opposite of the Lambertian assumption may be considered to be a mirror-like surface represented by a smooth, plane dielectric interface (such as a smooth, flat ice surface or a calm water surface). For such an interface the backward scattered radiation will be solely in the direction of specular reflection, it will have the same polarization as the incident radiation, and the reflected power is $|r_b(\theta_i)|^2$, where $r_b(\theta_i)$ is the *Fresnel reflection coefficient* for polarization direction $\hat{b}$ and incident direction θ_i. From Eq. (289)

$$\mathcal{E}_b(\theta_i, \phi_i) = 1 - \underbrace{\frac{1}{4\pi} \sum_{a=\perp,\|} \int_0^{2\pi} d\phi \int_0^{\pi/2} d\theta \sin\theta \rho_{ab}(\theta, \phi; \theta_i, \phi_i)}_{|r_b(\theta_i)|^2} = \mathcal{A}_b(\theta_i, \phi_i)$$

we obtain

$$|r_b(\theta_i)|^2 = \frac{1}{4\pi} \sum_{a=\perp,\|} \int d\omega \rho_{ab}(\theta, \phi; \theta_i, \phi_i) \tag{290}$$

and hence

$$\rho_{ab}(\theta_s, \phi_s; \theta_i, \phi_i) = |r_b(\theta_i)|^2 4\pi\delta(\cos\theta_s - \cos\theta_i)\delta(\phi_s - \phi_i)\delta_{ba}. \tag{291}$$

2.11.4 Scattering, Emission, and Transmission by a Random Rough Surface – Kirchhoff Approximation

The Kirchhoff approach is based on the *tangent plane approximation*, according to which the fields at any point on a surface can be approximated by the field that would be present at that point if the actual surface were replaced locally by the tangent plane. This approximation requires a large radius of curvature relative to the incident wavelength at every point on the surface.

2.11.4.1 Rough Dielectric Interface

The backscattered and transmitted fields are derived by using a combination of the *Kirchhoff (tangent plane) approximation* and geometrical optics (for details see Appendix C). Note that, unlike the case of a plane interface where there is no coupling between parallel and perpendicular polarization components and coupling only to the specular reflection and transmission directions, the incident radiance is coupled to all reflected and transmitted directions. In Figure 15, a *plane wave* is incident from medium 1 on to the interface with medium 2 in direction $\hat{k}_i$ with electric field

$$\mathbf{E}_i = \hat{e}_i E_0 e^{i\mathbf{k}_i \cdot \mathbf{r}} \tag{292}$$

where $\mathbf{k}_i$ denotes the incident wave vector and $\hat{e}_i$ the polarization of the electric field vector. The incident field will generate a reflected and a transmitted field in

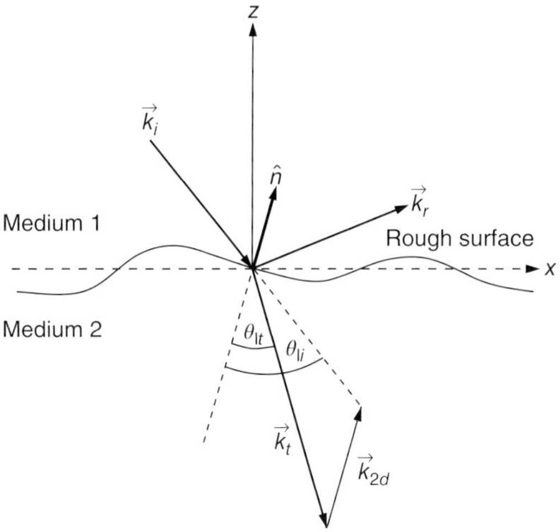

Figure 15 Schematic illustration of a plane wave incident on a rough surface characterized by a Gaussian random height distribution $z = f(x, y) = f(\mathbf{r}_\perp)$, where $f(\mathbf{r}_\perp) = f(x, y)$ with mean height $\langle z \rangle = \langle f(x, y) \rangle \equiv \langle f(\mathbf{r}_\perp) \rangle = 0$. Medium 1 above the surface has *permeability* μ_0 and *permittivity* ϵ_1, whereas medium 2 below the surface has permeability μ_0 and permittivity ϵ_2.

medium 1 and medium 2, respectively. The *local surface slopes* in the x and y directions for the reflected fields are

$$\alpha_1 = -\frac{k_{1dz}}{k_{1dx}} \equiv -\frac{k_{rdz}}{k_{rdx}} \qquad \beta_1 = -\frac{k_{1dz}}{k_{1dy}} \equiv -\frac{k_{rdz}}{k_{rdy}} \tag{293}$$

and for the transmitted fields

$$\alpha_2 = -\frac{k_{2dz}}{k_{2dx}} \equiv -\frac{k_{tdz}}{k_{tdx}} \qquad \beta_2 = -\frac{k_{2dz}}{k_{2dy}} \equiv -\frac{k_{tdz}}{k_{tdy}} \tag{294}$$

where (see Figure 15)

$$\mathbf{k}_{1d} = \mathbf{k}_i - \mathbf{k}_{1r} = \hat{x}k_{1dx} + \hat{y}k_{1dy} + \hat{z}k_{1dz} \tag{295}$$

$$\mathbf{k}_{2d} = \mathbf{k}_i - \mathbf{k}_{2t} = \hat{x}k_{2dx} + \hat{y}k_{2dy} + \hat{z}k_{2dz}. \tag{296}$$

The reflection and transmission conditions at a rough interface are (see Appendix C)

$$\tilde{\mathbf{I}}_1(\hat{k}_s) = \int_0^{2\pi} d\phi_i \int_0^{\pi/2} d\theta_i \sin\theta_i \mathbf{R}_{rs}(\theta_s, \phi_s; \theta_i, \phi_i)\tilde{\mathbf{I}}_1(\hat{k}_i) \tag{297}$$

$$\tilde{\mathbf{I}}_2(\hat{k}_t) = \int_0^{2\pi} d\phi_i \int_0^{\pi/2} d\theta_i \sin\theta_i \mathbf{T}_{rs}(\theta_t, \phi_t; \theta_i, \phi_i)\tilde{\mathbf{I}}_1(\hat{k}_i) \tag{298}$$

where $\tilde{\mathbf{I}}_1$ and $\tilde{\mathbf{I}}_2$ are the Stokes vectors for reflected and transmitted radiation, respectively:

$$\tilde{\mathbf{I}}_p \equiv [I_{\|p}, I_{\perp p}, U_p, V_p]^T \qquad p = 1, 2. \tag{299}$$

The *reflectance matrix* is related to the *bidirectional polarized reflectance distribution function (BP$_r$DF)* $\vec{\rho}_{rs}(\theta_s, \phi_s; \theta_i, \phi_i)$ by

$$\mathbf{R}_{rs}(\theta_s, \phi_s; \theta_i, \phi_i) = \vec{\rho}_{rs}(\theta_s, \phi_s; \theta_i, \phi_i) \times \cos\theta_i. \tag{300}$$

Similarly, the *transmittance matrix* is related to the *bidirectional polarized transmittance distribution function (BP$_t$DF)* $\vec{t}_{rs}(\theta_t, \phi_t; \theta_i, \phi_i)$ by

$$\mathbf{T}_{rs}(\theta_t, \phi_t; \theta_i, \phi_i) = \vec{t}_{rs}(\theta_t, \phi_t; \theta_i, \phi_i) \times \cos\theta_i. \tag{301}$$

The backscattered or the transmitted field in the direction $\hat{k}_s$ or $\hat{k}_t$, respectively, consists of contributions from all the scattered fields that are coupled to the direction $\hat{k}_s$ or $\hat{k}_t$, respectively.

Explicit expressions for the reflectance matrix $\mathbf{R}_{rs}$ and the transmittance matrix $\mathbf{T}_{rs}$ at the rough surface interface are given by (Appendix C)

$$\mathbf{R}_{rs}(\theta_s, \phi_s; \theta_i, \phi_i) = \frac{1}{\cos\theta_s} \frac{|\mathbf{k}_{1d}|^4}{4|\hat{k}_i \times \hat{k}_s|^4 k_{1dz}^4} \times \frac{1}{2\pi\sigma^2} \exp\left(-\frac{k_{1dx}^2 + k_{1dy}^2}{2k_{1dz}^2\sigma^2}\right) \mathbf{C}_{rs}^r(\theta_s, \phi_s; \theta_i, \phi_i) \tag{302}$$

$$\mathbf{T}_{rs}(\theta_t, \phi_t; \theta_i, \phi_i) = \frac{1}{\cos\theta_t} \frac{k_2^2|\mathbf{k}_{2d}|^2(\hat{n}\cdot\hat{k}_t)^2}{|\hat{k}_i \times \hat{k}_s|^4 k_{2dz}^4} \frac{\eta_1}{\eta_2} \times \frac{1}{2\pi\sigma^2} \exp\left(-\frac{k_{2dx}^2 + k_{2dy}^2}{2k_{2dz}^2\sigma^2}\right) \mathbf{C}_{rs}^t(\theta_t, \phi_t; \theta_i, \phi_i) \tag{303}$$

where η_1 and η_2 are the impedances of medium 1 and medium 2, respectively. The *impedance* is defined as $\eta = \sqrt{\mu/\epsilon}$, where μ is the *permeability*, and ϵ the *permittivity* of the medium. The matrices $\mathbf{C}_{rs}^\alpha$, $\alpha = r, t$ are defined in Appendix C. The factor

$$p(\alpha, \beta) \equiv \frac{1}{2\pi\sigma^2} \exp\left(-\frac{\alpha^2 + \beta^2}{2\sigma^2}\right) \qquad \alpha = \frac{k_{pdx}^2}{k_{pdz}^2}, \qquad \beta = \frac{k_{pdy}^2}{k_{pdz}^2}, \quad p = 1, 2$$

is the probability density for the surface slopes resulting from the assumption of a *Gaussian rough surface*, and σ^2 is the mean square *surface slope*.

Unpolarized Radiation

In the case of unpolarized incident radiation, the *reflection and transmittance matrices* become scalar quantities. For the reflected radiation, the matrix $\mathbf{C}_{rs}^r$ in Eq. (302) reduces to

$$C_{rs}^r(\theta_s, \phi_s; \theta_i, \phi_i) = |\hat{k}_i \times \hat{k}_s|^4 \frac{R_\perp^2 + R_\parallel^2}{2} \equiv |\hat{k}_i \times \hat{k}_s|^4 \, R_{\text{unpol}} \tag{304}$$

where R_{unpol} is given by Eq. (260), so that Eq. (302) becomes

$$R_{rs}(\theta_s, \phi_s; \theta_i, \phi_i) = \frac{1}{\cos\theta_s} \frac{1}{4} \left(\frac{|\mathbf{k}_{1d}|}{k_{1dz}}\right)^4 \frac{1}{2\pi\sigma^2} \exp\left[-\frac{k_{1dx}^2 + k_{1dy}^2}{2k_{1dz}^2\sigma^2}\right] R_{\text{unpol}}. \tag{305}$$

Since $\frac{k_{1dz}}{|\mathbf{k}_{1d}|} = \cos\theta_n$, where θ_n is the tilt of the surface facet normal from the vertical direction, and

$$P(\alpha, \beta) = \frac{1}{2\pi\sigma^2}\exp\left[-\frac{k_{1dx}^2 + k_{1dy}^2}{2k_{1dz}^2\sigma^2}\right] = \frac{1}{\pi\sigma^2}\exp\left[-\frac{\tan^2\theta_n}{\sigma^2}\right] \tag{306}$$

we find

$$R_{rs}(\theta_s, \phi_s; \theta_i, \phi_i) = \frac{1}{4\cos\theta_s\cos^4\theta_n}\frac{1}{\pi\sigma^2}\exp\left[-\frac{\tan^2\theta_n}{\sigma^2}\right] R_{\text{unpol}}. \tag{307}$$

Similarly, in the case of unpolarized radiation, the matrix $\mathbf{C}_{rs}^t$ in Eq. (303) for the transmitted radiation reduces to

$$C_{rs}^t(\theta_t, \phi_t; \theta_i, \phi_i) = |\hat{k}_i \times \hat{k}_s|^4 \frac{T_\perp^2 + T_\parallel^2}{2} \equiv |\hat{k}_i \times \hat{k}_s|^4\, T_{\text{unpol}} \tag{308}$$

where T_{unpol} is given by Eq. (261), and Eq. (303) becomes

$$T_{rs}(\theta_t, \phi_t; \theta_i, \phi_i) = \frac{1}{\cos\theta_t}\frac{k_2^2|\mathbf{k}_{2d}|^2(\hat{n}\cdot\hat{k}_t)^2}{k_{2dz}^4}\frac{\eta_1}{\eta_2}\frac{1}{\pi\sigma^2}\exp\left[-\frac{\tan^2\theta_n}{\sigma^2}\right] T_{\text{unpol}}. \tag{309}$$

Since $\eta_1/\eta_2 = n_2/n_1$ and

$$\frac{|\mathbf{k}_{2d}|^2}{k_{2dz}^4} = \left(\frac{|\mathbf{k}_{2d}|}{k_{2dz}}\right)^4 \frac{1}{|\mathbf{k}_{2d}|^2} = \frac{1}{\cos^4\theta_n}\frac{1}{|\mathbf{k}_{2d}|^2}$$

we have

$$T_{rs}(\theta_t, \phi_t; \theta_i, \phi_i) = \frac{1}{\cos\theta_t}\frac{1}{\cos^4\theta_n}\frac{k_2^2}{|\mathbf{k}_{2d}|^2}\frac{n_2}{n_1}(\hat{n}\cdot\hat{k}_t)^2\frac{1}{\pi\sigma^2}\exp\left[-\frac{\tan^2\theta_n}{\sigma^2}\right] T_{\text{unpol}}. \tag{310}$$

Furthermore (see lower part of Figure 15)

$$\mathbf{k}_{2d} = \hat{\mathbf{n}} = \mathbf{k}_2 - \mathbf{k}_1 = \mathbf{k}_t - \mathbf{k}_i$$

which implies

$$|\mathbf{k}_{2d}| = |k_2\cos\theta_{lt} - k_1\cos\theta_{li}| = \left|k_2(\hat{n}\cdot\hat{k}_t) - k_1(\hat{n}\cdot\hat{k}_i)\right| = k_2\left|(\hat{n}\cdot\hat{k}_t) - \frac{n_1}{n_2}(\hat{n}\cdot\hat{k}_i)\right|$$

and hence

$$\frac{k_2^2}{|\mathbf{k}_{2d}|^2} = \left(\frac{1}{(\hat{n}\cdot\hat{k}_t) - \frac{n_1}{n_2}(\hat{n}\cdot\hat{k}_i)}\right)^2.$$

Insertion in Eq. (310) yields

$$T_{rs}(\theta_t, \phi_t; \theta_i, \phi_i) = \frac{1}{\cos\theta_t}\frac{1}{\cos^4\theta_n}\left(\frac{(\hat{n}\cdot\hat{k}_t)}{(\hat{n}\cdot\hat{k}_t) - \frac{n_1}{n_2}(\hat{n}\cdot\hat{k}_i)}\right)^2\frac{n_2}{n_1}$$
$$\times\frac{1}{\pi\sigma^2}\exp\left[-\frac{\tan^2\theta_n}{\sigma^2}\right] T_{\text{unpol}}. \tag{311}$$

2.11.5 Slope Statistics for a Wind-Roughened Water Surface

Based on photographs of *sunglint* obtained from an airplane, Cox and Munk [145] expressed the slope distribution of the ocean waves by a *Gram–Charlier series*

$$
\begin{aligned}
P(z_x, z_y) = {} & \frac{1}{2\pi\sigma_x\sigma_y} \exp\left[-\frac{1}{2}\left(\frac{z_x^{\,2}}{\sigma_x^2} + \frac{z_y^{\,2}}{\sigma_y^2}\right)\right]\Big\{1 - \frac{1}{2}C_{21}(\xi^2 - 1) - \frac{1}{6}C_{03}(\eta^3 - 3\eta) \\
& + \frac{1}{24}C_{40}(\xi^4 - 6\xi^2 + 3) + \frac{1}{4}C_{22}(\xi^2 - 1)(\eta^2 - 1) \\
& + \frac{1}{24}C_{04}(\eta^4 - 6\eta^2 + 3)\Big\}
\end{aligned}
\qquad (312)
$$

where C_{21} and C_{03} are *skewness coefficients*; C_{40}, C_{22}, and C_{04} are *peakedness coefficients*; and σ_x and σ_y are the root-mean-square (rms) values of the *surface slope components* z_x and z_y, defined by

$$
z_x = \frac{\partial z}{\partial x} = \frac{\partial f(x,y)}{\partial x} = \sin\phi \tan\theta_n \qquad z_y = \frac{\partial z}{\partial y} = \frac{\partial f(x,y)}{\partial y} = \cos\phi \tan\theta_n
$$

where θ_n is the tilt angle and ϕ the relative azimuth angle of the surface facet as illustrated in Figure 16. In the absence of *skewness* ($C_{21} = C_{03} = 0$), and *peakedness* ($C_{40} = C_{22} = C_{04} = 0$), the Gram–Charlier series reduces to a two-dimensional Gaussian:

$$
P(z_x, z_y) = \frac{1}{2\pi\sigma_x\sigma_y} \exp\left[-\frac{1}{2}\left(\frac{z_x^{\,2}}{\sigma_x^2} + \frac{z_y^{\,2}}{\sigma_y^2}\right)\right] \qquad (313)
$$

which for an *isotropic slope distribution* ($\sigma_x = \sigma_y$) becomes (see Problem 2.12)

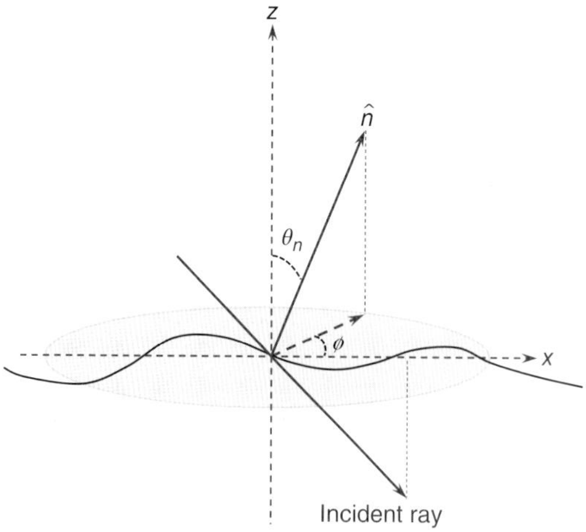

Figure 16 Illustration of tilt angle (θ_n) and relative azimuth angle ($\phi \equiv \Delta\phi$) in rough surface scattering. Here, $\hat{\mathbf{n}}$ is the normal to the facet, θ_n is the tilt angle, and ϕ is the azimuth angle relative to the glint direction ($\phi = 0$).

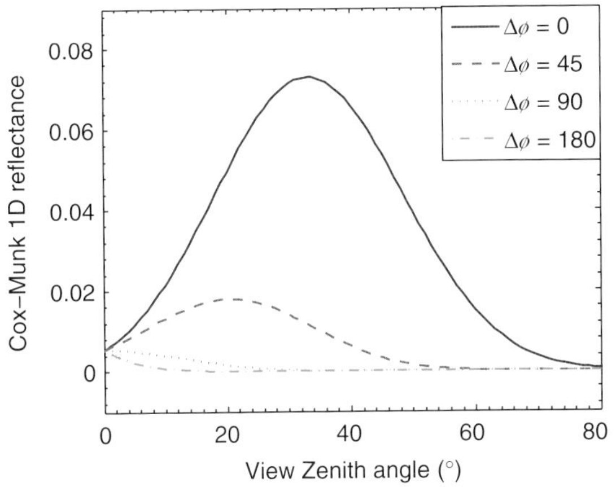

Figure 17 Reflectance $R(\theta) = \rho(\theta', \theta) \times \cos\theta'$ [Eq. (307)] for a Cox–Munk 1-D Gaussian BRDF $\rho(\theta', \theta)$ for a solar zenith angle of $\theta' = 30°$ and a wind speed of 5 m s^{-1} versus zenith angle θ. The relative azimuth angle $\Delta\phi = 0°$ corresponds to the glint direction (see Figure 16).

$$P(z_x, z_y) = \frac{1}{\pi\sigma^2} \exp\left(-\frac{\tan^2\theta_n}{\sigma^2}\right) \tag{314}$$

which is identical to Eq. (306). This probability distribution is widely used to represent the slope statistics of water waves with the numerical value of the slope variance $\sigma^2 = (0.003 + 0.512) \times \text{ws}$, ws stands for wind speeding in 5 m s^{-1} [145]. An example of a *"Cox-Munk" BRDF* for a wind speed of 5 m s^{-1} is shown in Figure 17.

2.12 Land Surfaces

The reflectance of land surfaces is usually described in terms of the *bidirectional reflectance distribution function (BRDF),* defined as [18]

$$\rho(\hat{\mathbf{\Omega}}', \hat{\mathbf{\Omega}}) = \frac{dI_r^+(\hat{\mathbf{\Omega}})}{I^-(\hat{\mathbf{\Omega}}') \cos\theta' d\omega'} \tag{315}$$

where $dI_r^+(\hat{\mathbf{\Omega}})$ is the radiance reflected in the direction $\hat{\mathbf{\Omega}} = (\theta, \phi)$, while $I^-(\hat{\mathbf{\Omega}}')$ is the incident radiance in direction $\hat{\mathbf{\Omega}}' = (\theta', \phi')$.

The interaction of radiation with land surfaces typically involves two distinct physical processes: *surface scattering* and *volume scattering.* Surface scattering can lead to a *"hot spot" effect,* which in plant canopies is caused by the absence of shadows in the direction of illumination, leading to enhanced brightness when a scene is viewed from the direction of the sun. This "hot spot" effect, also referred to as *retroreflectance,* or the *opposition effect,* is typically modeled using geometrical optics. Volume scattering, on the other hand, is due to multiple scattering within the plant canopy, which requires application of RT theory [146].

Before proceeding, let us explain the scattering geometry that is illustrated in Figure 18. For the sake of illustration, we assume that the Sun supplies a parallel beam of light incident on the surface (ignoring atmospheric scattering). The vector $\hat{k}_i$ in Figure 18 describes the illumination in direction $(-\mu_i, \phi_i) \equiv (-\mu', \phi')$, and the scattered (reflected) light is described by the vector $\hat{k}_r$ in direction $(\mu_r, \phi_r) \equiv (\mu, \phi)$. The relative azimuth between the incident beam in direction $\hat{k}_i$ and the scattered beam in direction $\hat{k}_r$ is $\Delta\phi = \phi' - \phi$ if we use the azimuth angle of $\hat{k}_i$ as the reference ($\phi = 0°$ in Figure 18), while the relative azimuth between $-\hat{k}_i$ and $\hat{k}_r$ is $\Delta\phi' = \pi - \Delta\phi$ as illustrated in Figure 18. The *phase (or backscattering)* angle α (see Figure 18) is usually defined as

$$\cos\alpha = \hat{k}_{\text{sun}} \cdot \hat{k}_r = -\hat{k}_i \cdot \hat{k}_r = \mu'\mu + \sqrt{1-\mu'^2}\sqrt{1-\mu^2}\cos(\Delta\phi')$$
$$= \mu'\mu - \sqrt{1-\mu'^2}\sqrt{1-\mu^2}\cos(\Delta\phi). \quad (316)$$

The *scattering angle* $\Theta = \pi - \alpha$ is given by

$$\cos\Theta = \hat{k}_i \cdot \hat{k}_r = \cos(\pi - \alpha) = -\mu'\mu + \sqrt{1-\mu'^2}\sqrt{1-\mu^2}\cos(\Delta\phi). \quad (317)$$

2.12.1 Unpolarized Light

The Hapke BRDF

The *Hapke BRDF*, $\rho_H(\theta', \theta, \phi' - \phi)$, is an analytic function used to model the reflectance from planetary and lunar surfaces [146]:

$$\rho_H(-\mu', \mu, \Delta\phi) = \frac{\varpi}{4}\left[\frac{(1+b)p(\cos\Theta) + h_0 h - 1}{\mu' + \mu}\right] \quad (318)$$

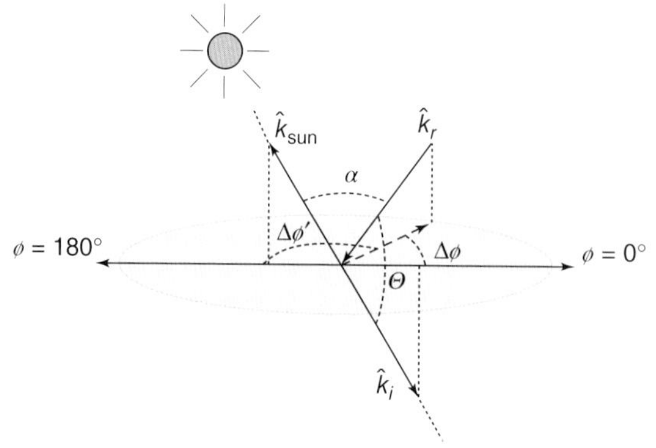

Figure 18 Illustration of the geometry involved in the description of the BRDF. The phase angle α is the supplementary angle of the scattering angle Θ, that is, $\alpha = \pi - \Theta$. The relative azimuth angle $\Delta\phi = 0°$ corresponds to the glint direction, and $\Delta\phi = 180°$ to the backscattering (hot-spot) direction.

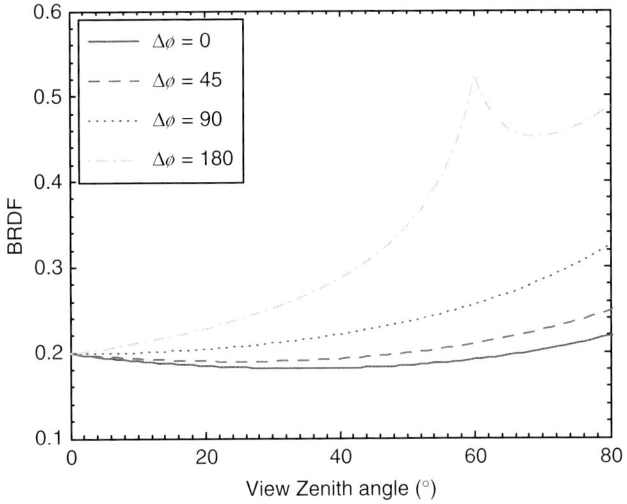

Figure 19 Hapke BRDF [Eq. (318)] for a solar zenith angle of 60°, $\Delta\phi = 0$, $\varpi = 0.6$, $b_0 = 1.0$, and $h_h = 0.06$. The relative azimuth angle $\Delta\phi = 0°$ corresponds to the glint direction, and $\Delta\phi = 180°$ to the backscattering (hot-spot) direction.

where $\mu' = \cos\theta'$, $\mu = \cos\theta$, $\Delta\phi = \phi' - \phi$ and ϖ is the *single-scattering albedo*. The *scattering phase function* $p(\cos\Theta)$ is approximated by

$$p(\cos\Theta) \approx 1 + \frac{\cos\Theta}{2} \tag{319}$$

where the cosine of the *scattering angle* Θ given by Eq. (317). The parameters h_0 and h are determined by the "albedo factor" $\sqrt{1-\varpi}$ and the angles of incidence and observation, respectively:

$$h_0 = \frac{1+2\mu'}{1+2\mu'\sqrt{1-\varpi}} \quad \text{and} \quad h = \frac{1+2\mu}{1+2\mu\sqrt{1-\varpi}}. \tag{320}$$

The term b accounts for the *opposition effect*, also known as *retroreflectance*, and depends on the angular width h_h of the retroreflectance as well as an empirical factor b_0 to account for the finite size of the Particles:

$$b = \frac{b_0 h_h}{h_h + \tan(\frac{\Theta}{2})}. \tag{321}$$

An example of the Hapke BRDF is provided in Figure 19. This kind of BRDF model has been used to simulate the reflectance of planetary and lunar surfaces [146].

The Rahman–Pinty–Verstraete (RPV) BRDF Model

In terrestrial remote sensing applications, the BRDF of land surfaces is often described by a few parameters. The *Rahman–Pinty–Verstraete (RPV) BRDF* model [147] has been widely used to compute the radiance reflected from land surfaces. The RPV BRDF can be written as [148]:

$$\rho_{\mathrm{RPV}}(-\mu', \mu, \phi' - \phi) = \rho_0 \left[\mu' \mu(\mu' + \mu)\right]^{k-1} F_{\mathrm{HG}}(g, \alpha) h(h_0, G) \tag{322}$$

where ρ_0 is an empirical parameter, and $F_{\mathrm{HG}}(g, \alpha)$ is the HG scattering function [see Eq. (24)] given by

$$F_{\mathrm{HG}}(g, \alpha) = \frac{1 - g^2}{[1 + g^2 - 2g\cos(\pi - \alpha)]^{3/2}} \tag{323}$$

and the *phase angle* is given by Eq. (316). The *hot spot* is approximated by the function

$$h(h_0, G) = 1 + \frac{1 - h_0}{1 + G} \tag{324}$$

where the geometric factor G is given by

$$G = [\tan^2\theta' + \tan^2\theta - 2\tan\theta' \tan\theta \times \cos(\Delta\phi)]^{1/2}. \tag{325}$$

Note that in the hot-spot direction, for which $\theta' = \theta$, $\phi' = \phi$, and $G = 0$, $h(h_0, G)$ is a maximum and the total reflectance also increases. Note also that the anisotropy of the surface is described by only four independent parameters:

- ρ_0 characterizing the intensity of the surface reflectance;
- k indicating the level of surface anisotropy. When $k = 1$, all the cosine functions become equal to 1 and the resulting anisotropy stems solely from the function $F_{\mathrm{HG}}(g, \alpha)$ and the hot spot function $h(h_0, G)$;
- g controlling the relative amount of forward ($0 \leq g \leq +1$) and backward ($-1 \leq g \leq 0$) scattering;
- h_0, a *hot spot* parameter.

As noted by Rahman *et al.* [147] and Dubovik *et al.* [149], all four parameters could potentially be used to discriminate between different land surfaces. An example of a RVP BRDF for a vegetated surface is given in Figure 20.

The Ross-Li BRDF Model

Ross–Li refers to a kernel-based BRDF model [150–152] described by the following equation:

$$\rho_{\mathrm{RL}}(\mu', \mu, \Delta\phi) = c_0 + c_s F_s(\mu', \mu, \Delta\phi) + c_v F_v(\mu', \mu, \Delta\phi) \tag{326}$$

where $F_s(\mu', \mu, \Delta\phi)$ represents the surface scattering (geometrical optics) kernel and $F_v(\mu', \mu, \Delta\phi)$ represents the volume (multiple) scattering kernel. The surface scattering kernel, designed to account for mutual shadowing of protrusions, is given by (Maignan *et al.* [153])

$$F_s(\mu', \mu, \Delta\phi) = \frac{\mu' + \mu}{\pi\mu'\mu}(t - \sin t\cos t - \pi) + \frac{1 + \cos\alpha}{2\mu'\mu} \tag{327}$$

$$\cos t = \frac{2\mu'\mu}{\mu' + \mu}\left(G^2 + (\tan\theta' \tan\theta \sin\Delta\phi)^2\right)^{1/2} \tag{328}$$

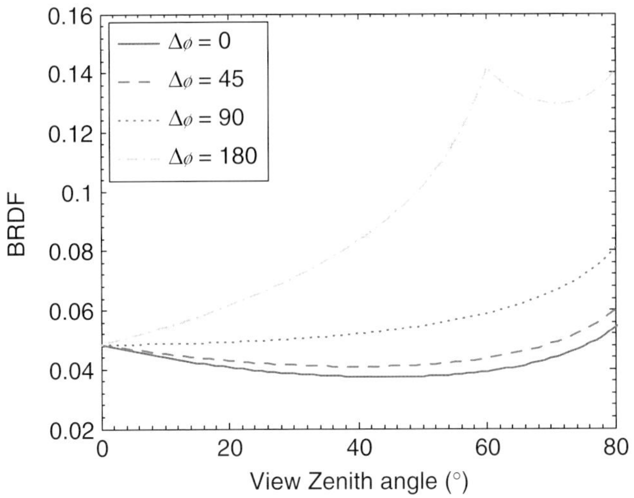

Figure 20 RPV BRDF [Eqs. (322)–(325)] for a polar incidence angle of 60°, $\rho_0 = 0.027$, $k = 0.647$, $g = -0.169$, and $h_0 = 0.1$. The relative azimuth angle $\Delta\phi = 0°$ corresponds to the glint direction, and $\Delta\phi = 180°$ to the backscattering (hot-spot) direction.

where G is given by Eq. (325) and

$$t = \begin{cases} \cos^{-1} t, & \cos t \leq 1, \\ 0, & \cos t > 1. \end{cases} \tag{329}$$

The volume scattering kernel is given by

$$F_v(\mu', \mu, \Delta\phi) = \frac{4C(\alpha, \alpha_0)}{3\pi(\mu' + \mu)}\left[(\frac{\pi}{2} - \alpha)\cos\alpha + \sin\alpha\right] - \frac{1}{3}. \tag{330}$$

The *phase angle* α is given by Eq. (316), and the factor $C(\alpha, \alpha_0)$, which is included to improve the hot-spot reflection, is given by (Maignan *et al.* [153])

$$C(\alpha, \alpha_0) = 1 + \left(1 + \frac{\alpha}{\alpha_0}\right)^{-1} \tag{331}$$

where the angle α_0 has been found to lie in the range 1–2° for most land surfaces [153].

The Ross–Li model [150–152] has been used to retrieve BRDF effects from MODIS data (Figure 21) [154], and the Rahman–Pinty–Verstraete (RPV) BRDF model [147] has been used to compute the radiance reflected from land surfaces. Comparisons of various BRDF models with satellite [153] and aircraft [155, 156] data have shown that both the Ross–Li and RPV models are capable of reproducing results that agree with multiangle observations, but the RVP model has been applied more extensively to interpret multidirectional images [149].

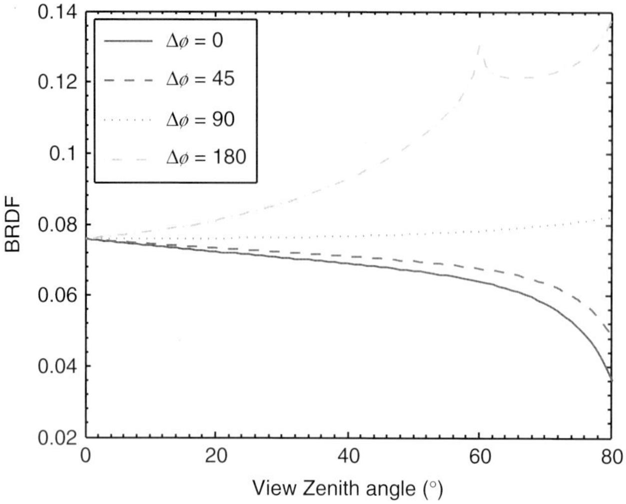

Figure 21 Ross–Li BRDF [Eqs. (326)–(331)] for a polar incidence angle of 60°, $c_0 = 0.091$, $c_v = 0.069$, and $\alpha_0 = 1.0$. The relative azimuth angle $\Delta\phi = 0°$ corresponds to the glint direction, and $\Delta\phi = 180°$ to the backscattering (hot-spot) direction.

2.12.2
Polarized Light

As discussed by Cairns *et al.* [157], accurate retrieval of aerosol properties over land surfaces requires adequate representation of the *polarized reflectance* of the underlying surface. Theoretical and empirical findings indicate that the polarization of light reflected by a surface (soil or vegetation) is generated primarily by specular reflection from suitably oriented surface facets, such as mineral surfaces or leaves. In addition to the illumination $[\hat{\Omega}' = (\theta', \phi')]$ and observation $[\hat{\Omega} = (\theta, \phi)]$ directions, two angles are needed to model the polarized reflectance. These are the angle of incidence $\alpha_i = \alpha/2$, which is the angle between the incoming light and the normal to the reflecting facet (the angle α is shown in Figure 18), and the tilt angle θ_n shown in Figure 16.

For the *polarized reflectance*, the model proposed by Maignan *et al.* [158] has proven to be very useful [149]. In this model, the *bidirectional polarized reflectance distribution function (BP$_r$DF)* is described as

$$\rho_p(\theta', \phi', \theta, \phi) = \rho_M F_p(\alpha_i, n) \tag{332}$$

where $F_p(\alpha_i, n)$ is the *polarized Fresnel reflectance* given by [see Eq. (262)]

$$F_p(\alpha_I, n) = \frac{1}{2}\left(\frac{\sin^2(\alpha_t - \alpha_i)}{\sin^2(\alpha_t + \alpha_i)} - \frac{\tan^2(\alpha_t - \alpha_i)}{\tan^2(\alpha_t + \alpha_i)}\right) \tag{333}$$

where the angle of refraction α_t is related to the angle of incidence α_i through Snell's law

$$\sin\alpha_t = \sin\alpha_i / n \tag{334}$$

and $n = n_{veg}/n_{air}$ is the refractive index of vegetation relative to air. The function ρ_M is given by

$$\rho_\mathrm{M} = \frac{C \exp(-\tan\alpha_i)\exp(-\nu)}{4(\mu' + \mu)} \tag{335}$$

where ν is the *normalized difference vegetation index (NDVI)*, defined as a normalized difference between the reflectance in two bands:

$$\nu = \frac{R_\mathrm{vis} - R_\mathrm{nir}}{R_\mathrm{vis} + R_\mathrm{nir}} \tag{336}$$

where R_vis is the observed reflectance of a channel in the visible spectral range, and R_nir is the observed reflectance of a channel in the near-infrared spectral range.

The attenuation term $\exp(-\nu)$ reflects the observed tendency of the polarized reflectance to decrease with an increase in vegetation cover. The NDVI value ν can be obtained from reflectance measurements obtained simultaneously with the polarization observations, and the factor C is a free parameter that can be chosen to fit the observed angular dependence of the $\mathrm{BP}_r\mathrm{DF}$.

Problems

2.1 Show that the Legendre expansion coefficients for the *Rayleigh scattering phase function* are given by $\chi_0 = 1$, $\chi_1 = 0$, $\chi_2 = \frac{2f}{5(3+f)}$, and $\chi_\ell = 0$ for $\ell > 2$.
2.2 Derive Eq. (25).
2.3 Derive Eq. (40).
2.4 Derive Eqs. (43)–(45).
2.5 Show that in the *Stokes vector* representation $\mathbf{I} = [I_{||}, I_{\perp}, U, V]^T$, and the *Stokes scattering matrix* for *Rayleigh scattering* is given by Eq. (60).
2.6 Verify that the normalization given by Eq. (103) is correct.
2.7 Derive Eqs. (109) and (110).
2.8 Derive Eqs. (136)–(151).
2.9 Verify that Eqs. (216) and (217) are correct.
2.10 Derive Eq. (289).
2.11 Derive Eq. (314).

3
Basic Radiative Transfer Theory

3.1
Derivation of the Radiative Transfer Equation (RTE)

3.1.1
RTE for Unpolarized Radiation

The radiant energy passing through a surface element dA in direction $\hat{\Omega}$ within a solid angle $d\omega$ in a time interval dt and a frequency interval $d\nu$ is given by (Thomas and Stamnes [18])

$$dE = I_\nu \cos\theta dA dt d\omega d\nu \tag{337}$$

where θ is the angle between the normal $\hat{n}$ to dA and the direction of propagation $\hat{\Omega}$, that is, $\cos\theta = \hat{n} \cdot \hat{\Omega}$. It follows from Eq. (337) that we may define the spectral radiance as

$$I_\nu = \frac{dE}{\cos\theta dA dt d\omega d\nu}. \tag{338}$$

According to Eqs. (1)–(3), an incident beam of light with radiance I_ν entering a volume element $dV = dA\, ds$ of a slab of cross sectional area dA and length ds will suffer an extinction dI_ν^{ext} described by

$$dI_\nu^{\text{ext}} = -\gamma(\nu) I_\nu ds = -[\alpha(\nu) + \beta(\nu)] I_\nu ds. \tag{339}$$

If the volume dV contains an optically active material emitting radiant energy, the emitted radiance will be given by

$$dI_\nu^{\text{em}} = j_\nu ds \tag{340}$$

where the *emission coefficient* is defined as

$$j_\nu = \frac{dE}{dA ds dt d\omega d\nu} = \frac{dE}{dV dt d\omega d\nu}. \tag{341}$$

The dependence of $I_\nu, \alpha(\nu), \beta(\nu), \gamma(\nu)$, and j_ν on position s is suppressed, but implied in Eqs. (337)–(341). Combining Eqs. (339) and (340), we obtain

$$dI_\nu = dI_\nu^{\text{ext}} + dI_\nu^{\text{em}} = -\gamma(\nu) I_\nu ds + j_\nu ds, \tag{342}$$

Radiative Transfer in Coupled Environmental Systems: An Introduction to Forward and Inverse Modeling,
First Edition. Knut Stamnes and Jakob J. Stamnes.

and dividing by $d\tau_s = \gamma(\nu)ds = [\alpha(\nu) + \beta(\nu)]ds$, we find that the *radiative transfer equation (RTE)* becomes

$$\frac{dI_\nu}{d\tau_s} = -I_\nu + S_\nu, \tag{343}$$

where the *source function* is defined as

$$S_\nu \equiv j_\nu/\gamma(\nu). \tag{344}$$

For a medium in local thermodynamic equilibrium at temperature T, the *emission coefficient* is given by $j_\nu^t = \alpha(\nu)B_\nu(T)$, where $B_\nu(T)$ is the *Planck function*. Hence, the thermal contribution to the source function becomes

$$S_\nu^t = \frac{j_\nu^t}{\gamma(\nu)} = \frac{\alpha(\nu)}{\gamma(\nu)}B_\nu(T) = [1 - \varpi(\nu)]B_\nu(T) \tag{345}$$

where $\varpi = \beta(\nu)/\gamma(\nu)$ is the *single-scattering albedo*. The *source function* due to *multiple scattering* by particles within dV is given by (Thomas and Stamnes [18]) (see Problem 3.1)

$$S_\nu^{\text{sca}} = \frac{\beta(\nu)}{\gamma(\nu)} \int_{4\pi} d\omega' \frac{p(\hat{\Omega}', \hat{\Omega})}{4\pi} I_\nu(\Omega') = \frac{\varpi(\nu)}{4\pi} \int_{4\pi} d\omega' p(\hat{\Omega}', \hat{\Omega}) I_\nu(\Omega') \tag{346}$$

where $p(\hat{\Omega}', \hat{\Omega})$ is the *scattering phase function* [see Eq. (8)] Hence, the complete *RTE* for unpolarized radiation, which includes both *thermal emission* and *multiple scattering*, becomes

$$\frac{dI_\nu}{d\tau_s} = -I_\nu + [1 - \varpi(\nu)]B_\nu(T) + \frac{\varpi(\nu)}{4\pi} \int_{4\pi} d\omega' p(\hat{\Omega}', \hat{\Omega}) I_\nu(\Omega'). \tag{347}$$

In the following, we will restrict our attention to a *plane-parallel medium* in which the *IOPs* are assumed to vary only in the vertical direction (denoted by z), increasing upwards, so that the corresponding *vertical optical depth*, denoted by $\tau(z)$, is defined by

$$\tau(z) = \int_z^\infty [\alpha(z') + \beta(z')]dz' \tag{348}$$

and hence

$$d\tau(z) = -[\alpha(\tau) + \beta(\tau)]dz \tag{349}$$

where the minus sign indicates that τ increases in the downward direction, whereas z increases in the upward direction. Equation (347) pertains to the total radiation field. In a *plane-parallel geometry*, with a collimated beam incident at the top of the upper slab, one may invoke the usual *diffuse-direct splitting* to obtain for the diffuse component of the radiation field (omitting the subscript ν and the functional dependence on ν for notational convenience)

$$u\frac{dI(\tau, u, \phi)}{d\tau} = I(\tau, u, \phi) - S(\tau, u, \phi) \tag{350}$$

where

$$S(\tau, u, \phi) = S_b(\tau, u, \phi) + [1 - \varpi(\tau)]B(\tau) + \frac{\varpi(\tau)}{4\pi} \int_0^{2\pi} d\phi' \int_{-1}^{1} p(\tau, u', \phi'; u, \phi) I(\tau, u', \phi') du. \tag{351}$$

Here, u is the cosine of the polar angle θ, ϕ is the azimuth angle, $\varpi(\tau) = \beta(\tau)/[\alpha(\tau) + \beta(\tau)]$ is the *single-scattering albedo*, $p(\tau, u', \phi'; u, \phi)$ is the *scattering phase function* defined by Eq. (8), and $B(\tau)$ is the *thermal radiation* field given by the Planck function. The term $S_b(\tau, u, \phi)$ in Eq. (351) is the single-scattering source function due to a collimated beam incident at the top of the upper slab, to be specified below.

3.1.2 RTE for Polarized Radiation

To generalize Eq. (350) to apply to polarized radiation, we note that the multiple scattering term in Eq. (351) must be replaced by (Hovenier *et al.* [16])

$$\mathbf{S}(\tau, u, \phi) = \frac{\varpi(\tau)}{4\pi} \int_0^{2\pi} d\phi' \int_{-1}^{1} \mathbf{P}(\tau, u', \phi'; u, \phi) \mathbf{I}(\tau, u', \phi') du \tag{352}$$

where $\mathbf{I}(\tau, u', \phi') \equiv \mathbf{I}_S(\tau, u', \phi') = [I, Q, U, V]^T$ is the Stokes vector, $\mathbf{P}(\tau, u', \phi'; u, \phi)$ is the scattering phase matrix [see Eq. (40)], and the first element of the vector $\mathbf{S}(\tau, u, \phi)$ represents the energy per unit solid angle per unit frequency interval per unit time that is scattered by a unit volume in the direction (θ, ϕ). Hence, in a plane-parallel (slab) geometry, the integro-differential equation for polarized radiative transfer is expressed in terms of a Stokes vector $\mathbf{I}(\tau, u, \phi)$ as

$$u \frac{d\,\mathbf{I}(\tau, u, \phi)}{d\tau} = \mathbf{I}(\tau, u, \phi) - \mathbf{S}(\tau, u, \phi) \tag{353}$$

where the source function is

$$\mathbf{S}(\tau, u, \phi) = \frac{\varpi(\tau)}{4\pi} \int_0^{2\pi} d\phi' \int_{-1}^{1} du' \, \mathbf{P}(\tau, u', \phi'; u, \phi) \mathbf{I}(\tau, u', \phi') + \mathbf{Q}(\tau, u, \phi). \tag{354}$$

The term $\mathbf{Q}(\tau, u, \phi)$, due to thermal and beam sources, will be specified below.

We note that the traditional heuristic derivation of the RTE given above, based on energy conservation arguments, is simple, easy to understand, and sufficient for most applications of radiative transfer in coupled environmental systems. As pointed out by Mandel and Wolf [24], the relationship between the classical radiative transfer theory and electromagnetic theory has been obscure for a long time. However, the connection between radiative transfer theory and Maxwell's equations has been much clarified by the work of Mishchenko [25, 159, 160], who provided a self-consistent microphysical derivation of the RTE including polarization using methods of statistical electromagnetics.

3.2 Radiative Transfer of Unpolarized Radiation in Coupled Systems

Consider a *coupled* system consisting of two adjacent slabs separated by a plane, horizontal interface across which the refractive index changes abruptly from a value m_1 in one of the slabs to a value m_2 in the other. If the IOPs in each of the two slabs vary only in the vertical direction denoted by z, where z increases upward, and both slabs are assumed to be in local thermodynamic equilibrium so that they emit radiation according to the local temperature $T(\tau(z))$, then the diffuse radiance distribution $I(\tau, \mu, \phi)$ can be described by the *RTE* given by Eq. (350). The *scattering angle* Θ and the polar and azimuth angles are related by Eq. (16). By definition, $\theta = 180°$ is directed toward the nadir (straight down) and $\theta = 0°$ toward the zenith (straight up). Thus, u varies in the range $[-1, 1]$ (from nadir to zenith). For cases of oblique illumination of the medium, $\phi = 180°$ is defined to be the azimuth angle of the incident light. The *vertical optical depth* τ is defined to increase downward with depth from $\tau = 0$ at the top of the upper slab (slab$_1$).

The *single-scattering source term* $S_b(\tau, u, \phi)$ in Eq. (351) in slab$_1$ (with *complex refractive index* $m_1 = n_1 + in'_1$) is different from that in the lower slab (slab$_2$, with refractive index $m_2 = n_2 + in'_2$). In slab$_1$, it is given by

$$S_{b1}(\tau, u, \phi) = \frac{\varpi(\tau)F_0}{4\pi}\Big\{p(\tau, -\mu_0, \phi_0; u, \phi)e^{-\tau/\mu_0} + R_F(-\mu_0; m_1, m_2)p(\tau, \mu_0, \phi_0; u, \phi)e^{-(2\tau_a-\tau)/\mu_0}\Big\} \quad (355)$$

where τ_a is the vertical optical depth of the upper slab, $R_F(-\mu_0; m_1, m_2)$ is the *Fresnel reflectance* at the slab$_1$–slab$_2$ interface, $\mu_0 = \cos\theta_0$, with θ_0 being the polar angle of the incident beam of illumination, and where $n_2 > n_1$. Note that the real part of the *refractive index* of the medium in slab$_1$ has been assumed to be smaller than that of the medium in slab$_2$, as would be the case for air overlying a water body. The first term on the right-hand side of Eq. (355) is due to *first-order scattering* of the attenuated incident beam of irradiance F_0 (normal to the beam), while the second term is due to first-order scattering of the attenuated incident beam that is reflected at the slab$_1$–slab$_2$ interface. In slab$_2$, the *single-scattering source term* consists of the attenuated incident beam that is refracted through the interface, that is

$$S_{b2}(\tau, u, \phi) = \frac{\varpi(\tau)F_0}{4\pi}\frac{\mu_0}{\mu_0^w}T_F(-\mu_0; m_1, m_2) \times p(\tau, -\mu_0^w, \phi_0; u, \phi)e^{-\tau_a/\mu_0}e^{-(\tau-\tau_a)/\mu_0^w} \quad (356)$$

where $T_F(-\mu_0; m_1, m_2)$ is the *Fresnel transmittance* through the interface, and μ_0^w is the cosine of the polar angle θ_0^w in slab$_2$, which is related to $\theta_0 = \arccos\mu_0$ by Snell's law.

For a two-slab system with *source terms* given by Eqs. (355) and (356), a solution based on the discrete-ordinate method [161, 162] of the RTE in Eq. (350) subject to appropriate *boundary conditions* at the top of slab_1, at the bottom of slab_2, and at the slab_1–slab_2 interface, was first developed by Jin and Stamnes [39] (see also Thomas and Stamnes [18]).

3.2.1 Isolation of Azimuth Dependence

The azimuth dependence in Eq. (350) may be isolated by expanding the *scattering phase function* in *Legendre polynomials* $P_\ell(\cos\Theta)$, and making use of the *addition theorem* for spherical harmonics. As shown in Section 2.1.2, the scattering phase function can be expressed in a *Fourier cosine series* as [see Eq. (18)]

$$p(\tau, u', \phi'; u, \phi) = \sum_{m=0}^{M} (2 - \delta_{0m}) p^m(\tau, u', u) \cos m(\phi' - \phi) \tag{357}$$

where δ_{0m} is the Kronecker delta function, that is, $\delta_{0m} = 1$ for $m = 0$ and $\delta_{0m} = 0$ for $m \neq 0$, and

$$p^m(\tau, u', u) = \sum_{\ell=m}^{M} (2\ell + 1)\chi_\ell(\tau)\Lambda_\ell^m(u')\Lambda_\ell^m(u). \tag{358}$$

Here, $\chi_\ell(\tau) = \frac{1}{2}\int_{-1}^{1} d(\cos\Theta)P_\ell(\cos\Theta)p(\tau, \cos\Theta)$ is an *expansion coefficient* and $\Lambda_\ell^m(u)$ is given by $\Lambda_\ell^m(u) \equiv \sqrt{(\ell - m)!}/\sqrt{(\ell + m)!}P_\ell^m(u)$, where $P_\ell^m(u)$ is an *associated Legendre polynomial* of order m. Expanding the radiance in a similar way, that is

$$I(\tau,u,\phi) = \sum_{m=0}^{M} I^m(\tau,u) \cos m(\phi - \phi_0) \tag{359}$$

where ϕ_0 is the azimuth angle of the incident light beam, one finds that each Fourier component satisfies the following *RTE* (see Thomas and Stamnes [18] for details):

$$u\frac{dI^m(\tau, u)}{d\tau} = I^m(\tau, u) - \frac{\varpi(\tau)}{2}\int_{-1}^{1} p^m(\tau, u', u)\, I^m(\tau, u)\, du' - Q^m(\tau, u) \tag{360}$$

where

$$Q^m(\tau, u) = S_b^m(\tau, u) + \delta_{0,m}[1 - \varpi(\tau)]B(\tau) \tag{361}$$

$m = 0, 1, 2, \ldots, 2N - 1$, and $\delta_{0,m} = 1$ if $m = 0$ and 0 otherwise. The (solar) beam source term in slab_1 (upper slab) is given by

$$S_{b1}^m(\tau, u) = \frac{\varpi(\tau)F_0}{4\pi}(2 - \delta_{0m})\Big\{ p^m(\tau, -\mu_0, u)e^{-\tau/\mu_0} + R_F(-\mu_0; m_1, m_2)p^m(\tau, -\mu_0, u)e^{-(2\tau_1-\tau)/\mu_0} \Big\} \tag{362}$$

and in slab$_2$ (lower slab) by

$$S^m_{b2}(\tau, u) = \frac{\varpi(\tau)F_0}{4\pi}(2 - \delta_{0m})\frac{\mu_0}{\mu_0^w} T_F(-\mu_0; m_1, m_2) \\ \times p^m(\tau, \mu_0^w, u)e^{-\tau_1/\mu_0}e^{-(\tau-\tau_a)/\mu_0^w} \tag{363}$$

where $p^m(\tau, u', u)$ is given by Eq. (358).

3.3 Radiative Transfer of Polarized Radiation in Coupled Systems

In a plane-parallel (slab) geometry, the *RTE* is given by Eq. (353), where the Stokes vector $\mathbf{I}_S(\tau, u, \phi) = [I, Q, U, V]^T$, and the superscript T denotes the transpose. Alternatively, the *Stokes vector* may be represented by $\mathbf{I}(\tau, u, \phi) = [I_{\|}, I_{\perp}, U, V]^T$. Then $I_{\|}$ and $I_{\perp}$ are the intensity components that are parallel and perpendicular, respectively, to the *scattering plane*, which is spanned by the directions of the incident and the scattered radiation. In either representation, U is the *degree of linear polarization* in the 45°/135° plane, and V is the *degree of circular polarization*. This Stokes vector is related to the more common one, $\mathbf{I}_S = [I, Q, U, V]^T$, by $I = I_{\|} + I_{\perp}$ and $Q = I_{\|} - I_{\perp}$, where Q is the degree of linear polarization in the 0°/90° plane.

In Eq. (353), $\mathbf{I}(\tau, u, \phi) = [I_{\|}, I_{\perp}, U, V]^T$ or $\mathbf{I}(\tau, u, \phi) = \mathbf{I}_S(\tau, u, \phi) = [I, Q, U, V]^T$ denotes the Stokes vector of the diffuse radiation, u is the cosine of the polar angle θ, ϕ denotes the azimuth angle, and τ denotes the optical depth in a plane-parallel, vertically inhomogeneous medium. The first term in the *source function* $\mathbf{S}(\tau, u, \phi)$ in Eq. (354) is the *multiple scattering* term, where $\mathbf{P}(\tau, u', \phi'; u, \phi)$ is the phase matrix and $\varpi(\tau) = \beta(\tau)/(\alpha(\tau) + \beta(\tau))$ denotes the *single-scattering albedo*.

As in the unpolarized case, for a coupled air–water system, the second term in Eq. (354) is different in the atmosphere and in water due to the change in the refractive index across the interface between the two media. In the upper slab (slab$_1$, air), it is given by

$$\mathbf{Q}_1(\tau, u, \phi) = \frac{\varpi(\tau)}{4\pi}\mathbf{P}(\tau, -\mu_0, \phi_0; u, \phi)\mathbf{S}_b e^{-\tau/\mu_0} + [1 - \varpi(\tau)]\,\mathbf{S}_t(\tau) \\ + \frac{\varpi(\tau)}{4\pi}\mathbf{P}(\tau, \mu_0, \phi_0; u, \phi)\mathbf{R}_F(-\mu_0, m_{\rm rel})\mathbf{S}_b e^{-(2\tau_a-\tau)/\mu_0}. \tag{364}$$

The first term on the right-hand side of Eq. (364) describes the incident beam $\mathbf{S}_b$ in direction $(-\mu_0, \phi_0)$, which is attenuated at depth τ by a factor $e^{-\tau/\mu_0}$ and undergoes single scattering into the direction (u, ϕ). For an unpolarized incident beam, $\mathbf{S}_b$ has the form

$$\mathbf{S}_b = [I_0/2, I_0/2, 0, 0]^T \;\; or \;\; [I_0, 0, 0, 0]^T \tag{365}$$

where the first or second expression corresponds to the choice of *Stokes vector* representation, $[I_{\|}, I_{\perp}, U, V]^T$ or $[I, Q, U, V]^T$. The second term on the right-hand side of Eq. (364) is due to *thermal emission*, which is unpolarized and given by

$$\mathbf{S}_t(\tau) = [B(T(\tau))/2, B(T(\tau))/2, 0, 0]^T \quad \text{or} \quad [B(T(\tau)), 0, 0, 0]^T \tag{366}$$

where B is the *Planck function*, and the first or second expression corresponds to the choice of Stokes vector representation. We have set $\mu_0 \equiv |u_0| \equiv |\cos\theta_0|$, where θ_0 is the polar angle of the incident light beam. The third term on the right-hand side of Eq. (364) describes radiation due to the incident beam $\mathbf{S}_b$ that has been attenuated by the factor $e^{-\tau_a/\mu_0}$ before reaching the air–water interface, undergoing *Fresnel reflection* given by the reflection matrix $\mathbf{R}_F(-\mu_0, m_{\text{rel}})$ [defined above, see Eq. (263)], attenuated by the factor $e^{-(\tau_a-\tau)/\mu_0}$ to reach the level τ in the atmosphere, and finally singly scattered from direction (μ_0, ϕ_0) into direction (u, ϕ) described by the factor $\frac{\varpi(\tau)}{4\pi}\mathbf{P}(\tau, \mu_0, \phi_0; u, \phi)$. Thus, the incident beam propagates through the entire atmosphere and a portion of it is reflected upwards by the interface to reach depth τ in the atmosphere, which explains the factor $e^{-(2\tau_a-\tau)/\mu_0}$. In the lower slab (slab$_2$, water), the *source term* becomes

$$\begin{aligned}\mathbf{Q}_2(\tau, u, \phi) = {} & \frac{\varpi(\tau)}{4\pi}\mathbf{P}(\tau, -\mu_0^w, \phi_0; u, \phi)\mathbf{S}_b\, e^{-\tau_a/\mu_0} \\ & \times \mathbf{T}_F(-\mu_0, m_{\text{rel}})\frac{\mu_0}{\mu_0^w} e^{-(\tau-\tau_a)/\mu_0^w} \\ & + [1 - \varpi(\tau)]\,\mathbf{S}_t(\tau)\end{aligned} \tag{367}$$

where $\mathbf{T}_F(-\mu_0, m_{\text{rel}})$ is the *Fresnel transmission* matrix. The first term in Eq. (367) is due to the incident beam $\mathbf{S}_b$ that has been attenuated through the atmosphere by the factor $e^{-\tau_a/\mu_0}$, transmitted into the water by the factor $\mathbf{T}_F(-\mu_0, m_{\text{rel}})\frac{\mu_0}{\mu_0^w}$, further attenuated by the factor $e^{-(\tau-\tau_a)/\mu_0^w}$ to reach depth τ in the water, and singly scattered from the direction $(-\mu_0^w, \phi_0)$ into the direction (u, ϕ), which explains the factor $\frac{\varpi(\tau)}{4\pi}\mathbf{P}(\tau, -\mu_0^w, \phi_0; u, \phi)$. The second term in Eq. (367) is due to *thermal emission* in water.

3.3.1 Isolation of Azimuth Dependence

We start by expanding the scattering *phase matrix* in a *Fourier series* [see Eq. (64)]:

$$\mathbf{P}(u', u; \phi' - \phi) = \sum_{m=0}^{M} \left\{\mathbf{P}_c^m(u', u)\cos m(\phi' - \phi) + \mathbf{P}_s^m(u', u)\sin m(\phi' - \phi)\right\}. \tag{368}$$

To isolate the azimuth dependence of the radiation field, we expand the *Stokes vector* $\mathbf{I}(\tau, u, \phi)$ in Eq. (353) and the source term $\mathbf{Q}_1(\tau, u, \phi)$ in Eq. (364) or $\mathbf{Q}_2(\tau, u, \phi)$ in Eq. (367) in a Fourier series in a manner similar to the expansion of the scattering *phase matrix* in Eq. (368):

$$\mathbf{I}(\tau, u, \phi) = \sum_{m=0}^{M} \left\{ \mathbf{I}_c^m(\tau, u) \cos m(\phi_0 - \phi) \right.$$
$$\left. + \mathbf{I}_s^m(\tau, u) \sin m(\phi_0 - \phi) \right\} \tag{369}$$

$$\mathbf{Q}_p(\tau, u, \phi) = \sum_{m=0}^{M} \left\{ \mathbf{Q}_{cp}^m(\tau, u) \cos m(\phi_0 - \phi) \right.$$
$$\left. + \mathbf{Q}_{sp}^m(\tau, u) \sin m(\phi_0 - \phi) \right\} \qquad p = 1, 2 \tag{370}$$

where the subscript s or c denotes sine or cosine mode, and the subscript p indicates the slab, $p = 1$ for slab$_1$ and $p = 2$ for slab$_2$. Substitution of Eqs. (368) into Eq. (364) and comparison with Eq. (370) yields for the upper slab ($u = u^a$)

$$\mathbf{Q}_{c1}^m(\tau, u^a) = \mathbf{X}_{c1,0}^m(\tau, u^a)\, e^{-\tau/\mu_0} + \mathbf{X}_{c1,1}^m(\tau, u^a)\, e^{\tau/\mu_0}$$
$$+ \delta_{0m}\, [1 - \varpi(\tau)]\, \mathbf{S}_t(\tau)$$
$$\mathbf{Q}_{s1}^m(\tau, u^a) = \mathbf{X}_{s1,0}^m(\tau, u^a)\, e^{-\tau/\mu_0} + \mathbf{X}_{s1,1}^m(\tau, u^a)\, e^{\tau/\mu_0} \tag{371}$$

where

$$\mathbf{X}_{c1,0}^m(\tau, u^a) = \frac{\varpi(\tau)}{4\pi}\, \mathbf{P}_c^m(\tau, -\mu_0, u^a)\, \mathbf{S}_b \tag{372}$$

$$\mathbf{X}_{c1,1}^m(\tau, u^a) = \frac{\varpi(\tau)}{4\pi}\, \mathbf{P}_c^m(\tau, \mu_0, u^a)\, \mathbf{S}_b$$
$$\times\, \mathbf{R}_F(-\mu_0, m_{\mathrm{rel}})\, e^{-2\tau_a/\mu_0} \tag{373}$$

$$\mathbf{X}_{s1,0}^m(\tau, u^a) = \frac{\varpi(\tau)}{4\pi}\, \mathbf{P}_s^m(\tau, -\mu_0, u^a)\, \mathbf{S}_b \tag{374}$$

$$\mathbf{X}_{s1,1}^m(\tau, u^a) = \frac{\varpi(\tau)}{4\pi}\, \mathbf{P}_s^m(\tau, \mu_0, u^a) \mathbf{R}_F(-\mu_0, m_{\mathrm{rel}})\, \mathbf{S}_b\, e^{-2\tau_a/\mu_0} \tag{375}$$

which can be written more compactly as (letting $\alpha = c, s$ represent the cosine or sine mode)

$$\mathbf{Q}_{\alpha 1}^m(\tau, u^a) = \mathbf{X}_{\alpha 1,0}^m(\tau, u^a)\, e^{-\tau/\mu_0} + \mathbf{X}_{\alpha 1,1}^m(\tau, u^a)\, e^{\tau/\mu_0}$$
$$+ \delta_{c\alpha}\, \delta_{0m}\, [1 - \varpi(\tau)]\, \mathbf{S}_t(\tau) \tag{376}$$

where $\delta_{c\alpha} = 1$ if $\alpha = c$, and $\delta_{c\alpha} = 0$ if $\alpha \neq c$. Repeating this procedure for the lower slab ($u = u^w$), we find

$$\mathbf{Q}_{c2}^m(\tau, u^w) = \mathbf{X}_{c2,2}^m(\tau, u^w)\, e^{-\tau/\mu_0^w} + \delta_{0m}\, [1 - \varpi(\tau)]\, \mathbf{S}_t(\tau) \tag{377}$$

$$\mathbf{Q}_{s2}^m(\tau, u^w) = \mathbf{X}_{s2,2}^m(\tau, u^w)\, e^{-\tau/\mu_0^w} \tag{378}$$

where

$$\mathbf{X}^m_{c2,2}(\tau,u^w) = \frac{\varpi(\tau)}{4\pi}\,\mathbf{P}^m_c(\tau,-\mu_0^w,u^w) \times \mathbf{T}(-\mu_0,m_{\rm rel})\,\mathbf{S}_b\,e^{-\tau_a(1/\mu_0-1/\mu_0^w)} \tag{379}$$

$$\mathbf{X}^m_{s2,2}(\tau,u^w) = \frac{\varpi(\tau)}{4\pi}\,\mathbf{P}^m_s(\tau,-\mu_0^w,u^w) \times \mathbf{T}(-\mu_0,m_{\rm rel})\mathbf{S}_b\,e^{-\tau_a(1/\mu_0-1/\mu_0^w)}. \tag{380}$$

Equations (377) and (378) can be written more compactly as ($\alpha = c, s$)

$$\mathbf{Q}^m_{\alpha 2}(\tau,u^w) = \mathbf{X}^m_{\alpha 2,2}(\tau,u^w)\,e^{-\tau/\mu_0^w} + \delta_{c\alpha}\delta_{0m}\,[1-\varpi(\tau)]\,\mathbf{S}_t(\tau). \tag{381}$$

By substituting Eqs. (368), (369), and (370) into Eqs. (353) and (354), performing the integration over ϕ' in Eq. (354), and comparing terms of equal order in the resulting *Fourier series*, one obtains a transfer equation for each Fourier component of the *Stokes vector*:

$$u\,\frac{d\mathbf{I}^m_{\alpha p}(\tau,u)}{d\tau} = \mathbf{I}^m_{\alpha p}(\tau,u) - \mathbf{S}^m_{\alpha p}(\tau,u) \qquad \alpha = c,s, \qquad p = 1,2 \tag{382}$$

where

$$\mathbf{S}^m_{\alpha p}(\tau,u) = \tilde{\mathbf{S}}^m_{\alpha p}(\tau,u) + \mathbf{Q}^m_{\alpha p}(\tau,u) \tag{383}$$

and the contributions due to *multiple scattering* are

$$\tilde{\mathbf{S}}^m_c(\tau,u) = \frac{\varpi(\tau)}{4}\int_{-1}^{1} du' \left\{ \mathbf{P}^m_c(\tau,u',u)\,\mathbf{I}^m_c(\tau,u')\,(1+\delta_{0m}) - \mathbf{P}^m_s(\tau,u',u)\,\mathbf{I}^m_s(\tau,u')\,(1-\delta_{0m}) \right\} \tag{384}$$

$$\tilde{\mathbf{S}}^m_s(\tau,u) = \frac{\varpi(\tau)}{4}\int_{-1}^{1} du' \left\{ \mathbf{P}^m_c(\tau,u',u)\,\mathbf{I}^m_s(\tau,u')\,(1+\delta_{0m}) + \mathbf{P}^m_s(\tau,u',u)\,\mathbf{I}^m_c(\tau,u')\,(1-\delta_{0m}) \right\}. \tag{385}$$

While the cosine modes start at $m = 0$, the sine modes start at $m = 1$, that is, $\mathbf{I}^m_s(\tau,u)$ and $\mathbf{P}^m_s(\tau,u',u)$ vanish for $m = 0$.

3.4 Methods of Solution of the RTE

Half-Range Quantities in Slab Geometry

In slab geometry it is convenient to introduce the half-range Stokes vectors [18]

$$\begin{aligned}\mathbf{I}^+_{\alpha p}(\tau,\theta,\phi) &\equiv \mathbf{I}^+_{\alpha p}(\tau,\theta \le \pi/2,\phi) = \mathbf{I}^+_{\alpha p}(\tau,\mu,\phi)\\ \mathbf{I}^-_{\alpha p}(\tau,\theta,\phi) &\equiv \mathbf{I}^-_{\alpha p}(\tau,\theta > \pi/2,\phi) = \mathbf{I}^-_{\alpha p}(\tau,\mu,\phi)\end{aligned} \tag{386}$$

where $\alpha = c, s$, $p = 1, 2$, and $\mu = |u| = |\cos\theta|$. Note that the sign of μ is given by the superscript of the Stokes vector. For the Fourier components [see Eq. (369)], we have similarly

$$\begin{aligned}\mathbf{I}_{\alpha p}^{m+}(\tau,\theta) &\equiv \mathbf{I}_{\alpha p}^{m+}(\tau,\theta \le \pi/2) = \mathbf{I}_{\alpha p}^{m+}(\tau,\mu)\\ \mathbf{I}_{\alpha p}^{m-}(\tau,\theta) &\equiv \mathbf{I}_{\alpha p}^{m-}(\tau,\theta > \pi/2) = \mathbf{I}_{\alpha p}^{m-}(\tau,\mu).\end{aligned} \quad (387)$$

3.4.1 Formal Solutions

In terms of these half-range quantities, Eq. (382) may written as ($\mu \equiv |u|$)

$$\mu \frac{d\mathbf{I}_{\alpha p}^{m+}(\tau,\mu)}{d\tau} = \mathbf{I}_{\alpha p}^{m+}(\tau,\mu) - \mathbf{S}_{\alpha p}^{m+}(\tau,\mu) \quad (388)$$

$$-\mu \frac{d\mathbf{I}_{\alpha p}^{m-}(\tau,\mu)}{d\tau} = \mathbf{I}_{\alpha p}^{m-}(\tau,\mu) - \mathbf{S}_{\alpha p}^{m-}(\tau,\mu). \quad (389)$$

Using an integrating factor, we readily obtain from Eqs. (388) and (389)

$$\begin{aligned}\frac{d}{d\tau}[\mathbf{I}_{\alpha p}^{m+}(\tau,\mu)e^{-\tau/\mu}] &= \left[\frac{d\mathbf{I}_{\alpha p}^{m+}(\tau,\mu)}{d\tau} - \frac{1}{\mu}\mathbf{I}_{\alpha p}^{m+}(\tau,\mu)\right]e^{-\tau/\mu}\\ &= -\frac{\mathbf{S}_{\alpha p}^{m+}(\tau,\mu)}{\mu}e^{-\tau/\mu}\end{aligned} \quad (390)$$

$$\begin{aligned}\frac{d}{d\tau}[\mathbf{I}_{\alpha p}^{m-}(\tau,\mu)e^{\tau/\mu}] &= \left[\frac{d\mathbf{I}_{\alpha p}^{m-}(\tau,\mu)}{d\tau} + \frac{1}{\mu}\mathbf{I}_{\alpha p}^{m-}(\tau,\mu)\right]e^{\tau/\mu}\\ &= \frac{\mathbf{S}_{\alpha p}^{m-}(\tau,\mu)}{\mu}e^{\tau/\mu}.\end{aligned} \quad (391)$$

Single-Slab Medium

For a single slab with constant *refractive index*, we adopt the following notation:

1) τ_0 is the optical depth at the upper boundary (top of the slab);
2) τ_b is the optical depth at the lower boundary (bottom of the slab).

Integrating Eq. (390) from τ_b to τ and Eq. (391) from τ_0 to τ ($\tau_0 \le \tau \le \tau_b$), and solving for $\mathbf{I}_{\alpha}^{m\pm}(\tau,\mu)$, we obtain (see Problem 3.2)

$$\mathbf{I}_{\alpha}^{m+}(\tau,\mu) = \mathbf{I}_{\alpha}^{m+}(\tau_b,\mu)e^{-(\tau_b-\tau)/\mu} + \int_{\tau}^{\tau_b}\frac{dt}{\mu}\mathbf{S}_{\alpha}^{m+}(t,\mu)e^{-(t-\tau)/\mu} \quad (392)$$

$$\mathbf{I}_{\alpha}^{m-}(\tau,\mu) = \mathbf{I}_{\alpha}^{m-}(\tau_0,\mu)e^{-(\tau-\tau_0)/\mu} + \int_{\tau_0}^{\tau}\frac{dt}{\mu}\mathbf{S}_{\alpha}^{m-}(t,\mu)e^{-(\tau-t)/\mu} \quad (393)$$

where we have dropped the subscript p since we are considering a single slab. Equations (392) and (393) show that for cases in which the source function $\mathbf{S}_{\alpha}^{m\pm}(t,\mu)$ is known, one can obtain a solution to the radiative transfer problem by numerical integration.

Single Homogeneous Slab – Thermal Emission

For a homogeneous, isothermal slab with *thermal emission* but no scattering ($\varpi = 0$), the source function is given by the (unpolarized) *Planck function*, which is isotropic and independent of the position. Then

$$\mathbf{S}_\alpha^{m\pm}(t,\mu) = \mathbf{B}^\pm \quad \text{independent of } t, \text{ and } \mu$$

implying that the integrals in Eqs. (392) and (393) can be evaluated to yield

$$\int_\tau^{\tau_b} \frac{dt}{\mu} \mathbf{S}_\alpha^{m+}(t,\mu) e^{-(t-\tau)/\mu} = \mathbf{B}^+ \left[1 - e^{-(\tau_b-\tau)/\mu}\right] \tag{394}$$

$$\int_\tau^{\tau_b} \frac{dt}{\mu} \mathbf{S}_\alpha^{m-}(t,\mu) e^{-(t-\tau)/\mu} = \mathbf{B}^- \left[1 - e^{-(\tau-\tau_0)/\mu}\right]. \tag{395}$$

Single Inhomogeneous Slab – Multilayered Medium

The vertical variation of the *IOPs* in a slab may be dealt with by dividing it into a number of adjacent horizontal layers in which the IOPs are taken to be constant within each layer, but allowed to vary from layer to layer. The number of layers should be large enough to resolve the vertical variation in the IOPs. In such a multilayered medium, consisting of a total of L layers, we may evaluate the integrals in Eqs. (392) and (393) by integrating layer by layer as follows ($\tau_{\ell-1} \le \tau \le \tau_\ell$ and $\mu > 0$, $\tau_b = \tau_L$, $\tau_0 = 0$):

$$\int_\tau^{\tau_L} \frac{dt}{\mu} \mathbf{S}_\alpha^{m+}(t,\mu) e^{-(t-\tau)/\mu} = \int_\tau^{\tau_\ell} \frac{dt}{\mu} \mathbf{S}_{\alpha,\ell}^{m+}(t,\mu) e^{-(t-\tau)/\mu} + \sum_{n=\ell+1}^{L} \int_{\tau_{n-1}}^{\tau_n} \frac{dt}{\mu} \mathbf{S}_{\alpha,n}^{m+}(t,\mu) e^{-(t-\tau)/\mu} \tag{396}$$

$$\int_0^{\tau} \frac{dt}{\mu} \mathbf{S}_\alpha^{m-}(t,\mu) e^{-(\tau-t)/\mu} = \sum_{n=1}^{\ell-1} \int_{\tau_{n-1}}^{\tau_n} \frac{dt}{\mu} \mathbf{S}_{\alpha,n}^{m-}(t,\mu) e^{-(\tau-t)/\mu} + \int_{\tau_{\ell-1}}^{\tau} \frac{dt}{\mu} \mathbf{S}_{\alpha,\ell}^{m-}(t,\mu)) e^{-(\tau-t)/\mu}. \tag{397}$$

We can evaluate the integrals in Eqs. (396) and (397) either numerically or analytically if the source function is known. Analytical integration is possible in the case of the *discrete-ordinate method* to be discussed below.

Coupled Two-Slab (Atmosphere–Water) System

We use the following notations:

1) τ_0 is the optical depth at the upper boundary (top of the atmosphere);
2) τ_a is the optical depth just *above* the air–water interface;
3) τ_a^+ is the optical depth just *below* the air–water interface where $\tau_a^+ = \tau_a + \epsilon$ ($\epsilon \to 0$);
4) τ_b is the optical depth at the lower boundary (bottom of the water);
5) $\mathbf{R}_F(\mu, m_{\rm rel})$ and $\mathbf{T}_F(\mu, m_{\rm rel})$ are the *Fresnel reflectance* and transmittance matrices given by Eqs. (263) and (269), respectively, with m_r replaced by $m_{\rm rel}$,

$\tilde{\mathbf{I}}_\alpha^{m\pm} = \mathbf{I}_\alpha^{m\pm}/n_p^2$, where n_p is the real part of the local refractive index, which has the value $n_1 = n_{ra} \approx 1$ in the air (slab$_1$, $p = 1$), and $n_2 = n_{rw} \approx 1.34$ in the water (slab$_2$, $p = 2$);

6) the *relative refractive index* is $m_{\rm rel} = n_2/n_1 = n_{rw}/n_{ra}$;
7) $\mu = \mu^a$ in the atmosphere;

$$\mu = \mu^w = \sqrt{1 - [1 - (\mu^a)^2]/m_{\rm rel}^2} \tag{398}$$

in the water, and

8) μ_0^w is the cosine of the solar zenith angle in the water evaluated using Eq. (398) with $\mu^a = \mu_0^a$.

Integrating Eq. (390) from τ_b to τ and Eq. (391) from τ_0 to τ ($\tau_0 \le \tau \le \tau_b$), we obtain

$$[\mathbf{I}_\alpha^{m+}(t,\mu)e^{-t/\mu}]\Big|_{\tau_b}^{\tau} = \int_{\tau_b}^{\tau} [-\frac{\mathbf{S}_\alpha^{m+}(t,\mu)}{\mu} e^{-t/\mu}]dt \tag{399}$$

and

$$[\mathbf{I}_\alpha^{m-}(t,\mu)e^{t/\mu}]\Big|_{\tau_0}^{\tau} = \int_{\tau_0}^{\tau} [\frac{\mathbf{S}_\alpha^{m-}(t,\mu)}{\mu} e^{t/\mu}]dt. \tag{400}$$

Since the integrals in Eqs. (399) and (400) apply to a coupled atmosphere–water system, where τ can have any value in the range $\tau_0 \le \tau \le \tau_b$, we need expressions for the *source function*, $\mathbf{S}_\alpha^{m\pm}(t,\mu)$, in the atmosphere and water.

The upper limit of integration in Eq. (399) or (400) applies to an arbitrary optical depth τ in the range $\tau_0 \le \tau \le \tau_b$ in the atmosphere–water system, but the light beam changes direction across the interface, which must be dealt with.

3.4.2
Single-Scattering Approximation

In the *single-scattering approximation*, we ignore the *multiple scattering* contributions by setting $\tilde{\mathbf{S}}_\alpha^{m\pm}(t,\mu) = \mathbf{0}$ in Eq. (383), so that the *source function* becomes

$$\mathbf{S}_{\alpha p}^{m\pm}(t,\mu) = \mathbf{Q}_{\alpha p}^{m\pm}(t,\mu) \qquad \alpha = c, s; \qquad p = 1, 2 \tag{401}$$

where in the upper slab [see Eq. (376)]

$$\mathbf{Q}_{\alpha 1}^{m\pm}(t,\mu^a) = \mathbf{X}_{\alpha 1,0}^{m\pm}(t,\mu^a)\, e^{-t/\mu_0} + \mathbf{X}_{\alpha 1,1}^{m\pm}(t,\mu^a)\, e^{t/\mu_0} + \delta_{c\alpha}\, \mathbf{B}_{0m}(t) \tag{402}$$

where

$$\mathbf{B}_{0m}(t) = \delta_{0m}\,[1 - \varpi(t)]\, \mathbf{S}_t(t), \tag{403}$$

and [see Eqs. (372)–(375)]

$$\mathbf{X}^{m\pm}_{c1,0}(\tau,\mu^a) = \frac{\varpi(\tau)}{4\pi}\,\mathbf{P}^m_c(\tau,-\mu_0,\pm\mu^a)\,\mathbf{S}_b \tag{404}$$

$$\mathbf{X}^{m\pm}_{c1,1}(\tau,\mu^a) = \frac{\varpi(\tau)}{4\pi}\,\mathbf{P}^m_c(\tau,\mu_0,\pm\mu^a)\,\mathbf{S}_b \times \mathbf{R}_F(-\mu_0,m_{\text{rel}})\,e^{-2\tau_a/\mu_0} \tag{405}$$

$$\mathbf{X}^{m\pm}_{s1,0}(\tau,\mu^a) = \frac{\varpi(\tau)}{4\pi}\,\mathbf{P}^m_s(\tau,-\mu_0,\pm\mu^a)\,\mathbf{S}_b \tag{406}$$

$$\mathbf{X}^{m\pm}_{s1,1}(\tau,\mu^a) = \frac{\varpi(\tau)}{4\pi}\,\mathbf{P}^m_s(\tau,\mu_0,\pm\mu^a)\mathbf{R}_F(-\mu_0,m_{\text{rel}})\,\mathbf{S}_b\,e^{-2\tau_a/\mu_0}. \tag{407}$$

In the lower slab [see Eq. (381)]

$$\mathbf{Q}^{m\pm}_{\alpha 2}(t,\mu^w) = \mathbf{X}^{m\pm}_{\alpha 2,2}(t,\mu^w)\,e^{-t/\mu_0^w} + \delta_{c\alpha}\,\mathbf{B}_{0m}(t) \tag{408}$$

where [see Eqs. (379)–(380)]

$$\mathbf{X}^{m\pm}_{c2,2}(\tau,\mu^w) = \frac{\varpi(\tau)}{4\pi}\,\mathbf{P}^m_c(\tau,-\mu_0^w,\pm\mu^w) \times \mathbf{T}(-\mu_0,m_{\text{rel}})\,\mathbf{S}_b\,e^{-\tau_a(1/\mu_0-1/\mu_0^w)} \tag{409}$$

$$\mathbf{X}^{m\pm}_{s2,2}(\tau,\mu^w) = \frac{\varpi(\tau)}{4\pi}\,\mathbf{P}^m_s(\tau,-\mu_0^w,\pm\mu^w) \times \mathbf{T}(-\mu_0,m_{\text{rel}})\mathbf{S}_b\,e^{-\tau_a(1/\mu_0-1/\mu_0^w)}. \tag{410}$$

Single Homogeneous Slab

For a single homogenous slab, $\varpi(t) = \varpi$, $\mathbf{P}^m_\alpha(t,\mu',\pm\mu) = \mathbf{P}^m_\alpha(\mu',\pm\mu)$, and $\mathbf{S}_t(t) = \mathbf{S}_t = [B/2, B/2, 0, 0]^T$ (or $\mathbf{S}_t = [B, 0, 0, 0]^T$), where B is the Planck function. Then, $\mathbf{S}^{m\pm}_{\alpha 2}(t,\mu) = \mathbf{0}$, $\mathbf{S}^{m\pm}_{\alpha 1}(t,\mu) \equiv \mathbf{S}^{m\pm}_\alpha(t,\mu)$, and Eq. (401) becomes

$$\mathbf{S}^{m\pm}_\alpha(t,\mu) = \mathbf{Q}^{m\pm}_\alpha(t,\mu) = \mathbf{B}_{0m} + \mathbf{X}^{m\pm}_\alpha(\mu)\,e^{-t/\mu_0} \tag{411}$$

where $\mathbf{X}^{m\pm}_\alpha(\mu) \equiv \mathbf{X}^{m\pm}_{\alpha 1,0}(\mu)$. Assuming that there is no diffuse radiation incident at the top of the slab, and that the lower boundary is black (no reflection) and non-emitting, the *boundary conditions* are as follows:

$$\mathbf{I}^{m-}_\alpha(\tau_0,\mu) = 0 \quad \leftarrow \quad \text{upper boundary condition} \tag{412}$$

$$\mathbf{I}^{m+}_\alpha(\tau_b,\mu) = 0 \quad \leftarrow \quad \text{lower boundary condition.} \tag{413}$$

Substitution of Eq. (411) in Eqs. (392) and (393) yields (see Problem 3.3)

$$\mathbf{I}^{m+}_\alpha(\tau,\mu) = \mathbf{B}_{0m}\left[1 - e^{-(\tau_b-\tau)/\mu}\right] + \frac{\mu_0\mathbf{X}^{m+}_\alpha(\mu)}{(\mu_0+\mu)}\left[e^{-\tau/\mu_0} - e^{-[(\tau_b-\tau)/\mu+\tau_b/\mu_0]}\right] \tag{414}$$

$$\mathbf{I}^{m-}_\alpha(\tau,\mu) = \mathbf{B}_{0m}\left[1 - e^{-(\tau-\tau_0)/\mu}\right] + \frac{\mu_0\mathbf{X}^{m-}_\alpha(\mu)}{(\mu_0-\mu)}\left[e^{-\tau/\mu_0} - e^{-[(\tau-\tau_0)/\mu+\tau_0/\mu_0]}\right]. \tag{415}$$

Single Multilayered Slab

For a multilayered slab, the corresponding solutions are (setting $\mathbf{B}_{0m} = \mathbf{0}$)

$$\mathbf{I}^{m+}_{\alpha,\ell}(\tau,\mu) = \frac{\mu_0}{(\mu_0+\mu)} \sum_{n=\ell}^{L} \mathbf{X}^{m+}_{\alpha,n}(\mu) \times \left[e^{-[\tau_{n-1}/\mu_0 + (\tau_{n-1}-\tau)/\mu]} - e^{-[\tau_n/\mu_0 + (\tau_n-\tau)/\mu]} \right] \tag{416}$$

with τ_{n-1} replaced by τ for $n = \ell$,

$$\mathbf{I}^{m-}_{\alpha,\ell}(\tau,\mu) = \frac{\mu_0}{(\mu_0-\mu)} \sum_{n=1}^{\ell} \mathbf{X}^{m-}_{\alpha,n}(\mu) \times \left[e^{-[\tau_n/\mu_0 + (\tau-\tau_n)/\mu]} - e^{-[\tau_{n+1}/\mu_0 + (\tau-\tau_{n+1})/\mu]} \right] \tag{417}$$

with τ_n replaced by τ for $n = \ell$.

Coupled Two-Slab (Atmosphere–Water) System

For the beam source in the atmosphere and water, the *source functions* are [163]

$$\mathbf{S}^{m\pm}_{\alpha p}(t,\mu) = \mathbf{Q}^{m\pm}_{\alpha p}(t,\mu) \qquad \alpha = c, s; \;\; p = 1, 2$$

$$\mathbf{S}^{m\pm}_{\alpha 1}(t,\mu^a) = \mathbf{Q}^{m\pm}_{\alpha 1}(t,\mu^a) = \mathbf{X}^{m\pm}_{\alpha 1,0}(\mu^a) e^{-t/\mu_0} + \mathbf{X}^{m\pm}_{\alpha 1,1}(\mu^a) e^{t/\mu_0} \tag{418}$$

$$\mathbf{S}^{m\pm}_{\alpha 2}(t,\mu^w) = \mathbf{Q}^{m\pm}_{\alpha 2}(t,\mu^w) = \mathbf{X}^{m\pm}_{\alpha 2,2}(\mu^w) e^{-t/\mu_0^w} \tag{419}$$

where μ^w and μ_0^w are given by Eq. (398). The *reflection matrix* $\mathbf{R}_F(-\mu_0, m_{\rm rel})$ and the *transmission matrix* $\mathbf{T}_F(-\mu_0, m_{\rm rel})$ are given by Eqs. (263) and (269), respectively, with m_r replaced by $m_{\rm rel}$. $\mathbf{X}^{m\pm}_{\alpha 1,0}(\mu^a)$ and $\mathbf{X}^{m\pm}_{\alpha,1}(\mu^a)$ are given by Eqs. (372)–(375), and $\mathbf{X}^{m\pm}_{\alpha 2,2}(\mu^w)$ by Eqs. (379)–(380).

Upward radiances in the water ($\tau_a^+ \le \tau \le \tau_b$)

Since there is no change in the direction of an upward light beam when $\tau \ge \tau_a^+$, we have

$$\mathbf{I}^{m+}_{\alpha}(\tau,\mu^w) = \mathbf{I}^{m+}_{\alpha}(\tau_b,\mu^w) e^{-(\tau_b-\tau)/\mu^w} + \frac{\mu_0^w \mathbf{X}^{m+}_{\alpha 2,2}(\mu^w)}{(\mu_0^w + \mu^w)} [e^{-\tau/\mu_0^w} - e^{-[\tau_b/\mu_0^w + (\tau_b-\tau)/\mu^w]}]. \tag{420}$$

Upward radiances in the atmosphere ($\tau_0 \le \tau \le \tau_a$)

When $\tau_0 \le \tau \le \tau_a$, the integral from τ_b to τ in Eq. (399) is completed by considering three subintervals: (1) τ_b to τ_a^+, (2) τ_a^+ to τ_a, and (3) τ_a to τ. The result is

$$\mathbf{I}^{m+}_{\alpha}(\tau,\mu^a) = \mathbf{I}^{m+}_{\alpha}(\tau_a,\mu^a) e^{-(\tau_a-\tau)/\mu^a} + \mathbf{I}^{m+}_{\alpha,\rm air}(\tau,\mu^a) \tag{421}$$

where

$$\mathbf{I}^{m+}_{\alpha}(\tau_a,\mu^a) = \mathbf{I}^{m-}_{\alpha}(\tau_a,\mu^a) \mathbf{R}_F(-\mu^a, m_{\rm rel}) + [\frac{\mathbf{I}^{m+}_{\alpha}(\tau_a^+,\mu^w)}{m^2_{\rm rel}}] \mathbf{T}_F(+\mu^w, m_{\rm rel}) \tag{422}$$

where $\mathbf{R}_F(-\mu^a, m_{\text{rel}})$ and $\mathbf{T}_F(+\mu^w, m_{\text{rel}})$ are given by Eqs. (263) and (269), respectively, with m_r replaced by m_{rel} and

$$\mathbf{I}^{m+}_{\alpha,\text{air}}(\tau, \mu^a) = e^{\tau/\mu^a} \int_{\tau}^{\tau_a} \frac{dt}{\mu^a} e^{-t/\mu^a} \mathbf{S}^{m+}_{\alpha}(t, \mu^a). \tag{423}$$

By substituting Eq. (418) into Eq. (423), we find after some algebra (see Problem 3.4)

$$\begin{aligned}\mathbf{I}^{m+}_{\alpha,\text{air}}(\tau, \mu^a) = & \frac{\mu_0 \mathbf{X}^{m+}_{\alpha 1,0}(\mu^a)}{(\mu_0 + \mu^a)} \left\{ e^{-\tau/\mu_0} - e^{-[\tau_a/\mu_0 + (\tau_a - \tau)/\mu^a]} \right\} \\ & + \frac{\mu_0 X^{m+}_{\alpha 1,1}(\mu^a)}{(\mu^a - \mu_0)} \left\{ e^{-[(\tau_a - \tau)/\mu^a - \tau_a/\mu_0]} - e^{\tau/\mu_0} \right\}\end{aligned} \tag{424}$$

where the second term on the right-hand side is due to specular reflection of the attenuated solar beam by the water surface.

Downward Radiances

The upper limit of integration in Eq. (400) applies to an arbitrary optical depth τ in the range $\tau_0 \leq \tau \leq \tau_b$ in an atmosphere–water system, but a downward light beam in the atmosphere changes direction when crossing the interface, and that change must be dealt with in the integral when $\tau \geq \tau_a^+$.

Downward Radiances in the Atmosphere ($\tau_0 \leq \tau \leq \tau_a$)

Since $\tau \leq \tau_a$, there is no change in the direction of the light beam, and integration yields

$$\begin{aligned}\mathbf{I}^{m-}_{\alpha}(\tau, \mu^a) = & \mathbf{I}^{m-}_{\alpha}(\tau_0, \mu^a) e^{-(\tau - \tau_0)/\mu^a} \\ & + \frac{\mu_0 \mathbf{X}^{m-}_{\alpha 1,0}(\mu^a)}{(\mu_0 - \mu^a)} \left\{ e^{-\tau/\mu_0} - e^{-[\tau_0/\mu_0 + (\tau - \tau_0)/\mu^a]} \right\} \\ & + \frac{\mu_0 \mathbf{X}^{m-}_{\alpha 1,1}(\mu^a)}{(\mu_0 + \mu^a)} \left\{ e^{\tau/\mu_0} - e^{\tau_0/\mu_0 - (\tau - \tau_0)/\mu^a} \right\}.\end{aligned} \tag{425}$$

Downward Radiances in the Water ($\tau_a^+ \leq \tau \leq \tau_b$)

Since there is a change in the direction of the light beam across the interface, we integrate in three steps as follows: (1) from τ_0 to τ_a, (2) from τ_a to τ_a^+, and (3) from τ_a^+ to τ ($\tau_a^+ \leq \tau \leq \tau_b$). The result is

$$\mathbf{I}^{m-}_{\alpha}(\tau, \mu^w) = \mathbf{I}^{m-}_{\alpha}(\tau_a^+, \mu^w) e^{(\tau_a^+ - \tau)/\mu^w} + \mathbf{I}^{m-}_{\alpha,\text{wat}}(\tau, \mu) \tag{426}$$

where in the *refraction region* ($\mu \geq \mu_{\text{crit}}$)

$$\frac{\mathbf{I}^{m-}_{\alpha}(\tau_a^+, \mu^w)}{m_{\text{rel}}^2} = \frac{\mathbf{I}^{m+}_{\alpha}(\tau_a^+, \mu^w)}{m_{\text{rel}}^2} \mathbf{R}_F(+\mu^w, m_{\text{rel}}) + \mathbf{I}^{m-}_{\alpha}(\tau_a, \mu^a) \mathbf{T}_F(-\mu^a, m_{\text{rel}}) \tag{427}$$

and in the region of *total reflection* ($\mu < \mu_{\text{crit}}$)

$$\mathbf{I}^{m-}_{\alpha}(\tau_a^+) = \mathbf{I}^{m+}_{\alpha}(\tau_a^+). \tag{428}$$

To evaluate the second term on the right-hand side of Eq. (426)

$$\mathbf{I}^{m-}_{\alpha,\mathrm{wat}}(\tau,\mu) = e^{-\tau/\mu^w} \int_{\tau_a^+}^{\tau} \frac{dt}{\mu^w} e^{t/\mu^w} \mathbf{S}^{m-}_{\alpha}(t,\mu^w) \tag{429}$$

we substitute Eq. (419) $\mathbf{S}^{m-}_{\alpha}(\tau,\mu^w) = \mathbf{X}^{m-}_{\alpha,2}(\mu^w)e^{-\tau/\mu_0^w}$ into Eq. (429) to obtain

$$\mathbf{I}^{m-}_{\alpha,\mathrm{wat}}(\tau,\mu) = \frac{\mu_0^w \mathbf{X}^{m-}_{\alpha 2,2}(\mu^w)}{(\mu_0^w - \mu^w)} \left\{ e^{-\tau/\mu_0^w} - e^{-[\tau_a^+/\mu_0^w + (\tau-\tau_a^+)/\mu^w]} \right\}. \tag{430}$$

Note:

- The expressions above can be used to compute the singly scattered *Stokes vector* in the coupled atmosphere–water system with a calm (flat) interface.
- The atmosphere and water are each assumed to be homogeneous slabs in which the inherent optical properties (IOPs) (the *single-scattering albedo* and the *scattering phase matrix*) are taken to be constant.
- To approximate a vertically inhomogeneous atmosphere–water system, the derivations above can be extended to apply to two adjacent slabs, each of which is divided into an appropriate number of homogeneous layers.

3.4.3 Successive Order of Scattering (SOS) Method

While solutions to the *scalar RTE*, which involves only the first component of the Stokes vector (the radiance or intensity I in the representation $\mathbf{I}_S = [I, Q, U, V]^T$), are well developed, modern RT models provide solutions of the vector RTE that fully accounts for polarization effects described by the second, third, and fourth component of the Stokes vector. Even if one's interest lies primarily in the *radiance*, it is important to realize that solutions of the *scalar RTE*, which are obtained by ignoring *polarization effects*, introduce errors in the computed radiances [41, 42, 164].

The *SOS method* has been used extensively to solve the scalar RT problem for computation of radiances at microwave frequencies [165–167] as well as for atmospheric correction of TOA radiances in the visible and thermal spectral ranges [168, 169]. The SOS method has been extended to solve the vector RTE in the atmosphere [49, 170] as well as to solve the vector RTE in a coupled atmosphere–ocean system [43, 48, 92].

In the SOS method, the radiation is assumed to be a sum of contributions from all orders of scattering, so that for each Fourier component of the *Stokes vector*

$$\mathbf{I}^{m\pm}_{\alpha}(\tau,\mu) = \sum_{i=1}^{M_s} \mathbf{I}^{m\pm}_{\alpha,i}(\tau,\mu) \tag{431}$$

where $\mathbf{I}^{m\pm}_{\alpha,1}(\tau,\mu)$ is the solution obtained in the *single-scattering approximation.* The integer M_s is chosen to be so large that contributions for $i > M_s$ are negligible. To illustrate the method, let us consider a single homogeneous slab with

$\varpi(t) = \varpi$, $\mathbf{P}_\alpha^m(t, \mu', \pm\mu) = \mathbf{P}_\alpha^m(\mu', \pm\mu)$, and $\mathbf{S}_t(t) = 0$ (no *thermal emission*), so that the *source function* is given by [see Eq. (411)]

$$\mathbf{S}_\alpha^{m\pm}(t, \mu) = \mathbf{X}_\alpha^{m\pm}(\mu)\, e^{-t/\mu_0}. \tag{432}$$

With no diffuse radiation incident at the top of the slab, and a black lower boundary (no reflection), we have [see Eqs. (412) and (413)]

$$\mathbf{I}_\alpha^{m-}(\tau_0, \mu) = 0, \qquad \mathbf{I}_\alpha^{m+}(\tau_b, \mu) = 0. \tag{433}$$

The first-order (single-scattering) solutions are [see Eqs. (414) and (415)]

$$\mathbf{I}_{\alpha,1}^{m+}(\tau, \mu) = \frac{\mu_0 \mathbf{X}_\alpha^{m+}(\mu)}{(\mu_0 + \mu)} \left[e^{-\tau/\mu_0} - e^{-[(\tau_b-\tau)/\mu + \tau_b/\mu_0]}\right] \tag{434}$$

$$\mathbf{I}_{\alpha,1}^{m-}(\tau, \mu) = \frac{\mu_0 \mathbf{X}_\alpha^{m-}(\mu)}{(\mu_0 - \mu)} \left[e^{-\tau/\mu_0} - e^{-[(\tau-\tau_0)/\mu + \tau_0/\mu_0]}\right]. \tag{435}$$

The contributions from second and higher orders of scattering are obtained from solutions of the *RTE*:

$$u \frac{\mathbf{I}_{\alpha,i}^m(\tau, u)}{d\tau} = \mathbf{I}_{\alpha,i}^m(\tau, u) - \mathbf{S}_{\alpha,i}^m(\tau, u) \qquad \alpha = c, s \tag{436}$$

where the *source function* is given by [see Eq. (383)]

$$\mathbf{S}_{\alpha,i}^m(\tau, u) = \tilde{\mathbf{S}}_{\alpha,i}^m(\tau, u) \qquad \alpha = c, s \tag{437}$$

with [see Eqs. (384) and (385)]

$$\tilde{\mathbf{S}}_{c,i}^m(\tau, u) = \frac{\varpi(\tau)}{4} \int_{-1}^{1} du' \left\{ (1 + \delta_{0m})\, \mathbf{P}_c^m(\tau, u', u)\, \mathbf{I}_{c,i-1}^m(\tau, u') \right.$$
$$\left. -\, (1 - \delta_{0m})\, \mathbf{P}_{s,i-1}^m(\tau, u', u)\, \mathbf{I}_{s,i-1}^m(\tau, u') \right\} \tag{438}$$

$$\tilde{\mathbf{S}}_{s,i}^m(\tau, u) = \frac{\varpi(\tau)}{4} \int_{-1}^{1} du' \left\{ (1 + \delta_{0m})\, \mathbf{P}_c^m(\tau, u', u)\, \mathbf{I}_{s,i-1}^m(\tau, u') \right.$$
$$\left. +\, (1 - \delta_{0m})\, \mathbf{P}_s^m(\tau, u', u)\, \mathbf{I}_{c,i-1}^m(\tau, u') \right\}. \tag{439}$$

Since we know $\mathbf{I}_{\alpha,i-1}^m(\tau, u')$ from the previous step, we can compute the source function from Eqs. (438) and (439), and use Eqs. (392) and (393) to solve Eq. (436) by numerical integration. An advantage of this method is that Eqs. (434)–(439) provide an explicit solution that can be executed in a successive manner. It can be applied to solve practical problems with vertical variations in the IOPs, but the speed of convergence depends on the *single-scattering albedo* and the *optical thickness* of the medium.

3.4.4 Discrete-Ordinate Method

Numerical realizations based on the *discrete-ordinate method* developed by Chandrasekhar and others in the 1940s [11] for the scalar RT problem were first explored in the early 1970s [171]. The scalar RT method was further developed in the 1980s [172–174], and eventually implemented numerically in a robust, well-tested, and freely available computer code (*DISORT*) in 1988 [161]. An upgraded and improved version of DISORT has recently been released [175]. The DISORT code provides the radiances and irradiances in a vertically inhomogenous medium that is divided into homogenous layers to resolve the vertical variation in the *IOPs*. The method was extended to be applicable to two adjacent media with different refractive indices, such as a coupled atmosphere–water system [39] or a coupled air–tissue system [33], and the performance of the resulting *C-DISORT* codes was tested against *Monte Carlo* (MC) computations [27, 30].

For a layered medium with a constant refractive index, solutions of the *vector RTE* [see Eq. (353)] were developed and implemented in the *VDISORT* code [176–179]. Discrete-ordinate solutions to the vector RTE have also been developed by Siewert [83], and this approach has been implemented in the *VLIDORT* [90] and *SCIATRAN* [55] vector RT codes. For a coupled system consisting of two media with different refractive indices, the coupled vector discrete-ordinate (*C-VDISORT*) RTM [180] was developed and tested for pure Rayleigh scattering [87]. Further, it was extended for applications to scattering by particles in the Mie regime and tested against *Monte Carlo* simulations and benchmark results [181] by Cohen *et al.* [84]. The SCIATRAN software package [55] is also applicable to a coupled atmosphere–ocean system.

Equations (382)–(385) can be rewritten as

$$u\,\frac{d\mathbf{I}_c^m(\tau,u)}{d\tau} = \mathbf{I}_c^m(\tau,u) - \frac{\varpi(\tau)}{4}\int_{-1}^{1} du' \left\{ \mathbf{P}_c^m(\tau,u',u)\,\mathbf{I}_c^m(\tau,u')\,(1+\delta_{0m}) - \mathbf{P}_s^m(\tau,u',u)\,\mathbf{I}_s^m(\tau,u') \right\} - \mathbf{Q}_c^m(\tau,u) \tag{440}$$

$$u\,\frac{d\mathbf{I}_s^m(\tau,u)}{d\tau} = \mathbf{I}_s^m(\tau,u) - \frac{\varpi(\tau)}{4}\int_{-1}^{1} du' \left\{ \mathbf{P}_c^m(\tau,u',u)\,\mathbf{I}_s^m(\tau,u') + \mathbf{P}_s^m(\tau,u',u)\,\mathbf{I}_c^m(\tau,u') \right\} - \mathbf{Q}_s^m(\tau,u) \tag{441}$$

where we have omitted the Kronecker deltas in Eq. (385) and the second Kronecker delta in Eq. (384) since the cosine modes start at $m = 0$, whereas the sine modes start at $m = 1$, i.e., $\mathbf{I}_s^m$ and $\mathbf{P}_s^m$ vanish for $m = 0$. The discrete-ordinate method consists of replacing the integration over u' by a discrete sum by introducing the *Gaussian quadrature* points u_j (the discrete ordinates) and corresponding weights w_j. One obtains for each Fourier component

$$u_i \frac{d\mathbf{I}_c^m(\tau, u_i)}{d\tau} = \mathbf{I}_c^m(\tau, u_i) - \frac{\varpi(\tau)}{4} \sum_{\substack{j=-N \\ j\neq 0}}^{N} w_j \left\{ (1+\delta_{0m}) \mathbf{P}_c^m(\tau, u_j, u_i) \mathbf{I}_c^m(\tau, u_j) - \mathbf{P}_s^m(\tau, u_j, u_i) \mathbf{I}_s^m(\tau, u_j) \right\} - \mathbf{Q}_c^m(\tau, u_i), \qquad i = \pm 1, \ldots, \pm N \tag{442}$$

$$u_i \frac{d\mathbf{I}_s^m(\tau, u_i)}{d\tau} = \mathbf{I}_s^m(\tau, u_i) - \frac{\varpi(\tau)}{4} \sum_{\substack{j=-N \\ j\neq 0}}^{N} w_j \left\{ \mathbf{P}_c^m(\tau, u_j, u_i) \mathbf{I}_s^m(\tau, u_j) + \mathbf{P}_s^m(\tau, u_j, u_i) \mathbf{I}_c^m(\tau, u_j) \right\} - \mathbf{Q}_s^m(\tau, u_i), \qquad i = \pm 1, \ldots, \pm N. \tag{443}$$

The convention for the indices of the quadrature points is such that $u_j < 0$ for $j < 0$, and $u_j > 0$ for $j > 0$. These points are distributed symmetrically about zero, that is, $u_{-j} = -u_j$. The corresponding weights are equal, that is, $w_{-j} = w_j$. In a plane-parallel medium, represented by a horizontal slab, it is convenient to consider the two hemispheres defined by $u < 0$ (downward radiation) and $u > 0$ (upward radiation) separately and to make use of the half-range quantity $\mu \equiv |u|$ introduced at the beginning of this chapter.

It should be emphasized that the essence of the *discrete-ordinate method* is to convert a pair of coupled integro-differential equations [Eqs. (440) and (441)] into a system of coupled linear differential equations [Eqs. (442) and (443)] by replacing the integrals with discrete sums. Combining cosine modes of the first two *Stokes parameters* ($I_\parallel$ and $I_\perp$) with the sine modes of the third and fourth Stokes parameters (U and V) in Eqs. (442) and (443), one obtains a set of coupled differential equations for the half-range $4N$ vectors $\tilde{\mathbf{I}}_c^{m\pm}(\tau, \mu_i) \equiv [I_{\parallel c}^m(\tau, \pm\mu_i), I_{\perp c}^m(\tau, \pm\mu_i), U_s^m(\tau, \pm\mu_i), V_s^m(\tau, \pm\mu_i)]^T$, $i = 1, \ldots, N$, which may be written in matrix form as [178, 182]

$$-\mathbf{U}\frac{d\tilde{\mathbf{I}}_c^{m-}}{d\tau} = (\mathbf{1} + \mathbf{A}_{11c}^m)\tilde{\mathbf{I}}_c^{m-} + \mathbf{A}_{12c}^m \tilde{\mathbf{I}}_c^{m+} - \mathbf{S}_c^{m-} \tag{444}$$

$$\mathbf{U}\frac{d\tilde{\mathbf{I}}_c^{m+}}{d\tau} = \mathbf{A}_{21c}^m \tilde{\mathbf{I}}_c^{m-} + (\mathbf{1} + \mathbf{A}_{22c}^m)\tilde{\mathbf{I}}_c^{m+} - \mathbf{S}_c^{m+} \tag{445}$$

where $\mathbf{1}$ denotes the $4N \times 4N$ identity matrix,

$$\mathbf{S}_c^{m\pm} = \begin{pmatrix} Q_{\parallel c}^m(\tau, \pm\mu_i) \\ Q_{\perp c}^m(\tau, \pm\mu_i) \\ Q_{Us}^m(\tau, \pm\mu_i) \\ Q_{Vs}^m(\tau, \pm\mu_i) \end{pmatrix}_{4N\times 1} \tag{446}$$

and $\mathbf{U}$ is the $4N \times 4N$ diagonal matrix

$$\mathbf{U} = \begin{pmatrix} \mathbf{U}_\perp & & & \\ & \mathbf{U}_\parallel & & \\ & & \mathbf{U}_U & \\ & & & \mathbf{U}_V \end{pmatrix} \tag{447}$$

where the $N \times N$ diagonal submatrices are given by

$$\mathbf{U}_\perp = \mathbf{U}_\parallel = \mathbf{U}_U = \mathbf{U}_V = \mathrm{diag}(\mu_1, \mu_2, \ldots, \mu_N)^T.$$

The matrices $\mathbf{A}^m_{ijc}$ are defined in Chapter 4.

In a similar manner, by combining sine modes of the first two Stokes parameters ($I_\parallel$ and $I_\perp$) with the cosine modes of the third and fourth Stokes parameters (U and V) in Eqs. (442) and (443), one obtains a set of coupled differential equations for the $4N$ vectors $\tilde{\mathbf{I}}^{m\pm}_s(\tau, \mu_i) \equiv [I^m_{\parallel s}(\tau, \pm\mu_i), I^m_{\perp s}(\tau, \pm\mu_i), U^m_c(\tau, \pm\mu_i), V^m_c(\tau, \pm\mu_i)]^T$, $i = 1, \ldots, N$, which may be written in matrix form as [178, 182]

$$-\mathbf{U}\frac{d\tilde{\mathbf{I}}^{m-}_s}{d\tau} = (\mathbf{1} + \mathbf{A}^m_{11s})\,\tilde{\mathbf{I}}^{m-}_s + \mathbf{A}^m_{12s}\,\tilde{\mathbf{I}}^{m+}_s - \mathbf{S}^{m-}_s \tag{448}$$

$$\mathbf{U}\frac{d\tilde{\mathbf{I}}^{m+}_s}{d\tau} = \mathbf{A}^m_{21s}\,\tilde{\mathbf{I}}^{m-}_s + (\mathbf{1} + \mathbf{A}^m_{22s})\,\tilde{\mathbf{I}}^{m+}_s - \mathbf{S}^{m+}_s \tag{449}$$

where

$$\mathbf{S}^{m\pm}_s = \begin{pmatrix} Q^m_{\parallel s}(\tau, \pm\mu_i) \\ Q^m_{\perp s}(\tau, \pm\mu_i) \\ Q^m_{Uc}(\tau, \pm\mu_i) \\ Q^m_{Vc}(\tau, \pm\mu_i) \end{pmatrix}_{4N\times 1} \tag{450}$$

and the matrices $\mathbf{A}^m_{ijs}$ are defined in Chapter 4.

Equations (444) and (445) or Eqs. (448) and (449) may be rewritten as follows:

$$\frac{d\tilde{\mathbf{I}}^{m-}_\alpha}{d\tau} = \tilde{\mathbf{A}}^m_{11\alpha}\,\tilde{\mathbf{I}}^{m-}_\alpha + \tilde{\mathbf{A}}^m_{12\alpha}\,\tilde{\mathbf{I}}^{m+}_\alpha - \tilde{\mathbf{S}}^{m-}_\alpha \qquad \alpha = c, s \tag{451}$$

$$\frac{d\tilde{\mathbf{I}}^{m+}_\alpha}{d\tau} = \tilde{\mathbf{A}}^m_{21\alpha}\,\tilde{\mathbf{I}}^{m-}_\alpha + \tilde{\mathbf{A}}^m_{22\alpha}\,\tilde{\mathbf{I}}^{m+}_\alpha - \tilde{\mathbf{S}}^{m+}_\alpha \qquad \alpha = c, s \tag{452}$$

or as

$$\frac{d}{d\tau}\begin{bmatrix} \tilde{\mathbf{I}}^{m-}_\alpha \\ \tilde{\mathbf{I}}^{m+}_\alpha \end{bmatrix} = \begin{pmatrix} \tilde{\mathbf{A}}^m_{11\alpha} & \tilde{\mathbf{A}}^m_{12\alpha} \\ \tilde{\mathbf{A}}^m_{21\alpha} & \tilde{\mathbf{A}}^m_{22\alpha} \end{pmatrix}\begin{bmatrix} \tilde{\mathbf{I}}^{m-}_\alpha \\ \tilde{\mathbf{I}}^{m+}_\alpha \end{bmatrix} - \begin{bmatrix} \tilde{\mathbf{S}}^{m-}_\alpha \\ \tilde{\mathbf{S}}^{m+}_\alpha \end{bmatrix} \qquad \alpha = c, s \tag{453}$$

where $\tilde{\mathbf{A}}^m_{11\alpha} = -\mathbf{U}^{-1}(\mathbf{1} + \mathbf{A}^m_{11\alpha})$, $\tilde{\mathbf{A}}^m_{12\alpha} = -\mathbf{U}^{-1}\mathbf{A}^m_{12\alpha}$, $\tilde{\mathbf{A}}^m_{21\alpha} = \mathbf{U}^{-1}\mathbf{A}^m_{21\alpha}$, $\tilde{\mathbf{A}}^m_{22\alpha} = \mathbf{U}^{-1}(\mathbf{1} + \mathbf{A}^m_{22\alpha})$, and $\tilde{\mathbf{S}}^{m\pm}_\alpha = \mathbf{U}^{-1}\mathbf{S}^{m\pm}_\alpha$.

Combining the vectors for the downward and upward directions by defining the $8N$ vectors ($i = 1, \ldots, N$) $\tilde{\mathbf{I}}^m_\alpha \equiv [\tilde{\mathbf{I}}^{m-}_\alpha(\mu_i), \tilde{\mathbf{I}}^{m+}_\alpha(\mu_i)]^T$, and $\tilde{\mathbf{S}}^m_\alpha \equiv [\tilde{\mathbf{S}}^{m-}_\alpha(\mu_i), \tilde{\mathbf{S}}^{m+}_\alpha(\mu_i)]^T$, we may write the coupled system of differential equations in the following compact form [178, 182]:

$$\frac{d\tilde{\mathbf{I}}^m_\alpha}{d\tau} = \tilde{\mathbf{A}}^m_\alpha\tilde{\mathbf{I}}^m_\alpha - \tilde{\mathbf{S}}^m_\alpha \qquad \alpha = c, s \tag{454}$$

where the matrix $\tilde{\mathbf{A}}_\alpha^m$ is given by

$$\tilde{\mathbf{A}}_\alpha^m \equiv \begin{pmatrix} \tilde{\mathbf{A}}_{11\alpha}^m & \tilde{\mathbf{A}}_{12\alpha}^m \\ \tilde{\mathbf{A}}_{21\alpha}^m & \tilde{\mathbf{A}}_{22\alpha}^m \end{pmatrix} \qquad \alpha = c, s. \tag{455}$$

Each of the two media (atmosphere and water) is divided into a number of adjacent layers, large enough to resolve vertical changes in the *IOPs* of each medium. Equation (454) applies in each layer in the atmosphere or water. As described in some detail in Chapter 4, the solution involves the following steps:

1) The homogeneous version of Eq. (454) with $\tilde{\mathbf{S}}_\alpha^m = 0$ yields a linear combination of exponential solutions (with unknown coefficients) obtained by solving an algebraic eigenvalue problem;
2) Particular analytic solutions are found by solving a system of linear algebraic equations;
3) The general solution is obtained by adding the homogeneous and particular solutions;
4) The solution is completed by imposing *boundary conditions* at the top of the atmosphere and the bottom of the water;
5) The solutions are required to satisfy continuity conditions across layer interfaces in the atmosphere and the water and, finally, to satisfy *Fresnel's equations* and *Snell's law* at the atmosphere–water interface, where there is an abrupt change in the *refractive index*;
6) The application of boundary, layer interface, and atmosphere–water interface conditions leads to a system of linear algebraic equations, and the numerical solution of this system of equations yields the unknown coefficients in the homogenous solutions.

3.4.5
Doubling-Adding and Matrix Operator Methods

The *doubling-adding method* has been widely used to solve radiative transfer problems. In this method, *doubling* refers to how one finds the *reflection and transmission matrices* of *two* layers with *identical* optical properties from those of the individual layers, while *adding* refers to the combination of two or more layers with *different* optical properties.

The doubling concept appears to have originated with Stokes in 1862 [183]. It was "rediscovered" and introduced into atmospheric physics one century later [184–186]. The theoretical aspects as well as the numerical techniques have since been developed by a number of investigators (for references, e.g., Wiscombe [187]). These methods or slight variants thereof are also referred to as *discrete space theory* [188, 189] or *matrix operator theory* [190–192]. The *doubling method* as commonly practiced today uses the known reflection and transmission properties of a single homogeneous layer to derive the resulting properties of two identical layers. To start the doubling procedure, the initial layer is frequently taken to be thin enough so that its reflection and transmission properties can be computed from single scattering. Repeated "doublings" are then applied to reach the desired optical thickness. The division of an inhomogeneous slab into a series of adjacent sublayers, each of which is homogeneous, but in principle different from all the others, is usually taken to be identical to that discussed previously for the *discrete-ordinate method*. The solution proceeds by first applying doubling to find the reflection and transmission matrices for each of the homogeneous layers, whereupon adding is subsequently used to find the solution for all the different layers combined.

Derivation of Doubling Rules – Matrix-Exponential Solution

In the *doubling-adding method*, one usually adopts the same discretization in angle as in the discrete-ordinate method. The basic problem is therefore to find solutions to the homogeneous version of Eq. (453), which we may rewrite as

$$\frac{d}{d\tau}\begin{bmatrix}\tilde{\mathbf{I}}_\alpha^{m-}\\ \tilde{\mathbf{I}}_\alpha^{m+}\end{bmatrix}=\tilde{\mathbf{A}}_\alpha^{m}\begin{bmatrix}\tilde{\mathbf{I}}_\alpha^{m-}\\ \tilde{\mathbf{I}}_\alpha^{m+}\end{bmatrix} \tag{456}$$

where the $8N\times 8N$ matrix $\tilde{\mathbf{A}}_\alpha^{m}$ defined in Eq. (455) consists of the $4N\times 4N$ matrices defined after Eq. (453). We may solve this equation formally for a homogeneous slab of optical thickness τ_b to obtain

$$\begin{bmatrix}\tilde{\mathbf{I}}_\alpha^{m-}(\tau_b)\\ \tilde{\mathbf{I}}_\alpha^{m+}(\tau_b)\end{bmatrix}=e^{-\tilde{\mathbf{A}}_\alpha^{m}\tau_b}\begin{bmatrix}\tilde{\mathbf{I}}_\alpha^{m-}(0)\\ \tilde{\mathbf{I}}_\alpha^{m+}(0)\end{bmatrix}. \tag{457}$$

It should be emphasized that this *matrix-exponential solution* is only a formal solution since we are dealing with a two-point boundary-value problem in which the *incident Stokes vectors* $\tilde{\mathbf{I}}_\alpha^{m-}(0)$ and $\tilde{\mathbf{I}}_\alpha^{m+}(\tau_b)$ constitute the *boundary conditions* and the *emergent* Stokes vectors $\tilde{\mathbf{I}}_\alpha^{m+}(0)$ and $\tilde{\mathbf{I}}_\alpha^{m-}(\tau_b)$ are to be determined. We notice the formal similarity with an *initial-value problem* in which the left side could be computed if the "initial" values (i.e., the right side) were known (which they are *not* in our problem).

Basically, the doubling concept starts with the notion that the emergent Stokes vectors $\tilde{\mathbf{I}}_\alpha^{m+}(0)$ and $\tilde{\mathbf{I}}_\alpha^{m-}(\tau_b)$ in Eq. (457) are determined by a *reflection matrix* $\mathbf{R}$ and a *transmission matrix* $\mathbf{T}$ through the following relations:

$$\tilde{\mathbf{I}}_\alpha^{m-}(\tau_b)=\mathbf{T}\,\tilde{\mathbf{I}}_\alpha^{m-}(0)+\mathbf{R}\,\tilde{\mathbf{I}}_\alpha^{m+}(\tau_b)$$
$$\tilde{\mathbf{I}}_\alpha^{m+}(0)=\mathbf{T}\,\tilde{\mathbf{I}}_\alpha^{+}(\tau_b)+\mathbf{R}\,\tilde{\mathbf{I}}_\alpha^{m-}(0)$$

for a homogeneous slab of thickness τ_b. These two relations, which are frequently referred to as the *interaction principle*, may be rewritten in matrix form as (see Problem 3.5)

$$\begin{bmatrix}\tilde{\mathbf{I}}_\alpha^{m-}(\tau_b)\\ \tilde{\mathbf{I}}_\alpha^{m+}(\tau_b)\end{bmatrix}=\begin{pmatrix}\mathbf{T}-\mathbf{R}\mathbf{T}^{-1}\mathbf{R} & \mathbf{R}\mathbf{T}^{-1}\\ -\mathbf{T}^{-1}\mathbf{R} & \mathbf{T}^{-1}\end{pmatrix}\begin{bmatrix}\tilde{\mathbf{I}}_\alpha^{m-}(0)\\ \tilde{\mathbf{I}}_\alpha^{m+}(0)\end{bmatrix} \tag{458}$$

where the superscript -1 denotes matrix inversion. By comparing Eqs. (457) and (458), we find

$$e^{-\tilde{\mathbf{A}}_\alpha^{m}\tau_b}=\begin{pmatrix}\mathbf{T}-\mathbf{R}\mathbf{T}^{-1}\mathbf{R} & \mathbf{R}\mathbf{T}^{-1}\\ -\mathbf{T}^{-1}\mathbf{R} & \mathbf{T}^{-1}\end{pmatrix}. \tag{459}$$

If we now let $\mathbf{T}_1 = T(\tau_b)$, $\mathbf{R}_1 = \mathbf{R}(\tau_b)$ and $\mathbf{T}_2 = \mathbf{T}(2\tau_b)$, $\mathbf{R}_2 = \mathbf{R}(2\tau_b)$, then the identity $e^{-\tilde{\mathbf{A}}_\alpha^m 2\tau_b} = (e^{-\tilde{\mathbf{A}}_\alpha^m \tau_b})^2$ implies [using Eq. (459)]

$$\begin{pmatrix} \mathbf{T}_2 - \mathbf{R}_2\mathbf{T}_2^{-1}\mathbf{R}_2 & \mathbf{R}_2\mathbf{T}_2^{-1} \\ -\mathbf{T}_2^{-1}\mathbf{R}_2 & \mathbf{T}_2^{-1} \end{pmatrix} = \begin{pmatrix} \mathbf{T}_1 - \mathbf{R}_1\mathbf{T}_1^{-1}\mathbf{R}_1 & \mathbf{R}_1\mathbf{T}_1^{-1} \\ -\mathbf{T}_1^{-1}\mathbf{R}_1 & \mathbf{T}_1^{-1} \end{pmatrix}^2 .$$

Solving for $\mathbf{T}_2$ and $\mathbf{R}_2$, we find

$$\mathbf{T}_2 = \mathbf{T}_1(\mathbf{1} - \mathbf{R}_1^2)^{-1}\mathbf{T}_1 \qquad (460)$$

$$\mathbf{R}_2 = \mathbf{R}_1 + \mathbf{T}_1\mathbf{R}_1(\mathbf{1} - \mathbf{R}_1^2)^{-1}\mathbf{T}_1 \qquad (461)$$

where 1 is the identity matrix. Equations (460) and (461) constitute the basic *doubling rules* from which the reflection and transmission matrices for a layer of thickness $2\tau_b$ are obtained from those of half the thickness, τ_b.

Connection Between Doubling and Discrete-Ordinate Methods

As already mentioned, in practical numerical implementations of the doubling method it is common to start with an infinitesimally thin layer so that multiple scattering can be ignored. The “starting values” for the **R** and **T** matrices are then simply determined from single scattering. Since the computational effort is directly proportional to the number of “doublings” required to obtain results for a given thickness, it would be useful to start the procedure with thicker layers. This improvement could be achieved by evaluating the left side of Eq. (459) requiring the eigenvalues and eigenvectors of $\tilde{\mathbf{A}}_\alpha^m$ to be determined by standard procedures as discussed previously. Inspection of the right side of Eq. (459) shows that inversion of the lower right quadrant of the matrix yields **T** and postmultiplication of the upper right quadrant with **T** yields **R** [193, 194].

The discussion above shows that the discrete-ordinate method and the doubling method, which are conceptually very different, are, in fact, intimately related. In particular, we note that the formal *matrix-exponential* (discrete-ordinate) solution can be used to derive the *doubling rules*. Moreover, the eigenvalues and eigenvectors that will be used to construct the basic solutions in the *discrete-ordinate method* (see Chapter 4) can also be used to compute the reflection and transmission matrices occurring in the *doubling method*.

Intuitive Derivation of the Doubling Rules – Adding of Dissimilar Layers

The doubling concept is illustrated in Figure 22 in which the two sublayers are taken to be identical and homogeneous. From this figure, the doubling rules can be derived in a more intuitive way. Writing $\mathbf{R}_1$ and $\mathbf{T}_1$ for the individual layers and $\mathbf{R}_2$ and $\mathbf{T}_2$ for the combined layers, we find

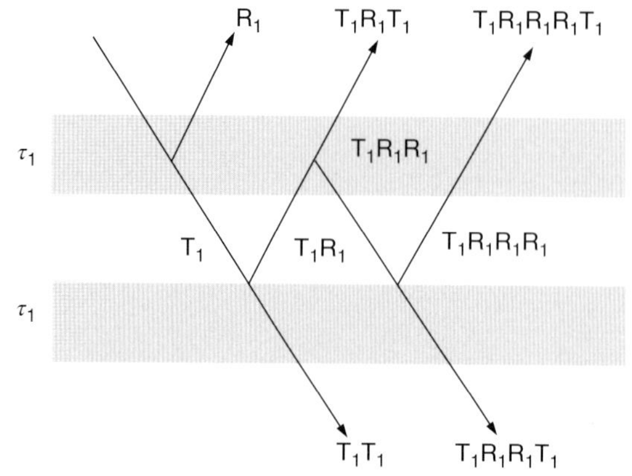

Figure 22 Illustration of the doubling concept. Two similar layers, each of optical thickness τ_1, are combined to give the reflection and transmittance matrices of a layer of twice the optical thickness $\tau_2 = 2\tau_1$ ([see Eqs. (462) and (463)]).

$$\begin{aligned}\mathbf{R}_2 &= \mathbf{R}_1 + \mathbf{T}_1\mathbf{R}_1\mathbf{T}_1 + \mathbf{T}_1\mathbf{R}_1\mathbf{R}_1\mathbf{R}_1\mathbf{T}_1 + \mathbf{T}_1\mathbf{R}_1\mathbf{R}_1\mathbf{R}_1\mathbf{R}_1\mathbf{R}_1\mathbf{T}_1 + \cdots \\ &= \mathbf{R}_1 + \mathbf{T}_1\mathbf{R}_1(\mathbf{1} + \mathbf{R}_1\mathbf{R}_1 + \mathbf{R}_1\mathbf{R}_1\mathbf{R}_1\mathbf{R}_1 + \cdots)\mathbf{T}_1 \\ &= \mathbf{R}_1 + \mathbf{T}_1\mathbf{R}_1(\mathbf{1} - \mathbf{R}_1^2)^{-1}\mathbf{T}_1 \end{aligned} \tag{462}$$

$$\begin{aligned}\mathbf{T}_2 &= \mathbf{T}_1\mathbf{T}_1 + \mathbf{T}_1\mathbf{R}_1\mathbf{R}_1\mathbf{T}_1 + \mathbf{T}_1\mathbf{R}_1\mathbf{R}_1\mathbf{R}_1\mathbf{R}_1\mathbf{T}_1 + \cdots \\ &= \mathbf{T}_1(\mathbf{1} + \mathbf{R}_1\mathbf{R}_1 + \mathbf{R}_1\mathbf{R}_1\mathbf{R}_1\mathbf{R}_1 + \cdots)\mathbf{T}_1 \\ &= \mathbf{T}_1(\mathbf{1} - \mathbf{R}_1^2)^{-1}\mathbf{T}_1 \end{aligned} \tag{463}$$

where we have used matrix inversion to sum the infinite series, that is,

$$\mathbf{1} + \mathbf{RR} + \mathbf{RRRR} + \cdots = (\mathbf{1} - \mathbf{RR})^{-1}.$$

We note that these expressions are identical to Eqs. (460) and (461) as they should be.

For two dissimilar, vertically inhomogeneous layers, we must generalize the expressions to account for the fact that the transmission and reflection matrices will in general be different for illumination from the bottom of the layer than from the top. This difference is illustrated in Figure 23, where the reflection and transmission matrices pertinent for illumination from below are denoted by the overbar symbol ($\overline{\mathbf{R}}$ and $\overline{\mathbf{T}}$), the individual layers are denoted by subscripts 1 and 2, and the combined layers are without subscript. Referring to Figure 23, we have

$$\begin{aligned}\mathbf{R} &= \mathbf{R}_1 + \mathbf{T}_1\mathbf{R}_2\overline{\mathbf{T}}_1 + \mathbf{T}_1\mathbf{R}_2\overline{\mathbf{R}}_1\mathbf{R}_2\overline{\mathbf{T}}_1 + \cdots \\ &= \mathbf{R}_1 + \mathbf{T}_1\mathbf{R}_2(\mathbf{1} - \overline{\mathbf{R}}_1\mathbf{R}_2)^{-1}\overline{\mathbf{T}}_1 \end{aligned} \tag{464}$$

$$\begin{aligned}\mathbf{T} &= \mathbf{T}_1\mathbf{T}_2 + \mathbf{T}_1\mathbf{R}_2\overline{\mathbf{R}}_1\mathbf{T}_2 + \mathbf{T}_1\mathbf{R}_2\overline{\mathbf{R}}_1\mathbf{R}_2\overline{\mathbf{R}}_1\mathbf{T}_2 + \cdots \\ &= \mathbf{T}_1(\mathbf{1} - \mathbf{R}_2\overline{\mathbf{R}}_1)^{-1}\mathbf{T}_2 \end{aligned} \tag{465}$$

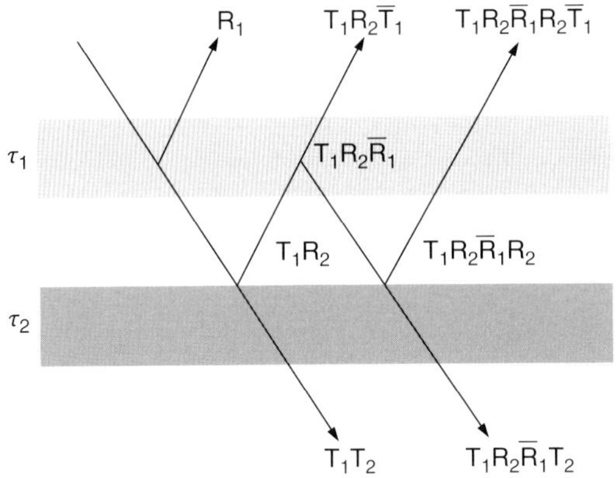

Figure 23 Illustration of the adding concept. Two dissimilar *inhomogeneous* layers are combined (added) to give the reflection and transmission matrices for two layers with different IOPs. Illumination from the top of the two layers yields Eq. (464) for the reflection matrix and Eq. (465) for the transmittance matrix. Illumination from the bottom of the two layers yields Eqs. (466) and (467).

which are the reflection and transmission matrices for the combined layers for illumination from the top boundary, as illustrated in Figure 23.

Illumination from the bottom boundary leads to the following expressions:

$$\begin{aligned}\overline{\mathbf{R}} &= \overline{\mathbf{R}}_2 + \overline{\mathbf{T}}_2\overline{\mathbf{R}}_1\mathbf{T}_2 + \overline{\mathbf{T}}_2\overline{\mathbf{R}}_1\mathbf{R}_2\overline{\mathbf{R}}_1\mathbf{T}_2 + \cdots \\ &= \overline{\mathbf{R}}_2 + \overline{\mathbf{T}}_2\overline{\mathbf{R}}_1(1 - \mathbf{R}_2\overline{\mathbf{R}}_1)^{-1}\mathbf{T}_2 \end{aligned} \tag{466}$$

$$\begin{aligned}\overline{\mathbf{T}} &= \overline{\mathbf{T}}_2\overline{\mathbf{T}}_1 + \overline{\mathbf{T}}_2\overline{\mathbf{R}}_1\mathbf{R}_2\overline{\mathbf{T}}_1 + \overline{\mathbf{T}}_2\overline{\mathbf{R}}_1\mathbf{R}_2\overline{\mathbf{R}}_1\mathbf{R}_2\overline{\mathbf{T}}_1 + \cdots \\ &= \overline{\mathbf{T}}_2(\mathbf{1} - \overline{\mathbf{R}}_1\mathbf{R}_2)^{-1}\overline{\mathbf{T}}_1. \end{aligned} \tag{467}$$

Equations (464)–(467) are the required of *doubling-adding formulas* for external illumination. To deal with internal sources, additional formulas are required. Readers who are interested in further pursuing this topic should consult the references given previously.

3.4.6
Monte Carlo Method

Monte Carlo simulations have been used for more than four decades to study light scattering in clouds [195], polarized radiative transfer in hazy atmospheres [196], and a variety of different problems in atmospheric optics [197–202] as well as in coupled atmosphere–ocean systems [52, 84, 202, 203]. One advantage of statistical Monte Carlo methods compared to deterministic methods for solving the *RTE* is that they can easily be extended to a three-dimensional (3-D) geometry.

The *Monte Carlo method* is based on statistics, and it makes use of simulations to determine probabilistic paths or *random walks* of a large number of beams (typically referred to as "photons") propagating through the medium. Thus, for the solar illumination problem, one considers a large number of beams that are incident at the top of the first medium (TOA for a coupled atmosphere–water system) and proceed until each of the beams is either absorbed or scattered back to space. For each beam incident at the TOA, a random number ρ between 0 and 1 is generated to determine its initial optical path length τ. To compute the new path length τ, one generates a new random number ρ between 0 and 1, and τ is computed from the formula

$$\tau = \sum_{j=1}^{\ell} d_j \gamma_j = \ln \rho \tag{468}$$

where d_j is the geometric path length, $\gamma_j = \alpha_j + \beta_j$ is the *attenuation coefficient*, and α_j and β_j are the *absorption and scattering coefficients*, respectively. The summation is carried out until the sum becomes larger than $\ln \rho$, and then d_ℓ is adjusted in layer ℓ so that the sum in Eq. (468) becomes equal to $\ln \rho$. This procedure determines the new vertical position of the beam. Once the new vertical position of the beam has been calculated, the attenuation coefficient, given by $\gamma = \alpha + \beta$, determines whether or not the beam is scattered or absorbed. If $\rho > \beta/\gamma$, the beam is absorbed; otherwise it is scattered. Next, a new random number ρ between 0 and 1 is generated and used to determine whether the scattering is of the Rayleigh or non-Rayleigh type, each defined by an appropriate *Stokes scattering matrix* [see Eq. (36)].

At the interface between the two media, where the refractive index changes abruptly, a random number ρ is used to determine whether the beam is reflected or refracted by comparing it with the reflectance R_{unpol} for unpolarized light given by [see Eq. (260)]

$$R_{\mathrm{unpol}} = \frac{1}{2}\left[\frac{\sin^2(\theta_i - \theta_t)}{\sin^2(\theta_i + \theta_t)} + \frac{\tan^2(\theta_i - \theta_t)}{\tan^2(\theta_i + \theta_t)}\right]. \tag{469}$$

If $\rho < R_{\mathrm{unpol}}$, the beam is reflected; otherwise the beam is refracted according to *Snell's law*, for details see [52, 92, 196–198, 200, 201]. In Eq. (469), $R_{\mathrm{unpol}} = (R_{||}^2 + R_{\perp}^2)/2$, where $R_{||}$ and $R_{\perp}$ are the *Fresnel reflection coefficients* given by Eqs. (255) and (257).

In *Monte Carlo* simulations, each beam contains information about the full *Stokes vector*, but two rotations are required to relate the beam's initial and final Stokes vectors, as discussed in Section 2.3.

3.5 Calculation of Weighting Functions – Jacobians

3.5.1 Linearized Radiative Transfer

To solve the inverse problem, it is useful to compute the response of the ***Stokes parameters*** (let us call them $y_i, i = 1, \ldots 4$) to changes in the system (state) parameters (let us call them $x_j, j = 1, \ldots n$) one seeks to retrieve. In general, this response will depend nonlinearly on the change in a state parameter, but if the change is sufficiently small, a perturbation approach can be used, so that we are only seeking $\partial y_i/\partial x_j$. Hence the parlance *linearized radiative transfer*, which may sound a little bit strange because the radiative transfer equation itself is linear. To illustrate this approach, let us look at the *discrete-ordinate method*. Once layer-by-layer solutions have been found in each of the two slabs comprising a coupled system, the application of boundary, layer interface, and slab_1–slab_2 interface conditions leads to a system of linear algebraic equations (see Appendix D)

$$\mathbf{Ac} = \mathbf{B} \tag{470}$$

which has solution $\mathbf{c} = \mathbf{A}^{-1}\mathbf{B}$. The matrix $\mathbf{A}$ contains information about the system optical properties (layer by layer in each slab) as well as the lower boundary BRDF properties. The column vector $\mathbf{c}$ contains all the unknown coefficients $\tilde{C}_{\pm j\ell}$ in the homogeneous solutions, and the column vector $\mathbf{B}$ contains information about the particular solutions as well as the lower boundary emissivity. For a coupled two-slab system, the order of the matrix $\mathbf{A}$ is $[8N_1 \times L_1 + 8N_2 \times L_2]$, which is the same as the number of unknown coefficients $\tilde{C}_{\pm j\ell}$. Here, N_1 is the number of discrete ordinates (quadrature points) and L_1 the number of layers in slab_1, and N_2 and L_2 the corresponding values in slab_2.

The "linearization" concept is based on the recognition that, if the optical properties of, say, one of the layers in the system is perturbed, there is *no need to repeat the entire computation* from scratch. For example, a perturbation in the scattering and/or absorption coefficients of one particular layer may translate into small changes in the *single-scattering albedo* and the optical depth of that layer. Based on a perturbation analysis of the *discrete-ordinate solution*, it has been shown [204] that the solution of the problem can be cast in a matrix form similar to Eq. (470):

$$\mathbf{Ac}' = \mathbf{B}' \tag{471}$$

where the matrix $\mathbf{A}$ is identical to that appearing in the unperturbed problem [Eq. (470)]. The column vector $\mathbf{B}'$ depends on the layer parameters that have been changed. The *LU-factorization* of the matrix $\mathbf{A}$ into the product of a lower diagonal matrix (L) and an upper diagonal matrix (U) has already been done for the unperturbed problem. Thus, the solution $\mathbf{c}' = \mathbf{A}^{-1}\mathbf{B}'$ can be found by back-substitution using the LU-factorized form of $\mathbf{A}$. Perhaps the most significant advancement is that, in this perturbation (or "linearized") approach, not only radiances but also the associated *Jacobians* or *weighting functions*, required to solve the *inverse problem*, can be computed at insignificant additional computational cost. For *polarized* RT, a vectorized version of the "linearized approach" [90] has been developed for a single slab, and for the coupled atmosphere–ocean system a linearized model exists for the unpolarized case [34].

3.5.2 Neural Network Forward Models

The linearized forward modeling approach described above has been used to compute *Jacobians* required to solve the inverse problem to retrieve atmospheric and water parameters in a *coupled* atmosphere–water *system* [38]. Although this method works quite well, it has two limitations: (i) repeated calls to the forward model required in iterative optimization schemes makes the method slower than desired for analysis of large amounts of satellite data, and (ii) if the atmospheric and water inherent optical property models are changed, then the analytic formulas for the linearization will have to be redone from scratch.

We would like to find a way both to (i) speed up the forward model computation and (ii) avoid having to redo tedious analytic "linearization" derivations if we change the models describing the *IOPs*. For example, for a coupled atmosphere–water system, we may be using a particular atmospheric aerosol model and a hydro-optical model to link desired *retrieval parameters*, such as *aerosol particle size* and *loading* as well as the amounts of organic and inorganic material in the water, to the IOPs that are the required inputs to the *radiative transfer model for the coupled system (CRTM).*

One way to achieve this goal is to store the computed *Stokes parameters* from the CRTM in a *look-up table (LUT)* for the expected range of values of the *retrieval parameters* and sun/satellite viewing geometries. But in order to get good accuracy, the LUT may have to be quite big, and the time it takes to search in this table may become longer than desired. Also, computing the derivatives $\partial y_i/\partial x_j$ to obtain the weighting functions from the LUT results may not be sufficiently accurate. An alternative approach is to use a *radial basis function neural network (RBF-NN)* for this purpose. An advantage of this approach is that the RBF-NN consists of analytical functions that can be differentiated to yield the required *Jacobians* or *weighting functions.*

Radial Basis Functions

In the *RBF-NN* approach, we first generate a set of *synthetic (model) data* by simulating the transport of radiation in the coupled system using the *CRTM*. Thus, the CRTM is used to compute the TOA *Stokes parameters* as a function of sun/satellite geometry for the expected range of variation of the desired retrieval or state parameters.

The RBF-NN was proposed by Broomhead and Lowe [205] in 1988. It employs RBFs as its *activation functions.* In most cases, an RBF is a taken to be a *Gaussian distribution*

$$f(x) = \exp\left[-\frac{(x-c)^2}{2\sigma^2}\right] \qquad (472)$$

where c is the central point or mean of the Gaussian distribution, also called the RBF weight, and σ is the width (or standard deviation). The center and the width of the RBF are the parameters of the *neural network* [206]. The main feature of an RBF is that its response decreases with increasing distance from its center: the smaller the distance from its center, the larger the activation. The architecture of an RBF-NN is illustrated in Figure 24. It consists of three strata: an input layer, hidden layers, and an output layer [207]. An RBF-NN has two hidden layers: an RBF layer and a linear layer, as illustrated in Figure 25. The output after the RBF layer is commonly written as

$$O_{1i} = \exp[-((x_{1i} - w_{1i})b_1)^2] \tag{473}$$

where x_{1i} is the input, w_{1i} is the weight, b_1 is called the bias, and O_{1i} is the output of the RBF layer. If the bias b_1 is selected to be

$$b_1 = \sqrt{\ln 2}/r = \sqrt{-\ln(0.5)}/r = 0.8326/r \tag{474}$$

where r is referred to as the spread, then the output is 0.5, the full-width at half-maximum (FWHM), when the weighted input is equal to the spread, that is, when $\|x_{1i} - w_{1i}\| = r$. Hence, Eq. (473) becomes

$$O_{1i} = \exp[-(\sqrt{\ln 2}\,(x_{1i} - w_{1i})/r)^2]. \tag{475}$$

The selectivity of the *RBF* is determined by the spread. If the spread r is too small, the RBF is too selective, and if it is too large, the selectivity is too low. More specifically, the selectivity is proportional to the reciprocal of the spread. If $(x_{1i} - w_{1i})$ is larger than the spread, the output O_{1i} of the RBF is larger than 0.5, and if $(x_{1i} - w_{1i})$ is less than the spread, then O_{1i} is smaller than 0.5. Thus, the relationship between

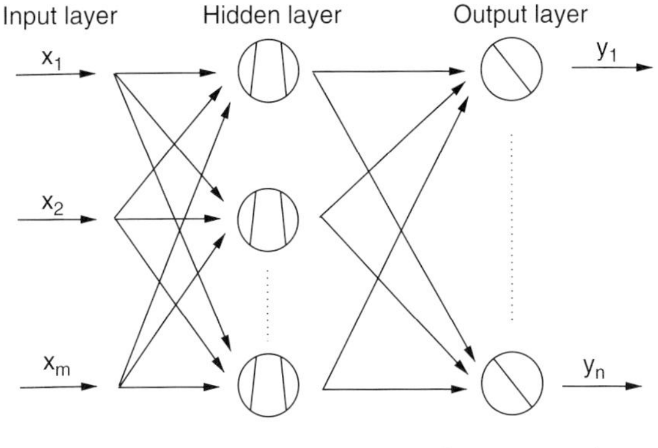

Figure 24 Architecture of a radial basis function neural network (RBF-NN).

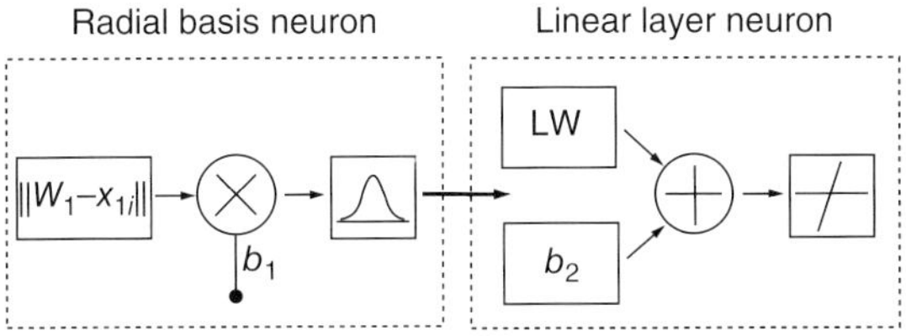

Figure 25 Radial basis function neural network (RBF-NN).

the sensitivity of the selection and the spread is given by

$$O_{1i} = \begin{Bmatrix} > 0.5 & \textit{if} \ \|x_{1i} - w_{1i}\| > r \\ = 0.5 & \textit{if} \ \|x_{1i} - w_{1i}\| = r \\ < 0.5 & \textit{if} \ \|x_{1i} - w_{1i}\| < r \end{Bmatrix}. \tag{476}$$

After the RBF layer, there is a linear layer with a transfer function given by

$$O_{2i} = w_{Li}x_{2i} + b_{2i} \tag{477}$$

where w_{Li} is a weight and b_{2i} is the bias of the second layer. The final output of the whole *RBF-NN* is given by O_{2i}. We need to combine the two hidden layers. If there are N_{in} input parameters, the equation for the first layer becomes

$$O_{1i} = \exp\left[-b_1^2 \sum_{k=1}^{N_{\text{in}}} (w_{1k} - x_{1k})^2\right]. \tag{478}$$

The input to the linear layer is the output of the RBF layer, defined as

$$x_{2i} = O_{1i} = \exp\left[-b_1^2 \sum_{k=1}^{N_{\text{in}}} (w_{1k} - x_{1k})^2\right] \tag{479}$$

which upon substitution in Eq. (477) yields

$$O_{2i} = w_{Li}x_{2i} + b_{2i} = w_{Li} \exp\left[-b_1^2 \sum_{k=1}^{N_{\text{in}}} (w_{1k} - x_{1k})^2\right] + b_{2i}. \tag{480}$$

If there is a total of N *neurons*, the ith output of the *RBF-NN* becomes

$$O_{2i} = \sum_{j=1}^{N} w_{Lij} \exp\left[-b_1^2 \sum_{k=1}^{N_{\text{in}}} (w_{1jk} - x_{1k})^2\right] + b_{2i}. \tag{481}$$

To simplify the notation, we set $a_{ij} = w_{Lij}$, $b = b_1$, $d_i = b_{2i}$, $c_{jk} = w_{1jk}$, $R_k = x_{1k}$, and set the ith output of the whole RBF-NN to $p_i = O_{2i}$, so that the complete RBF-NN function becomes

$$p_i = \sum_{j=1}^{N} a_{ij} \exp\left[-b^2 \sum_{k=1}^{N_{\text{in}}} (c_{jk} - R_k)^2\right] + d_i \tag{482}$$

where N is the total number of neurons and N_{in} is the number of input parameters. The *Jacobians* are obtained by calculating the derivative with respect to the retrieval parameter R_k:

$$J(k) = \frac{\partial p_i}{\partial R_k} = -2b^2(c_{jk} - R_k) \sum_{j=1}^{N} a_{ij} \exp\left[-b^2 \sum_{k=1}^{N_{\text{in}}} (c_{jk} - R_k)^2\right]. \tag{483}$$

The purpose of the training of the *RBF-NN* is to determine the coefficients a_{ij}, b, c_{jk}, and d_i appearing in Eq. (482).

Finally, it is important to note that, if the goal is to make a retrieval of state parameters directly from say TOA radiation measurements, then the input parameters R_k in Eq. (482) are the TOA *Stokes parameters* at the desired wavelengths plus the solar/viewing geometry, and the output parameters p_i are the desired *retrieval (state) parameters*. If, on the other hand, the goal is to use the RBF-NN as a fast interpolator to obtain the TOA Stokes parameters and associated *Jacobians*, then the input parameters R_k should be the state parameters and the solar/viewing geometry, and the output parameters p_i should be the TOA Stokes parameters.

Problems

3.1 Derive Eq. (346) for the multiple scattering source term.
3.2 Derive Eqs. (392)–(393).
3.3 Derive Eqs. (414)–(415).
3.4 Verify that Eqs. (421)–(424) are correct.
3.5 Derive Eq. (458).

4
Forward Radiative Transfer Modeling

4.1
Quadrature Rule – The *Double-Gauss* Method

The sums in Eqs. (442) and (443) were obtained by evaluating the integrals in Eqs. (440) and (441) over u' in the range $[-1, 1]$ using a numerical quadrature method. Carl Friedrich Gauss showed that, in the absence of information about the function to be integrated, optimum accuracy would be obtained by choosing the u' values to be roots of the *Legendre polynomials* $P_\ell(u)$.

It is convenient to use the *even-order* Legendre polynomials because the roots appear in *pairs*. Thus, if we use a negative subscript to label values of u corresponding to directions in the downward hemisphere and a positive subscript for values of u in the upward hemisphere, then $u_{-i} = -u_{+i}$, and the quadrature weights are the same in both hemispheres, that is, $w'_i = w'_{-i}$.

The "full-range" is based on the assumption that $\mathbf{I}(\tau, u)$ is a smoothly varying function of u $(-1 \leq u \leq +1)$ for all values of τ, which is problematic because, at least for small τ, the radiation field changes rather rapidly as u passes through zero, that is, as the observation direction passes through the horizontal [18]. To deal with this problem, the *Double-Gauss method* was devised, in which the two hemispheres are treated separately. Instead of approximating $\int_{-1}^{1} du\mathbf{I}(u) \approx \sum_{i=-M}^{+M} w'_i\mathbf{I}(u_i)$, where w'_i and u_i are the weights and roots of the even-order *Legendre polynomial* $P_{2M}(u)$, one may break the angular integration into two hemispheres, and approximate each integral separately:

$$\int_{-1}^{1} du\, \mathbf{I} = \int_{0}^{1} d\mu\, \mathbf{I}^{+} + \int_{0}^{1} d\mu\, \mathbf{I}^{-} \approx \sum_{j=1}^{N} w_j \mathbf{I}^{+}(\mu_j) + \sum_{j=1}^{N} w_j \mathbf{I}^{-}(\mu_j)$$

where w_j and μ_j are the weights and roots for the half-range. To obtain the highest accuracy, one must again use *Gaussian quadrature*, but the new interval is $(0 \leq \mu \leq 1)$ instead of $(-1 \leq u \leq 1)$. Algorithms to compute the roots and weights are usually based on the full range, but the half-range weights and roots can easily be found in terms of the weights w'_j and points u_j for the full range. Since the linear transformation $t = (2x - x_1 - x_2)/(x_2 - x_1)$ will map any interval $[x_1, x_2]$ into

Radiative Transfer in Coupled Environmental Systems: An Introduction to Forward and Inverse Modeling,
First Edition. Knut Stamnes and Jakob J. Stamnes.

$[-1, 1]$ provided $x_2 > x_1$, Gaussian quadrature can be used to numerically integrate a function $F(x)$:

$$\int_{x_1}^{x_2} dx F(x) = \int_{-1}^{1} dt\, F\left(\frac{(x_2 - x_1)t + x_2 + x_1}{2}\right) \frac{(x_2 - x_1)}{2}. \tag{484}$$

Choosing $x_1 = 0, x_2 = 1, x = \mu$, and $t = u$, one finds

$$\int_0^1 d\mu F(\mu) = \frac{1}{2} \int_{-1}^{1} du\, F\left(\frac{u+1}{2}\right)$$

and by applying Gaussian quadrature to each integral, one finds on setting $N = 2M$ for the half-range

$$\int_0^1 d\mu F(\mu) = \sum_{j=1}^{2M} w_j F(\mu_j) = \frac{1}{2} \int_{-1}^{1} du F\left(\frac{u+1}{2}\right) = \frac{1}{2} \sum_{\substack{j=-M \\ j\neq 0}}^{M} w'_j F\left(\frac{u_j+1}{2}\right). \tag{485}$$

Thus, for even orders the half-range points (μ-values) and weights are related to the full-range points u_j and weights w'_j by (Thomas and Stamnes [18])

$$\mu_j = \frac{u_j + 1}{2}; \qquad w_j = \frac{1}{2} w'_j. \tag{486}$$

Hence, the Double-Gauss weights for even orders are half the Gaussian weights for half the order, and each pair of roots $\pm|u_j|$ for any order M (full range) generates two positive roots $\mu_j = (-|u_j| + 1)/2$ and $\mu_{2N+1-j} = (|u_j| + 1)/2$ of order $N = 2M$ (half range). We have chosen to label the roots so that they appear in ascending order, that is, $\mu_1 < \mu_2 < \mu_3 < \cdots < \mu_N$. Note that the quadrature points (for even orders) are distributed symmetrically around $|u| = 0.5$ and clustered both toward $|u| = 1$ and $|u| = 0$, whereas in the Gaussian scheme for the complete range, $-1 < u < 1$, so they are clustered toward $u = \pm 1$. The clustering toward $|u| = 0$ will give superior results in the range where the intensity varies rapidly around $|u| = 0$. A half-range scheme is also preferable since the intensity is discontinuous at the boundaries. Another advantage is that half-range quantities such as upward and downward fluxes and average intensities are obtained immediately without any further approximations, whereas computation of *half-range* quantities by use of a *full-range* quadrature scheme is *not* self-consistent.

For a coupled two-slab (atmosphere–water) system, one may use the Double-Gauss rule to determine *quadrature points* and weights μ_i^a and w_i^a $(i = 1, \ldots, N_1)$ in the atmosphere, as well as the quadrature points and weights μ_i^w and w_i^w $(i = N_1 + 1, \ldots, N_2)$ in region **I** of *total reflection* in the water. A quadrature point in the *refractive region* **II** of the water with cosine of the polar angle given by μ_i^w is obtained by refraction of an associated downward beam in the atmosphere into the water with corresponding cosine of the polar angle given by μ_i^a, as illustrated

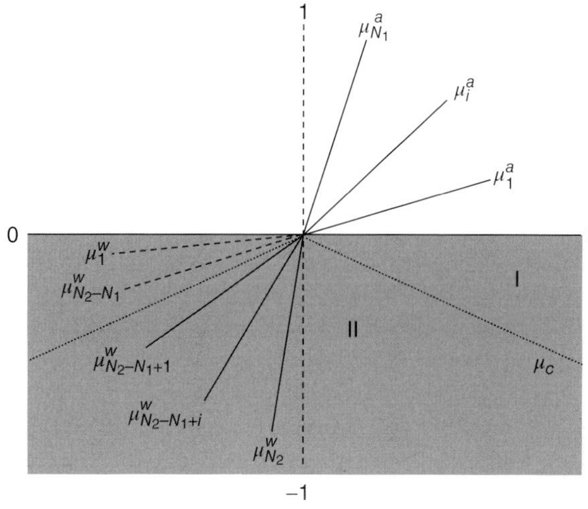

Figure 26 Schematic illustration of the quadrature adopted for a *coupled* atmosphere-water *system*. The dotted line labeled μ_c marks the separation between region **II**, in which light is refracted from the atmosphere into the water, and region **I** of total reflection in the water, in which the upward directed light beam in the water undergoes total internal reflection at the water–air interface. The quadrature angles in region **II** are connected to those in the atmosphere through *Snell's law*. Additional quadrature points are added to represent quadrature angles in region **I** as indicated. Note that for "bookkeeping purposes", μ^a_i in the air corresponds to $\mu^w_{N_2-N_1+i}$ in the water.

schematically in Figure 26. Thus, in the refractive region **II**, μ^w_i is related to μ^a_i by Snell's law:

$$\mu^w_i = S(\mu^a_i) = \sqrt{1 - [1 - (\mu^a_i)^2]/m^2_{\text{rel}}}$$

and from this relation the weights for this region are derived as

$$w^w_i = w^a_i \frac{dS(\mu^a_i)}{d\mu^a_i} = \frac{\mu^a_i}{m^2_{\text{rel}}\mu^w_i} w^a_i \qquad i = 1, \ldots, N_1.$$

The advantage of this choice of quadrature is that the points are clustered toward $\mu = 0$ both in the atmosphere and the ocean and, in addition, toward the direction associated with the *critical angle* (represented by μ_c) in the ocean. This clustering will give superior results near these directions where the radiation field varies rapidly. Also, the *scattering phase function* will remain correctly normalized.

The quadrature points and weights in the total reflection region $[0, \pm\mu_c]$ are obtained from Eq. (484) with $x_2 = \mu_c$, $x_1 = 0$, $x = \mu$, and $t = u$, which yields

$$\mu^w_i = \frac{\mu_c(u_i + 1)}{2}; \qquad w^w_i = \frac{\mu_c}{2} w'_i \qquad i = 1, \ldots, \frac{(N_2 - N_1)}{2}. \tag{487}$$

4.2
Discrete Ordinate Equations – Compact Matrix Formulation

As shown in Section 3.4.4 each Fourier component of the RTE satisfies the following equations ($i = \pm 1, \dots, \pm N$):

$$u_i \frac{d\mathbf{I}_c^m(\tau, u_i)}{d\tau} = \mathbf{I}_c^m(\tau, u_i) - \frac{\varpi(\tau)}{4} \sum_{\substack{j=-N \\ j\neq 0}}^{N} w_j \Big\{ (1+\delta_{0m}) \mathbf{P}_c^m(\tau, u_j, u_i) \mathbf{I}_c^m(\tau, u_j)$$
$$- \mathbf{P}_s^m(\tau, u_j, u_i) \mathbf{I}_s^m(\tau, u_j) \Big\} - \mathbf{Q}_c^m(\tau, u_i) \tag{488}$$

$$u_i \frac{d\mathbf{I}_s^m(\tau, u_i)}{d\tau} = \mathbf{I}_s^m(\tau, u_i) - \frac{\varpi(\tau)}{4} \sum_{\substack{j=-N \\ j\neq 0}}^{N} w_j \Big\{ \mathbf{P}_c^m(\tau, u_j, u_i) \mathbf{I}_s^m(\tau, u_j)$$
$$+ \mathbf{P}_s^m(\tau, u_j, u_i) \mathbf{I}_c^m(\tau, u_j) \Big\} - \mathbf{Q}_s^m(\tau, u_i). \tag{489}$$

Equations (488) and (489) apply to either of the slabs except that the source terms $\mathbf{Q}_\alpha^m(\tau, u_i)$ $\alpha = c, s$ are different in the upper and lower slabs [see Eqs. (376) and (381). In the following, we shall denote by N_1 the number of quadrature points per hemisphere in the upper slab, and by $N_2 > N_1$ those in the lower slab, as indicated in Section 4.1. Thus, there are a total of $2N_1$ *quadrature points* or "streams" in the upper slab and $2N_2$ streams in the lower slab.

As a consequence of certain symmetry relations, the Fourier components of the phase matrix have the following form [81]:

$$\mathbf{P}_c^m = \begin{pmatrix} P_{11c}^m & P_{12c}^m & 0 & 0 \\ P_{21c}^m & P_{22c}^m & 0 & 0 \\ 0 & 0 & P_{33c}^m & P_{34c}^m \\ 0 & 0 & P_{43c}^m & P_{44c}^m \end{pmatrix} \equiv \begin{pmatrix} \mathbf{P}_{1c}^m & \mathbf{0} \\ \mathbf{0} & \mathbf{P}_{2c}^m \end{pmatrix} \tag{490}$$

$$\mathbf{P}_s^m = \begin{pmatrix} 0 & 0 & P_{13s}^m & P_{14s}^m \\ 0 & 0 & P_{23s}^m & P_{24s}^m \\ P_{31s}^m & P_{32s}^m & 0 & 0 \\ P_{41s}^m & P_{42s}^m & 0 & 0 \end{pmatrix} \equiv \begin{pmatrix} \mathbf{0} & \mathbf{P}_{1s}^m \\ \mathbf{P}_{2s}^m & \mathbf{0} \end{pmatrix} \tag{491}$$

where we have introduced (2×2) block matrices for notational convenience.

4.2.1
"Cosine" Solutions

As discussed in Section 3.4.4, by combining cosine modes of the first two *Stokes parameters* ($I_\parallel$ and $I_\perp$) with the sine modes of the third and fourth Stokes parameters (U and V) in Eqs. (488) and (489), one obtains a set of coupled differential equations for the half-range $4N$ vectors $\tilde{I}_c^{m\pm}(\tau, \mu_i)$, which may be written in matrix form as

$$\frac{d\tilde{\mathbf{I}}_c^{m-}}{d\tau} = \tilde{\mathbf{A}}_{11c}^m \tilde{\mathbf{I}}_c^{m-} + \tilde{\mathbf{A}}_{12c}^m \tilde{\mathbf{I}}_c^{m+} - \tilde{\mathbf{S}}_c^{m-} \tag{492}$$

$$\frac{d\tilde{\mathbf{I}}_c^{m+}}{d\tau} = \tilde{\mathbf{A}}_{21c}^m \tilde{\mathbf{I}}_c^{m-} + \tilde{\mathbf{A}}_{22c}^m \tilde{\mathbf{I}}_c^{m+} - \tilde{\mathbf{S}}_c^{m+} \tag{493}$$

where $\tilde{\mathbf{A}}_{11c}^m = -\mathbf{U}^{-1}(\mathbf{1} + \mathbf{A}_{11c}^m)$, $\tilde{\mathbf{A}}_{12c}^m = -\mathbf{U}^{-1}\mathbf{A}_{12c}^m$, $\tilde{\mathbf{A}}_{21c}^m = \mathbf{U}^{-1}\mathbf{A}_{21c}^m$, $\tilde{\mathbf{A}}_{22c}^m = \mathbf{U}^{-1}(\mathbf{1} + \mathbf{A}_{22c}^m)$, and $\tilde{\mathbf{S}}_c^{m\pm} = \mathbf{U}^{-1}\mathbf{S}_c^{m\pm}$. Here, $\mathbf{U}$ is a $4N \times 4N$ diagonal block matrix [see Eq. (447)], given by

$$\mathbf{U} = \text{diag}(\mathbf{U}_\perp, \mathbf{U}_\parallel, \mathbf{U}_U, \mathbf{U}_V)^T \tag{494}$$

where the $N \times N$ diagonal submatrices are given by

$$\mathbf{U}_N \equiv \mathbf{U}_\perp = \mathbf{U}_\parallel = \mathbf{U}_U = \mathbf{U}_V = \text{d}iag(\mu_1, \mu_2, \ldots, \mu_N)^T \tag{495}$$

and

$$\tilde{\mathbf{I}}_c^{m\pm} = \begin{pmatrix} I_{\parallel c}^m(\tau, \pm\mu_i) \\ I_{\perp c}^m(\tau, \pm\mu_i) \\ U_s^m(\tau, \pm\mu_i) \\ V_s^m(\tau, \pm\mu_i) \end{pmatrix}_{4N\times 1} \qquad \mathbf{S}_c^{m\pm} = \begin{pmatrix} Q_{\parallel c}^m(\tau, \pm\mu_i) \\ Q_{\perp c}^m(\tau, \pm\mu_i) \\ Q_{Us}^m(\tau, \pm\mu_i) \\ Q_{Vs}^m(\tau, \pm\mu_i) \end{pmatrix}_{4N\times 1} \tag{496}$$

and the $4N \times 4N$ matrices $\mathbf{A}_{11c}^m$, $\mathbf{A}_{12c}^m$, $\mathbf{A}_{21c}^m$ and $\mathbf{A}_{22c}^m$ are identical expressions evaluated at different pairs of angles:

$$\mathbf{A}_{11c}^m(\tau) \equiv -w_j \frac{\varpi(\tau)}{4} \begin{pmatrix} (1+\delta_{0m})\mathbf{P}_{1c}^m & -\mathbf{P}_{1s}^m \\ \mathbf{P}_{2s}^m & \mathbf{P}_{2c}^m \end{pmatrix} \Bigg|_{-\mu_i, -\mu_j;\ i,j = 1,\ldots,N} \tag{497}$$

$$\mathbf{A}_{12c}^m(\tau) \equiv -w_j \frac{\varpi(\tau)}{4} \begin{pmatrix} (1+\delta_{0m})\mathbf{P}_{1c}^m & -\mathbf{P}_{1s}^m \\ \mathbf{P}_{2s}^m & \mathbf{P}_{2c}^m \end{pmatrix} \Bigg|_{-\mu_i, +\mu_j;\ i,j = 1,\ldots,N} \tag{498}$$

$$\mathbf{A}_{21c}^m(\tau) \equiv -w_j \frac{\varpi(\tau)}{4} \begin{pmatrix} (1+\delta_{0m})\mathbf{P}_{1c}^m & -\mathbf{P}_{1s}^m \\ \mathbf{P}_{2s}^m & \mathbf{P}_{2c}^m \end{pmatrix} \Bigg|_{+\mu_i, -\mu_j;\ i,j = 1,\ldots,N} \tag{499}$$

$$\mathbf{A}_{22c}^m(\tau) \equiv -w_j \frac{\varpi(\tau)}{4} \begin{pmatrix} (1+\delta_{0m})\mathbf{P}_{1c}^m & -\mathbf{P}_{1s}^m \\ \mathbf{P}_{2s}^m & \mathbf{P}_{2c}^m \end{pmatrix} \Bigg|_{+\mu_i, +\mu_j;\ i,j = 1,\ldots,N} \tag{500}$$

where we have used $w_{-j} = w_j$. Again, we should emphasize that the development so far is generic so that Eqs. (492) and (493) apply in both slabs except that $N = N_1$ in the upper slab and $N = N_2 > N_1$ in the lower slab. To further distinguish between the two slabs, we shall denote the *quadrature points* in the upper slab by μ^a, where the superscript a stands for atmosphere, and those in the lower slab by μ^w, where the superscript w stands for water, as already indicated in Section 4.1.

We also showed in Section 3.4.4 that we may combine the vectors for the downward and upward directions in Eqs. (492) and (493) by defining the $8N$

vectors, $i = 1, \ldots, N$, $\tilde{\mathbf{I}}_c^m \equiv [\tilde{\mathbf{I}}_c^{m-}(\mu_i), \tilde{\mathbf{I}}_c^{m+}(\mu_i)]^T$, and $\tilde{\mathbf{S}}_c^m \equiv [\tilde{\mathbf{S}}_c^{m-}(\mu_i), \tilde{\mathbf{S}}_c^{m+}(\mu_i)]^T$, so that we may write the coupled system of differential equations for these "cosine" solutions as

$$\frac{d\tilde{\mathbf{I}}_c^m}{d\tau} = \tilde{\mathbf{A}}_c^m \tilde{\mathbf{I}}_c^m - \tilde{\mathbf{S}}_c^m \tag{501}$$

where the matrix $\tilde{\mathbf{A}}_c^m$ is given by

$$\tilde{\mathbf{A}}_c^m \equiv \begin{pmatrix} \tilde{\mathbf{A}}_{11c}^m & \tilde{\mathbf{A}}_{12c}^m \\ \tilde{\mathbf{A}}_{21c}^m & \tilde{\mathbf{A}}_{22c}^m \end{pmatrix}_{8N\times 8N} . \tag{502}$$

We arrived at Eq. (501) by combining the cosine modes of the first two *Stokes parameters* ($I_\|$ and $I_\perp$) with the sine modes of the third and fourth Stokes parameters (U and V) in Eqs. (488) and (489), and then by combining the resulting Eqs. (492) and (493) for the upper and lower hemispheres into a single equation for the $8N$ Stokes vector.

Remark: If the incident beam has no U, V components, as for the disk-integrated solar pseudo-source, then the homogeneous "cosine" solutions obtained by solving Eq. (501) completely describe the homogeneous solutions because the "sine" solutions described below vanish.

4.2.2
"Sine" Solutions

In a similar manner, by combining the sine modes of the first two Stokes parameters ($I_\|$ and $I_\perp$) with the cosine modes of the third and fourth Stokes parameters (U and V) in Eqs. (488) and (489), one obtains a set of coupled differential equations for the $4N$ vectors $\tilde{\mathbf{I}}_s^{m\pm} \equiv [I_{\|s}^m(\tau, \pm\mu_i), I_{\perp s}^m(\tau, \pm\mu_i), U_c^m(\tau, \pm\mu_i), V_c^m(\tau, \pm\mu_i)]^T$ given by [see Eqs. (448) and (449) and Eqs. (451) and (452)]

$$\frac{d\tilde{\mathbf{I}}_s^{m-}}{d\tau} = \tilde{\mathbf{A}}_{11s}^m \tilde{\mathbf{I}}_s^{m-} + \tilde{\mathbf{A}}_{12s}^m \tilde{\mathbf{I}}_s^{m+} - \tilde{\mathbf{S}}_s^{m-} \tag{503}$$

$$\frac{d\tilde{\mathbf{I}}_s^{m+}}{d\tau} = \tilde{\mathbf{A}}_{21s}^m \tilde{\mathbf{I}}_s^{m-} + \mathbf{A}_{22s}^m \tilde{\mathbf{I}}_s^{m+} - \tilde{\mathbf{S}}_s^{m+} \tag{504}$$

where $\tilde{\mathbf{A}}_{11s}^m = -\mathbf{U}^{-1}(1 + \mathbf{A}_{11s}^m)$, $\tilde{\mathbf{A}}_{12s}^m = -\mathbf{U}^{-1}\mathbf{A}_{12s}^m$, $\tilde{\mathbf{A}}_{21s}^m = \mathbf{U}^{-1}\mathbf{A}_{21s}^m$, $\tilde{\mathbf{A}}_{22s}^m = \mathbf{U}^{-1}(1 + \mathbf{A}_{22s}^m)$, and $\tilde{\mathbf{S}}_s^{m\pm} = \mathbf{U}^{-1}\mathbf{S}_s^{m\pm}$. The matrix $\mathbf{U}$ is the $4N \times 4N$ diagonal matrix given by Eq. (494) and

$$\tilde{\mathbf{I}}_s^{m\pm} = \begin{pmatrix} I_{\|s}^m(\tau, \pm\mu_i) \\ I_{\perp s}^m(\tau, \pm\mu_i) \\ U_c^m(\tau, \pm\mu_i) \\ V_c^m(\tau, \pm\mu_i) \end{pmatrix}_{4N\times 1} \quad \mathbf{S}_s^{m\pm} = \begin{pmatrix} Q_{\|s}^m(\tau, \pm\mu_i) \\ Q_{\perp s}^m(\tau, \pm\mu_i) \\ Q_{Uc}^m(\tau, \pm\mu_i) \\ Q_{Vc}^m(\tau, \pm\mu_i) \end{pmatrix}_{4N\times 1} . \tag{505}$$

The $4N \times 4N$ matrices $\mathbf{A}_{11s}^m$, $\mathbf{A}_{12s}^m$, $\mathbf{A}_{21s}^m$, and $\mathbf{A}_{22s}^m$ are identical expressions evaluated at different pairs of angles:

$$\mathbf{A}_{11s}^m(\tau) \equiv -w_j \frac{\varpi(\tau)}{4} \begin{pmatrix} \mathbf{P}_{1c}^m & \mathbf{P}_{1s}^m \\ -\mathbf{P}_{2s}^m & (1 + \delta_{0m})\mathbf{P}_{2c}^m \end{pmatrix} \Bigg|_{-\mu_i, -\mu_j;\ i,j = 1,\ldots,N} \tag{506}$$

$$\mathbf{A}^m_{12s}(\tau) \equiv -w_j \frac{\varpi(\tau)}{4} \begin{pmatrix} \mathbf{P}^m_{1c} & \mathbf{P}^m_{1s} \\ -\mathbf{P}^m_{2s} & (1+\delta_{0m})\mathbf{P}^m_{2c} \end{pmatrix} \Bigg|_{-\mu_i,+\mu_j;\ i,j=1,\ldots,N} \tag{507}$$

$$\mathbf{A}^m_{21s}(\tau) \equiv -w_j \frac{\varpi(\tau)}{4} \begin{pmatrix} \mathbf{P}^m_{1c} & \mathbf{P}^m_{1s} \\ -\mathbf{P}^m_{2s} & (1+\delta_{0m})\mathbf{P}^m_{2c} \end{pmatrix} \Bigg|_{+\mu_i,-\mu_j;\ i,j=1,\ldots,N} \tag{508}$$

$$\mathbf{A}^m_{22s}(\tau) \equiv -w_j \frac{\varpi(\tau)}{4} \begin{pmatrix} \mathbf{P}^m_{1c} & \mathbf{P}^m_{1s} \\ -\mathbf{P}^m_{2s} & (1+\delta_{0m})\mathbf{P}^m_{2c} \end{pmatrix} \Bigg|_{+\mu_i,+\mu_j;\ i,j=1,\ldots,N} . \tag{509}$$

Combining Eqs. (503) and (504), we obtain the following equation for the $8N$ *Stokes vector*:

$$\frac{d\tilde{\mathbf{I}}^m_s}{d\tau} = \tilde{\mathbf{A}}^m_s \tilde{\mathbf{I}}^m_s - \tilde{\mathbf{S}}^m_s \tag{510}$$

which is of a form identical to Eq. (501). Here,

$$\tilde{\mathbf{A}}^m_s \equiv \begin{pmatrix} \tilde{\mathbf{A}}^m_{11s} & \tilde{\mathbf{A}}^m_{12s} \\ \tilde{\mathbf{A}}^m_{21s} & \tilde{\mathbf{A}}^m_{22s} \end{pmatrix}_{8N\times 8N} ; \quad \tilde{\mathbf{I}}^m_s \equiv \begin{pmatrix} \tilde{\mathbf{I}}^{m-}_s \\ \tilde{\mathbf{I}}^{m+}_s \end{pmatrix}_{8N\times 1} ; \quad \tilde{\mathbf{S}}^m_s \equiv \begin{pmatrix} \tilde{\mathbf{S}}^{m-}_s \\ \tilde{\mathbf{S}}^{m+}_s \end{pmatrix}_{8N\times 1} .$$

The "sine" solutions found by solving Eq. (510) are necessary for beam sources with nonzero U, V components.

4.3 Discrete-Ordinate Solutions

The *vector RTEs* [Eqs. (501) and (510)] have the same form in the upper slab (atmosphere) and the lower slab (water). Thus, we may rewrite Eqs. (501) and (510) as

$$\frac{d\tilde{\mathbf{I}}^m_{\alpha p}(\tau)}{d\tau} = \tilde{\mathbf{A}}^m_{\alpha p}(\tau)\tilde{\mathbf{I}}^m_{\alpha p}(\tau) - \tilde{\mathbf{S}}^m_{\alpha p}(\tau) \qquad \alpha = c, s, \qquad p = 1, 2. \tag{511}$$

4.3.1 Homogeneous Solution

For a homogeneous slab with a constant *single-scattering albedo* $\varpi(\tau) = \varpi$ and phase matrix $\mathbf{P}(\tau, u', u) = \mathbf{P}(u', u)$, the matrix $\tilde{\mathbf{A}}^m_{\alpha p}(\tau)$ will also be independent of position τ in the slab. Therefore, we may seek solutions of the homogeneous version of Eq. (511) (setting $\tilde{\mathbf{S}}^m_\alpha(\tau) = 0$) of the following form (dropping the subscript p for convenience):

$$\tilde{\mathbf{I}}^m_\alpha(\tau) = \mathbf{G}^m_\alpha e^{-\lambda^m \tau}. \tag{512}$$

Substituting this expression into Eq. (511) with $\tilde{\mathbf{S}}^m_\alpha(\tau) = 0$, we find

$$\tilde{\mathbf{A}}^m_\alpha \mathbf{G}^m_\alpha = -\lambda^m \mathbf{G}^m_\alpha . \tag{513}$$

Solving this *algebraic eigenvalue problem*, we get $8N$ eigenvalues $\lambda^m_{\alpha 1}, \ldots, \lambda^m_{\alpha 8N}$ with corresponding eigenvectors $\mathbf{G}^m_{\alpha 1}, \ldots, \mathbf{G}^m_{\alpha 8N}$. The general solution to the homogeneous version of Eq. (511) is a linear combination of these $8N$ linearly independent solutions of the form given by Eq. (512).

Reduction of Order

It is possible to reduce the size of the system of equations so that one may solve an algebraic eigenvalue problem for a matrix having half the dimension of the original system. For the *scalar RTE* pertinent to *unpolarized radiation*, the reduction in size is from $2N$ to N, where N is the number of *quadrature points* in each hemisphere, while for the *vector RTE* the corresponding reduction is from $8N$ to $4N$. From a computational point of view, this reduction implies that we will save a factor of about 8 in computing time because the computational effort required to solve an algebraic eigenvalue problem scales as N^3 for a square matrix of dimension N. Thus, a reduction in size by a factor of 2 leads to a computational savings of order $2^3 = 8$. For the unpolarized (scalar) case, such a *reduction of order* was first accomplished by Stamnes and Swanson [172], while for the polarized (vector) case it was developed by Siewert [83]. Here we review this approach, starting with the unpolarized case, and then proceed to generalize it to include polarization.

Unpolarized Case

The homogeneous version of the *unpolarized (scalar) RTE* can be written as (dropping the m-superscripts)

$$-\mu_i \frac{dI^-(\tau, \mu_i)}{d\tau} = I^-(\tau, \mu_i) - \sum_{j=1}^{N} w_j [D(\tau, -\mu_j, -\mu_i) I^-(\tau, \mu_j) + D(\tau, \mu_j, -\mu_i) I^+(\tau, \mu_j)] \tag{514}$$

$$\mu_i \frac{dI^+(\tau, \mu_i)}{d\tau} = I^+(\tau, \mu_i) - \sum_{j=1}^{N} w_j [D(\tau, -\mu_j, \mu_i) I^-(\tau, \mu_j) + D(\tau, \mu_j, \mu_i) I^+(\tau, \mu_j)] \tag{515}$$

where we have defined ($i, j = 1, \ldots, N$):

$$D(\tau, \pm\mu_j, \pm\mu_i) \equiv D^m(\tau, \pm\mu_j, \pm\mu_i) = \frac{\varpi(\tau)}{2} p^m(\tau, \pm\mu_j, \pm\mu_i) = \frac{\varpi(\tau)}{2} \sum_{\ell=m}^{2N-1} (2\ell + 1) \chi_\ell(\tau) \Lambda^m_\ell(\pm\mu_j) \Lambda^m_\ell(\pm\mu_i). \tag{516}$$

Equations (515) and (514) may be written in matrix form as

$$\frac{d\mathbf{I}^-}{d\tau} = \mathbf{A}_{11}\mathbf{I}^- + \mathbf{A}_{12}\mathbf{I}^+ \tag{517}$$

$$\frac{d\mathbf{I}^+}{d\tau} = -\mathbf{A}_{12}\mathbf{I}^- - \mathbf{A}_{11}\mathbf{I}^+ \tag{518}$$

where $\mathbf{A}_{11} = -\mathbf{U}_N^{-1}(\mathbf{1} + \mathbf{W}\mathbf{D}^+)$, $\mathbf{A}_{12} = -\mathbf{U}_N^{-1}\mathbf{W}\mathbf{D}^-$, $\mathbf{A}_{21} = -\mathbf{U}_N^{-1}\mathbf{W}\mathbf{D}^- = -\mathbf{A}_{12}$, $\mathbf{A}_{22} = \mathbf{U}_N^{-1}(\mathbf{1} + \mathbf{W}\mathbf{D}^+) = -\mathbf{A}_{11}$, and where the relations $\mathbf{A}_{22} = -\mathbf{A}_{11}$ and $\mathbf{A}_{21} = -\mathbf{A}_{12}$ are due to the symmetry of the matrices $\mathbf{D}^{\pm}$ in Eqs. (520) and (521) below, which in turn reflects a corresponding symmetry of the *scattering phase function*. Here, $\mathbf{1}$ is the $N \times N$ identity matrix, $\mathbf{U}_N$ is the $N \times N$ diagonal matrix containing the *quadrature points* given by Eq. (495), and

$$\mathbf{W} = \text{diag}\{w_1, \dots, w_N\} \tag{519}$$

$$\mathbf{D}^+ = [D(\mu_i, \mu_j)] = [D(-\mu_i, -\mu_j)] \qquad i, j = 1, \dots, N \tag{520}$$

$$\mathbf{D}^- = [D(\mu_i, -\mu_j)] = [D(-\mu_i, \mu_j)] \qquad i, j = 1, \dots, N. \tag{521}$$

Seeking solutions to Eqs. (517) and (518) of the form

$$\mathbf{I}^{\pm} = \mathbf{I}(\tau, \pm\mu_i) = \mathbf{g}^{\pm}(k, \mu_i)e^{-k\tau} \qquad i = 1, \dots, N \tag{522}$$

one finds

$$-k\mathbf{g}^- = \mathbf{A}_{11}\,\mathbf{g}^- + \mathbf{A}_{12}\,\mathbf{g}^+ \tag{523}$$

$$-k\mathbf{g}^+ = -\mathbf{A}_{12}\,\mathbf{g}^- - \mathbf{A}_{11}\,\mathbf{g}^+ \tag{524}$$

which upon addition and subtraction of Eqs. (523) and (524) yields

$$-k(\mathbf{g}^- + \mathbf{g}^+) = (\mathbf{A}_{11} - \mathbf{A}_{12})(\mathbf{g}^- - \mathbf{g}^+) \tag{525}$$

$$-k(\mathbf{g}^- - \mathbf{g}^+) = (\mathbf{A}_{11} + \mathbf{A}_{12})(\mathbf{g}^- + \mathbf{g}^+). \tag{526}$$

Next, one may combine Eqs. (525) and (526) to obtain

$$(\mathbf{A}_{11} - \mathbf{A}_{12})(\mathbf{A}_{11} + \mathbf{A}_{12})(\mathbf{g}^- + \mathbf{g}^+) = k^2(\mathbf{g}^- + \mathbf{g}^+) \tag{527}$$

$$(\mathbf{A}_{11} + \mathbf{A}_{12})(\mathbf{A}_{11} - \mathbf{A}_{12})(\mathbf{g}^- - \mathbf{g}^+) = k^2(\mathbf{g}^- - \mathbf{g}^+). \tag{528}$$

Solving the eigenvalue problem given by Eq. (527), one obtains the eigenvalues k^2 and the corresponding eigenvectors

$$\mathbf{g}^+ + \mathbf{g}^- \equiv \tilde{\gamma} \tag{529}$$

and using Eq. (526) we have

$$\mathbf{g}^+ - \mathbf{g}^- = \frac{1}{k}(\mathbf{A}_{11} + \mathbf{A}_{12})\tilde{\gamma}. \tag{530}$$

Combining Eqs. (529) and (530), one finds

$$\mathbf{g}^+ = \frac{1}{2}\left[\tilde{\gamma} + \frac{1}{k}(\mathbf{A}_{11} + \mathbf{A}_{12})\tilde{\gamma}\right] \tag{531}$$

$$\mathbf{g}^- = \frac{1}{2}\left[\tilde{\gamma} - \frac{1}{k}(\mathbf{A}_{11} + \mathbf{A}_{12})\tilde{\gamma}\right]. \tag{532}$$

There is a total of N eigenvalues $\lambda_i = k_i^2, i = 1, \dots, N$ and hence a total of $2N$ eigenvalues occurring in positive/negative pairs:

$$k_j = \sqrt{\lambda_j} > 0, \quad j = 1, \dots, N \tag{533}$$

$$k_{-j} = -k_j, \quad j = 1, \dots, N. \tag{534}$$

Since the real matrix $(\mathbf{A}_{11} - \mathbf{A}_{12})(\mathbf{A}_{11} + \mathbf{A}_{12})$ in Eq. (527) or $(\mathbf{A}_{11} + \mathbf{A}_{12})(\mathbf{A}_{11} - \mathbf{A}_{12})$ in Eq. (528) can be made symmetric, the eigenvalues for the scalar case are real, as discussed by Stamnes *et al.* [208].

Polarized Case

In the vector case, one may start with Eqs. (492) and (493) for the cosine modes and Eqs. (503) and (504) for the sine modes, and rewrite the homogenous version of these equations as

$$\frac{d\tilde{\mathbf{I}}_{\alpha}^{m-}}{d\tau} = \tilde{\mathbf{A}}_{11\alpha}^{m}\,\tilde{\mathbf{I}}_{\alpha}^{m-} + \tilde{\mathbf{A}}_{12\alpha}^{m}\,\tilde{\mathbf{I}}_{\alpha}^{m+} \qquad \alpha = c, s \tag{535}$$

$$\frac{d\tilde{\mathbf{I}}_{\alpha}^{m+}}{d\tau} = \tilde{\mathbf{A}}_{21\alpha}^{m}\,\tilde{\mathbf{I}}_{\alpha}^{m-} + \tilde{\mathbf{A}}_{22\alpha}^{m}\,\tilde{\mathbf{I}}_{\alpha}^{m+} \qquad \alpha = c, s. \tag{536}$$

Dropping the superscript m and the subscript α, and seeking solutions to Eqs. (535) and (536) of the form

$$\tilde{\mathbf{I}}^{\pm}(\tau, \pm\mu_i) = \mathbf{G}^{\pm} e^{-k\tau} \qquad \mathbf{G}^{\pm} = \mathbf{G}(k, \pm\mu_i) \tag{537}$$

one finds, upon invoking the symmetry relations $\tilde{\mathbf{A}}_{21} = -\tilde{\mathbf{A}}_{12}$ and $\tilde{\mathbf{A}}_{22} = -\tilde{\mathbf{A}}_{11}$ (see Problem 4.1) implied by the symmetry of the phase matrix summarized in Section 2.3.3 ($i = 1, \ldots, N$)

$$-k\mathbf{G}^{-} = \tilde{\mathbf{A}}_{11}\,\mathbf{G}^{-} + \tilde{\mathbf{A}}_{12}\,\mathbf{G}^{+} \tag{538}$$

$$-k\mathbf{G}^{+} = -\tilde{\mathbf{A}}_{12}\,\mathbf{G}^{-} - \tilde{\mathbf{A}}_{11}\,\mathbf{G}^{+}. \tag{539}$$

Addition and subtraction of Eqs. (538) and (539) yields

$$-k(\mathbf{G}^{-} + \mathbf{G}^{+}) = (\tilde{\mathbf{A}}_{11} - \tilde{\mathbf{A}}_{12})(\mathbf{G}^{-} - \mathbf{G}^{+}) \tag{540}$$

$$-k(\mathbf{G}^{-} - \mathbf{G}^{+}) = (\tilde{\mathbf{A}}_{11} + \tilde{\mathbf{A}}_{12})(\mathbf{G}^{-} + \mathbf{G}^{+}) \tag{541}$$

which in combination gives

$$(\tilde{\mathbf{A}}_{11} - \tilde{\mathbf{A}}_{12})(\tilde{\mathbf{A}}_{11} + \tilde{\mathbf{A}}_{12})(\mathbf{G}^{-} + \mathbf{G}^{+}) = k^2(\mathbf{G}^{-} + \mathbf{G}^{+}) \tag{542}$$

$$(\tilde{\mathbf{A}}_{11} + \tilde{\mathbf{A}}_{12})(\tilde{\mathbf{A}}_{11} - \tilde{\mathbf{A}}_{12})(\mathbf{G}^{-} - \mathbf{G}^{+}) = k^2(\mathbf{G}^{-} - \mathbf{G}^{+}). \tag{543}$$

By solving the eigenvalue problem given by Eq. (542), one obtains the eigenvalues k^2 and the corresponding eigenvectors

$$\mathbf{G}^{+} + \mathbf{G}^{-} \equiv \mathbf{\Gamma} \tag{544}$$

so that Eq. (541) yields

$$\mathbf{G}^{+} - \mathbf{G}^{-} = \frac{1}{k}(\tilde{\mathbf{A}}_{11} + \tilde{\mathbf{A}}_{12})\mathbf{\Gamma}. \tag{545}$$

Combining Eqs. (544) and (545), one finds

$$\mathbf{G}^{+} = \frac{1}{2}\left[\mathbf{\Gamma} + \frac{1}{k}(\tilde{\mathbf{A}}_{11} + \tilde{\mathbf{A}}_{12})\mathbf{\Gamma}\right] \tag{546}$$

$$\mathbf{G}^{-} = \frac{1}{2}\left[\mathbf{\Gamma} - \frac{1}{k}(\tilde{\mathbf{A}}_{11} + \tilde{\mathbf{A}}_{12})\mathbf{\Gamma}\right] \tag{547}$$

which, on noting that

$$\mathbf{G}^{+}_{-j} = \mathbf{G}^{+}(-k_j) = \mathbf{G}^{-}(k_j) = \mathbf{G}^{-}_{j}, \qquad j = 1, 2, \ldots, 4N \tag{548}$$

so that $\mathbf{G}^{\pm}(+k_j) = \mathbf{G}^{\mp}(-k_j)$, completes the reduction of the order and provides all the information needed for the solution of Eqs. (535) and (536).

There are a total of $4N$ eigenvalues $\lambda_i = k_i^2, i = 1, \ldots, 4N$ and hence a total of $8N$ eigenvalues occurring in positive/negative pairs:

$$k_j = \sqrt{\lambda_j} > 0, \quad j = 1, \ldots, 4N \tag{549}$$

$$k_{-j} = -k_j, \quad j = 1, \ldots, 4N. \tag{550}$$

Since both real and complex *eigensolutions* may be present, the complex-variable eigensolver DGEEV of LAPACK (http://www.netlib.org/lapack/lapack-3.5.0.html) may be used to get the solutions. If the eigensolutions are known to be real (as for *Rayleigh scattering*), the faster routine ASYMTX available in *DISORT* [161, 162] may be used.

The solution to Eqs. (535) and (536) can be written as ($i = 1, \ldots, N$)

$$\mathbf{I}_h^{+}(\tau) = \mathbf{I}_h(\tau, +\mu_i) = \sum_{\substack{j=-4N \\ j\neq 0}}^{4N} C_j \mathbf{G}_j^{+} e^{-k_j\tau} = \sum_{j=1}^{4N} [C_j \mathbf{G}_j^{+} e^{-k_j\tau} + C_{-j}\mathbf{G}_j^{-} e^{+k_j\tau}] \tag{551}$$

and

$$\mathbf{I}_h^{-}(\tau) = \mathbf{I}_h(\tau, -\mu_i) = \sum_{\substack{j=-4N \\ j\neq 0}}^{4N} C_j \mathbf{G}_j^{-} e^{-k_j\tau} = \sum_{j=1}^{4N} [C_j \mathbf{G}_j^{-} e^{-k_j\tau} + C_{-j}\mathbf{G}_j^{+} e^{+k_j\tau}] \tag{552}$$

or more compactly as

$$\mathbf{I}_h^{\pm}(\tau) = \mathbf{I}_h(\tau, \pm\mu_i) = \sum_{j=1}^{4N} [C_j \mathbf{G}_j^{\pm} e^{-k_j\tau} + C_{-j}\mathbf{G}_j^{\mp} e^{+k_j\tau}] \tag{553}$$

where $\{C_j\}$ and $\{C_{-j}\}$ are constants of integration to be determined by the *boundary conditions*.

Since both the eigenvalues $\{k_j\}$ and the eigenvectors $\mathbf{G}^{\pm}(k_j) = \mathbf{G}^{\mp}(-k_j)$ may be complex, it is desirable to rewrite Eqs. (551) and (552) in terms of real quantities. To that end, one may distinguish between real and complex eigenvalues $\{k_j\}$. The eigenvalues (and the associated eigenvectors) of a real matrix are either real or they occur in complex conjugate pairs [209]. Thus, letting J_r denote the number of real eigenvalues and J_x denote the number of complex conjugate pairs of eigenvalues, one may write Eqs. (551) and (552) as (see Problem 4.2)

$$\mathbf{I}_h^{\pm}(\tau) = \mathbf{I}_{r,h}^{\pm}(\tau) + \Re\{\mathbf{I}_{x,h}^{\pm}(\tau)\} \tag{554}$$

where the real-valued solutions are

$$\mathbf{I}_{r,h}^{+}(\tau) = \sum_{j=1}^{J_r} [C_j \mathbf{G}_j^{+} e^{-k_j\tau} + C_{-j}\mathbf{G}_j^{-} e^{+k_j\tau}] \tag{555}$$

$$\mathbf{I}^-_{r,h}(\tau) = \sum_{j=1}^{J_r}[C_j\mathbf{G}^-_j e^{-k_j\tau} + C_{-j}\mathbf{G}^+_j e^{+k_j\tau}] \tag{556}$$

and the complex-valued solutions are

$$\mathbf{I}^+_{x,h}(\tau) = \sum_{j=1}^{J_x}[C_{x,j}\mathbf{G}^+_{x,j}e^{-k_{x,j}\tau} + C_{x,-j}\mathbf{G}^-_{x,j}e^{+k_{x,j}\tau}] \tag{557}$$

$$\mathbf{I}^-_{x,h}(\tau) = \sum_{j=1}^{J_x}[C_{x,j}\mathbf{G}^-_{x,j}e^{-k_{x,j}\tau} + C_{x,-j}\mathbf{G}^+_{x,j}e^{+k_{x,j}\tau}] \tag{558}$$

where the subscript x (last letter in the word complex) is used to indicate that a quantity is a complex number or eigenvector. Taking the real parts of Eqs. (557) and (558), we find ($C_{x,j} = C_{R,j} + iC_{I,j}$, see Problem 4.2)

$$\begin{aligned}\Re\{\mathbf{I}^+_{x,h}(\tau)\} = \sum_{j=1}^{J_x}\Big\{ & C_{R,j}\mathbf{F}^+_R(\tau,k_j) + C_{R,-j}\mathbf{F}^-_R(\tau,-k_j) \\ & - C_{I,j}\mathbf{F}^+_I(\tau,k_j) - C_{I,-j}\mathbf{F}^-_I(\tau,-k_j)\Big\}\end{aligned} \tag{559}$$

$$\begin{aligned}\Re\{\mathbf{I}^-_{x,h}(\tau)\} = \sum_{j=1}^{J_x}\Big\{ & C_{R,j}\mathbf{F}^-_R(\tau,k_j) + C_{R,-j}\mathbf{F}^+_R(\tau,-k_j) \\ & - C_{I,j}\mathbf{F}^-_I(\tau,k_j) - C_{I,-j}\mathbf{F}^+_I(\tau,-k_j)\Big\}\end{aligned} \tag{560}$$

where

$$\mathbf{F}^\pm_R(\tau,\pm k_j) \equiv \left[\Re\{\mathbf{G}^\pm_{x,j}\}\Re\{e^{\mp k_{x,j}\tau}\} - \Im\{\mathbf{G}^\pm_{x,j}\}\Im\{e^{\mp k_{x,j}\tau}\}\right] \tag{561}$$

$$\mathbf{F}^\pm_I(\tau,\pm k_j) \equiv \left[\Re\{\mathbf{G}^\pm_{x,j}\}\Im\{e^{\mp k_{x,j}\tau}\} + \Im\{\mathbf{G}^\pm_{x,j}\}\Re\{e^{\mp k_{x,j}\tau}\}\right]. \tag{562}$$

Note that all the $8N = 2(J_r + 2J_x)$ constants ($2J_r$ values of $C_{\pm j}$, and $2J_x$ values each of $C_{R,\pm j}$ and $C_{I,\pm j}$) in Eqs. (555)–(562) are to be determined by applying the *boundary conditions*.

4.3.2 Vertically Inhomogeneous Media

For *vertically inhomogeneous media*, one may divide the upper slab (atmosphere) into L_1 adjacent homogeneous layers labeled by the index $\ell = 1, \ldots, L_1$. Likewise, one may divide the lower slab (water) into L_2 homogeneous layers with index $\ell = L_1 + 1, \ldots, L_1 + L_2$. In both media, the IOPs (the phase matrix and the single scattering albedo) are constant within each layer, but they are allowed to vary from layer to layer as required to adequately resolve the vertical variation in the *IOPs* within each slab, as illustrated in Figure 27. Then, the solution arrived at above [Eq. (512)] will be applicable to each layer separately. The exposition may

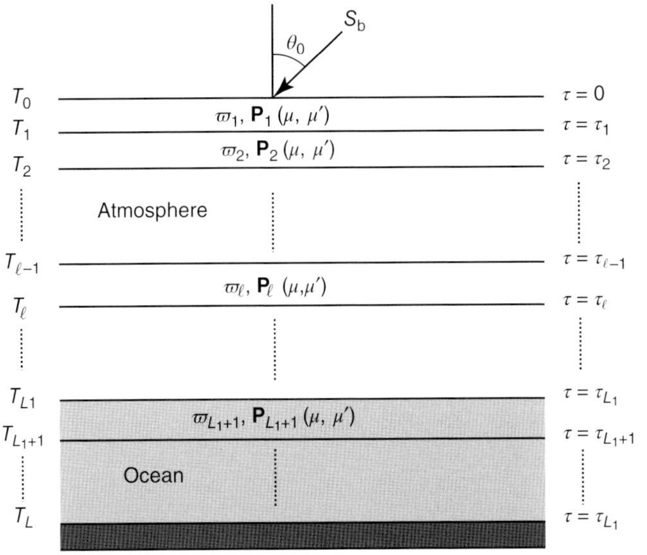

Figure 27 Schematic illustration of two vertically inhomogeneous slabs separated by an interface across which the refractive index changes abruptly like in an atmosphere–water system.

be simplified by rewriting the homogeneous solution given by Eq. (553) for mode $\alpha = c$ or s and layer denoted by ℓ as

$$\mathbf{I}_{\alpha\ell}^{m\pm,h}(\tau) = \sum_{j=1}^{4N} [C_{-\alpha j\ell}^{m} \mathbf{G}_{-\alpha j\ell}^{m\pm} e^{+k_{\alpha j\ell}^{m}\tau} + C_{\alpha j\ell}^{m} \mathbf{G}_{\alpha j\ell}^{m\pm} e^{-k_{\alpha j\ell}^{m}\tau}]. \tag{563}$$

Defining $\tilde{\mathbf{I}}_{\alpha\ell}^{m}(\tau) \equiv [\mathbf{I}_{\alpha\ell}^{m-}(\tau), \mathbf{I}_{\alpha\ell}^{m+}(\tau)]^T$, $\tilde{C}_{\alpha j\ell} \equiv [C_{-\alpha j\ell}, C_{\alpha j\ell}]^T$, $\tilde{\mathbf{G}}_{\alpha j\ell} \equiv [\mathbf{G}_{-\alpha j\ell}, \mathbf{G}_{\alpha j\ell}]^T$, one may write Eq. (563) representing the solution to the homogeneous part of the RTE for a Fourier mode denoted by the subscript α and a component denoted by the superscript m in a layer denoted by subscript ℓ more compactly as

$$\tilde{\mathbf{I}}_{\alpha\ell}^{m,h}(\tau) = \sum_{j=1}^{8N} \tilde{C}_{\alpha j\ell}^{m} \tilde{\mathbf{G}}_{\alpha j\ell}^{m} e^{-k_{\alpha j\ell}^{m}\tau} \qquad \ell = 1, \ldots, L_1 + L_2. \tag{564}$$

Note that this *generic homogeneous solution* applies in any layer of each slab, but that in the upper slab $N = N_1$, while in the lower slab $N = N_2 > N_1$.

4.3.3
Particular Solution – Upper Slab

The source term $\tilde{\mathbf{S}}_{\alpha p}^{m}(\tau)$ in the VRTE given by Eq. (511) is different in the upper slab ($p = 1$) and the lower slab ($p = 2$). For the cosine modes in the upper slab, the source term is (dropping the subscript $p = 1$)

$$\tilde{\mathbf{S}}_c^m \equiv \begin{pmatrix} \tilde{\mathbf{S}}_c^{m-} \\ \tilde{\mathbf{S}}_c^{m+} \end{pmatrix}_{8N\times 1} \quad ; \quad \tilde{\mathbf{S}}_c^{m\pm} = \begin{pmatrix} \tilde{Q}_{\|c}^m(\tau, \pm\mu_i) \\ \tilde{Q}_{\perp c}^m(\tau, \pm\mu_i) \\ \tilde{Q}_{Us}^m(\tau, \pm\mu_i) \\ \tilde{Q}_{Vs}^m(\tau, \pm\mu_i) \end{pmatrix}_{4N\times 1}$$

and for the sine modes

$$\tilde{\mathbf{S}}_s^m \equiv \begin{pmatrix} \tilde{\mathbf{S}}_s^{m-} \\ \tilde{\mathbf{S}}_s^{m+} \end{pmatrix}_{8N\times 1} \quad ; \quad \tilde{\mathbf{S}}_s^{m\pm} = \begin{pmatrix} \tilde{Q}_{\|s}^m(\tau, \pm\mu_i) \\ \tilde{Q}_{\perp s}^m(\tau, \pm\mu_i) \\ \tilde{Q}_{Uc}^m(\tau, \pm\mu_i) \\ \tilde{Q}_{Vc}^m(\tau, \pm\mu_i) \end{pmatrix}_{4N\times 1}$$

where (see Eq. (402)$\alpha = c, s$) $\tilde{\mathbf{Q}}_\alpha^{m\pm}(\tau, \mu_i) = \mathbf{U}^{-1}\mathbf{Q}_\alpha^{m\pm}(\tau, \mu_i)$ and

$$\mathbf{Q}_\alpha^{m\pm}(\tau, \mu^a) = \mathbf{X}_{\alpha,0}^{m\pm}(\tau, \mu^a)\, e^{-\tau/\mu_0} + \mathbf{X}_{\alpha,1}^{m\pm}(\tau, \mu^a)\, e^{\tau/\mu_0} + \delta_{c\alpha}\, \mathbf{B}_{0m}(\tau). \tag{565}$$

Here [see Eqs. (404)–(407)] $\mathbf{X}_{\alpha,0}^{m\pm}(\tau, \mu^a) = \frac{\varpi(\tau)}{4\pi}\, \mathbf{P}_\alpha^m(\tau, -\mu_0, \pm\mu^a)\, \mathbf{S}_b$ and $\mathbf{X}_{\alpha,1}^{m\pm}(\tau, \mu^a) = \frac{\varpi(\tau)}{4\pi}\, \mathbf{P}_\alpha^m(\tau, -\mu_0, \pm\mu^a)\mathbf{R}_F(-\mu_0, m_{\mathrm{rel}})\, \mathbf{S}_b\, e^{-2\tau_a/\mu_0}$. Thus, the source term $\tilde{\mathbf{S}}_{\alpha p}^m(\tau)$ in Eq. (511) can be written (dropping the subscript p) as

$$\tilde{\mathbf{S}}_\alpha^m(\tau) = \mathbf{U}^{-1}[\tilde{\mathbf{X}}_{\alpha,0}^m(\tau)\, e^{-\tau/\mu_0} + \tilde{\mathbf{X}}_{\alpha,1}^m(\tau)\, e^{\tau/\mu_0} + \delta_{c\alpha}\, \mathbf{B}_{0m}(\tau)] \tag{566}$$

where $\tilde{\mathbf{X}}_{\alpha,i}^m(\tau) = [\mathbf{X}_{\alpha,i}^{m-}(\tau, \mu^a), \mathbf{X}_{\alpha,i}^{m+}(\tau, \mu^a)]^T$, $i = 0, 1$, and where $i = 0$ represents the downward beam, and $i = 1$ represents the upward reflected beam (see Section 3.3.1).

Cosine Modes

Combining the two upper rows of the matrix $\mathbf{P}_c^m$ and the two lower rows of the matrix $\mathbf{P}_s^m$ given by [see Eqs. (490) and (491)]

$$\mathbf{P}_c^m = \begin{pmatrix} P_{11c}^m & P_{12c}^m & 0 & 0 \\ P_{21c}^m & P_{22c}^m & 0 & 0 \\ 0 & 0 & P_{33c}^m & P_{34c}^m \\ 0 & 0 & P_{43c}^m & P_{44c}^m \end{pmatrix} \; ; \; \mathbf{P}_s^m = \begin{pmatrix} 0 & 0 & P_{13s}^m & P_{14s}^m \\ 0 & 0 & P_{23s}^m & P_{24s}^m \\ P_{31s}^m & P_{32s}^m & 0 & 0 \\ P_{41s}^m & P_{42s}^m & 0 & 0 \end{pmatrix}$$

one obtains a matrix $\tilde{\mathbf{P}}_c^m$ and a corresponding vector $\tilde{\mathbf{S}}_b$ defined as

$$\tilde{\mathbf{P}}_c^m = \begin{pmatrix} P_{11c}^m & P_{12c}^m & 0 & 0 \\ P_{21c}^m & P_{22c}^m & 0 & 0 \\ P_{31s}^m & P_{32s}^m & 0 & 0 \\ P_{41s}^m & P_{42s}^m & 0 & 0 \end{pmatrix}_{8N_1\times 8N_1} \quad ; \quad \tilde{\mathbf{S}}_b = \begin{pmatrix} S_{\|b} \\ S_{\perp b} \\ S_{Ub} \\ S_{Vb} \end{pmatrix}_{8N_1\times 1} . \tag{567}$$

Hence, for the cosine modes we obtain

$$\tilde{\mathbf{X}}^m_{c,0} = \frac{\varpi(\tau)}{4\pi}\begin{pmatrix} P^m_{11c} & P^m_{12c} & 0 & 0 \\ P^m_{21c} & P^m_{22c} & 0 & 0 \\ P^m_{31s} & P^m_{32s} & 0 & 0 \\ P^m_{41s} & P^m_{42s} & 0 & 0 \end{pmatrix}_{8N_1\times 8N_1}\begin{pmatrix} S_{\|b} \\ S_{\perp b} \\ S_{Ub} \\ S_{Vb}\end{pmatrix}_{8N_1\times 1}$$

$$= \frac{\varpi(\tau)}{4\pi}\tilde{\mathbf{P}}^m_c\tilde{\mathbf{S}}_b = \frac{\varpi(\tau)}{4\pi}\tilde{\mathbf{P}}^m_{c,b} \tag{568}$$

and

$$\tilde{\mathbf{X}}^m_{c,1} = \frac{\varpi(\tau)\mathbf{R}_F}{4\pi}\tilde{\mathbf{P}}^m_{c,b}\, e^{-2\tau_a/\mu_0} \tag{569}$$

where the vector $\tilde{\mathbf{P}}^m_{c,b} \equiv \tilde{\mathbf{P}}^m_c\tilde{\mathbf{S}}_b$ is given by

$$\tilde{\mathbf{P}}^m_{b,c} = \begin{pmatrix} P^m_{11c}(\tau,-\mu_0,-\mu_i^a)S_{\|b} + P^m_{12c}(\tau,-\mu_0,-\mu_i^a)S_{\perp b} \\ P^m_{21c}(\tau,-\mu_0,-\mu_i^a)S_{\|b} + P^m_{22c}(\tau,-\mu_0,-\mu_i^a)S_{\perp b} \\ P^m_{31s}(\tau,-\mu_0,-\mu_i^a)S_{\|b} + P^m_{32s}(\tau,-\mu_0,-\mu_i^a)S_{\perp b} \\ P^m_{41s}(\tau,-\mu_0,-\mu_i^a)S_{\|b} + P^m_{42s}(\tau,-\mu_0,-\mu_i^a)S_{\perp b} \\ P^m_{11c}(\tau,-\mu_0,+\mu_i^a)S_{\|b} + P^m_{12c}(\tau,-\mu_0,+\mu_i^a)S_{\perp b} \\ P^m_{21c}(\tau,-\mu_0,+\mu_i^a)S_{\|b} + P^m_{22c}(\tau,-\mu_0,+\mu_i^a)S_{\perp b} \\ P^m_{31s}(\tau,-\mu_0,+\mu_i^a)S_{\|b} + P^m_{32s}(\tau,-\mu_0,+\mu_i^a)S_{\perp b} \\ P^m_{41s}(\tau,-\mu_0,+\mu_i^a)S_{\|b} + P^m_{42s}(\tau,-\mu_0,+\mu_i^a)S_{\perp b} \end{pmatrix}_{8N_1\times 1}$$

and $\mathbf{R}_F \equiv \mathbf{R}_F(-\mu_0, m_{\text{rel}})$ is the *reflection matrix*. Since the reflection is assumed to be specular, the reflection matrix is diagonal, implying that the order of the multiplicative factors is irrelevant.

Sine Modes

For the sine modes, one may proceed in a similar manner by combining the two upper rows of the matrix $\mathbf{P}^m_s$ and the two lower rows of the matrix $\mathbf{P}^m_c$ to obtain

$$\tilde{\mathbf{P}}^m_s = \begin{pmatrix} 0 & 0 & P^m_{13s} & P^m_{14s} \\ 0 & 0 & P^m_{23s} & P^m_{24s} \\ 0 & 0 & P^m_{33c} & P^m_{34c} \\ 0 & 0 & P^m_{43c} & P^m_{44c} \end{pmatrix}_{8N_1\times 8N_1} \tag{570}$$

Hence, for the sine modes, one obtains

$$\tilde{\mathbf{X}}^m_{s,0} = \frac{\varpi(\tau)}{4\pi}\tilde{\mathbf{P}}^m_s\,\tilde{\mathbf{S}}_b = \frac{\varpi(\tau)}{4\pi}\tilde{\mathbf{P}}^m_{s,b} \tag{571}$$

and

$$\tilde{\mathbf{X}}^m_{s,1} = \frac{\varpi(\tau)\mathbf{R}_F}{4\pi}\tilde{\mathbf{P}}^m_{s,b}\, e^{-2\tau_a/\mu_0} \tag{572}$$

where the vector $\tilde{\mathbf{P}}^m_{s,b} = \tilde{\mathbf{P}}^m_s \tilde{\mathbf{S}}_b$ is given by

$$\tilde{\mathbf{P}}^m_{s,b} = \begin{pmatrix} P^m_{13s}(\tau,-\mu_0,-\mu_i^a)S_{Ub} + P^m_{14s}(\tau,-\mu_0,-\mu_i^a)S_{Vb} \\ P^m_{23s}(\tau,-\mu_0,-\mu_i^a)S_{Ub} + P^m_{24s}(\tau,-\mu_0,-\mu_i^a)S_{Vb} \\ P^m_{33c}(\tau,-\mu_0,-\mu_i^a)S_{Ub} + P^m_{34c}(\tau,-\mu_0,-\mu_i^a)S_{Vb} \\ P^m_{43c}(\tau,-\mu_0,-\mu_i^a)S_{Ub} + P^m_{44c}(\tau,-\mu_0,-\mu_i^a)S_{Vb} \\ P^m_{13s}(\tau,-\mu_0,+\mu_i^a)S_{Ub} + P^m_{14s}(\tau,-\mu_0,+\mu_i^a)S_{Vb} \\ P^m_{23s}(\tau,-\mu_0,+\mu_i^a)S_{Ub} + P^m_{24s}(\tau,-\mu_0,+\mu_i^a)S_{Vb} \\ P^m_{33c}(\tau,-\mu_0,+\mu_i^a)S_{Ub} + P^m_{34c}(\tau,-\mu_0,+\mu_i^a)S_{Vb} \\ P^m_{43c}(\tau,-\mu_0,+\mu_i^a)S_{Ub} + P^m_{44c}(\tau,-\mu_0,+\mu_i^a)S_{Vb} \end{pmatrix}_{8N_1\times 1}.$$

From this expression, it follows that if the incident beam source has no U and V components, then the sine modes of the $(I_\parallel, I_\perp)$ and the cosine modes of the (U, V) *Stokes vector* components vanish, in which case Eq. (501) provides the complete homogeneous solution as $\tilde{\mathbf{I}}^m_s = \mathbf{0}$.

Beam Source

Consider now solutions for each of the *source terms* in Eq. (566). Thus, in a layer denoted by ℓ for the first term $\tilde{\mathbf{X}}^m_{\alpha\ell,0}\, e^{-\tau/\mu_0}$, where $\tilde{\mathbf{X}}^m_{\alpha\ell,0}$ is assumed to be constant within layer ℓ, consider a *particular solution* of the form

$$\tilde{\mathbf{I}}^m_{\alpha\ell,0}(\tau) = \mathbf{Z}^m_{\alpha\ell,0}\, e^{-\tau/\mu_0}. \tag{573}$$

Substitution in Eq. (511) with $\tilde{\mathbf{S}}^m_\alpha = \tilde{\mathbf{X}}^m_{\alpha\ell,0}\, e^{-\tau/\mu_0}$ leads to the following system of linear algebraic equations [$\mathbf{1}$ is the $8N_1 \times 8N_1$ identity matrix]:

$$[\tilde{\mathbf{A}}^m_\alpha + (\frac{1}{\mu_0})\mathbf{1}]\mathbf{Z}^m_{\alpha\ell,0} = \tilde{\mathbf{X}}^m_{\alpha\ell,0} \tag{574}$$

which can be solved to yield a particular solution vector $\mathbf{Z}^m_{\alpha\ell,0}$. Similarly, for the second term $\tilde{\mathbf{X}}^m_{\alpha\ell,1}\, e^{\tau/\mu_0}$ in Eq. (566), we seek a solution of the form

$$\tilde{\mathbf{I}}^m_{\alpha\ell,1}(\tau) = \mathbf{Z}^m_{\alpha\ell,1}\, e^{\tau/\mu_0} \tag{575}$$

which upon substitution in Eq. (511) with $\tilde{\mathbf{S}}^m_\alpha = \tilde{\mathbf{X}}^m_{\alpha\ell,1}\, e^{\tau/\mu_0}$ leads to the system of algebraic equations

$$[\tilde{\mathbf{A}}^m_\alpha - (\frac{1}{\mu_0})\mathbf{1}]\mathbf{Z}^m_{\alpha\ell,1} = \tilde{\mathbf{X}}^m_{\alpha\ell,1} \tag{576}$$

which again can be solved to yield the *particular solution vector* $\mathbf{Z}^m_{\alpha\ell,1}$.

Thermal Source

From the thermal source term, there is a contribution only for the cosine mode $m = 0$ in the Fourier expansion; there is no contribution for any of the sine modes. To get an approximate *particular solution* to the inhomogeneous radiative transfer equation, one may make polynomial approximations to the *Planck function* and the source vector $\mathbf{S}_t(\tau)$:

$$B(T(\tau)) = \sum_{k=0}^{K} b_k \tau^k$$

$$\tilde{\mathbf{S}}_t = (1 - \varpi(\tau))\delta_{0m} \sum_{k=0}^{K} \mathbf{D}_k \tau^k$$

where

$$\mathbf{D}_k \equiv \begin{pmatrix} \mathbf{D}_{1k} \\ \mathbf{D}_{0k} \\ \mathbf{D}_{1k} \\ \mathbf{D}_{0k} \end{pmatrix}_{8N_1 \times 1} \qquad \mathbf{D}_{1k} \equiv \begin{pmatrix} b_k \\ \vdots \\ b_k \end{pmatrix}_{2N_1 \times 1} \qquad \boldsymbol{D}_{0k} \equiv \begin{pmatrix} 0 \\ \vdots \\ 0 \end{pmatrix}_{2N_1 \times 1} \tag{577}$$

where we have taken into account that the thermal source is unpolarized. Next, one may substitute into Eq. (511) the assumption

$$\mathbf{I}_c^m = \delta_{0m} \sum_{k=0}^{K} \mathbf{X}_k^t \tau^k. \tag{578}$$

Since the thermal source term is nonvanishing only for the cosine mode $m = 0$, $\tilde{\mathbf{S}}_c^m = \delta_{0m}\tilde{\mathbf{S}}_t$ in this case. Thus, by substitution in Eq. (511) $\frac{d\tilde{\mathbf{I}}_c^m(\tau)}{d\tau} = \tilde{\mathbf{A}}_c^m(\tau)\tilde{\mathbf{I}}_c^m(\tau) - \delta_{0m}\tilde{\mathbf{S}}_t(\tau)$, one obtains

$$\delta_{0m} \sum_{k=1}^{K} k\, \mathbf{X}_k^t \tau^{k-1} = \delta_{0m}\tilde{\mathbf{A}}_c^m \sum_{k=0}^{K} \mathbf{X}_k^t \tau^k - \delta_{0m}(1 - \varpi(\tau)) \sum_{k=0}^{K} \mathbf{D}_k \tau^k$$

$$= \delta_{0m} \left[\tilde{\mathbf{A}}_c^m \sum_{k=1}^{K+1} \mathbf{X}_{k-1}^t \tau^{k-1} - (1 - \varpi(\tau)) \sum_{k=1}^{K+1} \mathbf{D}_{k-1} \tau^{k-1} \right].$$

Comparison of terms having equal powers of τ leads to

$$k < K + 1 \quad : \quad k\,\mathbf{X}_k^t = \tilde{\mathbf{A}}_c^0 \mathbf{X}_{k-1}^t - (1 - \varpi(\tau))\mathbf{D}_{k-1}$$
$$k = K + 1 \quad : \quad 0 = (\tilde{\mathbf{A}}_c^0 \mathbf{X}_K^t - (1 - \varpi(\tau))\mathbf{D}_K.$$

The first-order approximation ($K = 1$) is usually sufficient and yields (see Problem 4.3)

$$\tilde{\mathbf{A}}_c^0 \mathbf{X}_1^t = (1 - \varpi(\tau))\mathbf{D}_1 \tag{579}$$

and then we find $\mathbf{X}_0^t$ from

$$\tilde{\mathbf{A}}_c^0 \mathbf{X}_0^t = (1 - \varpi(\tau))\mathbf{D}_0 + \mathbf{X}_1^t. \tag{580}$$

4.3.4
Particular Solution – Lower Slab

In the lower slab, the *source term* in Eq. (566) is given by (dropping subscript $p = 2$)

$$\tilde{\mathbf{S}}_\alpha^m(\tau) = \tilde{\mathbf{X}}_{\alpha,2}^m(\tau)\, e^{-\tau/\mu_0^w} + \delta_{c\alpha}\, \mathbf{B}_{0m}(\tau) \tag{581}$$

where $\tilde{\mathbf{X}}^m_{\alpha,2}(\tau) = [\mathbf{X}^{m-}_{\alpha,2}(\tau,\mu^w), \mathbf{X}^{m+}_{\alpha,2}(\tau,\mu^w)]^T$, and [see Eqs. (379)–(380)]

$$\mathbf{X}^{m\pm}_{\alpha,2}(\tau,\mu^w) = \frac{\varpi(\tau)}{4\pi}\mathbf{P}^m_\alpha(\tau,-\mu_0^w,\pm\mu^w)\mathbf{T}(-\mu_0,m_{\text{rel}})\,\mathbf{S}_b\,e^{-\tau_a(1/\mu_0-1/\mu_0^w)}. \tag{582}$$

The matrices $\mathbf{P}^m_\alpha(\tau,-\mu_0^w,\pm\mu^w)$, $\alpha = c,s$ are analogous to those in Eqs. (567) and (570) except that the dimensions are $8N_2 \times 8N_2$ where $N_2 > N_1$ is the number of *quadrature points* in the lower slab (water) including those associated with the region of *total reflection*.

Beam Source

For the beam term $\tilde{\mathbf{X}}^m_{\alpha\ell,2}\,e^{-\tau/\mu_0^w}$ in a layer denoted by ℓ in the water, where $\tilde{\mathbf{X}}^m_{\alpha\ell,2}$ is assumed to be constant within layer ℓ, consider a solution of the form

$$\tilde{\mathbf{I}}^m_{\alpha\ell,2} = \mathbf{Z}^m_{\alpha\ell,2}\,e^{-\tau/\mu_0^w}$$

which upon substitution into Eq. (566) with $\mathbf{S}_\alpha(\tau) = \tilde{\mathbf{X}}^m_{\alpha\ell,2}\,e^{-\tau/\mu_0^w}$ leads to the system of linear algebraic equations [$\mathbf{1}$ is the $8N_2 \times 8N_2$ identity matrix]

$$[\tilde{\mathbf{A}}^m_\alpha + (\frac{1}{\mu_0^w})\mathbf{1}]\mathbf{Z}^m_{\alpha\ell,2} = \tilde{\mathbf{X}}^m_{\alpha\ell,2}. \tag{583}$$

Solving this system of equations, one obtains a *particular solution vector* $\mathbf{Z}^m_{\alpha\ell,2}$.

Thermal Source

The thermal source is completely analogous to that in the upper slab, so there is no need to repeat the solution.

4.3.5
General Solution

For each Fourier component denoted by the superscript m, the *general solution* is a combination of the homogeneous solution and the particular solutions for beam and thermal sources. Thus, in the upper slab (atmosphere), the general solution is

$$\tilde{\mathbf{I}}^m_{\alpha\ell}(\tau) = \sum_{j=1}^{8N_1}\tilde{C}^m_{\alpha j\ell}\tilde{\mathbf{G}}^m_{\alpha j\ell}e^{-k^m_{\alpha j\ell}\tau} + \mathbf{Z}^m_{\alpha\ell,0}e^{-\tau/\mu_0} + \mathbf{Z}^m_{\alpha\ell,1}e^{\tau/\mu_0}$$
$$+\,\delta_{0m}\delta_{\alpha c}[\mathbf{X}^t_{0,\ell} + \mathbf{X}^t_{1,\ell}\tau], \qquad \ell = 1,\ldots,L_1. \tag{584}$$

Since each of the solution vectors in the upper slab is of dimension $8N_1$, for example, $\tilde{\mathbf{G}}^m_{\alpha j\ell} = [\tilde{G}^m_{\alpha j\ell}(1), \tilde{G}^m_{\alpha j\ell}(2), \ldots, \tilde{G}^m_{\alpha j\ell}(8N_1)]^T \equiv g^m_{\alpha j\ell}(i), i = 1,\ldots,8N_1$, where the lower case letter g has been used to denote the vector component, one may rewrite Eq. (584) in component form as follows ($i = 1,\ldots,8N_1$):

$$\tilde{i}^m_{\alpha\ell}(\tau,i) = \sum_{j=1}^{8N_1}\tilde{C}^m_{\alpha j\ell}g^m_{\alpha j\ell}(i)e^{-k^m_{\alpha j\ell}\tau} + z^m_{\alpha\ell,0}(i)e^{-\tau/\mu_0} + z^m_{\alpha\ell,1}(i)e^{\tau/\mu_0}$$
$$+\,\delta_{0m}\delta_{\alpha c}[x^t_{0,\ell}(i) + x^t_{1,\ell}(i)\tau], \quad \ell = 1,\ldots,L_1. \tag{585}$$

For the lower slab, the result is

$$\tilde{\mathbf{I}}^m_{\alpha\ell}(\tau) = \sum_{j=1}^{8N_2} \tilde{C}^m_{\alpha j\ell} \tilde{\mathbf{G}}^m_{\alpha j\ell} e^{-k^m_{\alpha j\ell}\tau} + \mathbf{Z}^m_{\alpha\ell,2} e^{-\tau/\mu^w_0}$$
$$+ \delta_{0m}\delta_{\alpha c}[\mathbf{X}^t_{0,\ell} + \mathbf{X}^t_{1,\ell}\tau], \qquad \ell = 1, \dots, L_2 \tag{586}$$

and in component form ($i = 1, \dots, 8N_2$)

$$\tilde{i}^m_{\alpha\ell}(\tau, i) = \sum_{j=1}^{8N_2} \tilde{C}^m_{\alpha j\ell} g^m_{\alpha j\ell}(i) e^{-k^m_{\alpha j\ell}\tau} + z^m_{\alpha p,2}(i) e^{-\tau/\mu^w_0}$$
$$+ \delta_{0m}\delta_{\alpha c}[x^t_{0,\ell}(i) + x^t_{1,\ell}(i)\tau], \qquad \ell = 1, \dots, L_2. \tag{587}$$

4.3.6
Boundary Conditions

To complete the solution, one must apply *boundary conditions* in order to determine the unknown coefficients, the $\tilde{C}^m_{\alpha j\ell}$'s, in Eqs. (584) and (586). Mathematically, one is faced with a *two-point boundary-value problem*, which requires a specification of the radiation field incident at the top of the upper slab and at the bottom of the lower slab. In addition, one must ensure that *Snell's law* and *Fresnel's equations* are satisfied at the interface between the two slabs where the real part of the *refractive index* changes from a fixed value in the upper slab (atmosphere) to a different fixed value in the lower slab (water). Finally, at layer interfaces within each of the two slabs, the radiation field is required to be continuous because the refractive index is assumed to be constant within each slab. So, in summary, the following conditions are required

1) At the top of the upper slab, the incident *Stokes vector* must be specified;
2) The laws of reflection and transmission must be satisfied at the interface between the two slabs (atmosphere-water interface);
3) The Stokes vector must be continuous across layer interfaces in each slab, $\tilde{\mathbf{I}}^m_{\alpha\ell+1}(\tau_\ell) = \tilde{\mathbf{I}}^m_{\alpha\ell}(\tau_\ell)$;
4) At the bottom boundary of the lower slab (bottom of the water), the Stokes vector must be specified in terms of the bidirectional polarized (reflectance) distribution function [see Eq. (333) for an example].

Implementation of these conditions leads to a system of linear equations, and its solution yields the desired coefficients, the $\tilde{C}^m_{\alpha j\ell}$'s. For each layer in the upper slab (atmosphere), there are $8N_1$ equations corresponding to the Stokes vector components at the $8N_1$ quadrature points. For L_1 layers in the upper slab, one obtains $8N_1 \times L_1$ equations. Likewise, $8N_2$ quadrature points and L_2 layers in the lower slab lead to $8N_2 \times L_2$ equations. Hence, one obtains a linear system of equations of dimension $(8N_1 \times L_1 + 8N_2 \times L_2) \times (8N_1 \times L_1 + 8N_2 \times L_2)$, and the solution of this linear system of equations will determine the $(8N_1 \times L_1 + 8N_2 \times L_2)$ unknown coefficients, the $\tilde{C}^m_{\alpha j\ell}$'s, required to complete the solution. The details are provided in Appendix D.

Problems

4.1 Show that the matrices appearing in Eqs. (535)–(536) satisfy the symmetry relations $\tilde{\mathbf{A}}_{21} = -\tilde{\mathbf{A}}_{12}$ and $\tilde{\mathbf{A}}_{22} = -\tilde{\mathbf{A}}_{11}$.
4.2 Verify Eqs. (559)–(562).
4.3 Derive Eqs. (579)–(580).

5
The Inverse Problem

In the previous chapters, the focus was on formulating the radiative transfer problem in coupled turbid media and obtaining solutions of the pertinent *radiative transfer equation (RTE)* for *unpolarized radiation* as well as *polarized radiation*. Solutions to the RTE can be used to simulate the transport of electromagnetic radiation in coupled turbid media, and such solutions have important applications in the development of forward–inverse methods used to quantify

- morphological and physiological parameters describing the *inherent optical properties (IOPs)* and hence tissue health for a coupled air–tissue system [32];
- types and concentrations of scattering and absorbing constituents such as *aerosol* and *cloud particles* as well as dissolved and particulate biogeochemical matter in a coupled atmosphere–water system [38].

While the previous chapters have discussed the *forward modeling* aspects of the forward–inverse problem, the purpose of this chapter is to provide a basic introduction to the corresponding *inverse problem*. The presentation in this chapter is based on material from three main sources: (i) *Inverse methods for atmospheric sounding: Theory and practice* by Rodgers [37], (ii) *Data Analysis: A Bayesian Tutorial* by Sivia and Skilling [210], and (iii) *Parameter Estimation and Inverse Problems* by Aster *et al.* [211].

5.1
Probability and Rules for Consistent Reasoning

Consider a proposition such as (i) it will be sunshine tomorrow or (ii) the flip of this coin is twice as likely to show head as tail. To consistently evaluate the validity of such a proposition, we may assign a real number to it and let its numerical value be larger the more we believe in the proposition. But these numbers must obey certain rules to be logically consistent. For example, if we specify how much we believe that a proposition is true, then consistency requires that we simultaneously have specified how much we believe it is false. Thus, if we first specify to what degree we believe that claim A is true, and then specify to what degree we believe that A is true given that B is true, we have at the same time specified to what degree we believe that both A and B are true.

Radiative Transfer in Coupled Environmental Systems: An Introduction to Forward and Inverse Modeling, First Edition. Knut Stamnes and Jakob J. Stamnes.

It can be shown that consistency requires that the real number we attach to our belief in various propositions obey the rules of *probability theory* [210]:

$$\text{prob}(A|I) + \text{prob}(\overline{A}|I) = 1 \tag{588}$$

$$\text{prob}(A, B|I) = \text{prob}(A|B, I) \times \text{prob}(B|I) \tag{589}$$

where $0 = \text{prob(false)}$, $1 = \text{prob(true)}$, and $\overline{A}$ denotes that A is false. The symbol | (vertical bar) means "given" (i.e., that all items to the right of | are taken to be true), the comma signifies the conjunction "and", and the probabilities are conditional on some background information I, implying that there is no such thing as an absolute probability. Note that

- The *sum rule*, expressed by Eq. (588), states that the sum of the probability that A is true and the probability that A is false equals 1, while
- the *product rule*, expressed by Eq. (589), states that the probability that both A and B are true is equal to the probability that A is true given that B is true times the probability that B is true.
- The sum and product rules form the basic rules of probability theory, but many other rules can be derived from them, including *Bayes' theorem* (see below).

Bayes' theorem follows directly from the product rule [Eq. (589)], which may be rewritten as (interchanging A and B)

$$\text{prob}(B, A|I) = \text{prob}(B|A, I) \times \text{prob}(A|I). \tag{590}$$

Since the statement "A and B are both true" [Eq. (589)] is equivalent to the statement "B and A are both true" [Eq. (590)], we can equate the right-hand sides of Eqs. (589) and (590) to obtain

$$\text{prob}(A|B, I) \times \text{prob}(B|I) = \text{prob}(B|A, I) \times \text{prob}(A|I) \tag{591}$$

from which Bayes' theorem, which relates $\text{prob}(A|B, I)$ and $\text{prob}(B|A, I)$, follows:

$$\text{prob}(A|B, I) = \frac{\text{prob}(B|A, I) \times \text{prob}(A|I)}{\text{prob}(B|I)}. \tag{592}$$

If we replace A and B with H = *hypothesis* and D = *data*, respectively, then Bayes' theorem asserts

$$\text{prob}(H|D, I) \propto \text{prob}(D|H, I) \times \text{prob}(H|I). \tag{593}$$

The power of Bayes' theorem is that it relates the quantity of interest, which is the probability $\text{prob}(H|D, I)$ that the hypothesis H is true given the data D, to the probability $\text{prob}(D|H, I)$ that we would have observed D if H was true. Note, in particular, that the *prior probability* = $\text{prob}(H|I)$, which represents our state of knowledge about the truth of the hypothesis before making measurements, is multiplied by the experimental data represented by the *likelihood function* $\text{prob}(D|H, I)$ to provide the *posterior probability* $\text{prob}(H|D, I)$. Note also that in Eq. (593) the equality in Eq. (592) has been replaced by proportionality because

- we have omitted the denominator, which is called the *evidence* prob($D|I$). This omission is fine for problems involving *parameter estimation*, but the evidence term plays a crucial role in model selection.

Integrating over the parameter B, we obtain

$$\text{prob}(A|I) = \int_{-\infty}^{\infty} \text{prob}(A, B|I) dB. \tag{594}$$

This result, referred to as *marginalization*, can be obtained by returning to the *product rule* [Eqs. (589) and 590], from which it follows

$$\text{prob}(A, B|I) = \text{prob}(B, A|I) = \text{prob}(B|A, I) \times \text{prob}(A|I) \tag{595}$$

and

$$\text{prob}(A, \overline{B}|I) = \text{prob}(\overline{B}, A|I) = \text{prob}(\overline{B}|A, I) \times \text{prob}(A|I). \tag{596}$$

Combining Eqs. (595) and (596), we have

$$\text{prob}(A, B|I) + \text{prob}(A, \overline{B}|I) = \underbrace{[\text{prob}(B|A, I) + \text{prob}(\overline{B}|A, I)]}_{=1} \times \text{prob}(A|I)$$

or

$$\text{prob}(A|I) = \text{prob}(A, B|I) + \text{prob}(A, \overline{B}|I) \tag{597}$$

which states that the probability that A is true irrespective of whether or not B is true is equal to the sum of the probability that both A and B are true and the probability that A is true and B is false. We may generalize Eq. (597) to the situation in which there is a whole set of possibilities: $B_1, B_2, \ldots, B_M$. Then the probability that A is true, irrespective of which one of the propositions B_i is true, is

$$\text{prob}(A|I) = \sum_{i=1}^{M} \text{prob}(A, B_i|I) \tag{598}$$

and since $\text{prob}(A, B_i|I) = \text{prob}(B_i|A, I) \times \text{prob}(A|I)$, we get

$$\sum_{i=1}^{M} \text{prob}(B_i|A, I) = 1. \tag{599}$$

Note that the normalization in Eq. (599) is valid only if the $\{B_i\}$'s form a *mutually exclusive and exhaustive* set of possibilities, so that **if one of the propositions B_i is true, then all the others must be false, but one of them has to be true.** Equation (594) is a continuous generalization of Eq. (598) for the case in which $M \to \infty$, B represents a numerical value of a parameter of interest, and the integrand prob($A, B|I$) is technically a *probability density* function (pdf) rather than a probability.

To preserve uniformity of notation between continuous and discrete cases we will use "prob" for anything related to probabilities, so that in the continuum limit we have

$$\text{prob}(A|I) = \sum_{i=1}^{M} \text{prob}(A, B_i|I) \quad \to \quad \int_{-\infty}^{\infty} \text{prob}(A, B|I) dB \tag{600}$$

and

$$\sum_{i=1}^{M} \text{prob}(B_i|A,I) = 1 \quad \rightarrow \quad \int_{-\infty}^{\infty} \text{prob}(B|A,I)dB = 1. \tag{601}$$

Marginalization is a powerful device to deal with *nuisance parameters*, that is, quantities that necessarily enter the data analysis but are of no intrinsic interest. Examples of such quantities include the unwanted background signal present in many experimental measurements, and instrumental parameters that are difficult to calibrate.

5.2
Parameter Estimation

We will start by considering how to estimate the value of a single parameter because it will serve as a good introduction to the use of *Bayes' theorem*, and allow for a discussion of error bars and confidence intervals.

5.2.1
Optimal Estimation, Error Bars and Confidence Intervals

The posterior pdf, denoted by $\mathcal{P}(x) = \text{prob}(x|y,I)$, contains our inference about the value of a parameter x, and the best estimate x_0 is the value of x for which the posterior pdf attains its maximum, that is,

$$\left.\frac{d\mathcal{P}(x)}{dx}\right|_{x=x_0} = 0 \;\; \mathit{and} \;\; \left.\frac{d^2\mathcal{P}(x)}{dx^2}\right|_{x=x_0} < 0. \tag{602}$$

The reliability of this estimate is given by the width of $\mathcal{P}(x)$ about x_0. Taking the natural logarithm of the posterior pdf, we obtain

$$L = \ln\left[\text{prob}(x|y,I)\right] = \ln \mathcal{P}(x) \tag{603}$$

and a *Taylor series expansion* about $x = x_0$ gives

$$\ln \mathcal{P}(x) = L = L(x_0) + \frac{1}{2}\left.\frac{d^2L}{dx^2}\right|_{x_0}(x-x_0)^2 + \ldots . \tag{604}$$

The best estimate of x is given by

$$\left.\frac{dL}{dx}\right|_{x=x_0} = 0 \tag{605}$$

which follows from Eq. (602) since L is a monotonic function of $\mathcal{P}(x)$. Taking the exponential of each side of Eq. (604), we find

$$\mathcal{P}(x) \approx C\exp\left[\frac{1}{2}\left.\frac{d^2L}{dx^2}\right|_{x_0}(x-x_0)^2\right] = C\exp\left[-\frac{(x-\mu)^2}{2\sigma^2}\right] \tag{606}$$

where C is a normalization constant, $\mu \equiv x_0$, and

$$\sigma = \left(-\left.\frac{d^2L}{dx^2}\right|_{x_0}\right)^{-1/2}. \tag{607}$$

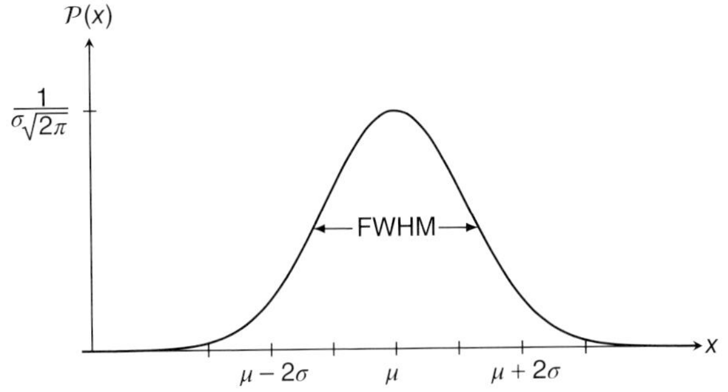

Figure 28 The Gaussian or normal distribution, given by $\mathcal{P}(x)$ in Eq. (608), is symmetric about its maximum value $1/\sigma\sqrt{2\pi}$ at $x = \mu$, and has a full-width at half-maximum (FWHM) of about 2.35 σ.

By requiring Eq. (606) to be normalized, so that $\int_{-\infty}^{\infty} \mathcal{P}(x)dx = 1$, we find $C = 1/\sigma\sqrt{2\pi}$, and hence we can rewrite Eq. (606) as

$$\mathcal{P}(x) = \text{prob}(x|\mu, \sigma) = \frac{1}{\sigma\sqrt{2\pi}}\, e^{-\frac{(x-\mu)^2}{2\sigma^2}} \tag{608}$$

which is a *Gaussian distribution* with a maximum at $x = \mu \equiv x_0$ and a width proportional to σ (see Figure 28). Our inference about the quantity of interest is conveyed concisely by

$$x = x_0 \pm \sigma \tag{609}$$

where x_0 is the best estimate of x, and σ, usually referred to as the *error bar*, is a measure of its reliability.

The probability that the true value of x lies within $\pm\sigma$ of $x = x_0$ is 67%, that is

$$\text{prob}(x_0 - \sigma \le x \le x_0 + \sigma|y, I) = \int_{x_0-\sigma}^{x_0+\sigma} \text{prob}(x, y|I)dx \approx 0.67 \tag{610}$$

and the probability that x lies within $\pm 2\sigma$ of x_0 is 95%, that is

$$\text{prob}(x_0 - 2\sigma \le x \le x_0 + 2\sigma|D, I) = \int_{x_0-2\sigma}^{x_0+2\sigma} \text{prob}(x, y|I)dx \approx 0.95. \tag{611}$$

Asymmetric Posterior pdfs

For a highly asymmetric posterior pdf, the "mean" value is different from the best estimate, defined as the most probable value where the pdf attains its maximum. For a normalized posterior pdf, we define the weighted average as

$$E(x) = \langle x \rangle = \int x\, \text{prob}(x|y, I)\, dx. \tag{612}$$

This mean or expectation value takes the skewness of the pdf into account, and is therefore more representative than x_0. If the posterior pdf is symmetric about the maximum, as in the *Gaussian distribution*, then $\langle x \rangle = x_0$.

The reliability at which x can be inferred may be expressed through a *confidence interval*. Since the area under the pdf between x_1 and x_2 is proportional to our expectation that x lies in that range, the *shortest* interval that encloses 95% of the total area represents a measure of the uncertainty of the estimate:

$$\text{prob}(x_1 \leq x \leq x_2|y, I) = \int_{x_1}^{x_2} \text{prob}(x|y, I)dx \approx 0.95 \tag{613}$$

where $x_2 - x_1$ is as small as possible. The region $x_1 \leq x \leq x_2$ is called the shortest 95% confidence interval. For the *Gaussian distribution* considered above, $x_1 = x_0 - 2\sigma$ and $x_2 = x_0 + 2\sigma$.

Multimodal Posterior pdfs

Sometimes we may encounter posterior pdfs that are *multimodal.*

- If one of the maxima is very much larger than the others, we can simply focus on the global maximum and ignore the others, but
- what do we do if there are several maxima of comparable magnitude? Then:
- what do we mean by a best estimate, and how should we quantify its reliability?

The posterior pdf gives a complete description of what we can infer about the desired parameter in light of the data, and the relevant prior knowledge. The idea of a best estimate and an error bar, or even a confidence interval, is simply an attempt to summarize the posterior pdf with just a few numbers. For a multimodal pdf, this characterization may not be possible, but the posterior pdf still exists, and we are free to draw from it whatever conclusions that may be appropriate. For a general *multimodal pdf,* the best approach may simply be to display the posterior pdf itself.

Gaussian Noise and Averages

The *normal distribution* is often used to describe noise associated with experimental data. The probability that a data point labeled k has the value x_k is given by

$$\text{prob}(x_k|\mu, \sigma) = \frac{1}{\sigma\sqrt{2\pi}} \exp\left[-\frac{(x_k - \mu)^2}{2\sigma^2}\right] \tag{614}$$

where μ is the true value of the parameter of interest, and σ is a measure of the error in the measurement. For a given set of data $\{x_k\}$, we would like to find the best estimate of μ and our confidence in its prediction. To estimate μ, we may use *Bayes' theorem* to calculate the posterior pdf:

$$\text{prob}(\mu|\{x_k\}, \sigma, I) \propto \text{prob}(\{x_k\}|\mu, \sigma, I) \times \text{prob}(\mu|\sigma, I). \tag{615}$$

If the data are independent, implying that the measurement of one data point does not influence what can be inferred about the outcome of another data point, the likelihood function is given by the product

$$\text{prob}(\{x_k\}|\mu, \sigma, I) = \prod_{k=1}^{N} \text{prob}(x_k|\mu, \sigma, I) \tag{616}$$

where the factors in the product are the probabilities of obtaining each of the N individual data points. Next let us assume a uniform pdf for the prior $[A = 1/(\mu_{\max} - \mu_{\min})]$

$$\text{prob}(\mu|\sigma, I) = A \quad \text{if} \quad \mu_{\min} \leq \mu \leq \mu_{\max}; \; 0 \text{ otherwise.} \tag{617}$$

Inserting this prior into the likelihood function in Eq. (615), we obtain using Eqs. (614) and (616)

$$\text{prob}(\mu|\{x_k\}, \sigma, I) \propto \prod_{k=1}^{N} \frac{1}{\sigma\sqrt{2\pi}} \exp\left[-\frac{(x_k - \mu)^2}{2\sigma^2}\right]. \tag{618}$$

Taking the logarithm of Eq. (618), we find

$$L = \ln\left[\text{prob}(\mu|\{x_k\}, \sigma, I)\right] = \text{constant} - \sum_{k=1}^{N} \frac{(x_k - \mu)^2}{2\sigma^2} \tag{619}$$

where the constant does not depend on μ. To find the best estimate $\mu = \mu_0$, we set the first derivative of L with respect to μ equal to zero [i.e., see Eq. (607)]:

$$\left.\frac{dL}{d\mu}\right|_{\mu_0} = \sum_{k=1}^{N} \frac{(x_k - \mu_0)}{\sigma^2} = 0 \tag{620}$$

which implies

$$\sum_{k=1}^{N} x_k = \sum_{k=1}^{N} \mu_0 = N\mu_0 \tag{621}$$

or

$$\mu_0 = \frac{1}{N} \sum_{k=1}^{N} x_k \tag{622}$$

which shows that μ_0 is the arithmetic mean. Evaluating the second derivative of L, we find [see Eq. (607)]

$$\frac{1}{\sigma_{\text{ave}}^2} = -\left.\frac{d^2L}{d\mu^2}\right|_{\mu_0} = \sum_{k=1}^{N} \frac{1}{\sigma^2} = \frac{N}{\sigma^2} \tag{623}$$

or

$$\sigma_{\text{ave}} = \frac{\sigma}{\sqrt{N}}. \tag{624}$$

Hence we may summarize our inference about the value of μ as follows:

$$\mu = \mu_0 \pm \frac{\sigma}{\sqrt{N}}. \tag{625}$$

Note that the reliability of the data is proportional to $\frac{1}{\sqrt{N}}$, and that the error bar analysis relies on the validity of the quadratic expansion in Eq. (604): $L = L(x_0) + \frac{1}{2}\frac{d^2L}{dx^2}\big|_{x_0}(x - x_0)^2 + \dots$. For a Gaussian noise distribution, all higher derivatives of L are zero, so the Gaussian approximation is exact. Thus, for Gaussian noise, the posterior pdf is completely defined by the parameters of Eq. (625). In the analysis above, it has been assumed that the magnitude of the error bar for each data point was the same, which may be reasonable if all measurements were made with the same experimental setup. But if the data were obtained from several instruments of varying quality, the question of how to combine evidence from data of differing reliability arises.

Data with Error Bars of Different Magnitudes

If one assumes again that the measurement error can be modeled by a Gaussian pdf, the probability of the kth data point having value x_k is given by

$$\text{prob}(x_k, \mu, \sigma_k) = \frac{1}{\sigma_k \sqrt{2\pi}} \exp[-\frac{(x_k - \mu)^2}{2\sigma_k^2}] \tag{626}$$

where the error bars σ_k for different data points may not be of the same magnitude. Proceeding as in the previous case, we find that the logarithm for the posterior pdf for μ is given by

$$L = \ln[\text{prob}(\mu|\{x_k\}, \{\sigma_k\}, I)] = \text{constant} - \sum_{k=1}^{N} \frac{(x_k - \mu)^2}{2\sigma_k^2}. \tag{627}$$

Setting the first derivative of L equal to zero and the corresponding value of μ equal to μ_0, we find

$$\mu_0 = \frac{\sum_{k=1}^{N} w_k x_k}{\sum_{k=1}^{N} w_k}, \quad \text{where} \quad w_k = \frac{1}{\sigma_k^2}. \tag{628}$$

The best estimate is now a weighted mean instead of the arithmetic mean, and less reliable data will have larger *error bars*, and correspondingly smaller weights w_k. The second derivative of L yields the error bar, and our inference about μ becomes

$$\mu = \mu_0 \pm \left(\sum_{k=1}^{N} w_k \right)^{-1/2}. \tag{629}$$

The Cauchy Distribution

Suppose a lighthouse, which is located somewhere off a straight coastline at a position a along the shore and a distance d out at sea, as illustrated in Figure 29, emits a series of short, highly collimated flashes at random intervals and hence at random angles θ_k. These pulses are detected at positions $\{x_k\}$ on the coast by photodetectors, which record that a flash has occurred but not the angle θ_k from which it came. Given that N flashes have been recorded at positions $\{x_k\}$, where is the lighthouse? It seems reasonable to assign a uniform pdf for the angle θ_k of the kth data point:

$$\text{prob}(\theta_k|a, d, I) = \frac{1}{\pi} \tag{630}$$

where the angle must lie between $\pm\pi/2$ ($\pm 90°$) to have been detected. Since the photodetectors are sensitive only to position along the coast and not direction,

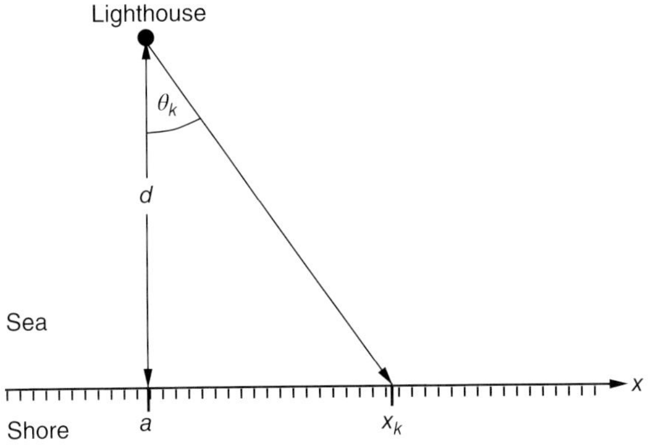

Figure 29 Schematic illustration of the geometry of the lighthouse problem.

we must relate the angle θ_k to the positions x_k. From the geometry in Figure 29, it follows that

$$d \tan \theta_k = x_k - a. \tag{631}$$

Differentiation of this equation with respect to x_k yields

$$\frac{d\theta_k}{dx_k} = \frac{\cos^2 \theta_k}{d}. \tag{632}$$

Using the trigonometric identity $\tan^2 \theta + 1 \equiv 1/\cos^2 \theta$, and substituting for $\tan \theta_k$ from Eq. (631), we obtain

$$\frac{d\theta_k}{dx_k} = \left[d(1 + \tan^2 \theta_k)\right]^{-1} = \frac{d}{d^2 + (x_k - a)^2}. \tag{633}$$

Hence, the probability that the kth flash will be recorded at position x_k, given the coordinates (a, d) of the lighthouse, is given by the *Cauchy distribution* shown in Figure 30:

$$\text{prob}(x_k|a, d, I) = \text{prob}(\theta_k|a, d, I) \times \frac{d\theta_k}{dx_k} = \frac{1}{\pi} \times \frac{d}{d^2 + (x_k - a)^2}. \tag{634}$$

This distribution is symmetric about the maximum, at $x_k = a$, and has a full-width at half-maximum (FWHM) of $2d$. Assuming that d (the distance out to sea) is known, we may use *Bayes' theorem* to estimate the posterior pdf required to infer the lighthouse position:

$$\text{prob}(a|\{x_k\}, d, I) \propto \text{prob}(\{x_k\}|a, d, I) \times \text{prob}(a|d, I). \tag{635}$$

Next, let us assume a uniform pdf for the prior $[A = 1/(a_{\text{max}} - a_{\text{min}})]$

$$\text{prob}(a|d, I) = \text{prob}(a|I) = A \;\; \text{if} \;\; a_{\text{min}} \leq \alpha \leq a_{\text{max}}; \;\; 0 \text{ otherwise}. \tag{636}$$

Since recording of a signal in one photodetector does not influence the recording of a signal in another detector, the likelihood function for these independent data

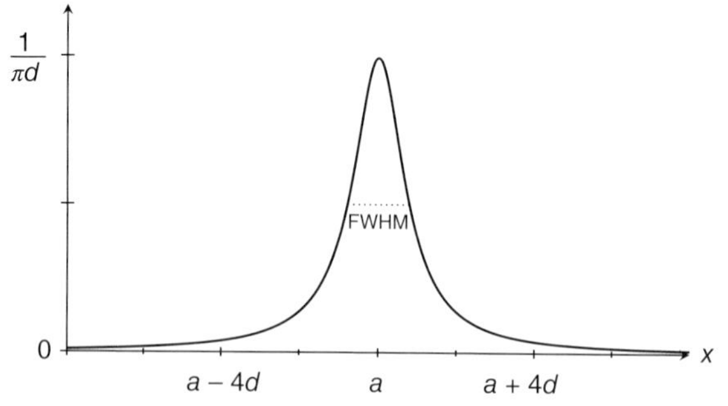

Figure 30 The Cauchy or Lorentzian distribution, given by Eq. (634), is symmetric with respect to its maximum at $x = a$, and has an FWHM of $2d$.

is given by

$$\text{prob}(\{x_k\}|a, d, I) = \prod_{k=1}^{N} \text{prob}(x_k|a, d, I). \tag{637}$$

Substituting Eqs. (636) and (637) into Eq. (635), we have

$$\text{prob}(a|\{x_k\}, d, I) \propto \prod_{k=1}^{N} \text{prob}(x_k|a, d, I) \times A = A \times \prod_{k=1}^{N} \frac{d}{\pi[d^2 + (x_k - a)^2]}$$

and taking the logarithm, we obtain

$$L = \ln[\text{prob}(a|\{x_k\}, d, I)] = \text{constant} - \sum_{k=1}^{N} \ln[d^2 + (x_k - a)^2] \tag{638}$$

where the constant is independent of a.

To determine the best estimate of the position $a = a_0$, we may differentiate Eq. (638) with respect to a, and set the result for $a = a_0$ equal to zero:

$$\left.\frac{dL}{da}\right|_{a_0} = 2 \sum_{k=1}^{N} \frac{x_k - a_0}{d^2 + (x_k - a_0)^2} = 0. \tag{639}$$

But this equation cannot be solved analytically for a_0. Instead, it is easier to evaluate Eq. (638) numerically for several values of a, and select the best estimate a_0 as the value of a that gives the largest L. Plotting $\exp(L)$ on the vertical axis as a function of a on the horizontal axis, we obtain the posterior pdf for the position of the lighthouse. This approach not only provides a complete visual representation of our inference but also has the advantage that we do not need to worry about the posterior pdf being asymmetric or multimodal.

5.2.2
Problems with More Than One Unknown Parameter

Now we will discuss problems with more than one unknown parameter, generalize the idea of *error bars* to include *correlations*, and use marginalization to deal with unwanted variables.

Optimal Estimation, Error Bars, and Correlations

As in the one-parameter case, the posterior pdf in the multiparameter case encodes our inference about the value of the parameters, given the data and the relevant background, and as before we may want to summarize the inference with just a few numbers: the best estimates and a measure of their reliabilities. Our *optimal estimate* is given by the maximum of the posterior pdf. If we denote the quantities of interest by $\{x_j\}$, with posterior pdf $\mathcal{P}(x) = \text{prob}(\{x_j\}|y, I)$, the best estimate of their values, $\{x_{0j}\}$, is given by the solution to the following set of *simultaneous equations*:

$$\left.\frac{\partial \mathcal{P}(x)}{\partial x_i}\right|_{\{x_{0j}\}} = 0, \quad i = 1, \ldots, M, \tag{640}$$

where M is the number of parameters to be inferred. As before, it is more convenient to work with the logarithm of $\mathcal{P}(x)$:

$$L = \ln \mathcal{P}(x) = \ln\left[\text{prob}(\{x_j\}|y, I)\right]. \tag{641}$$

Consider, first, the $M = 2$ case, and define $x \equiv x_1$ and $z \equiv x_2$ to simplify the notation. Then the pair of simultaneous equations to be solved is ($L = \ln\left[\text{prob}(x, z|y, I)\right]$)

$$\left.\frac{\partial L}{\partial x}\right|_{x_0, z_0} = 0 \quad \text{and} \quad \left.\frac{\partial L}{\partial z}\right|_{x_0, z_0} = 0. \tag{642}$$

Using a Taylor series expansion, we obtain the 2-D version of Eq. (604):

$$L = L(x_0, z_0) + \frac{1}{2} Q \tag{643}$$

where

$$Q = \left.\frac{\partial^2 L}{\partial x^2}\right|_{x_0, z_0} (x - x_0)^2 + \left.\frac{\partial^2 L}{\partial z^2}\right|_{x_0, z_0} (z - z_0)^2 + 2\left.\frac{\partial^2 L}{\partial x \partial z}\right|_{x_0, z_0} (x - x_0)(z - x_0) + \cdots$$

where x_0 and z_0 are determined by Eq. (642), and we have assumed $\frac{\partial^2 L}{\partial x \partial z} = \frac{\partial^2 L}{\partial z \partial x}$. The three quadratic terms, denoted by Q, can be written in matrix notation as ($\tilde{x} \equiv x - x_0, \tilde{z} \equiv z - z_0$, and $\mathbf{x}^T = [\tilde{x}\ \tilde{z}]$):

$$Q = [\tilde{x}\ \tilde{z}] \begin{bmatrix} A & C \\ C & B \end{bmatrix} \begin{bmatrix} \tilde{x} \\ \tilde{z} \end{bmatrix} = \mathbf{x}^T \begin{bmatrix} A & C \\ C & B \end{bmatrix} \mathbf{x} = \frac{1}{2} \mathbf{x}^T \mathbf{H} \mathbf{x}, \tag{644}$$

where

$$\mathbf{H} \equiv 2 \times \begin{bmatrix} A & C \\ C & B \end{bmatrix}, \quad A = \left.\frac{\partial^2 L}{\partial x^2}\right|_{x_0, z_0}, \quad B = \left.\frac{\partial^2 L}{\partial z^2}\right|_{x_0, z_0}, \quad C = \left.\frac{\partial^2 L}{\partial x \partial z}\right|_{x_0, z_0}. \tag{645}$$

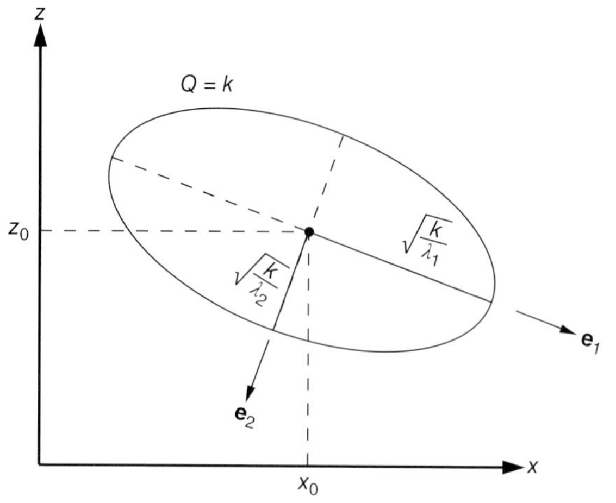

Figure 31 The contour in the *xz* parameter space along which $Q = k =$ constant is an ellipse centered at (x_0, z_0). The ellipse is characterized by the eigenvalues and eigenvectors defined by Eq. (646).

Figure 31 shows a contour of Q in the xz plane. The posterior pdf is constant along this contour, which is an ellipse centered at (x_0, z_0). The orientation and eccentricity of this ellipse are determined by the values of A, B, and C.

The directions of the principal axes are along the eigenvectors $\mathbf{e}_1$ and $\mathbf{e}_2$, obtained by solving the eigenvalue problem

$$\begin{bmatrix} A & C \\ C & B \end{bmatrix} \begin{bmatrix} \tilde{x} \\ \tilde{z} \end{bmatrix} = \lambda \begin{bmatrix} \tilde{x} \\ \tilde{z} \end{bmatrix} \tag{646}$$

and the two eigenvalues λ_1 and λ_2 determine the length of the axes of the ellipse: $\sqrt{k/|\lambda_1|}$ and $\sqrt{k/|\lambda_2|}$ for a given value of $k = Q$ (see Figure 31). For the point (x_0, z_0) to be a maximum (rather than a minimum or a saddle point), both λ_1 and λ_2 must be negative, which implies

$$A < 0, \qquad B < 0, \qquad AB > C^2.$$

The reliability of our best estimates would be easy to determine if $C = 0$, because the ellipse would then be aligned with the x and z axes, and the error bars for x_0 and z_0 will be proportional to $1/\sqrt{|\lambda_1|} = 1/\sqrt{|A|}$ and $1/\sqrt{|\lambda_2|} = 1/\sqrt{|B|}$. If the ellipse is skewed, and we are only interested in the value of x, we can integrate out z as a *nuisance parameter*:

$$\text{prob}(x|y, I) = \int_{-\infty}^{+\infty} \text{prob}(x, z|y, I) dz. \tag{647}$$

Within the *quadratic approximation*, that is, $\text{prob}(x, z|y, I) = \exp(L) \propto \exp(Q/2)$, this integral can be evaluated analytically to yield (see Problem 5.1)

$$\text{prob}(x|y, I) \propto \exp\left\{ \frac{1}{2}\left[\frac{AB - C^2}{B}\right](x - x_0)^2 \right\} = \exp\left\{ -\frac{(x - x_0)^2}{2\sigma_x^2} \right\} \tag{648}$$

which is the marginal distribution for x, a one-dimensional (1-D) Gaussian, with *marginal error bar*

$$\sigma_x = \sqrt{\frac{-B}{AB - C^2}}. \tag{649}$$

We obtain an analogous result for z by integrating out x as a *nuisance parameter* to obtain an expression for prob($z|y, I$)) given by Eq. (647) with x interchanged by z, and with marginal error bar (see Problem 5.1)

$$\sigma_z = \sqrt{\frac{-A}{AB - C^2}}. \tag{650}$$

These expressions for σ_x and σ_z provide useful error bars for our best estimates, but they do not provide a complete picture. The denominator $AB - C^2$ in the above expressions is the determinant of the matrix in Eq. (646), which for such a real symmetric matrix is equal to the product of the eigenvalues.

- If either λ_1 or λ_2 becomes very small, so that the ellipse in Figure 31 becomes extremely elongated in one of its principal directions, then $AB - C^2 \to 0$, and σ_x and σ_z will both be large, except when $C = 0$.
- Although neither x nor z can be reliably inferred in this case, the question is whether there could there be some joint aspect of the two parameters that could be well determined, because the posterior pdf might be very sharp in one direction but very broad in the other. To examine this question, we now compute the *variance* of the posterior pdf.

In accordance with Eq. (608) [$\text{prob}(x|\mu, \sigma) = \frac{1}{\sigma\sqrt{2\pi}} e^{-\frac{(x-\mu)^2}{2\sigma^2}}$, see also Figure 28], we have thought of the error bar as the FWHM $\approx 2.35 \times \sigma$, that is, the width of the Gaussian pdf. Alternatively, we may consider the variance, given by

$$Var(x) = \langle (x - \mu)^2 \rangle = \int (x - \mu)^2 \,\text{prob}(x|y, I)\, dx \tag{651}$$

where $\mu = \langle x \rangle$. For the 1-D Gaussian, we have

$$\langle (x - \mu)^2 \rangle = \sigma^2 \tag{652}$$

while for the 2-D case

$$\sigma_x^2 = \langle (x - x_0)^2 \rangle = \int\int (x - x_0)^2 \,\text{prob}(x, z|y, I)\, dx\, dz \tag{653}$$

$$\sigma_z^2 = \langle (z - z_0)^2 \rangle = \int\int (z - z_0)^2 \,\text{prob}(x, z|y, I)\, dx\, dz \tag{654}$$

where σ_x and σ_z are the same as in Eqs. (649) and (650) if we invoke the quadratic approximation $\text{prob}(x, z|y, I) = \exp(L) \propto \exp(Q/2)$. Considering the simultaneous deviations of both x and z, we get the *covariance*

$$\sigma_{xz}^2 = \langle (x - x_0)(z - z_0) \rangle = \int\int (x - x_0)(z - z_0) \,\text{prob}(x, z|y, I)\, dx\, dz \tag{655}$$

which is a measure of the *correlation* between the inferred parameters in the quadratic approximation

$$\sigma_{xz}^2 = \frac{C}{AB - C^2}. \tag{656}$$

Note that, if an overestimation of one parameter usually leads to a larger than average value of the other parameter, so that $z - z_0 > 0$ when $x - x_0 > 0$, and if the same is true for an underestimation so that $z - z_0 < 0$ when $x - x_0 < 0$, then the product of the deviations will be positive, that is, the covariance will be positive.

On the other hand, if there is an anticorrelation, so that an overestimation of one parameter is accompanied with an underestimation of the other, then the covariance will be negative. When our estimate of one parameter has little, or no, influence on the inferred value of the other, then the magnitude of the covariance will be negligible in comparison to the variance terms: $|\sigma_{xz}^2| << \sqrt{\sigma_x^2\sigma_z^2}$. In the quadratic approximation, we get the following *covariance matrix* (see Problem 5.2):

$$\begin{pmatrix} \sigma_x^2 & \sigma_{xz}^2 \\ \sigma_{xz}^2 & \sigma_z^2 \end{pmatrix} = \frac{1}{AB - C^2}\begin{pmatrix} -B & C \\ C & -A \end{pmatrix} = -\begin{pmatrix} A & C \\ C & B \end{pmatrix}^{-1}. \tag{657}$$

When $C = 0$, $\sigma_{xz}^2 = 0$: the inferred values of the parameters are uncorrelated, and the principal directions of the posterior pdf will be parallel to the coordinate axes (see Figure 32a). As the magnitude of C increases relative to A and B, the posterior pdf becomes more and more skewed and elongated, which reflects the growing strength of the correlation between our estimates of x and z (see Figure 32(b) and (c)). In the extreme case when $C = \pm\sqrt{AB}$, the elliptical contours will be infinitely wide in one direction and oriented at an angle $\pm\tan^{-1}\sqrt{AB}$ wrt the x-axis. Although the error bars σ_x and σ_z will be large (the individual x and z estimates are completely unreliable), the large off-diagonal elements of the covariance matrix imply that we can still reliably infer a linear combination of x and z. If the covariance is negative, then the posterior pdf will be very broad in the direction $z = -mx$, where $m = \sqrt{AB}$, but fairly narrow perpendicular to it, indicating that the data contain a lot of information about the sum $z + mx$ but little about the

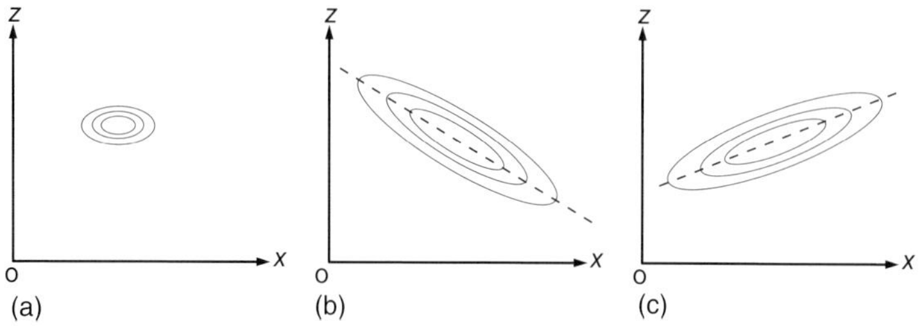

Figure 32 Schematic illustration of covariance and correlation. (a) Posterior pdf with zero covariance (inferred values of x and z are uncorrelated). (b) Large and negative covariance ($z + mx$ = constant along dotted line). (c) Large and positive covariance ($z - mx$ = constant along dotted line).

difference $z - x/m$. Similarly, when the covariance is positive, we can infer the difference $z - mx$ but not the sum $z + x/m$.

Generalization of the Quadratic Approximation

We will now generalize the *bivariate analysis* above to the case of several parameters M. As before, the best estimate will be chosen by maximizing the logarithm L of the posterior pdf for the set of M parameters $\{x_{0j}\} \equiv \mathbf{x}_0$:

$$\frac{\partial L}{\partial x_i}\bigg|_{\mathbf{x}_0} = 0, \qquad i = 1, \ldots, M. \tag{658}$$

For the *multivariate case*, the Taylor series expansion becomes

$$L = L(\mathbf{x}_0) + \frac{1}{2}\sum_{i=1}^{M}\sum_{j=1}^{M} \frac{\partial^2 L}{\partial x_i \partial x_j}\Big|_{x_0} (x_i - x_{0i})(x_j - x_{0j}) + \cdots. \tag{659}$$

Defining the $M \times 1$ column vector $\tilde{\mathbf{x}} \equiv \mathbf{x} - \mathbf{x}_0$, and $\mathbf{H} \equiv \nabla\nabla L(\mathbf{x}_0)$, this pdf is a *multivariate Gaussian*

$$\text{prob}(\tilde{\mathbf{x}}|y, I) = \frac{1}{(2\pi)^{M/2}\sqrt{\det(\mathbf{H})}} \exp\left[\frac{1}{2}\tilde{\mathbf{x}}^T \mathbf{H}\tilde{\mathbf{x}}\right] \tag{660}$$

where $\mathbf{H}$ is a symmetric $M \times M$ matrix with $\mathbf{H}_{ij} = [\nabla\nabla L(\mathbf{x}_0)]_{ij} = \frac{\partial^2 L}{\partial x_i \partial x_j}$, and the transpose $\tilde{\mathbf{x}}^T$ is a row vector. Since the quadratic exponent is just a generalization of Q in Eq. (643), we can think of Figure 31 as a 2-D slice through it. The *multivariate Gaussian* [Eq. (660)] is normalized so that $\int \text{prob}(\mathbf{x}|y, I)d\mathbf{x} = 1$, where the integration is carried out with respect to all parameters $\{x_j\}$, and has its maximum at $\mathbf{x}_0$ determined by Eq. (658): $\nabla L(\mathbf{x}_0) = 0$. By comparison with the 1-D Gaussian in Eq. (614), namely $\text{prob}(x_k, \mu, \sigma) = \frac{1}{\sigma\sqrt{2\pi}} \exp[-\frac{(x_k-\mu)^2}{2\sigma^2}]$, we see that $\mathbf{H} \equiv \nabla\nabla L(\mathbf{x}_0)$ is analogous to $-1/\sigma^2$, suggesting that the width of the posterior pdf should be related to the inverse of the second-derivative matrix. In fact, the measurement covariance matrix is given by

$$[\mathbf{S}_m]_{ij} = \langle (x_i - x_{0i})(x_j - x_{0j})\rangle = [-\mathbf{H}^{-1}]_{ij} = [-\nabla\nabla L(\mathbf{x}_0)^{-1}]_{ij} \tag{661}$$

which is a generalization of Eq. (657). The square root of the diagonal elements $(i = j)$ correspond to the error bars of the associated parameters, and the off-diagonal elements $(i \neq j)$ provide information about the correlations between the inferred values of x_i and x_j.

Example: Gaussian Noise Revisited

For the 1-D Gaussian case considered previously, we assumed that the magnitude σ of the *error-bar* was known. Now we want to relax this condition to estimate the posterior pdf $\text{prob}(\mu|\{x_k\}, I)$ rather than the posterior pdf $\text{prob}(\mu|\{x_k\}, \sigma, I)$, which is based on the assumption that σ is given. Thus, we must integrate out σ as a *nuisance parameter* from the joint posterior pdf:

$$\text{prob}(\mu|\{x_k\}, I) = \int_0^\infty \text{prob}(\mu|\{x_k\}, \sigma, I)d\sigma. \tag{662}$$

The integrand can be expressed as a product of the likelihood function and the prior using *Bayes' theorem*:

$$\text{prob}(\mu|\{x_k\},\sigma,I) = \text{prob}(\{x_k\}|\mu,\sigma,I) \times \text{prob}(\mu,\sigma|I). \tag{663}$$

If the data are independent, the likelihood function is simply the product of 1-D *Gaussian distributions*: $\text{prob}(x_k,\mu,\sigma) = 1/(\sigma\sqrt{2\pi})\exp[-(((x_k-\mu)^2)/(2\sigma^2))]$. Thus, for N data points

$$\text{prob}(\{x_k\},\mu,\sigma,I) = (\sigma\sqrt{2\pi})^{-N}\exp\left[-\frac{1}{2\sigma^2}\sum_{k=1}^{N}(x_k-\mu)^2\right]. \tag{664}$$

For simplicity, let us assume a *flat prior*

$$\text{prob}(\mu,\sigma|I) = \text{constant for } \sigma > 0; \quad 0 \text{ otherwise.} \tag{665}$$

Multiplying this prior with the likelihood function in Eq. (664), the marginal distribution of Eq. (662) becomes

$$\text{prob}(\mu|\{x_k\},I) \propto \int_0^\infty t^{N-2}\exp\left[-\frac{t^2}{2}\sum_{k=1}^{N}(x_k-\mu)^2\right]dt \tag{666}$$

where we have made the change of variable $\sigma = 1/t \rightarrow d\sigma = -dt/t^2$. Substitution of $\tau = t\sqrt{\sum(x_k-\mu)^2} \rightarrow d\tau = dt\sqrt{\sum(x_k-\mu)^2}$ yields

$$\text{prob}(\mu|\{x_k\},I) \propto \frac{1}{[\sqrt{\sum(x_k-\mu)^2}]^{N-2}}\int_0^\infty \tau^{N-2}\exp\left[-\frac{1}{2}\tau^2\right]\frac{d\tau}{\sqrt{\sum(x_k-\mu)^2}} \tag{667}$$

or

$$\text{prob}(\mu|\{x_k\},I) \propto \left[\sum(x_k-\mu)^2\right]^{-(N-1)/2} \tag{668}$$

where the integral over τ has been absorbed in the constant of proportionality. To find the best estimate μ_0, we differentiate $L = \ln[\text{prob}(\mu|\{x_k\},I)]$:

$$\left.\frac{dL}{d\mu}\right|_{\mu_0} = \frac{(N-1)\sum_{k=1}^{N}(x_k-\mu_0)}{\sum_{k=1}^{N}(x_k-\mu_0)^2} = 0 \text{ implying } \mu_0 = \frac{1}{N}\sum_{k=1}^{N}x_k. \tag{669}$$

Differentiating L once more, and evaluating it at the maximum $\mu = \mu_0$, we have

$$\left.\frac{d^2L}{d\mu^2}\right|_{\mu_0} = -\frac{N(N-1)}{\sum_{k=1}^{N}(x_k-\mu_0)^2}. \tag{670}$$

Since the *error bar* for the best estimate is given by

$$\left(-\left.\frac{d^2L}{d\mu^2}\right|_{\mu_0}\right)^{-1/2} = \frac{1}{\sqrt{N}}\left[\frac{1}{N-1}\sum_{k=1}^{N}(x_k-\mu_0)^2\right]^{1/2}$$

we can summarize our inference of the mean by

$$\mu = \mu_0 \pm \frac{S}{\sqrt{N}}, \quad \text{where } S^2 = \frac{1}{N-1}\sum_{k=1}^{N}(x_k-\mu_0)^2. \tag{671}$$

Comparing this result with the previous one [Eq. (625)]

$$\mu = \mu_0 \pm \frac{\sigma}{\sqrt{N}} \tag{672}$$

in which we assumed σ to be given, we see that the only difference is that σ has been replaced by the estimate S derived from the data.

The Student-*t* and the χ^2 Distributions

As noted in the 1-D case, when σ is known, the posterior pdf $\text{prob}(\mu|\{x_k\}, \sigma, I)$ is defined completely by the best estimate and the associated error bar ($\pm\sigma/\sqrt{N}$), because the quadratic termination of the Taylor series is exact in this case. But such is not the case when σ is not known beforehand. Then the summary of Eq. (671):

$$\mu = \mu_0 \pm \frac{S}{\sqrt{N}}, \quad \text{where} \quad S^2 = \frac{1}{N-1}\sum_{k=1}^{N}(x_k - \mu_0)^2 \text{ just represents}$$

a useful approximation to the (marginal) posterior pdf: $\text{prob}(\mu|\{x_k\}, \sigma, I)$, and the actual pdf is given by [Eq. (668)]

$$\text{prob}(\mu|\{x_k\}, I) \propto \left[\sum_{k=1}^{N}(x_k - \mu)^2\right]^{-(N-1)/2}. \tag{673}$$

If we rewrite the sum as

$$\sum_{k=1}^{N}(x_k - \mu)^2 = N(\bar{x} - \mu)^2 + V; \quad V = \sum_{k=1}^{N}(x_k - \bar{x})^2; \quad \bar{x} = \mu_0 = \sum_{k=1}^{N} x_k/N \tag{674}$$

then substitution into Eq. (673) yields the *Student-t distribution*

$$\text{prob}(\mu|\{x_k\}, I) \propto \left[N(\bar{x} - \mu)^2 + V\right]^{-(N-1)/2}. \tag{675}$$

When $N = 3$, this Student-*t* pdf has the same form as the *Cauchy distribution* (see Figure 30), $\text{prob}(\mu|\{x_k\}, I) \propto \left[N(\bar{x} - \mu)^2 + V\right]^{-1}$. It has a maximum at $\mu = \bar{x}$, an FWHM that is proportional to $\sqrt{V}$, and very long tails. As N increases, this function is multiplied by itself many times: the wide wings are killed off, leaving a more Gaussian-like distribution centered about $\bar{x}$. The optimal value is always $\bar{x} = \mu_0 = \sum_{k=1}^{N} x_k/N$, the sample average, and the error bar is given by Eq. (671): $\frac{S}{\sqrt{N}}$, $S^2 = 1/(N-1)\sum_{k=1}^{N}(x_k - \mu_0)^2 \equiv V/(N-1)$, which becomes more meaningful as N increases. So far, we have assumed a *flat prior* given by Eq. (665), but to express complete prior ignorance we should assign a pdf that is uniform in μ and $\ln\sigma$ because the position μ is associated with an additive uncertainty, whereas the width is a multiplicative scale factor. Assigning a pdf which is uniform in $\ln\sigma$ would lead to the following posterior pdf:

$$\text{prob}(\mu|\{x_k\}, I) \propto \left[N(\bar{x} - \mu)^2 + V\right]^{-N/2}. \tag{676}$$

Equation (676) is identical to Eq. (675) except that the power has increased from $(N-1)/2)$ to $N/2$. Thus, we have a *Student-t distribution* with $N-1$ degrees of freedom instead of $N-2$. Note also that their shapes are very similar, with a maximum at $\bar{x}$, but the pdf of Eq. (676) is a little narrower than that of Eq. (675): $S^2 \propto V/N$ rather than $S^2 \propto V/(N-1)$, but this difference is negligible when N is moderately large. We conclude that use of the flat prior [Eq. (665)] gives a slightly more conservative estimate of the error bar, but the results remain essentially unchanged.

What can we learn about the magnitude of the expected error in the measurements? Our inference about σ is described by the (marginal) posterior pdf

$$\mathrm{prob}(\sigma|\{x_k\},I) = \int_{-\infty}^{\infty} \mathrm{prob}(\mu,\sigma|\{x_k\},I)d\mu. \tag{677}$$

Using *Bayes' theorem* to rewrite the integrand in the above equation

$$\mathrm{prob}(\mu,\sigma|\{x_k\},I) = \mathrm{prob}(\{x_k\}|\mu,\sigma,I) \times \mathrm{prob}(\mu,\sigma|I)$$

and Eq. (664) $\mathrm{prob}(\{x_k\},\mu,\sigma,I) = (\sigma\sqrt{2\pi})^{-N}\exp[-\frac{1}{2\sigma^2}\sum_{k=1}^{N}(x_k-\mu)^2]$ for the likelihood function, a flat prior, and the substitution of $\sum_{k=1}^{N}(x_k-\mu)^2 = N(\bar{x}-\mu)^2 + V$, we obtain

$$\mathrm{prob}(\sigma|\{x_k\},I) \propto \sigma^{-N}\exp\left(-\frac{V}{2\sigma^2}\right)\int_{-\infty}^{\infty}\exp\left[-\frac{N(\bar{x}-\mu)^2}{2\sigma^2}\right]d\mu. \tag{678}$$

Making the change of variable $t = \frac{\sqrt{N}(\mu-\bar{x})}{\sqrt{2}\sigma}$, so that $d\mu = (\sqrt{2/N})\sigma dt$, we see that the integral in Eq. (678) is proportional to σ, so that we obtain

$$\mathrm{prob}(\sigma|\{x_k\},I) \propto \sigma^{1-N}\exp\left(-\frac{V}{2\sigma^2}\right). \tag{679}$$

This result is related to the χ^2 *distribution* through the substitution $x^2 = V/\sigma^2$. As usual, the best estimate and its *error bar* can be derived from the first and second derivatives of the logarithm of Eq. (679):

$$L = \ln[\mathrm{prob}(\sigma|\{x_k\},I)] \propto -(N-1)\sigma - \frac{V}{2\sigma^2} \tag{680}$$

$$\left.\frac{dL}{d\sigma}\right|_{\sigma_0} = -(N-1) + V\sigma_0^{-3} = 0, \text{ and } \sigma_0^3 = V/(N-1); \qquad \left.\frac{d^2L}{d\sigma^2}\right|_{\sigma_0} = -3V\sigma_0^{-4}.$$

Thus, we may summarize our inference about σ by

$$\sigma = \sigma_0 \pm \frac{3(N-1)}{\sigma_0}; \qquad \sigma_0 = \left(\frac{V}{N-1}\right)^{1/3}. \tag{681}$$

The Linear Problem

If a set of M parameters $\{x_j\}$ $(j = 1, 2, \ldots, M)$ are represented by the components of a column vector $\mathbf{x}$, the condition for the best estimate, denoted by $\mathbf{x}_0$, is

$$\nabla L(\mathbf{x}_0) = 0, \qquad [\nabla L(\mathbf{x}_0)]_j = \left.\frac{\partial L}{\partial x_j}\right|_{\mathbf{x}=\mathbf{x}_0} \tag{682}$$

where [see Eq. (641)] $L = \ln[\mathrm{prob}(\{x_j\}|y,I)]$. Equation (682) is a compact notation for M simultaneous equations, which are difficult to solve in general, but if they are linear, we have [see Eq. (659)]

$$\nabla L = \mathbf{H}\mathbf{x} + \mathbf{C}. \tag{683}$$

If each component of the vector $\mathbf{C}$ and the square matrix $\mathbf{H}$ is a constant, then the solution of Eq. (683) for $\nabla L = 0$ is

$$\mathbf{x}_0 = -\mathbf{H}^{-1}\mathbf{C}. \tag{684}$$

Differentiating Eq. (683), we obtain

$$\nabla\nabla L = \mathbf{H} \tag{685}$$

implying that all higher derivatives are zero. Therefore the *covariance matrix* is [see Eq. (661)]

$$[\mathbf{S}]_{ij} = \langle (x_i - x_{0i})(x_j - x_{0j}) \rangle = -[\mathbf{H}^{-1}]_{ij} \tag{686}$$

where $\{x_{0j}\}$ are proportional to the components of the vector $\mathbf{x}_0$ in Eq. (684). Since $\mathbf{H}^{-1}$ is proportional to $1/\det \mathbf{H}$, if the determinant of $\mathbf{H}$ is very small, the covariance matrix in Eq. (686) will be sensitive to small changes in the data, leading to large error bars. In this case, the ellipsoid of $Q = k$ (Figure 31), where

$$Q = (\mathbf{x} - \mathbf{x}_0)^T \nabla\nabla L(\mathbf{x}_0)(\mathbf{x} - \mathbf{x}_0)$$

will be very long in at least one of its principal directions. When $\det \mathbf{H} \approx 0$, it is useful to analyze the eigenvalues and eigenvectors of $\mathbf{H}$ to find its principal axes. They indicate which linear combinations of the parameters one may infer independently of one another, and the eigenvalues indicate how reliably they can be estimated. The only real cure in this case is to improve the posterior pdf by obtaining more relevant data, or by supplementing them with appropriate prior information.

We can use the linear approach also in cases when Eq. (683), with $\nabla L = \mathbf{H}\mathbf{x}_0 + \mathbf{C} = 0$, is not quite satisfied. Consider a *Taylor expansion* of L about an arbitrary point $\mathbf{x}_1$

$$L(\mathbf{x}) = L(\mathbf{x}_1) + (\mathbf{x} - \mathbf{x}_1)^T \nabla L(\mathbf{x}_1) + \frac{1}{2}(\mathbf{x} - \mathbf{x}_1)^T \nabla\nabla L(\mathbf{x}_1)(\mathbf{x} - \mathbf{x}_1) + \cdots \tag{687}$$

where, previously, the first derivative term was missing because we expanded about the *optimal estimate*. Differentiating with respect to $\{x_j\}$, we obtain a Taylor expansion of ∇L

$$\nabla L(\mathbf{x}) = \nabla L(\mathbf{x}_1) + \nabla\nabla L(\mathbf{x}_1)(\mathbf{x} - \mathbf{x}_1) + \cdots. \tag{688}$$

If we ignore higher order terms, the solution to $\nabla L(\mathbf{x}_2) = 0$ becomes

$$\mathbf{x}_2 \approx \mathbf{x}_1 - [\nabla\nabla L(\mathbf{x}_1)]^{-1} \nabla L(\mathbf{x}_1). \tag{689}$$

When $\mathbf{x}_1 = \mathbf{x}_2$, or if ∇L is linear, this relationship will be exact. But it will be a reasonable approximation as long as $\mathbf{x}_1$ is fairly close to the optimal estimate, giving rise to the following *iterative* algorithm:

- start with a good guess $\mathbf{x}_1$ of the optimal solution;
- evaluate the gradient vector ∇L, and the second-derivative matrix $\nabla\nabla L$ at $\mathbf{x} = \mathbf{x}_1$;
- calculate an improved estimate $\mathbf{x}_2$ by equating it to the right-hand side of Eq. (689);
- repeat the process until $\nabla L = 0$.

This *Newton–Raphson* algorithm is a generalization of a numerical method to find roots of a function $f(x_0) = 0$. For the multivariate case $\nabla L(\mathbf{x}) = 0$, the Newton–Raphson algorithm is given by the recursion formula

$$\mathbf{x}_{i+1} = \mathbf{x}_i - \left[\nabla\nabla L(\mathbf{x}_i)\right]^{-1} \nabla L(\mathbf{x}_i) \tag{690}$$

where $\mathbf{x}_i$ is the estimate of the solution after $i - 1$ iterations. Equation (690) will converge rapidly to $\mathbf{x}_0$ as long as the starting point is "close enough" to the solution.

The stability can be improved by adding a small (negative) number to the diagonal elements of $\nabla\nabla L$:

$$\mathbf{x}_{i+1} = \mathbf{x}_i - \left[\nabla\nabla L(\mathbf{x}_i) + c\mathbf{1}\right]^{-1} \nabla L(\mathbf{x}_i) \tag{691}$$

where $\mathbf{1}$ is the *identity matrix*. The matrix can be characterized by its eigenvalues $\{\lambda_j\}$ and eigenvectors $\{\mathbf{e}_j\}$:

$$[\nabla\nabla L]\mathbf{e}_j = \lambda_j \mathbf{e}_j; \qquad j = 1, \dots, M. \tag{692}$$

Adding $c\mathbf{1}$, one obtains a new matrix with the same eigenvectors, but different eigenvalues:

$$[\nabla\nabla L + c\mathbf{1}]\mathbf{e}_j = [\lambda_j + c]\mathbf{e}_j. \tag{693}$$

The modification of the diagonal does not affect the eigenvectors and hence not the orientation of the ellipsoid, and it leaves the correlations between the parameters unaltered. However, a suitable choice of the constant c causes a significant narrowing of the ellipsoid in those principal directions in which it was very elongated. Since small eigenvalues are associated with large uncertainties, Eq. (691) stabilizes the algorithm by reducing their influence. The inverse of a matrix is proportional to the reciprocal of its determinant, which is given by the product of the eigenvalues. Hence, small eigenvalues will cause $[\nabla\nabla L(\mathbf{x}_i)]^{-1}$ to become *ill conditioned*. Adding a small (negative) value times the identity matrix to $\nabla\nabla L(\mathbf{x}_i)$ ensures that the magnitude of the determinant is bounded away from zero.

The *Newton–Raphson* method will diverge if the initial guess $\mathbf{x}_1$ is not close enough to the optimal solution. In such cases, an "up-hill" *simplex* search algorithm can be employed. Such an algorithm works with the function L directly and is robust for unimodal pdfs. It can be used for up to a couple of dozen parameters, but lacks the efficiency of the Newton–Raphson method, and is therefore best used as a first step to get close enough to the optimal solution so that the Newton–Raphson will work. If the posterior pdf is multimodal, the optimization task becomes very difficult, and it is almost impossible to guarantee that the optimal solution has been found. It makes little sense to talk about a single "best" answer if the posterior pdf has several maxima of comparable magnitude. No algorithm seems to exist that can claim universal success or applicability if the posterior pdf is multimodal.

5.2.3 Approximations: Maximum Likelihood and Least Squares

The virtue of *Bayes' theorem* is that it helps one relate the required pdf to others that are easier to deal with. If the vector **x** denotes a set of unknown parameters M, and the vector **y** denotes N measured data, then according to Bayes' theorem

$$\text{prob}(\mathbf{x}|\mathbf{y},\mathbf{I}) \propto \text{prob}(\mathbf{y}|\mathbf{x},\mathbf{I}) \times \text{prob}(\mathbf{x}|\mathbf{I}) \tag{694}$$

where **I** represents all relevant background information, and the prior pdf should represent what we know about **x** before we analyze the current data. If we have little prior knowledge, we may indicate that by assigning a uniform prior pdf

$$\text{prob}(\mathbf{x}|\mathbf{I}) = \text{constant} \tag{695}$$

so that Eq. (694) becomes

$$\text{prob}(\mathbf{x}|\mathbf{y},\mathbf{I}) \propto \text{prob}(\mathbf{y}|\mathbf{x},\mathbf{I}). \tag{696}$$

Our *best estimate* $\mathbf{x}_0$, given by the maximum of the posterior pdf, $\text{prob}(\mathbf{x}|\mathbf{y},\mathbf{I})$, is the solution that yields the greatest value for the probability of the observed data $\text{prob}(\mathbf{y}|\mathbf{x},\mathbf{I})$, referred to as the *maximum likelihood estimate.* If we further assume that the observed data are independent, then

$$\text{prob}(\mathbf{y}|\mathbf{x},\mathbf{I}) = \prod_{k=1}^{N} \text{prob}(y_k|\mathbf{x},\mathbf{I}) \tag{697}$$

where y_k is the kth data point.

Although we have used this result several times before, we emphasize that it follows from the product rule:

$$\text{prob}(y_k, y_l|\mathbf{x},\mathbf{I}) = \text{prob}(y_k|y_l,\mathbf{x},\mathbf{I}) \times \text{prob}(y_l|\mathbf{x},\mathbf{I}) \tag{698}$$

and the assumption that our knowledge of one data point has no influence on our ability to predict the outcome of another data point, provided **x** is given:

$$\text{prob}(y_k|y_l,\mathbf{x},\mathbf{I}) = \text{prob}(y_k|\mathbf{x},\mathbf{I}). \tag{699}$$

If we also assume that the measurement noise can be represented by a Gaussian process, then

$$\text{prob}(y_k|\mathbf{x},\mathbf{I}) = \frac{1}{\sigma_k\sqrt{2\pi}} \exp\left[-\frac{(F_k - y_k)^2}{2\sigma_k^2}\right] \tag{700}$$

where **I** includes knowledge of both the expected size of the error bars $\{\sigma_k\}$ and a model of the functional relationship between the parameters **x** and the noiseless data **F**:

$$F_k = f(\mathbf{x}, k). \tag{701}$$

Equations (697) and (700) allow one to approximate the likelihood function by

$$\text{prob}(\mathbf{y}|\mathbf{x},\mathbf{I}) \propto \exp\left(-\frac{\chi^2}{2}\right) \tag{702}$$

where χ^2 is the sum of the squares of the normalized residuals $R_k = (F_k - y_k)/\sigma_k$:

$$\chi^2 = \sum_{k=1}^{N} \left(\frac{F_k - y_k}{\sigma_k}\right)^2. \tag{703}$$

With a uniform prior [Eq. (695)], the logarithm L of the posterior pdf is given by [see Eq. (702)]

$$L = \ln\,[\text{prob}(\mathbf{x}|\mathbf{y}, \mathbf{I})] = \text{constant} - \frac{\chi^2}{2} \tag{704}$$

which shows that the maximum of the posterior pdf will occur when χ^2 is smallest. Therefore, the corresponding *optimal solution* $\mathbf{x}_0$ is called the *least-squares estimate*. Note that the maximum likelihood procedure follows directly from *Bayes' theorem* by assuming **a uniform prior, and that the least-squares solution is based on the further assumptions of independent data and Gaussian measurement noise.** If these assumptions are not fulfilled, one should start again from Bayes' theorem and derive a more suitable statistical description. The main reason for the popularity of the *least-squares method* is that it is easy to use, especially when the functional relationship $F_k = f(\mathbf{x}, k)$ is linear, because then $\nabla L = \mathbf{Hx} + \mathbf{C}$, which makes the optimization problem easy. To see that such is the case, we write the equation for the kth noiseless data point as

$$F_k = \sum_{k=1}^{M} T_{kj}x_j + C_k \quad or \quad \mathbf{F} = \mathbf{Tx} + \mathbf{C} \ \text{ in matrix notation,} \tag{705}$$

where both T_{kj} and C_k are independent of the parameters $\{x_j\}$. Using the chain-rule of differentiation

$$\frac{\partial L}{\partial x_j} = -\frac{1}{2}\frac{\partial \chi^2}{\partial x_j} = -\sum_{k=1}^{N} \frac{(F_k - y_k)}{\sigma_k^2}\frac{\partial F_k}{\partial x_j} = -\sum_{k=1}^{N} \frac{(F_k - y_k)}{\sigma_k^2} T_{kj}. \tag{706}$$

To verify the linearity of ∇L, we differentiate again with respect to x_i, noting that the elements of the second-derivative matrix

$$\nabla\nabla L = \frac{\partial^2 L}{\partial x_i \partial x_j} = -\sum_{k=1}^{N} \frac{T_{ki}T_{kj}}{\sigma_k^2} \tag{707}$$

are all constant. Since all higher order derivatives of L are identically zero, the posterior pdf is defined completely by the optimal solution $\mathbf{x}_0$ and its *covariance matrix*, $\mathbf{S}_m$. The components of the covariance matrix are related to twice the inverse of the Hessian matrix, $\nabla\nabla\chi^2$:

$$[\mathbf{S}_m]_{ij} = \langle (x_i - x_{0i})(x_j - x_{0j}) \rangle = -[(\nabla\nabla L)^{-1}]_{ij} = 2[(\nabla\nabla\chi^2)^{-1}]_{ij}. \tag{708}$$

The property that the matrix $\nabla\nabla L =$ constant is not generally true unless the least squares solution, $L = \ln\,[\text{prob}(\mathbf{x}|\mathbf{y}, \mathbf{I})] = \text{constant} - (\chi^2/2)$, is valid, and it explains why the least-squares approximation is very convenient. For example

- the model $y_k = n_0[Ae^{-(x_k - x_0)^2/2w^2} + B]$ is linear wrt the amplitude A and the background B,

- but it is difficult to write down the optimal solution (A_0, B_0) analytically, because
- the gradient vector of the posterior pdf $L = \ln[\text{prob}(A, B|\{N_k\}, I)] = \text{constant} + \sum_{k=1}^{M}[N_k \ln(y_k) - y_k]$ cannot be rearranged into the linear form: $\nabla L = \mathbf{Hx} + \mathbf{C}$.

But for the *Poisson likelihood distribution*

$$\text{prob}(N_k|y_k) = \text{prob}(N_k|A, B, I) = \frac{y_k^{N_k}\, e^{-y_k}}{N_k!} \propto \exp\left[-\frac{(N_k - y_k)^2}{2y_k}\right] \tag{709}$$

where the Gaussian form applies for large N, we can get a good estimate using the *least-squares approximation*. We can summarize the result above as $N_k \approx y_k \pm \sqrt{y_k}$. With a *uniform prior*, the logarithm of the posterior pdf is approximated well with Eq. (704): $L = \ln[\text{prob}(\mathbf{x}|\mathbf{y}, \mathbf{I})] = \text{constant} - (\chi^2/2)$, and setting $F_k = N_k$, the χ^2 statistic is given by the l_2-norm:

$$\chi^2 = \sum_{k=1}^{N} \frac{(F_k - y_k)^2}{2y_k} \tag{710}$$

where y_k is the measured data point, F_k is our estimate of the measured data point based on the linear relationship $y_k = n_0[Ae^{-(x_k - x_0)^2/2w^2} + B]$, and the *error bar* in Eq. (703): $\chi^2 = \sum_{k=1}^{N}\left(\frac{F_k - y_k}{\sigma_k}\right)^2$ is replaced by the square root of the data point ($\sigma^2 = y_k$).

Despite the practical benefits of the least-squares solution, it should be emphasized again that **the real justification for its use hinges on the assignment of a Gaussian likelihood function, and a uniform prior.** What other reason do we have for using the sum of the squares of the residuals as a misfit statistic as opposed to anything else? For example, the l_1-norm

$$l_1 - \text{norm} = \sum_{k=1}^{N} \left|\frac{F_k - D_k}{\sigma_k}\right| \tag{711}$$

has the advantage that it is less susceptible to "freak" data than the l_2-norm. The maximum entropy principle can be used to assign the relevant uniform, Gaussian, and exponential pdfs needed to justify the maximum likelihood, least-squares, and l_1-norm estimates.

The term "maximum likelihood estimate" suggests that we have obtained the most probable values of the parameters of interest, which is not exactly true. Instead, by maximizing the likelihood function we have found values that make the measured data most probable. Although we expect these values to be relevant to our real question, "What are the most probable values of the parameters, given the data"?, the *"most probable values of the parameters of interest"* are not the same as *"the values that make the measured data most probable"* since, in general, $\text{prob}(A|B) \neq \text{prob}(B|A)$. But as noted in Eq. (696): $\text{prob}(\mathbf{x}|\mathbf{y}, \mathbf{I}) \propto \text{prob}(\mathbf{y}|\mathbf{x}, \mathbf{I})$ if we assign a uniform prior. However, even with a flat prior, due consideration of the range over which it is valid can lead to a better estimate of the parameters involved and enable us to tackle problems that would otherwise be inaccessible.

5.2.4 Error Propagation: Changing Variables

How will uncertainties in our estimate of a set of parameters translate into reliabilities of quantities derived from them? For example, if we are told $x = 10 \pm 3$ and $y = 7 \pm 2$, what can we say about the difference $x - y$, or the ratio x/y, or the sum of their squares $x^2 + y^2$, and so on? The problem simply involves a change of variables. Given $\text{prob}(x, y|I)$, we need to determine $\text{prob}(z|I)$ where $z = x - y$, or $z = x/y$, and so on. Consider a single variable and some function of it. Given $y = f(x)$, how is $\text{prob}(x|I)$ related to $\text{prob}(y|I)$? The probability that x lies in the range between $x^* - \delta x/2$ and $x^* + \delta x/2$ about $x = x^*$ is given by

$$\text{prob}\left(x^* - \frac{\delta x}{2} \le x^* + \frac{\delta x}{2}|I\right) \approx \text{prob}(x = x^*|I)\delta x \tag{712}$$

where the approximation becomes exact in the limit $\delta x \to 0$. Suppose now that we view the pdf as a function of another variable y, related (monotonically) to x by $y = f(x)$. Then f will map the point $x = x^*$ (uniquely) to $y = y^* = f(x^*)$, and the interval δx to the corresponding region δy, as illustrated in Figure 33. Since the range of y-values spanned by $y \pm \delta y/2$ is equivalent to the range of x-values spanned by $x \pm \delta x/2$, the area under the pdf $\text{prob}(y|I)$ should equal the probability represented by Eq. (712). Thus, we require

$$\text{prob}(x = x^*|I)\delta x = \text{prob}(y = y^*|I)\delta y. \tag{713}$$

Since this relationship must be true for any point in the x-space, in the limit of infinitesimally small intervals

$$\text{prob}(x|I) = \text{prob}(y|I) \times \left|\frac{dy}{dx}\right| \tag{714}$$

where $\left|\frac{dy}{dx}\right|$ is called the *Jacobian*, and represents a ratio of lengths.

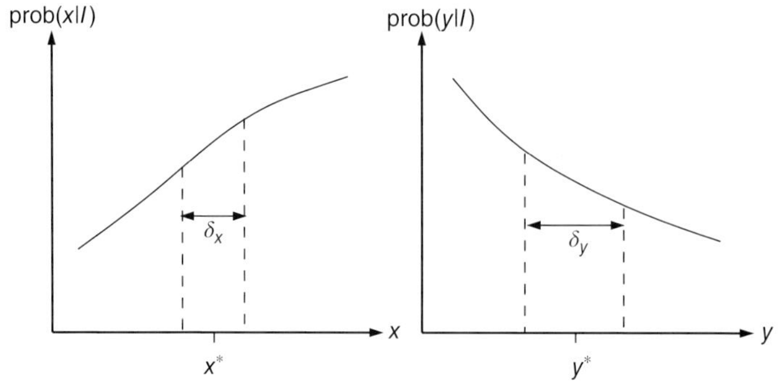

Figure 33 Change of variables in one dimension. The function f maps the point x^* to $y^* = f(x^*)$.

Example: Lighthouse Problem Revisited

As an example, we consider the lighthouse problem, where we assigned a uniform pdf for the angle of the kth data point θ_k:

$$\text{prob}(\theta_k|a, d, I) = \frac{1}{\pi}. \tag{715}$$

From Eq. (633): $d \tan\theta_k = x_k - a$, which follows from the geometry (see Figure 29), it can be shown using Eq. (714) that the pdf in θ_k can be transformed to its equivalent form in terms of x_k: (see Problem 5.3)

$$\text{prob}(x_k|a, d, I) = \frac{d}{\pi[d^2 + (x_k - a)^2]}. \tag{716}$$

Generalization

The result $\text{prob}(x|I) = \text{prob}(y|I) \times \left|dy/dx\right|$ can be generalized to apply to several variables:

- to express the pdf for M variables $\{x_j\}$ in terms of the same number of $\{y_j\}$ quantities related to them, we must ensure that

$$\text{prob}(\{x_j\}|I)\ \delta x_1 \delta x_2 \cdots \delta x_M = \text{prob}(\{y_j\}|I)\ \delta^M \text{Vol}(\{y_j\}) \tag{717}$$

where $\text{Vol}(\{y_j\})$ is the M-dimensional volume in y-space mapped out by the small hypercube region $\delta x_1 \delta x_2 \cdots \delta x_M$ in x-space. It can be shown that

$$\delta^M \text{Vol}(\{y_j\}) = \overbrace{\left|\frac{\partial(y_1, y_2, \ldots, y_M)}{\partial(x_1, x_2, \ldots, x_M)}\right|}^{\text{multivariate Jacobian}} \delta x_1 \delta x_2 \cdots \delta x_M, \tag{718}$$

where the *multivariate Jacobian* is given by the determinants of the $M \times M$ matrix of partial derivatives $\partial y_i/\partial x_j$. Thus, the general form of Eq. (714): $\text{prob}(x|I) = \text{prob}(y|I) \times \left|\frac{dy}{dx}\right|$ becomes

$$\text{prob}(\{x_j\}|I) = \text{prob}(\{y_j\}|I) \times \left|\frac{\partial(y_1, y_2, \ldots, y_M)}{\partial(x_1, x_2, \ldots, x_M)}\right|. \tag{719}$$

Illustration

Consider the transformation of a pdf defined on a two-dimensional Cartesian grid $\{y_j\} = (x, y)$ to its equivalent polar coordinates $\{x_j\} = (R, \theta)$. From Figure 34, we see that $x = R\cos\theta$ and $y = R\sin\theta$, and it is easy to show that the *Jacobian* becomes (see Problem 5.4)

$$\left|\frac{\partial(x, y)}{\partial(R, \theta)}\right| = R. \tag{720}$$

Therefore, according the Eq. (719)

$$\text{prob}(R, \theta|I) = \text{prob}(x, y|I) \times R. \tag{721}$$

Thus, if the pdf for x and y were an *isotropic* bivariate Gaussian

$$\text{prob}(x, y|I) = \frac{1}{2\pi\sigma^2} \exp\left[-\frac{(x^2 + y^2)}{2\sigma^2}\right] \tag{722}$$

then the corresponding pdf for R and θ would be (see Problem 5.4)

$$\mathrm{prob}(R,\theta|I) = \frac{R}{2\pi\sigma^2}\exp\left[-\frac{R^2}{2\sigma^2}\right]. \tag{723}$$

Alternative Derivation

Equation (721) can be obtained directly from a simple geometrical argument:

- the probability that the polar parameters lie in the small range $R \pm dR/2$ and $\theta \pm d\theta/2$ is given by
- the product of the value of $\mathrm{prob}(x,y|I)$ at the point (R,θ) and the element of area $Rd\theta \times dR$ in Figure 34;
- by the definition of the pdf for R and θ, this area is equal to $\mathrm{prob}(R,\theta|I)d\theta dR$; hence we obtain the desired result $\mathrm{prob}(x,y|I)Rd\theta \times dR = \mathrm{prob}(R,\theta|I)d\theta dR$ or

$$\mathrm{prob}(R,\theta|I) = \mathrm{prob}(x,y|I) \times R.$$

To obtain the pdf for the radius $R = \sqrt{x^2+y^2}$, we marginalize $\mathrm{prob}(R,\theta|I)$ over θ:

$$\mathrm{prob}(R|I) = \int_0^{2\pi} \mathrm{prob}(R,\theta|I)\, d\theta = \frac{R}{\sigma^2}\exp\left[-\frac{R^2}{2\sigma^2}\right] \tag{724}$$

which could have been derived directly from Eq. (722) by multiplication with the area of the circular shell in Figure 34: $2\pi RdR$, yielding $\mathrm{prob}(R|I)dR = \mathrm{prob}(x,y|I)2\pi RdR$ which yields Eq. (724), since $R^2 = x^2 + y^2$.

Multidimensional Generalization

The likelihood function of Eqs. (702): $\mathrm{prob}(\mathbf{y}|\mathbf{x},\mathbf{I}) \propto \exp(-\chi^2/2)$ and (703): $\chi^2 = \sum_{k=1}^{N}((F_k - y_k)/(\sigma_k))^2$ is an N-dimensional isotropic Gaussian when viewed in terms of the normalized residuals:

$$\mathrm{prob}(\mathbf{y}|\mathbf{x},\mathbf{I}) \propto \exp\left[-\frac{r_1^2 + r_2^2 + \cdots + r_N^2}{2}\right] = \exp\left[-\frac{R^2}{2}\right] \tag{725}$$

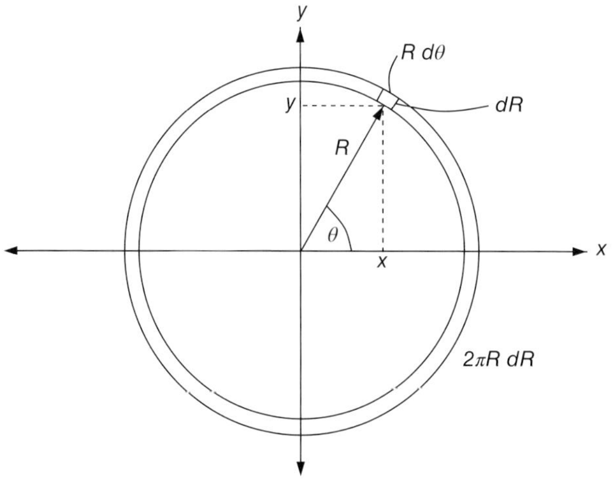

Figure 34 Change of variables from Cartesian to polar coordinates.

where $r_k = (F_k - y_k)/\sigma_k$, and $R = \sqrt{\sum r_k^2} = \chi$. The probability that R lies in a narrow range δR, $\text{prob}(R|\mathbf{x}, \mathbf{I})\delta R$, will be equal to the product of the magnitude of the likelihood function at that radius and the hypervolume of the associated spherical shell, which is proportional to R^{N-1}:

$$\text{prob}(R|\mathbf{x}, \mathbf{I}) \propto R^{N-1} \exp\left(-\frac{R^2}{2}\right). \tag{726}$$

We may use $\text{prob}(x|I) = \text{prob}(y|I) \times \left|(dy)/(dx)\right|$ to turn this result into a pdf for χ^2. Since $\chi^2 = R^2$, we obtain

$$\text{prob}(\chi^2|\mathbf{x}, \mathbf{I}) \propto (\chi^2)^{N/2-1} \exp\left(-\frac{\chi^2}{2}\right) \tag{727}$$

which is called a χ^2 *distribution* with N *degrees of freedom*:

- for $N \geq 2$, it has a maximum at $N - 2$;
- if the number of data is large, its shape is well described by a Gaussian pdf summarized by $\chi^2 \approx N \pm \sqrt{2N}$.

In summary: The basic ingredients required for the propagation of errors involves either

- a transformation in the sense of Eq. (719): $\text{prob}(\{x_j\}|I) = \text{prob}(\{y_j\}|I) \times \left|\frac{\partial(y_1, y_2, \ldots, y_M)}{\partial(x_1, x_2, \ldots, x_M)}\right|$, or an integration
- such as Eq. (724): $\text{prob}(R|I) = \int_0^{2\pi} \text{prob}(R, \theta|I)\, d\theta = (R/\sigma^2) \exp[-(R^2/(2\sigma^2))]$, or a combination of the two.

5.3 Model Selection or Hypothesis Testing

Suppose two models or theories are available: model A and a competing model B which has an adjustable parameter λ. Which model should we prefer on the basis of available data D? For real data, supposing we must decide whether a signal peak should be Gaussian or Lorentzian, which would be better? This type of question is called *model selection* (or comparison) or *hypothesis testing*. One might think that the choice between available model options could be based on how well a particular model fits the data, but there is a problem. More complex models with many adjustable parameters, will always be able to give better agreement with experimental data. However, although a high-order polynomial may fit one-dimensional graphical data better than a simple straight line, one would tend to prefer the straight line unless the discrepancy is very large. To judge the relative merits of models A and B, we need to evaluate the posterior probabilities for models A and B being correct by taking the ratio

$$\text{posterior ratio} = \frac{\text{prob}(A|D, I)}{\text{prob}(B|D, I)}. \tag{728}$$

If the ratio is much larger than one, we prefer model A; if it is much smaller than one, we prefer model B, and if it is close to one, we cannot make an informed judgement based on the available data.

Applying *Bayes' theorem* to both the numerator and denominator in Eq. (728), we have

$$\frac{\text{prob}(A|D,I)}{\text{prob}(B|D,I)} = \frac{\text{prob}(D|A,I)}{\text{prob}(D|B,I)} \times \frac{\text{prob}(A|I)}{\text{prob}(B|I)} \tag{729}$$

because the term $prob(D|I)$ cancels out. The answer to our question depends on our assessment of the models before analyzing the data. Taking the ratio of the prior terms to be unity, we still need to assign $\text{prob}(D|A,I)$ and $\text{prob}(D|B,I)$. We must compare the data with the predictions of models A and B; the larger the mismatch, the lower the probability. This calculation is straightforward for model A, but for model B we cannot make a prediction without a value of λ. To circumvent this problem, we can use the sum and product rules and *marginalization* to write

$$\text{prob}(D|B,I) = \int \text{prob}(D,\lambda|B,I)d\lambda = \int \text{prob}(D|\lambda,B,I) \times \text{prob}(\lambda|B,I)d\lambda \tag{730}$$

where $\text{prob}(D|\lambda,B,I)$ (λ given) is an ordinary likelihood function, while $\text{prob}(\lambda|B,I)$ is model B's prior pdf for λ. The knowledge (or ignorance) of λ in model B must be specified before it can be applied to the data. Assume now that in model B λ is only specified to lie in a certain range, so that a *uniform prior* may be assigned:

$$\text{prob}(\lambda|B,I) = \frac{1}{\lambda_{\text{max}} - \lambda_{\text{min}}} \quad \text{for } \lambda_{\text{min}} \le \lambda \le \lambda_{\text{max}}, \quad 0 \text{ otherwise.} \tag{731}$$

Let us further assume that a value λ_0 exists that yields the closest agreement with the data, and that the corresponding probability $\text{prob}(D,\lambda_0|B,I)$ will be the maximum of model B's likelihood function. As long as $\lambda \approx \lambda_0 \pm \sigma_\lambda$, we would expect a reasonable fit to the data to be given by the Gaussian pdf:

$$\text{prob}(D|\lambda,B,I) = \text{prob}(D|\lambda_0,B,I) \times \exp\left[-\frac{(\lambda-\lambda_0)^2}{2\sigma_\lambda^2}\right]. \tag{732}$$

The assignments of Eqs. (731) and (732) are illustrated in Figure 35.

Since the prior in Eq. (731) does not depend explicitly on λ, we may rewrite Eq. (730) as

$$\text{prob}(D|B,I) = \frac{1}{\lambda_{\text{max}} - \lambda_{\text{min}}} \int_{\lambda_{\text{min}}}^{\lambda_{\text{max}}} \text{prob}(D|\lambda,B,I)\, d\lambda. \tag{733}$$

Assuming that the cut-offs at λ_{min} and λ_{max} do not cause a significant truncation of the Gaussian in Eq. (732), so that its integral is $\sigma_\lambda\sqrt{2\pi}$, we get

$$\text{prob}(D|B,I) = \frac{\text{prob}(D|\lambda_0,B,I) \times \sigma_\lambda\sqrt{2\pi}}{\lambda_{\text{max}} - \lambda_{\text{min}}}. \tag{734}$$

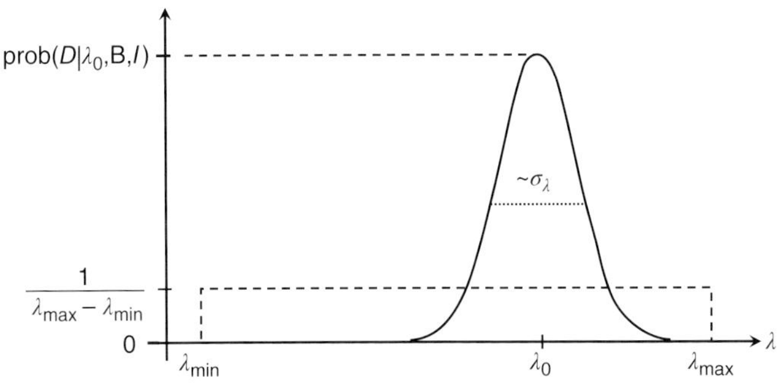

Figure 35 Schematic illustration of the prior pdf (dashed line) and the Gaussian likelihood function (solid line) for the parameter $\lambda \approx \lambda_0 \pm \sigma_\lambda$.

Substituting Eq. (734) into Eq. (729), we see that the ratio of the posteriors becomes

$$\frac{\text{prob}(A|D,I)}{\text{prob}(B|D,I)} = \frac{\text{prob}(A|I)}{\text{prob}(B|I)} \times \frac{\text{prob}(D|A,I)}{\text{prob}(D|\lambda_0,B,I)} \times \frac{\lambda_{\max} - \lambda_{\min}}{\sigma_\lambda\sqrt{2\pi}}. \tag{735}$$

- The first ratio reflects our relative preference for models A and B; to be fair, we set it to unity.
- The second ratio is a measure of how well the best predictions of each of the models agree with the data: with the added flexibility of model B's adjustable parameter, the second ratio can only favor model B, but
- the third ratio acts to penalize model B for the additional parameter, because the prior range $\lambda_{\max} - \lambda_{\min}$ will generally be much larger than the uncertainty $\pm\delta\lambda$ permitted by the data, which encompasses the spirit of
- *Ockham's razor: Frustra fit per plura quod potest fieri per pauciora* (**it is vain to do with more what can be done with fewer**).

According to Ockham, we should prefer the simplest theory that agrees with the empirical evidence, but

- what do we mean by the simpler theory if alternative models have the same number of adjustable parameters?
- In the choice between Gaussian and Lorentzian peak shapes, for example, each peak is defined by the position of the maximum and the width.

All that we are obliged to do in addressing such questions is to adhere to the rules of probability.

Consider now the case when model A also has an adjustable parameter, denoted by μ. Proceeding as in the case of model B, we find that Eq. (735) becomes

$$\frac{\text{prob}(A|D,I)}{\text{prob}(B|D,I)} = \frac{\text{prob}(A|I)}{\text{prob}(B|I)} \times \frac{\text{prob}(D|\mu_0,A,I)}{\text{prob}(D|\lambda_0,B,I)} \times \frac{\sigma_\mu(\lambda_{\max} - \lambda_{\min})}{\sigma_\lambda(\mu_{\max} - \mu_{\min})}. \tag{736}$$

As an example, Eq. (736) could represent a situation in which one had to choose between a Lorentzian and Gaussian shape for a signal peak. If the position of the maximum were fixed at the origin by theory, and the amplitude were constrained by the normalization of the data, then

- models A and B could be hypotheses favoring different lineshapes, where μ and λ are the FWHM values.
- If we were to give equal weight to models A and B before the analysis and assign similar prior ranges for μ and λ, then Eq. (736) would reduce to

$$\frac{\text{prob}(A|D,I)}{\text{prob}(B|D,I)} \approx \frac{\text{prob}(D|\mu_0,A,I)}{\text{prob}(D|\lambda_0,B,I)} \times \frac{\sigma_\mu}{\sigma_\lambda}. \tag{737}$$

For data of good quality, the dominant factor would be the best-fit likelihood ratio, but if both were to give comparable agreement with the data, then the shape with the larger *error bar* would be favored. **The reason behind this choice is that, in the context of model selection, a larger "error bar" means that more parameter values are consistent with the given hypothesis; hence its preferential treatment.**

If models A and B were based on the same physical theory, but were assigned different prior ranges, then with equal initial weightings for models A and B, Eq. (736) would give

$$\frac{\text{prob}(A|D,I)}{\text{prob}(B|D,I)} = \frac{(\lambda_{\max} - \lambda_{\min})}{(\mu_{\max} - \mu_{\min})} \tag{738}$$

because the best likelihood ratio would be unity ($\mu_0 = \lambda_0$) and $\sigma_\mu = \sigma_\lambda$. Thus, the analysis would lead us to prefer the model with a narrower prior range, which is reasonable because that model must be based on some additional insight, allowing it to assign the value of the parameter more accurately.

Comparison with Parameter Estimations

What is the difference between parameter estimation and model selection? To answer this question, we use *Bayes' theorem* to infer the value of λ from the data, given that model B is correct:

$$\text{prob}(\lambda|D,B,I) = \frac{\text{prob}(D|\lambda,B,I) \times \text{prob}(\lambda|B,I)}{\text{prob}(D|B,I)} \tag{739}$$

where the denominator, called the "evidence" for B, ascertains the merit of model B relative to a competing alternative. Since all the ingredients required for both parameter estimation and model selection appear in Eq. (739), we are not dealing with any new principles. The difference between parameter estimation and model selection is simply that we are asking different questions of the data:

- *parameter estimation* requires the location of the maximum of the likelihood function, whereas *model selection* entails the calculation of its average value.

As long as $\lambda_{\min}$ and $\lambda_{\max}$ encompass a significant region of prob$(D|\lambda,B,I)$ around λ_0, the precise bounds do not matter for optimal estimation, but since the prior range is employed in the computation of the average likelihood function, this range is important in model selection.

Hypothesis Testing

The preceding discussion of model selection suggests that we are dealing with a problem called *hypothesis testing.* Suppose we have a hypothesis H_1, which could be, for example, that the shape of a signal peak is Gaussian. To quantify our belief in this hypothesis based on available data D, and relevant background information I, we can use *Bayes' theorem* to evaluate the posterior pdf:

$$\text{prob}(H_1|D,I) = \frac{\text{prob}(D|H_1,I) \times \text{prob}(H_1|I)}{\text{prob}(D|I)} \tag{740}$$

where the denominator can be ignored if we are only interested in the relative merits of H_1 compared to another hypothesis H_2. In that case, we can apply Bayes' theorem to the second proposition, and divide by the expression above to arrive at

$$\frac{\text{prob}(H_1|D,I)}{\text{prob}(H_2|D,I)} = \frac{\text{prob}(D|H_1,I)}{\text{prob}(D|H_2,I)} \times \frac{\text{prob}(H_1|I)}{\text{prob}(H_2|I)} \tag{741}$$

which is the same as Eq. (729) with $H_1 = A$ and $H_2 = B$. But prob$(D|I)$ is important if we want to assess the intrinsic truth of H_1. Then we might let H_2 be the hypothesis that H_1 is false: $H_2 = \overline{H}_1$, and we could use *marginalization* and the product rule to obtain

$$\text{prob}(D|I) = \text{prob}(D|H_1,I) \times \text{prob}(H_1|I) + \text{prob}(D|\overline{H}_1,I) \times \text{prob}(\overline{H}_1|I) \tag{742}$$

where the two priors are related by the sum rule

$$\text{prob}(H_1|I) + \text{prob}(\overline{H}_1|I) = 1. \tag{743}$$

Considering Eq. (740) with H_1 replaced by $\overline{H}_1$, we see that the difficult term is the likelihood function prob$(D|\overline{H}_1,I)$ for $\overline{H}_1$:

- in general, we cannot compare the predictions of a hypothesis with the data given only that the hypothesis is false:
- we need well-defined alternatives. For example, if the shape of a peak is not Gaussian, could it be Lorentzian?
- With a specific set of possibilities $\{H_j\}$, the problem becomes one of model selection in which one has to compare the evidences given by prob$(D|H_j,I)$.

Even though agreement with data does not ensure the truth of a hypothesis, it is often said that it should be *rejected* in case of a poor fit to the data. Thus

- traditional hypothesis testing involves the use of procedures designed to asses the mismatch between theory and experiment. Often, the χ^2 statistic is employed. As shown by Eq. (727) [prob$(\chi^2|\mathbf{x},\mathbf{I}) \propto (\chi^2)^{N/2-1} \exp(-(\chi^2/2))$],
- if the data of interest are subject to Gaussian noise, the expected value of χ^2 will be approximately equal to the number of measurements N. Thus, deviations more than a few times $\sqrt{N}$ would not be expected. Nevertheless,
- one would be reluctant to reject a hypothesis because χ^2 is too large, the point being that:

- the misfit statistic is a measure of the likelihood function prob($D|H, I$). To reject or accept a hypothesis, one would need the posterior pdf prob($H|D, I$).

Even though a larger value of χ^2 would give a smaller value of prob($D|H, I$) likelihood for the data, one also needs prob($D|I$) and a value of the prior to determine the posterior pdf prob($H|D, I$).

- In spite of the conventional practice to use P-values as a criterion for rejection of a hypothesis, its basis appears to be doubtful.[2)]
- A misfit statistic can nonetheless serve a purpose if a poor quality of fit prompts one to think of alternative hypotheses, and probability theory then can provide the tools required to select the best one.
- **If there is no clearly stated alternative, and the null hypothesis is rejected, we are left with no rule at all, whereas the null hypothesis, though not satisfactory, may at any rate show some sort of correspondence with the facts (Jeffreys [212]).**
- While there was never a time when Newton's theory of gravity would not have failed a P-test, "The success of Newton was not that he explained all the variation of the observed positions of the planets, but that he explained most of it" (Jeffreys [212]).

5.4 Assigning Probabilities

5.4.1 Ignorance: Indifference, and Transformation Groups

Cox [213] showed that

- any method of plausible reasoning that satisfies elementary requirements of logical consistency must be equivalent to the use of probability theory;
- the sum and product rules specify the relationships between pdfs, but they do not tell us how to assign pdfs – are there any general principles or methods available for this purpose?

In 1713, Bernoulli proposed the *principle of insufficient reason* renamed by Keynes [214] as the *principle of indifference*. This principle states that **if we can enumerate a set of basic, mutually exclusive possibilities, and have no reason to believe that any one of them is more likely to be true than another, then we should assign the same probability to all of them.**

For an ordinary die, we can list the six potential outcomes of a roll as

$$x_i \equiv \text{the top face has } i \text{ dots, for } i = 1, 2, \dots, 6.$$

2) In statistical significance testing, the P-value is the probability of obtaining a test statistic result at least as extreme as the one that was actually observed, assuming that the null hypothesis is true.

According to Bernoulli's principle, we have

$$\text{prob}(x_i|I) = \frac{1}{6} \tag{744}$$

where the background information I only consists of the enumeration of the possibilities. Although this assignment is very reasonable, the question is whether it can be justified in a fundamental manner. Consistency demands that our probability assignment should not change if the order in which we list the outcomes are rearranged, and the only way to satisfy the requirement is through Eq. (744), which led Jaynes [215] to suggest that we should think of it as a consequence of the *desideratum consistency*, that is, our desire for consistency.

As an example, consider the case of colored balls that are being drawn randomly from an urn. If we knew that the urn contained only W white balls and R red balls, then according to the principle of indifference, we should assign a uniform pdf

$$\text{prob}(j|I) = \frac{1}{R+W} \tag{745}$$

for the proposition that any particular ball, denoted by j, would be drawn. Applying *marginalization* and the product rule, we can express the probability that the ball drawn will be red as [using Eq. (745)]

$$\begin{aligned}\text{prob}(\text{red}|I) &= \sum_{j=1}^{R+W} \text{prob}(\text{red}, j|I) = \sum_{j=1}^{R+W} \text{prob}(\text{red}|j, I) \times \text{prob}(j|I)\\ &= \frac{1}{R+W}\sum_{j=1}^{R+W} \text{prob}(\text{red}|j, I) = \frac{R}{R+W}\end{aligned} \tag{746}$$

since $\text{prob}(\text{red}, j|I) = 1$ if the jth ball is red and $\text{prob}(\text{red}, j|I) = 0$ if it is white, and we know the contents of the urn. This result justifies the common notion of probability as

$$\text{prob}(\text{red}|I) = \frac{\text{number of cases favorable to red}}{\text{total number of equally possible cases}}.$$

Consider next the result of repeating this ball-drawing procedure many times using *sampling with replacement* so that the contents of the urn are the same each time. Using *marginalization* and the *product rule*, we find that the probability that N such trials result in r red balls is given by

$$\text{prob}(r|N, I) = \sum_{k=1}^{2^N} \text{prob}(r, S_k|N, I) = \sum_{k=1}^{2^N} \text{prob}(r|S_k, N, I) \times \text{prob}(S_k|N, I) \tag{747}$$

since there are 2^N possible sequences of red–white outcomes $\{S_k\}$ among N trials. Now, $\text{prob}(r|S_k, N, I)$ will be equal to 1 if S_k contains exactly r red balls and 0 otherwise. Thus, we need only consider those sequences that have precisely r red outcomes for $\text{prob}(S_k|N, I)$.

Since I assumes a general ignorance about the situation, other than knowledge of the contents of the urn, the result of one draw does not influence what we can infer about the outcome of another. The probability of drawing any particle sequence S_k depends only on the total number of red (and complementary white) balls obtained, and not on their order:

$$\text{prob}(S_k|N,I) = [\text{prob}(\text{red}|I)]^r \times [\text{prob}(\text{white}|I)]^{N-r}.$$

Substituting for $\text{prob}(\text{red}|I) = R/(R+W)$ from Eq. (746), and for the corresponding probability of getting a white ball, $\text{prob}(\text{white}|I) = W/(R+W)$, we get

$$\text{prob}(S_k|N,I) = \left(\frac{R}{R+W}\right)^r \times \left(\frac{W}{R+W}\right)^{N-r} = \frac{R^r W^{N-r}}{(R+W)^N}. \tag{748}$$

Hence, the sum in Eq. (747) is the term above times the number of possible sequences of N-draws, $\text{prob}(r|S_k,N,I)$, which contains exactly r red balls, the evaluation of which requires a discussion of *permutations* and *combinations*.

First we ask the question: in how many ways can n objects be arranged in a straight line? To answer this question, we note that there are n choices for the first object; $n-1$ possibilities for the second object; $n-2$ for the third object, and so on. Thus, the number of permutations becomes

$$n \times (n-1) \times (n-2) \times \cdots 3 \times 2 \times 1 = n! \tag{749}$$

Next, we ask the related question: in how many ways can we sequentially pick m objects from n different ones? Answer: we just stop the product above when m items have been chosen, that is

$$n \times (n-1) \times (n-2) \times \cdots (n-m+2) \times (n-m+1) \equiv {}^nP_m = \frac{n!}{(n-m)!}. \tag{750}$$

Note that ${}^nP_n = n!$ because $0! = 1$, which follows from substitution of $n = 1$ in $n! = (n-1)! \times n$.

If the order in which the objects are picked is of no interest, then we should divide Eq. (750) by $m!$ to obtain the number of possible *permutations*:

$${}^nC_m = \frac{n!}{m!(n-m)!} \tag{751}$$

which we recognize as the coefficient in the *binomial expansion formula*

$$(a+b)^N = \sum_{j=0}^{N} {}^NC_j\, a^j\, b^{N-j} = \sum_{j=0}^{N} \frac{N!}{j!(N-j)!}\, a^j\, b^{N-j} \tag{752}$$

which for $a = b = 1$ yields $\sum_{m=0}^{n} {}^nC_m = 2^n$, and for $a = b = 1/2$ $1 = \sum_{m=0}^{n} {}^nC_m (1/2)^m (1/2)^{n-m} = \sum_{m=0}^{n} {}^nC_m (1/2)^n$.

To complete the evaluation of $\text{prob}(r|N,I)$, we need the number of different ways in which exactly r red balls can be drawn in N trials, given by $\text{prob}(r|S_k,N,I)$. We may think of the problem as follows: Imagine that integers 1 to N have been written on separate pieces of paper. If we select r of them, then the numbers chosen can be thought of as representing the draws in which a red ball was obtained. Thus, the sequences we require correspond to the number of ways of selecting r integers out of N when their order is irrelevant. This number is given by Eq. (751) with n replaced by N and m replaced by r. Then, there are ${}^NC_r = \frac{N!}{r!(N-r)!}$ such sequences, which is equal to $\text{prob}(r|S_k, N, I)$. Combining the result of Eqs. (748) with this result for $\text{prob}(r|S_k, N, I)$, we find that Eq. (747) gives

$$\text{prob}(r|N, I) = \frac{N!}{r!(N-r)!} \times \frac{R^r W^{N-r}}{(R+W)^N}. \tag{753}$$

This pdf is properly normalized because

$$\sum_{r=0}^{N} \text{prob}(r|N, I) = \sum_{r=0}^{N} \frac{N!}{r!(N-r)!} p^r q^{N-r} = (p+q)^N = 1 \tag{754}$$

$$\text{since } p = R/(R+W), \quad q = W/(R+W), \quad \text{so that } p + q = 1. \tag{755}$$

We can use the pdf in Eq. (753) to compute the frequency r/N at which we expect to observe red balls:

$$\left\langle \frac{r}{N} \right\rangle = \sum_{r=0}^{N} \frac{r}{N} \text{prob}(r|N, I) = \sum_{r=1}^{N} \frac{(N-1)!}{(r-1)!(N-r)!} p^r q^{N-r}$$

where the lower limit of the sum has been changed to $r = 1$ because there is no contribution from the term $r = 0$. Taking p outside the summation and letting $j = r - 1$, we obtain

$$\left\langle \frac{r}{N} \right\rangle = p \sum_{j=0}^{N-1} \frac{(N-1)!}{j!(N-1-j)!} p^j q^{N-1-j} = p(p+q)^{N-1}.$$

Substituting for $p = R/(R+W)$ and $q = W/(R+W)$, we get

$$\left\langle \frac{r}{N} \right\rangle = \frac{R}{R+W}. \tag{756}$$

Thus, **the expected frequency of red balls in repetitions of the urn "experiment" is equal to the probability of obtaining one red ball in a single trial.** A similar calculation of the *variance* of r/N yields

$$\left\langle \left(\frac{r}{N} - p \right)^2 \right\rangle = \frac{pq}{N} \to 0 \text{ as } N \to \infty \tag{757}$$

verifying that *Bernoulli's famous theorem* of large numbers is obeyed:

$$\lim_{N \to \infty} \left(\frac{r}{N} \right) = \text{prob}(\text{red}|I). \tag{758}$$

Bernoulli did not provide an answer to the question: what could one say about the probability of obtaining a red ball, in a single draw, given a finite number of observed outcomes? The answer to this question, which is essential in data analysis, had to wait for Bayes and Laplace.

The Binomial Distribution

If the outcome of an experiment can attain only one of two values, "success" or "failure", we may define

$$\text{prob}(\text{success}|I) = p \qquad \text{and} \qquad \text{prob}(\text{failure}|I) = q = 1 - p$$

for the probability p of success and q of failure in a single trial. The formulas of Eqs. (752)–(755) then give the pdf of obtaining r successes in N trials as the *binomial distribution*

$$\text{prob}(r|N,I) = \frac{N!}{r!(N-r)!}\,p^r\,(1-p)^{N-r} \qquad r = 0, 1, 2, \ldots, N. \tag{759}$$

The expected number of successes $\langle r\rangle$ and the *mean-square deviation* from this average value follow from Eqs. (756)–(758):

$$\langle r\rangle = Np \qquad \text{and} \qquad \left\langle (r-Np)^2\right\rangle = Np(1-p). \tag{760}$$

Location and Scale Parameters

Bernoulli's principle of insufficient reason can be used when we are able to enumerate a set of basic possibilities, assuming that the quantity of interest x is restricted to certain discrete values [see Eq. (744)]. For the case of continuous parameters, the probability that x lies in the infinitesimal range between $\tilde{x}$ and $\tilde{x} + \delta x$ is

$$\text{prob}(x = \tilde{x}|I) = \lim_{\delta x \to 0}\ \text{prob}(\tilde{x} \le x < \tilde{x} + \delta x|I).$$

If a mistake had been made so that the position $\tilde{x}$ was actually $\tilde{x} + x_0$, and if I indicated gross ignorance about the details, then consistency would demand that the pdf for x should change very little with the value of x_0:

$$\text{prob}(x|I)dx \approx \text{prob}(x + x_0|I)d(x + x_0) = \text{prob}(x + x_0|I)dx$$

which implies

$$\text{prob}(x|I) \approx \text{constant in the allowed range; 0 otherwise.} \tag{761}$$

Thus, **complete ignorance about a *location parameter* is ensured by the assignment of a uniform pdf.**

For quantities associated with a size or magnitude, the relative or fractional change is important (rather than the absolute change, as in the case of a location parameter). Such a *scale* parameter could be the length L of a biological molecule. The question is what to assign for prob$(L|I)$, for the case in which I represents gross ignorance about the value of L. If a mistake had been made about the unit of length (mm versus μm), then such a mistake should not make much difference, and consistency would require that

$$\text{prob}(L|I)dL \approx \text{prob}(\beta L|I)d(\beta L).$$

Here, β is a positive constant, and since $d(\beta L) = \beta dL$, the above requirement can be satisfied only if

$$\text{prob}(L|I) \propto 1/L \quad \text{in the allowed range, and 0 otherwise.} \tag{762}$$

This pdf represents complete ignorance about the value of the scale parameter, because it is equivalent to a uniform pdf for the logarithm of L: prob$(\ln L|I) =$ constant, which can be verified by a change of variables according to Eq. (714): prob$(x|I) =$ prob$(y|I) \times |(dy/dx)|$. Thus, with $x = L$, $y = \ln L$, we get prob$(L|I) =$ prob$(\ln L|I)(1/L) \propto (1/L)$.

5.4.2 Testable Information: The Principle of Maximum Entropy

Suppose a die is rolled a very large number of times, and we are told that the average result is

$$\sum_{i=1}^{6} i\,\text{prob}(x_i|I) = 4.5 \tag{763}$$

instead of $\sum_{i=1}^{6} i\,\text{prob}(x_i|I) = (1/6)\sum_{i=1}^{6} i = (1+2+3+4+5+6)/6 = 3.5$ predicted from the uniform pdf of Eq. (744). What probability should we assign for the various possible outcomes $\{x_i\}$ that the face on top had i dots? The constraint given by Eq. (763) constitutes an example of the so-called *testable information,* that is, a condition that can be used to either accept or reject a proposed pdf.

Jaynes [216, 217] suggested that the assignment should be made by using the *principle of maximum entropy* (MaxEnt):

$$S = -\sum_{i=1}^{6} p_i \ln p_i, \qquad p_i = \text{prob}(x_i|I) \tag{764}$$

subject to normalization and the condition of Eq. (763):

$$\sum_{i=1}^{6} p_i = 1 \qquad \text{and} \qquad \sum_{i=1}^{6} i\,p_i = 4.5.$$

We can perform such a constrained optimization using the method of *Lagrange multipliers,* but why is entropy the best selection criterion? Arguments in favor of using entropy range from information theory to logical consistency. We now consider two examples: (i) the *kangaroo problem* and (ii) a combinatorial argument, often phrased in terms of a hypothetical team of monkeys.

The Kangaroo Problem

The kangaroo problem is as follows:

- *Information:* A third of all kangaroos have blue eyes, and a quarter of all kangaroos are left-handed.
- *Question:* On the basis of this information alone, what proportions of kangaroos are both blue-eyed and left-handed?

For any given kangaroo, there are four distinct possibilities:

(1) blue-eyed and left-handed	p_1;	$p_1 + p_2 + p_3 + p_4 = 1$	$\rightarrow p_1 = x$
(2) blue-eyed and right-handed	p_2;	$p_1 + p_2 = 1/3$	$\rightarrow p_2 = 1/3 - x$
(3) not blue-eyed but left-handed	p_3;	$p_1 + p_3 = 1/4$	$\rightarrow p_3 = 1/4 - x$
(4) not blue-eyed and right-handed	p_4;		$\rightarrow p_4 = 5/12 + x.$

Following Bernoulli's law of large numbers, the expected values of the fraction of kangaroos with traits (1)–(4) will be equal to the probabilities we have assigned to each of these propositions, denoted by p_1, p_2, p_3, and p_4. All solutions where $0 \leq x \leq 1/4$ satisfy the constraint of the testable information, but which one is best? If we had to make a choice between permissible pdfs, then common sense would suggest $x = 1/12$ based on independence, because any other value would indicate that knowledge of the kangaroo's eye color could tell us something about its handedness.

Although in this case it was easy to decide which would be the most sensible pdf assignment, we may ask whether there is some function of the $\{p_i\}$ which, when maximized subject to the known constraints, would yield this common sense result. If so, it would be a good candidate for a general *variational principle* that one could use in situations too complicated for common sense to be applicable. Skilling [218] has shown that

- the only functions that give uncorrelated assignments in general are those related monotonically to the *entropy* $S = -\sum p_i \ln p_i$. In the present case, it yields the *optimal value* $x = 1/12$ and zero correlation between handedness and eye color.

The Monkey Argument

If there are M distinct possibilities $\{x_j\}$, then our task is to assign truth vales to them given some testable information I: $\text{prob}(x_i, I) = p_i$. To do so, we can play the following game:

- Let the various propositions be represented by different boxes, all of the same size, into which coins are thrown at random (unbiased monkey job). After a large number of coins have been thrown
- the fraction found in each of the boxes gives a possible assignment for the pdf of $\{x_j\}$;
- if the resulting pdf is inconsistent with the constraints I, it should be rejected; otherwise it is accepted as a viable option;
- now empty the boxes and repeat the scattering of coins.

After many such trials, some distributions will be found to occur more often than others, and the one that occurs most frequently (and satisfies I) would be a sensible choice for $\text{prob}(x_i, I) = p_i$, because it agrees with all the testable information available, but does it correspond to the pdf with the greatest value of $S = -\sum p_i \ln p_i$?

After all the coins have been thrown, suppose we find n_1 in the first box, n_2 in the second, and so on, and the total number of coins is a large number N, given by

$$N = \sum_{i=1}^{M} n_i \tag{765}$$

where N is much larger than the number M of boxes: $N >> M$. This distribution $\{n_i\}$ corresponds to a candidate pdf $\{p_i\}$ for the possibilities $\{x_i\}$:

$$p_i = n_i/N \qquad i = 1, 2, \ldots, M. \tag{766}$$

Since each of the N coins can land in any of the M boxes, there are M^N different ways of distributing the coins among the boxes, and each is equally likely to occur. But all of these basic sequences are not distinct because many of them will yield the same distribution $\{n_i\}$. Therefore, the expected frequency F with which a candidate pdf $\{p_i\}$ will arise is

$$F(\{p_i\}) = \frac{\text{number of ways of obtaining } \{n_i\}}{M^N}. \tag{767}$$

To evaluate the numerator, we use the results of permutations and combinations discussed previously and ask:

- For box 1: in how many ways can n_1 coins be chosen from a total of N coins? Answer: ${}^{N}C_{n_1}$.
- For box 2: in how many ways can n_2 coins be chosen from the remaining $N - n_1$ coins? Answer: ${}^{N-n_1}C_{n_2}$.
- For box 3: in how many ways can n_3 coins be chosen from the remaining $N - n_1 - n_2$ coins? Answer: ${}^{N-n_1-n_2}C_{n_3}$.

Continuing in this manner, we find that the numerator in Eq. (767) is a product of M such binomial terms:

$${}^{N}C_{n_1} \times {}^{N-n_1}C_{n_2} \times {}^{N-n_1-n_2}C_{n_3} \times \cdots \times {}^{n_M}C_{n_M} = \frac{N!}{n_1!\,n_2!\cdots n_M!}.$$

Here we have used Eq. (751): ${}^{n}C_m = \frac{n!}{m!(n-m)!}$ and Eq. (765): $N = \sum_i n_i$. Substituting this expression for the numerator in Eq. (767) and taking the logarithm, we find

$$\ln F = -N \ln M + \ln [N!] - \sum_{i=1}^{M} \ln [n_i!] \tag{768}$$

which can be simplified by using *Stirling's approximation*:

$$\ln [n!] = n \ln n - n$$

so that Eq. (768) becomes

$$\ln F = -N \ln M + N \ln N - \sum_{i=1}^{M} n_i \ln n_i.$$

Finally, we use $n_i = Np_i$ and $\sum p_i = 1$ to obtain

$$\ln F = -N \ln M - N \sum_{i=1}^{M} p_i \ln p_i \tag{769}$$

which is related monotonically to the frequency with which a candidate pdf $\{p_i\}$ will be produced. Therefore

- the assignment $\text{prob}(x_i, I) = p_i$ that best represents our state of knowledge is the one that gives the greatest value for $\ln F$ consistent with the testable information I.

Since M and N are constants, this requirement is equivalent to the constrained maximization of the entropy given by

$$S = -\sum_{i=1}^{M} p_i \ln p_i. \tag{770}$$

The Lebesque Measure

In the *monkey argument*, all boxes were assumed to have the same size. Likewise for an ordinary die, the principle of indifference suggests that we should assign equal *a priori* weight to the six possible outcomes. But suppose for some reason, the problem is posed in term of just three hypotheses:

$$x_i \equiv \text{the face on the top has} \begin{Bmatrix} i \text{ dots} & \text{for } i = 1, 2, \\ 3, 4, 5 \text{ or } 6 \text{ dots} & \text{for } i = 3. \end{Bmatrix}$$

Given the six-sided nature of the die, we would in this case be inclined to make the box for x_3 four times as large as that for x_1 and x_2, and the question is how such an unevenness would affect the preceding analysis.

To answer this question, let us return to the M distinct possibilities, but adjust the size of the boxes so that the chance that a coin will be thrown into the ith box is m_i. We have the condition that

$$\sum_{i=1}^{M} m_i = 1$$

but the m_i values are not necessarily equal, and if they were equal, then $m_i = 1/M$ would pertain for all i. The expected frequency F of finding n_1 coins in box 1, n_2 in box 2, and so on, will now be given by the number of different ways of distributing the coins, which yields the distribution $\{n_i\}$ times the probability of obtaining such a sequence of throws:

$$F(\{p_i\}) = \frac{N!}{n_1!\, n_2! \cdots n_m!} \times m_1^{n_1}\, m_2^{n_2} \cdots m_M^{n_M}. \tag{771}$$

This *multinomial distribution* is a generalization of its binomial ($M = 2$) counterpart given by Eq. (759): $\text{prob}(r|N, I) = ((N!)/(r!(N-r)!))\, p^r\, (1-p)^{N-r}$. Clearly, if all the M terms were equal ($m_i = 1/M$ for all i), then $m_1^{n_1} m_2^{n_2} \cdots m_M^{n_M} = (1/M)^{n_1+n_2+\cdots+n_M} = 1/(M^N)$, and Eq. (771) would lead to Eq. (767):

$$F(\{p_i\}) = \frac{\text{number of ways of obtaining } \{n_i\}}{M^N} = \frac{N!}{n_1!\, n_2! \cdots n_m!} \times \frac{1}{M^N}.$$

Taking the logarithm of Eq. (771), and using Stirling's approximation ($\ln[n!] = n \ln n - n$), we obtain

$$\ln F = \sum_{i=1}^{M} n_i \ln m_i - N \sum_{i=1}^{M} p_i \ln p_i$$

and using Eq. (766): $p_i = n_i/N$, we finally have

$$\frac{1}{N} \ln F = -\sum_{i=1}^{M} p_i \ln \left[\frac{p_i}{m_i}\right] \equiv S. \tag{772}$$

Thus, we see that the entropy of Eq. (770): $S = -\sum_{i=1}^{M} p_i \ln p_i$ is a special case (where $m_i = 1/M$) of the more general form. Equation (772) is known as the *Shannon–Jaynes entropy*, the Kullback number, or the *cross entropy*.

The generalization provided by Eq. (772) is necessary in the limit of continuous parameters:

$$S = -\int p(x) \ln \left[\frac{p(x)}{m(x)}\right] dx. \tag{773}$$

The *Lebesque measure* $m(x)$ is required to ensure that the entropy expression is invariant under a change of variables $x \to y = f(x)$. Invariance is obtained because both $p(x)$ and $m(x)$ transform in the same way. Thus,

- the measure $m(x)$ takes into account how (uniform) bin widths in x-space translate to a corresponding set of (variable) box sizes in an alternative y-space.

To get a better understanding of the nature of $m(x)$, we now maximize the entropy given by Eq. (773) subject to the normalization condition $\int p(x)dx = 1$. For the discrete case, we may use the method of *Lagrange multipliers* to maximize a function Q with respect to the $\{p_i\}$, treated as independent variables, where Q is given by

$$Q = -\sum_{i=1}^{M} p_i \ln \left[\frac{p_i}{m_i}\right] + \lambda \left(1 - \sum_{i=1}^{M} p_i\right).$$

Since the partial derivatives $\partial p_i/\partial p_j = 0$ if $i \neq j$, we have (for all j)

$$\frac{\partial Q}{\partial p_j} = -1 - \ln \left[\frac{p_j}{m_j}\right] - \lambda = 0 \quad \text{implying} \quad p_j = m_j e^{-(1+\lambda)}.$$

The normalization requirement $\sum_{i=1}^{M} p_i = 1$ determines λ, and in the continuum limit we have

$$p(x) = \text{prob}(x|\text{normalization}) \propto m(x). \tag{774}$$

Thus, $m(x)$ is any multiple of the pdf that expresses complete ignorance about the value of x, implying that the transformation-group (invariance) arguments discussed previously are appropriate for ascertaining the measure.

MaxEnt Examples: Some Common pdfs

We have just considered the simplest situation where the testable information consists purely of the normalization condition. If the measure is uniform, then according to Eq. (774) we should assign $\text{prob}(x|I) = \text{constant}$, which for the case of discrete possibilities reduces to $\text{prob}(x_i|I) = 1/M$ in accordance with Bernoulli's principle of insufficient reason. The uniform pdf gives rise to other pdfs by using the sum and product rules of probability. When considering the repeated drawing of red and white balls from an urn, we found that

- assuming the contents of the container and our state of ignorance to be the same in each trial, we obtained the *binomial distribution*.

- If we had considered sampling without replacement, we would have found the pdf to be a *hypergeometric distribution.*

For the continuous case of the lighthouse problem, we found that a uniform pdf for the angle of the emitted beam gave rise to a *Cauchy distribution* for the position of the flashes detected on the coast. Next, we will examine how MaxEnt gives rise to some commonly encountered pdfs.

Averages and Exponentials

Suppose the testable information was knowledge of the expectation value μ, which we can express as the constraint

$$\langle x \rangle = \int x \,\mathrm{prob}(x|I)dx = \mu. \tag{775}$$

According to the *MaxEnt principle,* we should seek that pdf which has the largest entropy while satisfying the normalization constraint given in Eq. (775). Thus, we need to maximize the function Q given by

$$Q = -\sum_{i=1}^{M} p_i \ln\left[\frac{p_i}{m_i}\right] + \lambda_0\left(1 - \sum_{i=1}^{M} p_i\right) + \lambda_1\left(1 - \sum_{i=1}^{M} x_i p_i\right)$$

where λ_0 and λ_1 are *Lagrange multipliers.* Setting $\partial Q/\partial p_j = 0$, we obtain

$$p_j = m_j e^{-(1+\lambda_0)} e^{-\lambda_1 x_j}. \tag{776}$$

For a uniform measure given by $m_i = 1/6$ and $x_i = i$, Eq. (776) is the solution to the die problem. Then λ_0 and λ_1 can be calculated numerically to ensure that the resultant pdf satisfies the requirements $\sum_i i p_i = 4.5$ and $\sum_i p_i = 1$.

Generalizing to the continuous case with a uniform measure, Eq. (776) becomes a simple exponential function

$$\mathrm{prob}(x|I) \propto \exp[-\lambda_1 x].$$

If the limits of integration are 0 and ∞, then $\langle x \rangle = \int_0^\infty x \exp[-\lambda_1 x]dx = 1/\lambda_1$ implying $\mu = 1/\lambda_1$, and hence

$$\mathrm{prob}(x|\mu) = \frac{1}{\mu}\exp\left[-\frac{x}{\mu}\right] \qquad \text{for} \qquad x \geq 0. \tag{777}$$

A uniform measure is not always the most appropriate as will be demonstrated below with the binomial and Poisson distributions. We will now explore what happens when we have other types of testable information.

Variance and the Gaussian Distribution

Suppose we know not only μ but also the *variance* σ^2 about that value

$$\left\langle (x-\mu)^2 \right\rangle = \int (x-\mu)^2 \,\mathrm{prob}(x|I)dx = \sigma^2. \tag{778}$$

To assign $\mathrm{prob}(x|I)$, we need to maximize its entropy subject to normalization and the constraint given by Eq. (778), which in the discrete case is equivalent to finding the extremum of

$$Q = -\sum_{i=1}^{M} p_i \ln\left[\frac{p_i}{m_i}\right] + \lambda_0\left(1 - \sum_i p_i\right) + \lambda_1\left(\sigma^2 - \sum_i (x-\mu)^2 p_i\right)$$

where λ_0 and λ_1 are Lagrange multipliers. Setting $\partial Q/\partial p_j = 0$, we obtain

$$p_j = m_j\, e^{-(1+\lambda_0)}\, e^{-\lambda_1 (x_j-\mu)^2}$$

which for a uniform measure generalizes in the continuum limit to

$$\text{prob}(x|I) \propto \exp[-\lambda_1 (x-\mu)^2]. \tag{779}$$

If the limits of integration are $\pm\infty$, we obtain the standard *Gaussian pdf*

$$\text{prob}(x|\mu, I) = \frac{1}{\sigma\sqrt{2\pi}} \exp\left[-\frac{(x-\mu)^2}{2\sigma^2}\right] \tag{780}$$

where σ is defined in Eq. (778), and the mean μ is defined in Eq. (775). Hence

- ***the normal (Gaussian) distribution is the most honest description of our state of knowledge, when all we know is the mean and the variance.***

The Multivariate Case

The *MaxEnt* analysis can easily be extended to the case of several parameters by expressing the entropy of Eq. (773) as a multidimensional integral:

$$S = -\int\int\cdots\int p(\mathbf{x}) \ln\left[\frac{p(\mathbf{x})}{m(\mathbf{x})}\right] d^N\mathbf{x} \tag{781}$$

where $p(\mathbf{x}) = \text{prob}(x_1, x_2, \ldots, x_N|I)$. If the testable information pertaining to the quantities $\{x_k\}$ represents knowledge of their individual variances

$$\left\langle (x_k-\mu)^2 \right\rangle = \int\int\cdots\int (x_k-\mu)^2\, p(\mathbf{x}) d^N\mathbf{x} = \sigma_k^2 \qquad k = 1, 2, \ldots, N \tag{782}$$

then one can show that the maximization of Eq. (781), with a uniform measure, yields a product of Gaussian pfds:

$$\text{prob}(\{x_k\}|\{\mu_k, \sigma_k\}) = \prod_{k=1}^{N} \frac{1}{\sigma_k\sqrt{2\pi}} \exp\left[-\frac{(x_k-\mu_k)^2}{2\sigma_k^2}\right]. \tag{783}$$

This pdf is the same as a *least-squares likelihood function*. Thus,

- if we identify $\{x_k\}$ as $\{D_k\}$, with error bars $\{\sigma_k\}$, and $\{\mu_k\}$ as the predictions $\{F_k\}$ based on some model, then Eq. (783) corresponds to Eqs. (702): $\text{prob}(\mathbf{D}|\mathbf{x}, \mathbf{I}) \propto \exp(-(\chi^2/2))$ and (703): $\chi^2 = \sum_{k=1}^{N} ((F_k - D_k)/(\sigma_k))^2$.

From a MaxEnt point of view

- the least-squares likelihood does not imply the existence (or assumption) of a mechanism for generating independent, additive Gaussian noise;
- it is just the pdf that best represents our state of knowledge given only Eq. (782);

- if we had convincing information about the covariance $\langle (x_i - \mu_i)(x_j - \mu_j) \rangle$, where $i \neq j$, then *MaxEnt* would assign a correlated multivariate Gaussian pdf for $\text{prob}(\{x_k\}|I)$.

In §5.2.3, we considered briefly the possibility of using the l_1-norm Eq. (711): $l_1 - \text{norm} = \sum_{k=1}^{N} |((F_k - D_k)/\sigma_k)|$ as a criterion for fitting functional models to data. Use of the l_1-norm involves a minimization of the moduli of the misfit residuals rather than their squares. This procedure follows naturally from *MaxEnt* if the *testable information* consists only of the expected value of the modulus of the discrepancy between theory and experiment for the individual data, given by

$$\langle |x_k - \mu_k| \rangle = \int\int \cdots \int |x_k - \mu_k| p(\mathbf{x}) d^N\mathbf{x} = \epsilon_k, \qquad k = 1, 2, \ldots, N. \quad (784)$$

One can show that maximization of Eq. (781), with a uniform measure, yields the product of symmetric exponential pdfs:

$$\text{prob}(\{x_k\}|\{\mu_k, \epsilon_k\}) = \prod_{k=1}^{N} \frac{1}{2\epsilon_k} \exp\left(- \frac{|x_k - \mu_k|}{\epsilon_k} \right). \quad (785)$$

The logarithm of the likelihood function is, therefore, given by the l_1-norm (plus an additive constant).

Having seen how the nature of the testable constraints can influence the pdf we assign, we will now look at an example where the measure plays an equally important role.

MaxEnt and the Binomial Distribution

To examine how the *binomial distribution* emerges from use of the *MaxEnt principle*, suppose we are given (only) the expected number of successes in M trials, $\langle N \rangle = \mu$. What should we assign for the probability of a specific number of favorable outcomes, $\text{prob}(N|M, \mu)$?

According to the *MaxEnt principle*, we must maximize the entropy of Eq. (772): $S = -\sum_{i=1}^{M} p_i \ln [p_i/m_i]$, subject to normalization and *testable information* given by

$$\langle N \rangle = \sum_{N=0}^{M} N \, \text{prob}(N|M, \mu) = \mu. \quad (786)$$

This optimization yields the pdf of Eq. (776): $p_j = m_j e^{-(1+\lambda_0)} e^{-\lambda_1 x_j}$, which leads to

$$\text{prob}(N|M, \mu) \propto m(N) e^{-\lambda N}. \quad (787)$$

In Eq. (787), λ is a Lagrangian multiplier, determined by the constraint on the mean, but how do we assign $m(N)$, the *Lebesque measure*? As discussed above, $m(N)$ is proportional to the pdf which reflects gross ignorance about the details of the situation:

- Given only that there are M trials, we should assign equal probability to each of the 2^M possible outcomes, according to the *principle of indifference.*

- The number of different ways of obtaining N successes in M trials, or ${}^{M}C_N = (M!/(N!(M-N)!)$ according to Eq. (751), is therefore an appropriate measure for this problem:

$$m(N) = \frac{M!}{N!(M-N)!}. \tag{788}$$

This result should be substituted into Eq. (787) before we impose the constraints of normalization and Eq. (786). The algebra is simplified by noting that

$$\sum_{N=0}^{M} m(N)e^{-\lambda N} = (e^{-\lambda}+1)^M \tag{789}$$

which follows from Eq. (752): $(a+b)^N = \sum_{j=0}^{N}(N!/(j!(N-j)!))\, a^j\, b^{N-j}$ on putting $a = e^{-\lambda}$ and $b = 1$. The reciprocal of Eq. (789) provides the constant of proportionality in Eq. (787): $\text{prob}(N|M,\mu) = m(N)e^{-\lambda N}/(e^{-\lambda}+1)^M$, and differentiation of Eq. (789) with respect to λ, which gives

$$\sum_{N=0}^{M} Nm(N)e^{-\lambda N} = M(e^{-\lambda}+1)^{M-1}\, e^{-\lambda}$$

which shows that Eq. (786) can be written as

$$\langle N\rangle = \sum_{N=0}^{M} N\text{prob}(N|M,\mu) = M(1+e^{\lambda})^{-1} = \mu$$

so that Eq. (787) becomes

$$\text{prob}(N|M,\mu) = \frac{M!}{N!(M-N)!}\left(\frac{\mu}{M}\right)^{N}\left(1-\frac{\mu}{M}\right)^{M-N} \tag{790}$$

which is a binomial pdf.

5.5 Generic Formulation of the Inverse Problem

We start by considering a set of measurements assembled into a measurement vector $\mathbf{y}$ with N elements, and the retrieval parameters (the unknowns) assembled into a state vector $\mathbf{x}$ with M elements. The *forward model* $\mathbf{F}(\mathbf{x})$ provides the relationship between these two vectors

$$\mathbf{y} = \mathbf{F}(\mathbf{x}) + \vec{\epsilon} \tag{791}$$

where $\vec{\epsilon}$ is the measurement error. The set of equations represented by Eq. (791) is generally nonlinear, and difficult to solve, but a least-squares solution can be obtained as shown above [see Eqs. (694)–(704)]. According to *Bayes' theorem*

$$\text{prob}(\mathbf{x}|\mathbf{y},\mathbf{I}) \propto \text{prob}(\mathbf{y}|\mathbf{x},\mathbf{I}) \times \text{prob}(\mathbf{x}|\mathbf{I}) \tag{792}$$

and if we assign a uniform or flat prior pdf, $\text{prob}(\mathbf{x}|\mathbf{I}) = \text{constant}$, then Eq. (792) becomes

$$\text{prob}(\mathbf{x}|\mathbf{y},\mathbf{I}) \propto \text{prob}(\mathbf{y}|\mathbf{x},\mathbf{I}). \tag{793}$$

Our best estimate $\mathbf{x}_0$, given by the maximum of the posterior $\text{prob}(\mathbf{x}|\mathbf{y},\mathbf{I})$, is the *maximum likelihood estimate*. If the data are independent, then

$$\text{prob}(\mathbf{y}|\mathbf{x},\mathbf{I}) = \prod_{k=1}^{N} \text{prob}(y_k|\mathbf{x},\mathbf{I}) \tag{794}$$

where y_k is the kth data point. If we also assume that the measurement noise can be represented by a Gaussian

$$\text{prob}(y_k|\mathbf{x},\mathbf{I}) = \frac{1}{\sigma_k\sqrt{2\pi}} \exp\left[-\frac{(F_k - y_k)^2}{2\sigma_k^2}\right] \tag{795}$$

and that an adequate model exists of the functional relationship between the parameters $\mathbf{x}$ and the noiseless data $\mathbf{F}$

$$F_k = f(\mathbf{x}, k) \tag{796}$$

then Eqs. (794) and (795) allow us to approximate the likelihood function by

$$\text{prob}(\mathbf{y}|\mathbf{x},\mathbf{I}) \propto \exp\left(-\frac{\chi^2}{2}\right), \qquad \chi^2 = \sum_{k=1}^{N}\left(\frac{F_k - y_k}{\sigma_k}\right)^2. \tag{797}$$

With a uniform prior the logarithm L of the posterior pdf is given by [see Eq. (797)]

$$L = \ln\left[\text{prob}(\mathbf{x}|\mathbf{y},\mathbf{I})\right] = \text{constant} - \frac{\chi^2}{2} \tag{798}$$

and the least-squares solution to this nonlinear problem is found by minimizing χ^2.

Before looking further into methods for solving nonlinear inversion problems, we will consider the easier *linear inverse problem.*

5.6 Linear Inverse Problems

Linearizing Eq. (791) about a reference state $\mathbf{x}_0$, we find

$$\mathbf{y} - \mathbf{F}(\mathbf{x}_0) = \frac{\partial \mathbf{F}(\mathbf{x})}{\partial \mathbf{x}}(\mathbf{x} - \mathbf{x}_0) + \vec{\epsilon} = \mathbf{J}(\mathbf{x} - \mathbf{x}_0) + \vec{\epsilon} \tag{799}$$

where the $M \times N$ *weighting function matrix* $\mathbf{J}$ is called the *Jacobian,* and its elements $J_{ij} = \partial F_i(x)/\partial x_j$ are called *Fréchet derivatives.*

If we assume that $\mathbf{F}(\mathbf{x}_0) = \mathbf{J}\,\mathbf{x}_0$, then we are left with the linear problem

$$\mathbf{y} = \mathbf{J}\mathbf{x} + \vec{\epsilon}. \tag{800}$$

5.6.1
Linear Problems without Measurement Errors

If we ignore the measurement error by setting $\vec{\epsilon} = 0$, the problem is reduced to the solution of a set of simultaneous linear algebraic equations

$$\mathbf{y} = \mathbf{J}\mathbf{x} \tag{801}$$

and we need to [37, 211]

- determine whether there are ***no*** solutions, ***one*** solution, or ***an infinite number*** of solutions,
- investigate what information can be extracted from the measurements $\mathbf{y}$ about $\mathbf{x}$ when there is ***no*** solution or ***no unique*** solution.

The N weighting function vectors $\mathbf{J}_j$ will span some subspace of state space, which will be of dimension p called the rank of $\mathbf{J}$, where p is the number of linearly independent rows (or columns). The rank $p \leq N$, and if $p < N$, the vectors are **not** linearly independent. If $N > M$ (more measurements than unknowns), then $p \leq M$. The subspace spanned by the rows of $\mathbf{J}$ is called the *row space* or *range* of $\mathbf{J}$, which may or may not comprise the whole of the state space.

As in the nonlinear case, Eq. (801) can be solved by minimizing the residual of the l_2 norm given by

$$||\mathbf{J}\mathbf{x} - \mathbf{y}||_2 = \left[\sum_{i=1}^{N}(y_i - (\mathbf{J}\mathbf{x})_i)^2\right]^{1/2} \tag{802}$$

which leads to the *least-squares solution* [211]

$$\mathbf{x} = (\mathbf{J}^T\mathbf{J})^{-1}\mathbf{J}^T\mathbf{y}. \tag{803}$$

If we assume a uniform prior so that the posterior pdf is proportional to the likelihood function [see Eq. (793)], then the χ^2 statistic is given by Eq. (797) as in the nonlinear problem, but with $F_k = (\mathbf{J}\mathbf{x})_k$, in the linear approximation. Hence, the least-squares solution is found by minimizing

$$\chi^2 = \sum_{k=1}^{N}\left(\frac{y_k - (\mathbf{J}\mathbf{x})_k}{\sigma_k}\right)^2. \tag{804}$$

We can imagine an orthogonal coordinate system or *basis* for the state space with p orthogonal *base vectors* in row space, and $(M - p)$ base vectors outside, which are orthogonal to row space and thus orthogonal to all $\mathbf{J}_j$ weighting function vectors. Only components of the state vector lying in row space will contribute to the measurement vector, implying that all other components (being orthogonal to it) will not contribute to the measurement: they are *unmeasurable*. This part of state space is called the *null space* of $\mathbf{J}$. The problem is underdetermined if $p < M$, so that a null space exists. In this case

- the solution is *nonunique* because there are components of state space which are not determined by the measurements, and which could thus take any value;

- if a state (to be retrieved) has components that lie in the null space, their values cannot be obtained from the measurements.

Consider just the components of the state vector in row space. They will be **overdetermined** if $N > p$ so that the number of measurements exceeds the rank of $\mathbf{J}$, and **well determined** if $N = p$. Thus, it is possible for a problem to be simultaneously **overdetermined** (in row space) and **underdetermined** (if there is a null space), implying a *mixed-determined* condition. It is even possible for there to be more measurements than unknowns ($N > M$), and for the problem to be underdetermined, if $p < M$. A problem is well determined only if $N = M = p$; then a unique solution can be found by solving a set of $p \times p$ equations. If a problem is overdetermined in row space and there is no measurement error, then either the measurements must be linearly related in the same way as the $\mathbf{J}_j$-vectors, or they are inconsistent, implying that no exact solution exists. In summary:

- *the measurement represented by* $\mathbf{J}$ *provides not more than p independent quantities or pieces of information with which to describe the state.*

Identifying the Null Space and the Row Space

The row space of a given $\mathbf{J}$ can be identified by finding a *basis* for the state space consisting of an orthogonal set of vectors in terms of which we can express every $\mathbf{j}_j$. This basis must be a linear combination of the $\mathbf{j}_j$. To find the basis, we may use *singular value decomposition*, that is, we express $\mathbf{J}$ as

$$\mathbf{J} = \mathbf{U}\mathbf{S}\mathbf{V}^T \tag{805}$$

where $\mathbf{U}$ ($N \times p$) and $\mathbf{V}$ ($M \times p$) are matrices composed of the **left** and **right** *singular vectors* of the ($N \times M$) matrix $\mathbf{J}$ of rank p, and $\mathbf{S}$ is a ($p \times p$) diagonal matrix of the nonzero *singular values* of $\mathbf{J}$. Note that

- The p columns of the ($M \times p$) matrix $\mathbf{V}$ form an orthogonal basis for the row (or state) space, while
- the p columns of the ($N \times p$) matrix $\mathbf{U}$ form a corresponding orthogonal basis for the column (or measurement) space.

Inserting Eq. (805) in the forward model $\mathbf{y} = \mathbf{J}\mathbf{x} = \mathbf{U}\mathbf{S}\mathbf{V}^T\mathbf{x}$, and multiplying by $\mathbf{U}^T$, yields

$$\mathbf{y}' = \mathbf{U}^T\mathbf{y} = \mathbf{S}\mathbf{V}^T\mathbf{x} = \mathbf{\Lambda}\mathbf{x}' \tag{806}$$

where $\mathbf{y}' = \mathbf{U}^T\mathbf{y}$ and $\mathbf{x}' = \mathbf{V}^T\mathbf{x}$ are a ($p \times 1$) column vectors. Hence

- The p transformed measurements $\mathbf{y}'$ in column space are each proportional to a component of a transformed state $\mathbf{x}'$ in p-dimensional row space;
- The ($M \times p$) matrix $\mathbf{V}$ forms a natural basis for row space, closely related to the ($N \times p$) matrix $\mathbf{U}$ for column space.

5.6.2
Linear Problems with Measurement Errors

Since all measurements have error or "noise", the proper treatment of experimental error is a very important consideration in the design of retrieval methods. Thus we need a **formalism** that can be used to express uncertainty in the measurements and the resulting uncertainty in the retrievals, and to ensure that the retrieval uncertainty is as small as possible. As we have already seen, a description of the error in terms of probability density functions (pdfs) based on the Bayesian approach is useful. A measurement has mean value $\bar{y}$ and error σ if our knowledge of the true value is described by a pdf $\mathcal{P}(y)$ with mean value $\bar{y}$ and variance σ^2 given by

$$\bar{y} = \int y\mathcal{P}(y)dy \qquad \sigma^2 = \int (y - \bar{y})^2 \mathcal{P}(y)dy \tag{807}$$

where $\mathcal{P}(y)dy$ is the probability that y lies in the range $(y, y + dy)$. If $\mathcal{P}(y)$ is a Gaussian, we have

$$\mathcal{P}(y) = \frac{1}{(2\pi)^{1/2}\sigma} \exp\left[-\frac{(y - \bar{y})^2}{2\sigma^2}\right] \tag{808}$$

which is usually a good approximation for the experimental error, and very convenient for algebraic manipulations. Also, as we have discussed previously, if the only information available about a pdf is its mean and variance, then the *Gaussian distribution* is the most honest description about the measured quantity.

If the measured quantity is a vector, then

- $\mathcal{P}(\mathbf{y})d\mathbf{y}$ ($d\mathbf{y} = dy_1 \cdots dy_m$) is the probability that the true value of the measurement lies in the multidimensional interval $(\mathbf{y}, \mathbf{y} + d\mathbf{y})$ in measurement space.

Different elements of a vector may be correlated, in the sense that the off-diagonal elements of the covariance matrix have finite values, that is

$$S_{ij} = \mathcal{E}\Big\{(y_i - \bar{y}_i)(y_j - \bar{y}_j)\Big\} \neq 0 \quad \rightarrow \quad \mathbf{S}_\mathbf{y} = \mathcal{E}\Big\{(\mathbf{y} - \bar{\mathbf{y}})(\mathbf{y} - \bar{\mathbf{y}})^T\Big\} \neq 0 \tag{809}$$

where S_{ij} is an element of the *covariance matrix* of y_i and y_j, and $\mathcal{E}$ is the *expected value operator*. The diagonal elements of the *covariance matrix* $\mathbf{S}_\mathbf{y}$ are the variances of the individual elements of $\mathbf{y}$, whereas the off-diagonal elements carry information about the correlations between elements. It should be noted that a covariance matrix is **symmetric** and **nonnegative definite**, and almost always *positive definite*.

The Gaussian distribution for the vector $\mathbf{y}$ is of the form

$$\mathcal{P}(\mathbf{y}) = \frac{1}{(2\pi)^{n/2}|\mathbf{S}_y|^{1/2}} \exp\Big\{-\frac{1}{2}(\mathbf{y} - \bar{\mathbf{y}})^T \mathbf{S}_y{}^{-1}(\mathbf{y} - \bar{\mathbf{y}})\Big\} \tag{810}$$

where $\bar{\mathbf{y}}$ is the mean value and $\mathbf{S}_y$ is the covariance matrix (assumed to be nonsingular).

Example

To see how Eq. (810) is related to the scalar Gaussian distribution, we may transform to a basis in which $\mathbf{S}_y$ is diagonal: $\mathbf{S}_y = \mathbf{L}\boldsymbol{\Lambda}\mathbf{L}^T$ (using eigenvector decomposition), implying $\mathbf{S}_y{}^{-1} = \mathbf{L}\boldsymbol{\Lambda}^{-1}\mathbf{L}^T$

$$\mathcal{P}(\mathbf{y}) = \frac{1}{(2\pi)^{n/2}|\mathbf{L}\boldsymbol{\Lambda}\mathbf{L}^T|^{1/2}} \exp\left\{ -\frac{1}{2}(\mathbf{y}-\overline{\mathbf{y}})^T\mathbf{L}\boldsymbol{\Lambda}^{-1}\mathbf{L}^T(\mathbf{y}-\overline{\mathbf{y}}) \right\}$$

$$P(\mathbf{z}) = \frac{1}{(2\pi)^{n/2}|\boldsymbol{\Lambda}|^{1/2}} \exp\left\{ -\frac{1}{2}\mathbf{z}^T\boldsymbol{\Lambda}^{-1}\mathbf{z} \right\} \qquad \mathbf{z} = \mathbf{L}^T(\mathbf{y}-\overline{\mathbf{y}}). \tag{811}$$

Thus, the pdf can be written as a product of the independent pdfs of each element of $\mathbf{z}$:

$$P(\mathbf{z}) = \prod_i \frac{1}{(2\pi\lambda_i)^{1/2}} \exp\left\{ -\frac{z_i^2}{2\lambda_i} \right\} \tag{812}$$

where the eigenvalue λ_i is the variance of z_i. The eigenvector transformation provides a basis for the measurement space in which

- the transformed measurements are statistically independent. Thus
- a singular covariance matrix would have one or more zero eigenvalues corresponding to elements of $\mathbf{z}$ that are known without error. **Such components would not correspond to physical measurements, and can be eliminated or ignored.**

Notice that surfaces of constant probability of the pdf are of the form

$$(\mathbf{y}-\overline{\mathbf{y}})^T\mathbf{S}_y{}^{-1}(\mathbf{y}-\overline{\mathbf{y}}) = \sum_i \frac{z_i^2}{\lambda_i} = \text{constant} \tag{813}$$

and are ellipsoids in measurement space, with principal axes (of lengths proportional to $\lambda_i^{1/2}$) corresponding to the eigenvectors of $\mathbf{S}_y$, as illustrated in Figure 31 in the 2-D case.

5.7 Bayesian Approach to the Inverse Problem

As already discussed, the Bayesian approach to inverse problems consists of asking the question: How do we relate the pdf of the measurement to the pdf of the state?

The act of measurement maps the state into the measurement space according to the forward model $\mathbf{y} = \mathbf{J}\mathbf{x} + \vec{\epsilon}$. We know $\vec{\epsilon}$ only statistically. Thus [37],

- even though $F(x) \approx \mathbf{J}\mathbf{x}$ is a deterministic mapping, **in the presence of measurement error, a point in state space maps into a region in measurement space determined by the pdf of $\vec{\epsilon}$.**

Conversely, if $\mathbf{y}$ is a given measurement, even in the absence of a *null space*, $\mathbf{y}$ could be the result of a mapping from anywhere inside a certain region of state space described by some pdf, rather than a single point. Furthermore, we may have some prior knowledge about the state, which can also be described by a pdf, and used to constrain the solution. We can think of such *prior knowledge* as a *virtual measurement* that provides an estimate of some function of the state. As we have already alluded to, the Bayesian approach is a very helpful way of looking at the noisy inversion problem, in which we have some prior understanding or expectation about some quantity, and want to update the understanding in the light of new information. Thus, imperfect prior knowledge can be quantified as a pdf over the state space. Similarly, a measurement (imperfect due to experimental error) can be quantified as a pdf over the measurement space.

We would like to know how the measurement pdf maps into state space and combines with prior knowledge: *Bayes' theorem*

$$\mathrm{prob}(\mathbf{x}|\mathbf{y},\mathbf{I}) = \frac{\mathrm{prob}(\mathbf{y}|\mathbf{x},\mathbf{I})\,\mathrm{prob}(\mathbf{x}|\mathbf{I})}{\mathrm{prob}(\mathbf{y}|\mathbf{I})} \tag{814}$$

tells us how. Here,

- $\mathrm{prob}(\mathbf{x}|\mathbf{y},\mathbf{I})$ is the posterior pdf of the state $\mathbf{x}$ when the measurement $\mathbf{y}$ is given.
- $\mathrm{prob}(\mathbf{y}|\mathbf{x},\mathbf{I})$ describes the knowledge of $\mathbf{y}$ that would be obtained if the state were $\mathbf{x}$; an explicit expression requires knowledge of a forward model and a statistical description of the measurement error.
- $\mathrm{prob}(\mathbf{y},\mathbf{I})$ can be obtained by integrating $\mathrm{prob}(\mathbf{x},\mathbf{y}|\mathbf{I}) = \mathrm{prob}(\mathbf{y}|\mathbf{x},\mathbf{I})\mathrm{prob}(\mathbf{x},\mathbf{I})$, but in practice it is only a normalizing factor and is often not needed.

We now have a conceptual approach to the inverse problem:

- Before making a measurement, we have knowledge in terms of a prior pdf given by $\mathrm{prob}(\mathbf{x}|\mathbf{I})$;
- The measurement process is expressed as a forward model, described by $\mathrm{prob}(\mathbf{y}|\mathbf{x},\mathbf{I})$, which maps the state space into measurement space.

To summarize

- Bayes' theorem provides **a formalism to invert this mapping** and calculate a posterior pdf $\mathrm{prob}(\mathbf{x}|\mathbf{y},\mathbf{I})$ by updating the prior pdf $\mathrm{prob}(\mathbf{x}|\mathbf{I})$ with a measurement pdf $\mathrm{prob}(\mathbf{y}|\mathbf{x},\mathbf{I})$.
- The Bayesian view is general, implying that **it is not just an inverse method** producing a solution that may be compared to other solutions; rather, **it encompasses all inverse methods** by providing a way of characterizing the class of possible solutions, considering all states, and assigning a probability density to each.
- The **forward model is never explicitly inverted** in this approach: an explicit "answer" is never produced. It provides some intuition about how the measurement improves knowledge of the state, but
- **to obtain an explicit retrieval we must choose one state from the ensemble** described by the posterior pdf $\mathrm{prob}(\mathbf{x}|\mathbf{y},\mathbf{I})$, perhaps
- **the expected** or **most probable value** of the state, together with some measure of **the width** of the pdf to quantify the accuracy of the retrieval. Thus
- further work is needed that results in **expressions mathematically equivalent to inverting the forward model.**

The Linear Problem with Gaussian Statistics

Consider a linear problem in which all the pdfs are Gaussian. The forward model is

$$\mathbf{y} = \mathbf{F}(\mathbf{x}) + \vec{\epsilon} \approx \mathbf{J}x + \vec{\epsilon}$$

and the Gaussian distribution for the vector $\mathbf{y}$ is given by

$$\mathcal{P}(\mathbf{y}) = \frac{1}{(2\pi)^{N/2}|\mathbf{S}_y|^{1/2}} \exp\left\{ -\frac{1}{2}(\mathbf{y}-\overline{\mathbf{y}})^T \mathbf{S}_y^{-1}(\mathbf{y}-\overline{\mathbf{y}})\right\}. \quad (815)$$

Taking the logarithm, we have

$$\ln \mathcal{P}(\mathbf{y}) \propto (\mathbf{y}-\overline{\mathbf{y}})^T \mathbf{S}_y^{-1}(\mathbf{y}-\overline{\mathbf{y}})$$

where $\overline{\mathbf{y}}$ is the mean value and $\mathbf{S}_y$ is the covariance matrix (assumed to be nonsingular). Assuming that prob($\mathbf{y}|\mathbf{x}, \mathbf{I}$) [i.e., the probability that $\mathbf{y}$ lies between $\mathbf{y}$ and ($\mathbf{y} + \mathbf{dy}$) given $\mathbf{x}$] can be described by a Gaussian, we have ($\mathbf{y} - \mathbf{Jx} = \vec{\epsilon}$):

$$\ln[\text{prob}(\mathbf{y}|\mathbf{x}, \mathbf{I})] = \vec{\epsilon}^{\,T}\mathbf{S}_\epsilon^{-1}\vec{\epsilon} + c_1 = (\mathbf{y}-\mathbf{Jx})^T\mathbf{S}_\epsilon^{-1}(\mathbf{y}-\mathbf{Jx}) + c_1 \quad (816)$$

where c_1 is a constant, and $\mathbf{S}_\epsilon$ is the measurement error covariance. If we assume that the prior knowledge of $\mathbf{x}$ is also a Gaussian pdf so that $\mathbf{x}_a$ is the *a priori* value of $\mathbf{x}$, and $\mathbf{S}_a$ is the associated covariance matrix), then

$$\ln[\text{prob}(\mathbf{x}, \mathbf{I})] = (\mathbf{x}-\mathbf{x}_a)^T\mathbf{S}_a^{-1}(\mathbf{x}-\mathbf{x}_a) + c_2 \quad (817)$$

where c_2 is another constant.

Substituting Eqs. (816) and (817) into Eq. (814) we obtain

$$\begin{aligned}\ln[\text{prob}(\mathbf{x}|\mathbf{y}, \mathbf{I})] &= (\mathbf{y}-\mathbf{Jx})^T\mathbf{S}_\epsilon^{-1}(\mathbf{y}-\mathbf{Jx}) \\ &\quad + (\mathbf{x}-\mathbf{x}_a)^T\mathbf{S}_a^{-1}(\mathbf{x}-\mathbf{x}_a) + c_3 \\ &= \vec{\epsilon}^{\,T}\mathbf{S}_\epsilon^{-1}\vec{\epsilon} + (\mathbf{x}-\mathbf{x}_a)^T\mathbf{S}_a^{-1}(\mathbf{x}-\mathbf{x}_a) + c_3, \end{aligned} \quad (818)$$

where c_3 is yet another constant. Since Eq. (818) is a quadratic form in $\mathbf{x}$, we can rewrite it as

$$\ln[\text{prob}(\mathbf{x}|\mathbf{y}, \mathbf{I})] = (\mathbf{x}-\hat{\mathbf{x}})^T\hat{\mathbf{S}}^{-1}(\mathbf{x}-\hat{\mathbf{x}}) + c_4 \qquad c_4 = \text{constant}. \quad (819)$$

Thus, the posterior pdf is also a *Gaussian distribution* with expected value $\hat{\mathbf{x}}$, and covariance $\hat{\mathbf{S}}$. Equating terms of Eqs. (818) and (819) that are quadratic in $\mathbf{x}$, we find

$$\mathbf{x}^T\mathbf{J}^T\mathbf{S}_\epsilon^{-1}\mathbf{Jx} + \mathbf{x}^T\mathbf{S}_a^{-1}\mathbf{x} = \mathbf{x}^T\hat{\mathbf{S}}^{-1}\mathbf{x} \quad (820)$$

and solving for $\hat{\mathbf{S}}^{-1}$ yields

$$\hat{\mathbf{S}}^{-1} = \mathbf{J}^T\mathbf{S}_\epsilon^{-1}\mathbf{J} + \mathbf{S}_a^{-1}. \quad (821)$$

Likewise, equating terms linear in $\mathbf{x}^T$, we obtain

$$(-\mathbf{Jx})^T\mathbf{S}_\epsilon^{-1}(\mathbf{y}) + (\mathbf{x})^T\mathbf{S}_a^{-1}(-\mathbf{x}_a) = (\mathbf{x})^T\hat{\mathbf{S}}^{-1}(-\hat{\mathbf{x}}) \quad (822)$$

or

$$(\mathbf{x})^T\mathbf{J}^T\mathbf{S}_\epsilon^{-1}\mathbf{y} + (\mathbf{x})^T\mathbf{S}_a^{-1}\mathbf{x}_a = (\mathbf{x})^T\hat{\mathbf{S}}^{-1}\hat{\mathbf{x}}. \quad (823)$$

Since Eq. (823) must be valid for arbitrary values of $\mathbf{x}$, we may cancel the $\mathbf{x}^T$'s. Thus

$$\mathbf{J}^T\mathbf{S}_\epsilon^{-1}\mathbf{y} + \mathbf{S}_a^{-1}\mathbf{x}_a = (\mathbf{J}^T\mathbf{S}_\epsilon^{-1}\mathbf{J} + \mathbf{S}_a^{-1})\hat{\mathbf{x}} \quad (824)$$

where we have used Eq. (821). Solving for $\hat{\mathbf{x}}$, we find

$$\begin{aligned}\hat{\mathbf{x}} &= (\mathbf{J}^T\mathbf{S}_\epsilon^{-1}\mathbf{J} + \mathbf{S}_a^{-1})^{-1}(\mathbf{J}^T\mathbf{S}_\epsilon^{-1}\mathbf{y} + \mathbf{S}_a^{-1}\mathbf{x}_a)\\ &= (\mathbf{J}^T\mathbf{S}_\epsilon^{-1}\mathbf{J} + \mathbf{S}_a^{-1})^{-1}(\mathbf{J}^T\mathbf{S}_\epsilon^{-1}\mathbf{y} - \mathbf{J}^T\mathbf{S}_\epsilon^{-1}\mathbf{J}\mathbf{x}_a + (\mathbf{J}^T\mathbf{S}_\epsilon^{-1}\mathbf{J} + \mathbf{S}_a^{-1})\mathbf{x}_a)\\ &= \mathbf{x}_a + (\mathbf{J}^T\mathbf{S}_\epsilon^{-1}\mathbf{J} + \mathbf{S}_a^{-1})^{-1}\mathbf{J}^T\mathbf{S}_\epsilon^{-1}(\mathbf{y} - \mathbf{J}\mathbf{x}_a).\end{aligned} \tag{825}$$

Because we have assumed Gaussian statistics

- the expected value $\hat{\mathbf{x}}$ is the same as the maximum probability value obtained by requiring $(\partial \ln \text{prob}(\mathbf{x}|\mathbf{I})/\partial\mathbf{x}) = (1/\text{prob}(\mathbf{x}|\mathbf{I}))(\partial\,\text{prob}(\mathbf{x}|\mathbf{I})/\partial\mathbf{x}) = \mathbf{0}$, or by evaluating $\int \mathbf{x}\,\text{prob}(\mathbf{x}|\mathbf{I})d\mathbf{x}$.

An alternate form for $\hat{\mathbf{x}}$ can be obtained from Eq. (825)

$$\hat{\mathbf{x}} = \mathbf{x}_a + \mathbf{S}_a\mathbf{J}^T(\mathbf{J}\mathbf{S}_a\mathbf{J}^T + \mathbf{S}_\epsilon)^{-1}(\mathbf{y} - \mathbf{J}\mathbf{x}_a). \tag{826}$$

- Recall that the Bayesian solution to the *inverse problem* is not $\hat{\mathbf{x}}$, but the Gaussian pdf $\text{prob}(\mathbf{x}|\mathbf{y},\mathbf{I})$, and that $\hat{\mathbf{x}}$ is the expected value and $\hat{\mathbf{S}}$ the covariance of $\text{prob}(\mathbf{x}|\mathbf{y},\mathbf{I})$. Note that
- $\hat{\mathbf{x}}$ is a linear function of the prior expected value and the measurement, and
- the *inverse* covariance matrix $\hat{\mathbf{S}}^{-1} = \mathbf{J}^T\mathbf{S}_\epsilon^{-1}\mathbf{J} + \mathbf{S}_a^{-1}$ is a linear function of the inverse measurement error *covariance matrix* $\mathbf{S}_\epsilon^{-1}$ and the inverse prior covariance matrix $\mathbf{S}_a^{-1}$.

5.7.1 Optimal Solution for Linear Problems

Given (i) **a measurement** together with a description of its error statistics, (ii) **a forward model** connecting the measurement and the unknown state, and (iii) any **a priori** information that might be available, we may identify the class of possible states consistent with available information, and assign a probability density function to them. But rather than the complete ensemble of possible solutions given by the posterior pdf $\text{prob}(\mathbf{x}|\mathbf{y},\mathbf{I})$, we wish to select

- just one of the possible states as the *optimal solution* to the inverse problem, and assign to it some *error estimate*. To select just one **optimal** state, we may

(1) use the posterior pdf of the state vector to select the solution as either **the expected value** or **the most likely** state, together with **the width of the distribution** as a measure of the uncertainty of the solution;

(2) use the **smoothing error**, the **measurement error**, and/or the **modeling error** as **quantities to be minimized** in seeking the solution.

Note that **the estimation theory aspects of nonlinear problems are similar to those of the linear problem**. The forward model for the linear problem is $\mathbf{y} = \mathbf{J}\mathbf{x} + \vec{\epsilon}$, with a constant $N \times M$ weighting function matrix $\mathbf{J}$. We anticipate a linear solution of the form

$$\hat{\mathbf{x}} = \mathbf{x}_o + \mathbf{G}\mathbf{y} \tag{827}$$

where $\mathbf{G}$ is a constant $M \times N$ matrix and $\mathbf{x}_o$ is some constant offset. Our task is to identify appropriate forms of $\mathbf{G}$ and $\mathbf{x}_o$ for various types of optimality.

The Maximum *a Posteriori* (MAP) Solution

The simplest way to select a state from an ensemble described by a pdf is to choose **the most likely state**, that is, the one for which $\text{prob}(\mathbf{x}|\mathbf{y},\mathbf{I})$ attains a maximum (the *maximum a posteriori (MAP) solution*), or the *expected value solution*:

$$\hat{\mathbf{x}} = \int \mathbf{x}\,\text{prob}(\mathbf{x}|\mathbf{y},\mathbf{I})d\mathbf{x}. \tag{828}$$

In either case, the width of the pdf provides the error estimate, and for the linear problem with Gaussian pdfs these two solutions are identical due to the symmetry of the pdf. For *non-Gaussian statistics*, the MAP and expected value solutions will give different results when the pdf is skew or asymmetric. Then the covariance provides insufficient information and we need higher order moments of the pdf.

The MAP solution given by Eq. (825) can be written in several different forms:

$$\hat{\mathbf{x}} = (\mathbf{J}^T\mathbf{S}_\epsilon^{-1}\mathbf{J} + \mathbf{S}_a^{-1})^{-1}(\mathbf{J}^T\mathbf{S}_\epsilon^{-1}\mathbf{y} + \mathbf{S}_a^{-1}\mathbf{x}_a) \tag{829}$$

$$= \mathbf{x}_a + (\mathbf{J}^T\mathbf{S}_\epsilon^{-1}\mathbf{J} + \mathbf{S}_a^{-1})^{-1}\mathbf{J}^T\mathbf{S}_\epsilon^{-1}(\mathbf{y} - \mathbf{J}\mathbf{x}_a) \quad M\text{-form} \tag{830}$$

$$= \mathbf{x}_a + \mathbf{S}_a\mathbf{J}^T(\mathbf{J}\mathbf{S}_a\mathbf{J}^T + \mathbf{S}_\epsilon)^{-1}(\mathbf{y} - \mathbf{J}\mathbf{x}_a) \quad N\text{-form}. \tag{831}$$

And its covariance given by Eq. (821) can be written as

$$\hat{\mathbf{S}}^{-1} = \mathbf{J}^T\mathbf{S}_\epsilon^{-1}\mathbf{J} + \mathbf{S}_a^{-1} = \mathbf{S}_a - \mathbf{S}_a\mathbf{J}^T(\mathbf{J}\mathbf{S}_a\mathbf{J}^T + \mathbf{S}_\epsilon^{-1})\mathbf{J}\mathbf{S}_a. \tag{832}$$

The two different forms are distinguished by the matrix to be inverted. In the N-form, it is an $N \times N$ matrix, while in the M-form it is an $M \times M$ matrix. Thus, the choice of formulation is determined in part by the relative sizes of the state vector and the measurement vector.

For a *well-posed problem*, an exact solution is possible. Then a matrix $\mathbf{G}$ exists such that $\mathbf{JG} = \mathbf{I}_N$, the unit matrix. For example, we could choose $\mathbf{G} = \mathbf{J}^T(\mathbf{J}\mathbf{J}^T)^{-1}$. Inserting $\mathbf{JG}$ in front of $\mathbf{y}$ in Eq. (829), we obtain

$$\hat{\mathbf{x}} = (\mathbf{J}^T\mathbf{S}_\epsilon^{-1}\mathbf{J} + \mathbf{S}_a^{-1})^{-1}(\mathbf{J}^T\mathbf{S}_\epsilon^{-1}\mathbf{J}(\mathbf{G}\mathbf{y}) + \mathbf{S}_a^{-1}\mathbf{x}_a) \tag{833}$$

which represents a weighted mean of $\mathbf{Gy}$ and $\mathbf{x}_a$ with weights $\mathbf{J}^T\mathbf{S}_\epsilon^{-1}\mathbf{J}$ and $\mathbf{S}_a^{-1}$, respectively, which looks like the familiar combination of two scalar measurements x_1 and x_2 of an unknown x, with variances σ_1 and σ_2, respectively, that is,

$$\hat{x} = \left(\frac{1}{\sigma_1^2} + \frac{1}{\sigma_2^2}\right)^{-1}\left(\frac{x_1}{\sigma_1^2} + \frac{x_2}{\sigma_2^2}\right). \tag{834}$$

Hence, the matrix version can be interpreted as

- the weighted mean of the *a priori* $\mathbf{x}_a$ and any exact retrieval $\mathbf{x}_e = \mathbf{Gy}$, exact in the sense that, if inserted in the forward model, the measured quantity $\mathbf{y}$ is recovered exactly.

An alternative approach to the MAP solution is based on the Bayesian solution pdf. Given prob($\mathbf{x}|\mathbf{y}, \mathbf{I}$), find the state $\hat{\mathbf{x}}$ such that the variance about $\hat{\mathbf{x}}$ is minimized:

$$\frac{\partial}{\partial \mathbf{x}} \int (\mathbf{x} - \hat{\mathbf{x}})^T (\mathbf{x} - \hat{\mathbf{x}}) \text{prob}(\mathbf{x}|\mathbf{y}, \mathbf{I}) d\mathbf{x} = 0 \qquad (835)$$

which gives

$$\hat{\mathbf{x}} = \int \mathbf{x} \, \text{prob}(\mathbf{x}|\mathbf{y}, \mathbf{I}) d\mathbf{x} \qquad (836)$$

that is, the minimum variance solution is the conditional expected value. This result is true for an arbitrary pdf.

5.8
Ill Posedness or Ill Conditioning

As stated in Section 5.5, we consider a set of measurements assembled into a measurement vector $\mathbf{y}$ with N elements, and a set of retrieval parameters assembled into a state vector $\mathbf{x}$ with M elements. The connection between these two vectors is provided by the forward model $\mathbf{F}(\mathbf{x})$

$$\mathbf{y} = \mathbf{F}(\mathbf{x}) + \vec{\epsilon} \qquad (837)$$

where $\vec{\epsilon}$ is the measurement error. The set of equations represented by Eq. (837) is generally nonlinear, and difficult to solve, but a least-squares solution can be obtained as shown above [see Eqs. (694)–(704)]. Since there may be many solutions (state vectors) that fit the data adequately, one must determine the quality of a given solution in terms of physical plausibility, and fit to the measurements and consistency with other constraints. Even in the linear case, it becomes essential to examine the solution in terms of [211, 219]

- *existence*: no solution may be found that exactly fits the measurements, because (i) the forward model is approximate or (ii) the measurements are noisy;
- *uniqueness*: if exact solutions exist, they may not be unique;
- *instability*: the inverse solution may be very sensitive to small changes in the measurements, so that the linear system becomes *ill posed* in the case of a continuous set of measurements, and *ill conditioned* for a discrete set of measurements.

Recall that for the linear noiseless version of Eq. (837), we have

$$\mathbf{J}\mathbf{x} = \mathbf{y} \qquad (838)$$

where the range of $\mathbf{J}$, $R(\mathbf{J})$ consists of all vectors $\mathbf{y}$ for which there is a model $\mathbf{x}$ such that $\mathbf{J}\mathbf{x} = \mathbf{y}$. If rank($\mathbf{J}$) $= M$, then $\mathbf{J}$ has full column rank, and no solution $\mathbf{x}$ may satisfy Eq. (838) because $dim\ R(\mathbf{J}) = M < N$ and a noisy measurement vector $\mathbf{y}$ can lie outside $R(\mathbf{J})$. A useful approximate solution is the model $\mathbf{x}$ that minimizes the **residual** vector

$$\mathbf{r} = \mathbf{y} - \mathbf{J}\mathbf{x} \qquad (839)$$

which leads to the least-squares solution [see Eq. (803)]

$$\mathbf{x} = (\mathbf{J}^T\mathbf{J})^{-1}\mathbf{J}^T\mathbf{y}. \tag{840}$$

It can be shown that $(\mathbf{J}^T\mathbf{J})^{-1}$ exists if $\mathbf{J}$ is of full rank.

5.8.1 SVD Solutions and Resolution Kernels

For least-squares problems that are ill conditioned and/or rank deficient, *singular value decomposition* (SVD) is very useful. In SVD, an $N \times M$ matrix $\mathbf{J}$ is factored into [219]

$$\mathbf{J} = \mathbf{U}\mathbf{S}\mathbf{V}^T \tag{841}$$

where $\mathbf{U}$ is an $N \times N$ **orthogonal** matrix with columns that are unit basis vectors spanning the **measurement space** R^N, $\mathbf{V}$ is an $M \times M$ **orthogonal** matrix with columns that are basis vectors spanning the **model space**, R^M, and $\mathbf{S}$ is an $N \times M$ diagonal matrix with diagonal elements called *singular values.* The nonzero singular values along the diagonal of $\mathbf{S}$ may be arranged in decreasing size, $s_1 \geq s_2 \geq \ldots \geq s_{\min} \geq 0$. If only the first p singular values are nonzero, we can rewrite the SVD of $\mathbf{J}$ in its compact form:

$$\mathbf{J} = \mathbf{U}_p\mathbf{S}_p\mathbf{V}_p^T, \tag{842}$$

where $\mathbf{U}_p$ and $\mathbf{V}_p$ denote the first p columns of $\mathbf{U}$ and $\mathbf{V}$, respectively. We denote by $\mathbf{U}_0$ the last $N - p$ columns of $\mathbf{U}$ and by $\mathbf{V}_0$ the last $M - p$ columns of $\mathbf{V}$. For any vector $\mathbf{y}'$ in the range of $\mathbf{J}$, use of Eq. (842) gives

$$\mathbf{y}' = \mathbf{J}\mathbf{x}' = \mathbf{U}_p(\mathbf{S}_p\mathbf{V}_p^T\mathbf{x}') = \mathbf{U}_p\mathbf{z}'. \tag{843}$$

Thus, any vector $\mathbf{y}'$ in $R(\mathbf{J})$ can be written $\mathbf{y}' = \mathbf{U}_p\mathbf{z}'$, where $\mathbf{z}' = \mathbf{S}_p\mathbf{V}_p^T\mathbf{x}'$, or

$$\mathbf{y}' = \sum_{i=1}^{p} z_i'\mathbf{U}_{.,i} \tag{844}$$

where $\mathbf{U}_{.,i}$ denotes the columns of $\mathbf{U}$. The columns of $\mathbf{U}_p$ span $R(\mathbf{J})$, are linearly independent, and form an orthonormal basis for $R(\mathbf{J})$. Because this orthonormal basis has p vectors, rank$(\mathbf{J}) = p$. The p columns of $\mathbf{U}_p$ form an orthonormal basis for R^N. Since $N(\mathbf{J}^T) + R(\mathbf{J}) = R^N$, the remaining $N - p$ columns of $\mathbf{U}_0$ form an orthonormal basis for $N(\mathbf{J}^T)$, the *data null space.*

Similarly, $\mathbf{J}^T = (\mathbf{U}_p\mathbf{S}_p\mathbf{V}_p^T)^T = \mathbf{V}_p\mathbf{S}_p\mathbf{U}_p^T$ implies $\tilde{\mathbf{y}} = \mathbf{J}^T\tilde{\mathbf{x}} = \mathbf{V}_p(\mathbf{S}_p\mathbf{U}_p^T\tilde{\mathbf{x}}) = \mathbf{V}_p\tilde{\mathbf{z}}$, and hence $\tilde{\mathbf{z}} = \mathbf{S}_p\mathbf{U}_p^T\tilde{\mathbf{x}}$, so that the p columns of $\mathbf{V}_p$ form an orthonormal basis for $R(\mathbf{J}^T)$ and the remaining $M - p$ columns of $\mathbf{V}_0$ form an orthonormal basis for $N(\mathbf{J})$, the *model null space.* The *singular values* of $\mathbf{J}$ and the eigenvalues of $\mathbf{J}\mathbf{J}^T$ and $\mathbf{J}^T\mathbf{J}$ are connected as

$$\mathbf{J}\mathbf{J}^T\mathbf{U}_{.,i} = s_i^2\mathbf{U}_{.,i} \tag{845}$$

and

$$\mathbf{J}^T\mathbf{J}\mathbf{V}_{.,i} = s_i^2\mathbf{V}_{.,i}. \tag{846}$$

We can use the SVD to compute a *generalized inverse* of $\mathbf{J}$ (recall $\mathbf{J} = \mathbf{U}_p\mathbf{S}_p\mathbf{V}_p^T$):

$$\mathbf{J}^\dagger = \mathbf{V}_p\mathbf{S}_p^{-1}\mathbf{U}_p^T \tag{847}$$

and using Eq. (847), we define the *pseudo inverse solution* to be

$$\mathbf{x}_\dagger = \mathbf{J}^\dagger\mathbf{y} = \mathbf{V}_p\mathbf{S}_p^{-1}\mathbf{U}_p^T\mathbf{y}. \tag{848}$$

This pseudo inverse solution [Eq. (848)] has the desirable property that $\mathbf{J}^\dagger$, and hence $\mathbf{x}_\dagger$, always exist. In contrast, the inverse of $\mathbf{J}^T\mathbf{J}$ that appears in the least-squares solution: $\mathbf{x} = (\mathbf{J}^T\mathbf{J})^{-1}\mathbf{J}^T\mathbf{y}$ does not exist unless $\mathbf{J}$ is of full column rank. To examine what the SVD tells us about our linear system, $\mathbf{J}$, and the corresponding generalized inverse system, $\mathbf{J}^\dagger$, we consider four cases:

(1) If $N = M = p$, then $N(\mathbf{G}) = \mathbf{0}$ and $N(\mathbf{J}^T) = \mathbf{0}$ so that $\mathbf{U}_p = \mathbf{U}$ and $\mathbf{V}_p = \mathbf{V}$ are square orthogonal matrices, implying that $\mathbf{U}_p^T = \mathbf{U}_p^{-1}$, and $\mathbf{V}_p^T = \mathbf{V}_p^{-1}$. Hence, Eq. (847) can be rewritten as (recall $\mathbf{J} = \mathbf{U}_p\mathbf{S}_p\mathbf{V}_p^T$)

$$\mathbf{J}^\dagger = \mathbf{V}_p\mathbf{S}_p^{-1}\mathbf{U}_p^T = (\mathbf{U}_p\mathbf{S}_p\mathbf{V}_p^T)^{-1} = \mathbf{J}^{-1} \tag{849}$$

which is the **matrix inverse of a square full rank matrix**. The solution is unique, and the measurements are fit exactly.

(2) If $p < M$ so that $N(\mathbf{J}) \neq \mathbf{0}$, but $N = p$ so that $N(\mathbf{J}^T) = \mathbf{0}$, then $\mathbf{U}_p^T = \mathbf{U}_p^{-1}$ and $\mathbf{V}_p^T\mathbf{V}_p = \mathbf{I}_p$, and $\mathbf{J}$ applied to the *generalized inverse* solution $\mathbf{x}_\dagger = \mathbf{J}^\dagger\mathbf{y}$ gives

$$\mathbf{J}\mathbf{x}_\dagger = \mathbf{J}\mathbf{J}^\dagger\mathbf{y} = \mathbf{U}_p\mathbf{S}_p\mathbf{V}_p^T\mathbf{V}_p\mathbf{S}_p^{-1}\mathbf{U}_p^T\mathbf{y} = \mathbf{U}_p\mathbf{S}_p\mathbf{I}_p\mathbf{S}_p^{-1}\mathbf{U}_p^T\mathbf{y} = \mathbf{y}. \tag{850}$$

Hence, the measurements are fitted exactly, but the solution is nonunique because a *model null space* exists $[N(\mathbf{J}) \neq \mathbf{0}]$. It can be shown that the square of the l_2 norm to the *least-squares solution* $\mathbf{J}\mathbf{x} = \mathbf{y}$ is given by

$$||\mathbf{x}||_2^2 = ||\mathbf{x}_\dagger||_2^2 + \sum_{i=p+1}^{M} \alpha_i^2 \geq ||\mathbf{x}_\dagger||_2^2 \tag{851}$$

where we have equality only if all of the model null space coefficients α_i are zero, implying that the generalized inverse solution is a *minimum length solution.*

(3) If $p = M$ so that $N(\mathbf{J}) = \mathbf{0}$, but $p < N$ so that $N(\mathbf{J}^T) \neq \mathbf{0}$, then $R(\mathbf{J})$ is a strict subset of R^N. Here, (recall $\mathbf{J} = \mathbf{U}_p\mathbf{S}_p\mathbf{V}_p^T$ and $\mathbf{J}^T = \mathbf{V}_p\mathbf{S}_p\mathbf{U}_p^T$)

$$\mathbf{J}\mathbf{x}_\dagger = \mathbf{U}_p\mathbf{S}_p\mathbf{V}_p^T\mathbf{V}_p\mathbf{S}_p^{-1}\mathbf{U}_p^T(\mathbf{U}_p\mathbf{U}_p^T\mathbf{y}) = \mathbf{U}_p\mathbf{U}_p^T\mathbf{y}. \tag{852}$$

The product $\mathbf{U}_p\mathbf{U}_p^T\mathbf{y}$ gives the projection of $\mathbf{y}$ onto $R(\mathbf{J})$. Thus, $\mathbf{J}\mathbf{x}_\dagger$ is the point in $R(\mathbf{J})$ that is closest to $\mathbf{y}$, and $\mathbf{x}_\dagger$ is a *least-squares solution* to $\mathbf{J}\mathbf{x} = \mathbf{y}$. This solution is unique, but cannot fit general data exactly.

(4) If $p < M$ and $p < N$, then both $N(\mathbf{G}^T) \neq \mathbf{0}$ and $N(\mathbf{G}) \neq \mathbf{0}$. In this case, the generalized inverse solution encapsulates the behavior of both of the two previous cases, minimizing both $||\mathbf{J}\mathbf{x} - \mathbf{y}||_2$ and $||\mathbf{x}||_2$. As in case 3,

$$\mathbf{J}\mathbf{x}_\dagger = \mathbf{U}_p\mathbf{S}_p\mathbf{V}_p^T\mathbf{V}_p\mathbf{S}_p^{-1}\mathbf{U}_p^T(\mathbf{U}_p\mathbf{U}_p^T\mathbf{y}) = \mathbf{U}_p\mathbf{U}_p^T\mathbf{y} = \mathrm{proj}_{R(\mathbf{G})}\mathbf{y}. \tag{853}$$

Thus, $\mathbf{x}_\dagger$ is a least-squares solution to $\mathbf{Jx} = \mathbf{y}$. As in case (2), we can write the model and its norm using Eq. (851): $||\mathbf{x}||_2^2 = ||\mathbf{x}_\dagger||_2^2 + \sum_{i=p+1}^{n} \alpha_i^2 \geq ||\mathbf{x}_\dagger||_2^2$. Thus, $\mathbf{x}_\dagger$ is the least-squares solution of minimum length.

In summary, the generalized inverse $\mathbf{G}^\dagger = \mathbf{V}_p\mathbf{S}_p^{-1}\mathbf{U}_p^T$ provides

- an inverse solution given by Eq. (848): $\mathbf{x}_\dagger = \mathbf{J}^\dagger\mathbf{y} = \mathbf{V}_p\mathbf{S}_p^{-1}\mathbf{U}_p^T\mathbf{y}$ that always exists, and is both least-squares and minimum length.

Resolution of the Generalized Inverse Solution

Although $\mathbf{x}_\dagger$ is a solution with well-determined properties, the question is how well does it represent the true situation. If the measurement errors are independent and normally distributed, the *least-squares solution* provides an unbiased estimate of the true model, and the estimated model parameters have a multivariate normal distribution with covariance

$$\mathbf{S}_\mathbf{x} = \sigma^2(\mathbf{J}^T\mathbf{J})^{-1}. \tag{854}$$

For the generalized inverse solution $\mathbf{x}_\dagger$, since $\mathbf{J}^\dagger = \mathbf{V}_p\mathbf{S}_p^{-1}\mathbf{U}_p^T$ and $(\mathbf{J}^\dagger)^T = \mathbf{U}_p\mathbf{S}_p^{-1}\mathbf{V}_p^T$, we have

$$\mathbf{S}_{\mathbf{x}_\dagger} = \sigma^2 \sum_{i=1}^{p} \frac{\mathbf{V}_{.,i}\mathbf{V}_{.,i}^T}{s_i^2}. \tag{855}$$

Unfortunately, unless $p = M$, the generalized inverse solution is **not** an **unbiased** estimator of the true solution, because

- the true solution may have nonzero projections onto basis vectors in $\mathbf{V}_0$.
- In practice, the **bias** introduced by restricting the solution to the subspace spanned by the columns of $\mathbf{V}_p$ is frequently far **larger than the uncertainty due to measurement errors**.

We can use the concept of *model resolution*

- to **examine how closely the generalized inverse solution matches a given model, assuming no errors in the data**.

By multiplying $\mathbf{J}$ times a given model $\mathbf{x}$, we find a corresponding data vector $\mathbf{y} = \mathbf{Jx}$. If we then multiply $\mathbf{J}^\dagger$ times $\mathbf{y}$, we get

$$\mathbf{x}_\dagger = \mathbf{J}^\dagger\mathbf{y} = \mathbf{J}^\dagger\mathbf{Jx} = \mathbf{R}_x\mathbf{x} \tag{856}$$

where the model resolution is given by

$$\mathbf{R}_x = \mathbf{J}^\dagger\mathbf{J} = \mathbf{V}\mathbf{S}^{-1}\mathbf{U}_p^T\mathbf{U}_p\mathbf{S}_p\mathbf{V}_p^T = \mathbf{V}_p\mathbf{V}_p^T. \tag{857}$$

If $N(\mathbf{G}) = \mathbf{0}$, then $\text{rank}(\mathbf{G}) = p = M$, and $\mathbf{R}_x = \mathbf{I}_M$, and thus $\mathbf{x}_\dagger = \mathbf{x}$. Hence, **the resolution is perfect** in the sense that the original model is recovered exactly.

If $N(\mathbf{J}) \neq \mathbf{0}$, then $p = \text{rank}(\mathbf{J}) < M$, so that $\mathbf{R}_x \neq \mathbf{I}_M$. Then

- if $p < M$, $\mathbf{R}_x$ is a symmetric matrix describing how $\mathbf{x}_\dagger$ ($= \mathbf{J}^\dagger\mathbf{y} = \mathbf{J}^\dagger\mathbf{Jx} = \mathbf{R}_x\mathbf{x}$) smears out the original model $\mathbf{x}$ into a recovered model $\mathbf{x}_\dagger$.

- The trace of $\mathbf{R}_x$ can be used as a simple quantitative measure of the resolution:
- If $\mathrm{Tr}(\mathbf{R}_x)$ is close to M, then $\mathbf{R}_x$ is relatively close to the identity matrix.

The model resolution matrix can be used to quantify the bias introduced by the pseudo inverse when $\mathbf{J}$ does not have full column rank ($p < M$).

We begin by showing that the expected value of $\mathbf{x}_\dagger$ is $\mathbf{R}_x\mathbf{x}_{\text{true}}$:

$$\mathcal{E}[\mathbf{x}_\dagger] = \mathcal{E}[\mathbf{J}^\dagger \mathbf{y}] = \mathbf{J}^\dagger \mathcal{E}[\mathbf{y}] = \mathbf{J}^\dagger \mathbf{J}\mathbf{x}_{\text{true}} = \mathbf{R}_x \mathbf{m}_{\text{true}}. \tag{858}$$

Thus the bias in the *generalized inverse* solution is

$$\mathcal{E}[\mathbf{x}_\dagger] - \mathbf{x}_{\text{true}} = \mathbf{R}_x\mathbf{x}_{\text{true}} - \mathbf{x}_{\text{true}} = (\mathbf{R}_x - \mathbf{1})\mathbf{x}_{\text{true}}, \tag{859}$$

where

$$\mathbf{R}_x - \mathbf{1} = \mathbf{V}_p\mathbf{V}_p^T - \mathbf{V}\mathbf{V}^T = -\mathbf{V}_0\mathbf{V}_0^T \quad \rightarrow \mathbf{0} \ \text{ as } p \ \text{ increases.} \tag{860}$$

Note that as p increases, $\mathbf{R}_x$ approaches the identity matrix $\mathbf{1}$, and Eq. (855): $\mathbf{S}_{\mathbf{x}_\dagger} = \sigma^2 \sum_{i=1}^{p}((\mathbf{V}_{.,i}\mathbf{V}_{.,i}^T)/(s_i^2))$ and Eq. (860) reveal an important trade-off associated with the value of p: **As p increases**, the **variance $\mathbf{S}_{\mathbf{x}_\dagger}$** in $\mathbf{x}_\dagger$ **increases**, but the **bias decreases**. In practice, the model resolution matrix is commonly used in two different ways. **First**, we can **examine the diagonal elements** of $\mathbf{R}_x$:

- Diagonal elements that are close to 1 correspond to parameters for which we can claim good resolution.
- Conversely, if any of the diagonal elements is small, then the corresponding model parameter will be poorly resolved.

We can multiply $\mathbf{J}^\dagger$ and $\mathbf{J}$ in the opposite order from Eq. (857): $\mathbf{R}_x = \mathbf{J}^\dagger\mathbf{J} = \mathbf{V}_p\mathbf{V}_p^T$ to obtain the **data space resolution matrix $\mathbf{R}_y$**

$$\mathbf{y}_\dagger = \mathbf{J}\mathbf{x}_\dagger = \mathbf{J}\mathbf{J}^\dagger\mathbf{y} = \mathbf{R}_y\mathbf{y} \tag{861}$$

where

$$\mathbf{R}_y = \mathbf{U}_p\mathbf{S}_p\mathbf{V}_p^T\mathbf{V}_p\mathbf{S}_p^{-1}\mathbf{U}_p^T = \mathbf{U}_p\mathbf{U}_p^T. \tag{862}$$

- If $N(\mathbf{J}^T) = \mathbf{0}$, then $p = N$, and $\mathbf{R}_y = \mathbf{1}_N$. In this case, $\mathbf{y}_\dagger = \mathbf{y}$, and the generalized inverse solution $\mathbf{x}_\dagger$ fits the data exactly.
- However, if $N(\mathbf{J}^T)$ is nontrivial, then $p < N$, and $\mathbf{R}_y \neq \mathbf{1}_N$. In this case, $\mathbf{x}_\dagger$ does not exactly fit the data.
- Note: the model space resolution matrix Eq. (857): $\mathbf{R}_x = \mathbf{J}^\dagger\mathbf{J} = \mathbf{V}_p\mathbf{V}_p^T$ and the data space resolution matrix Eq. (862): $\mathbf{R}_y = \mathbf{J}\mathbf{J}^\dagger = \mathbf{U}_p\mathbf{U}_p^T$ do not depend on specific data or models, but are exclusive properties of $\mathbf{J}$.
- They reflect the physics and geometry of a problem, and can thus be assessed during the design phase of an experiment.

Instability of the Generalized Inverse Solution

The *generalized inverse solution* $\mathbf{x}_\dagger$ has zero projection onto $N(\mathbf{J})$, but it may include terms involving column vectors in $\mathbf{V}_p$ with very small nonzero *singular values.* Small singular values cause the generalized inverse solution to be extremely sensitive to small amounts of noise in the data, and it can be **difficult to distinguish between zero and extremely small singular values.** To **quantify the instabilities** created by small singular values, we **recast** the generalized inverse **solution to make** the **effect of small singular values explicit**. The formula for the generalized inverse solution can be written as

$$\mathbf{x}_\dagger = \mathbf{V}_p \mathbf{S}_p^{-1} \mathbf{U}_p^T \mathbf{y} = \sum_{i=1}^{p} \frac{(\mathbf{U}_{.,i})^T \mathbf{y}}{s_i} \mathbf{V}_{.,i}. \tag{863}$$

In the presence of random noise, $\mathbf{y}$ will generally have a nonzero projection onto each of the directions specified by the columns of $\mathbf{U}$.

- The presence of a very small singular value s_i in Eq. (863) can lead to a very large coefficient for the corresponding model space basis vector $\mathbf{V}_{.,i}$.
- **In the worst case, the generalized inverse solution is just a noise amplifier, and the answer is practically useless.**
- A measure of the instability of the solution is the *condition number.*

It can be shown that for arbitrary values of p

$$\frac{||\mathbf{x}_\dagger - \mathbf{x}'_\dagger||_2}{||\mathbf{x}_\dagger||_2} \le \frac{s_1}{s_p} \frac{||\mathbf{y} - \mathbf{y}'||_2}{||\mathbf{d}||_2}. \tag{864}$$

- If we decrease p and thus eliminate model space vectors associated with small singular values, the solution becomes more stable.
- But the increased stability comes at the expense of reducing the dimension of the subspace of R^M where the solution lies.
- The model resolution matrix for the stabilized solution obtained by decreasing p becomes less like the identity matrix ($\mathbf{R}_y \neq \mathbf{1}_N$): the **fit** to the data **worsens**.

The condition number of $\mathbf{J}$ is the coefficient in Eq. (864)

$$\text{cond}(\mathbf{J}) = \frac{s_1}{s_k} \quad \text{where } k = \min(N, M). \tag{865}$$

- If $\mathbf{J}$ is of full rank, and we use all of the singular values in the pseudo inverse solution ($p = k$), then the condition number is exactly given by Eq. (865).
- If $\mathbf{J}$ is of **less than** full rank, then the condition number is effectively infinite.
- As with the model and data resolution matrices [Eq. (857); $\mathbf{R}_x = \mathbf{J}^\dagger \mathbf{J} = \mathbf{V}_p \mathbf{V}_p^T$ and (862): $\mathbf{R}_y = \mathbf{J}\mathbf{J}^\dagger = \mathbf{U}_p \mathbf{U}_p^T$], cond($\mathbf{J}$) is a property of $\mathbf{J}$ that can be computed in the design phase of an experiment before any data is collected.

The Truncated SVD or TSVD Solution

A useful solution may be obtained by truncating Eq. (863) at some highest term $p' < p$ to produce a *truncated SVD* (**TSVD**) solution. One way to decide when to truncate Eq. (863) is to apply the *discrepancy principle,* that is

- pick the smallest value of p' so that the model fits the data to some tolerance based on the length of the residual vector:

$$||\mathbf{Jx} - \mathbf{y}||_2 \leq \delta, \tag{866}$$

where $\mathbf{J}$ and $\mathbf{y}$ are the system matrix and data vector, respectively, and $\delta \approx \sqrt{N}$ is a reasonable choice because the approximate χ^2 distribution with N degrees of freedom is N. The TSVD solution is an example of *regularization,* in which we select solutions that sacrifice fit to the data in exchange for solution stability. Understanding the trade-off between fitting the data and solution stability involved in regularization is of fundamental importance.

5.8.2 Twomey–Tikhonov Regularization – TT-Reg

The *generalized inverse solution* given by Eq. (863) can become extremely unstable when one or more of the *singular values* s_i is small:

- to stabilize or regularize the solution, we may drop terms in the sum associated with small singular values, but this
- **regularized solution** has reduced resolution and is **not unbiased**.

The Twomey–Tikhonov regularization [220, 221] (TT-reg) is a widely used technique of regularizing discrete ill-posed problems:

- the **TT-reg solution** can be expressed in terms of the SVD of $\mathbf{J}$;
- it is a variant on the generalized inverse solution that effectively gives **greater weight to large singular values** in the SVD solution and **less weight to small singular values**.

In order to select a good solution using TT-reg, we consider all solutions with $||\mathbf{Jx} - \mathbf{y}||_2 \leq \delta$, and we select from these solutions the one that minimizes the norm of $\mathbf{x}$:

$$\min_{||\mathbf{Jx} - \mathbf{y}||_2 \leq \delta} \quad ||\mathbf{x}||_2 \tag{867}$$

Note that, as δ increases, the set of feasible models expands, and the minimum value of $||\mathbf{x}||_2$ decreases. We can thus trace out a curve of minimum values of $||\mathbf{x}||_2$ versus δ, as illustrated in the Figure 36(a).

It is also possible to trace out this curve by considering problems of the form

$$\min_{||\mathbf{x}||_2 \leq \varepsilon} \quad ||\mathbf{Jx} - \mathbf{y}||_2 \tag{868}$$

As ε decreases, the set of feasible solutions becomes smaller, and the minimum value of $||\mathbf{Jx} - \mathbf{y}||_2$ increases. Again, as we adjust ε, we trace out the curve of optimal values of $||\mathbf{x}||_2$ and $||\mathbf{Jx} - \mathbf{y}||$ (see Figure 36(b)).

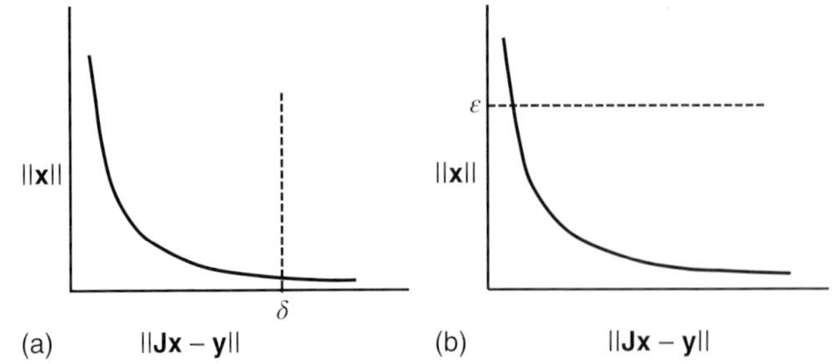

Figure 36 (a) A particular misfit norm, $\delta = ||\mathbf{Jx} - \mathbf{y}||_2$, and its associated model norm, $||\mathbf{x}||_2$. (b) A particular misfit norm, $\varepsilon = ||\mathbf{x}||_2$, and its associated model norm, $||\mathbf{Jx} - \mathbf{y}||_2$.

A third option is to consider the damped least-squares problem:

$$\min \quad ||\mathbf{Jx} - \mathbf{y}||_2^2 + \alpha^2||\mathbf{x}||_2^2 \tag{869}$$

which arises when we apply the method of Lagrange multipliers to Eq. (867). Here, α is a *regularization parameter*. For appropriate choices of δ, ε, and α, the three problems, namely Eqs. (867), (868), and (869), yield the same solution. Therefore, we may concentrate on solving the damped least-squares form of the problem given by Eq. (869), since solutions to Eqs. (867) and (868) can be obtained using Eq. (869) by adjusting the *regularization parameter* α until the constraints are satisfied. We note that

- when plotted on a log–log scale, the curve of optimal values of $||\mathbf{x}||_2$ versus $||\mathbf{Jx} - \mathbf{y}||_2$ often takes on a characteristic L shape, because
- $||\mathbf{x}||_2$ is a strictly decreasing function of α and $||\mathbf{Jx} - \mathbf{y}||_2$ is a strictly increasing function of α.
- The sharpness of the "corner" varies from problem to problem, but it is frequently well defined. For this reason, the curve is called an *L-curve*.
- In addition to the **discrepancy principle**, another popular criterion for picking the value of α is the *L*-**curve criterion**:
- select the value of α that gives the solution closest to the corner of the L-curve.

5.8.3 Implementation of the Twomey–Tikhonov Regularization

The damped least-squares problem in Eq. (869) is equivalent to the ordinary least-squares problem $\mathbf{Jx} = \mathbf{y}$ by augmenting it as follows:

$$\min \left\| \begin{bmatrix} \mathbf{J} \\ \alpha\mathbf{1} \end{bmatrix} \mathbf{x} - \begin{bmatrix} \mathbf{y} \\ \mathbf{0} \end{bmatrix} \right\|_2^2 .$$

or

$$(\mathbf{J}^T\mathbf{J} + \alpha^2\mathbf{1})\mathbf{x} = \mathbf{J}^T\mathbf{y} \tag{870}$$

which is a set of constraints for a **zeroth-order TT-reg** solution of $\mathbf{Jx} = \mathbf{y}$. Employing the SVD of $\mathbf{J}$, we can write Eq. (870) as

$$(\underbrace{\mathbf{VS}^T\mathbf{U}^T}_{\mathbf{J}^T}\underbrace{\mathbf{USV}^T}_{\mathbf{J}} + \alpha^2\mathbf{1})\mathbf{x} = (\mathbf{VS}^T\mathbf{SV}^T + \alpha^2\mathbf{I})\mathbf{x} = \underbrace{\mathbf{VS}^T\mathbf{U}^T}_{\mathbf{J}^T}\mathbf{y}. \tag{871}$$

Since Eq. (871) is nonsingular, it has a unique solution

$$\mathbf{x}_\alpha = \sum_{i=1}^{k} \frac{s_i^2}{s_i^2 + \alpha^2} \frac{(\mathbf{U}_{.,i})^T\mathbf{d}}{s_i} \mathbf{V}_{.,i} \tag{872}$$

where $k = \min(N, M)$, so that **all singular values are included**, and the filter factors are given by

$$f_i = \frac{s_i^2}{s_i^2 + \alpha^2}. \tag{873}$$

Note that

- for $s_i >> \alpha, f_i \approx 1$, and for $s_i << \alpha, f_i \approx 0$.
- for $0 < s_i < 1$, as s_i decrease, f_i produce a monotonically decreasing contribution of corresponding model space vectors, $\mathbf{V}_{.,i}$.

Resolution, Bias, and Uncertainty in the TT-reg Solution

As in the TSVD approach, we can compute a model resolution matrix for the TT-reg method. Using Eq. (870) $(\mathbf{J}^T\mathbf{J} + \alpha^2\mathbf{I})\mathbf{x} = \mathbf{J}^T\mathbf{y}$ and SVD, the solution can be written as

$$\mathbf{x}_\alpha = (\mathbf{J}^T\mathbf{J} + \alpha^2\mathbf{1})^{-1}\mathbf{J}^T\mathbf{y} = \mathbf{J}^\ddagger\mathbf{y} = \mathbf{VFS}^\dagger\mathbf{U}^T\mathbf{d}. \tag{874}$$

Here, $\mathbf{F}$ is an $M \times N$ diagonal matrix with the filter factors f_i on the diagonal, $\mathbf{S}^\dagger$ is the generalized inverse of $\mathbf{S}$, and $\mathbf{J}^\ddagger = (\mathbf{J}^T\mathbf{J} + \alpha^2\mathbf{I})^{-1}\mathbf{J}^T$ is a *generalized inverse* matrix that can be used to construct model data and data resolution matrices as was done for the SVD solution in Eq. (857): $\mathbf{R}_m = \mathbf{V}_p\mathbf{V}_p^T$ and Eq. (862): $\mathbf{R}_d = \mathbf{U}_p\mathbf{U}_p^T$. The resolution matrices are

$$\mathbf{R}_{x,\alpha} = \mathbf{J}^\ddagger\mathbf{J} = \mathbf{VFV}^T \tag{875}$$

and

$$\mathbf{R}_{y,\alpha} = \mathbf{J}\mathbf{J}^\ddagger = \mathbf{UFU}^T. \tag{876}$$

Note that $\mathbf{R}_x$ and $\mathbf{R}_y$ depend on the particular value of α used in Eq. (874).

Higher Order Twomey–Tikhonov Regularization

So far we have minimized an objective function involving $||\mathbf{x}||_2$, but

- In many situations, we would prefer to obtain a solution that minimizes some other measure of $\mathbf{x}$, such as the norm of the first or second derivative.

For example, if we have discretized our problem, then we can approximate the first derivative of the model by **Lm**, where

$$\mathbf{L} = \begin{bmatrix} -1 & 1 & & & & \\ & -1 & 1 & & & \\ & & \cdots & & & \\ & & & -1 & 1 & \\ & & & & -1 & 1 \end{bmatrix}. \tag{877}$$

Matrices that are used to differentiate **x** for the purposes of regularization are referred to as *roughening matrices.* In Eq. (877), **Lx** is a finite-difference approximation that is proportional to the **first** derivative of **x**.

- By minimizing $||\mathbf{Lx}||_2$, we will favor solutions that are relatively flat. Note:
- $||\mathbf{Lx}||_2$ is a seminorm: it is zero for any constant model, not just for $\mathbf{x} = \mathbf{0}$.

In **first-order TT-reg**, we solve the damped least-squares problem

$$\min ||\mathbf{Jx} - \mathbf{y}||_2^2 + \alpha^2||\mathbf{Lx}||_2^2 \tag{878}$$

using an **L** matrix like Eq. (877). In **second-order TT-reg**, we use

$$\mathbf{L} = \begin{bmatrix} 1 & -2 & 1 & & & & \\ & 1 & -2 & 1 & & & \\ & & & \cdots & & & \\ & & & & 1 & -2 & 1 \\ & & & & & 1 & -2 & 1 \end{bmatrix} \tag{879}$$

where **Lx** is a finite-difference approximation proportional to the **second** derivative of **x**. Minimizing the seminorm $||\mathbf{Lx}||_2$, one penalizes solutions that are rough in a second derivative sense.

We already applied zeroth-order TT-reg to solve Eq. (878): $\min ||\mathbf{Jx} - \mathbf{y}||_2^2 + \alpha^2||\mathbf{Lx}||_2^2$, for $\mathbf{L} = \mathbf{1}$, using the *singular value decomposition* Eq. (872): $\mathbf{x}_\alpha = \sum_{i=1}^{k}(s_i^2/(s_i^2 + \alpha^2))(((\mathbf{U}_{.,i})^T\mathbf{d})/(s_i)\mathbf{V}_{.,i})$.

5.9 Nonlinear Inverse Problems

The linearity of inverse problems may be classified as follows: [37]

- **Linear:** when the forward problem can be put in the form $\mathbf{y} = \mathbf{Jx}$, and any *a priori* is Gaussian: very few practical problems are truly linear.
- **Nearly linear:** problems which are nonlinear, but for which a **linearization** about some prior state is adequate to find a solution.
- **Moderately nonlinear:** problems where **linearization** is adequate for the error analysis, but not for finding the solution.
- **Grossly nonlinear:** problems which are nonlinear even within the range of the errors.

However, it should be noted that much of what has been said about linear problems applies directly to *moderately nonlinear* problems when they are appropriately linearized. But there are *no general, explicit expressions* for locating optimal solutions in the *moderately nonlinear* case; they must be found numerically and iteratively. Thus, the main topic of discussion is **numerical methods for finding solutions to *moderately nonlinear* problems**. If the problem is no worse than *moderately nonlinear*, and the measurement error is Gaussian, then the retrieval error will be Gaussian, and the **linear error analysis** will apply.

Formulation of the Inverse Problem

The primary task of a linear retrieval is to

- **select a state** satisfying some criterion of optimality **from an ensemble of states** which agrees with the measurement within experimental error. Locating the ensemble in state space is straightforward in the linear case, but
- **in the nonlinear case, an *explicit solution is not available.***

The expected value could be found in principle by integrating Eq. (836) [$\hat{\mathbf{x}} = \int \mathbf{x}\, \text{prob}(\mathbf{x}|\mathbf{y})d\mathbf{x}$] over all state space, but that would be exceedingly expensive. The optimality criterion may also need reconsideration in the nonlinear case:

- **if the Bayesian solution** $\text{prob}(\mathbf{x}|\mathbf{y}, \mathbf{I})$ is described by a **Gaussian** pdf (or at least a symmetric pdf), then the **maximum probability** and the **expected value** criteria still lead to the **same solution**;
- **if it is not symmetric**, then they will lead to **different solutions** in general, **but the expected value is still the minimum variance solution**, because Eq. (836) applies to any problem, however nonlinear.

5.9.1
Gauss–Newton Solution of the Nonlinear Inverse Problem

For nonlinear problems, we will consider the MAP approach. The **Bayesian solution to the linear problem**, Eq. (818)

$$\ln \text{prob}(\mathbf{x}|\mathbf{y}, \mathbf{I}) = (\mathbf{y} - \mathbf{J}\mathbf{x})^T \mathbf{S}_\epsilon^{-1}(\mathbf{y} - \mathbf{J}\mathbf{x}) + (\mathbf{x} - \mathbf{x}_a)^T \mathbf{S}_a^{-1}(\mathbf{x} - \mathbf{x}_a) + c_3 \qquad (880)$$

can be **modified for an *inverse problem*** in which

(i) the forward model is a general (nonlinear) function of the state,
(ii) the measurement error is Gaussian, and
(iii) there is a prior estimate with a Gaussian error

$$\ln \text{prob}(\mathbf{x}|\mathbf{y}, \mathbf{I}) = \underbrace{[\mathbf{y} - \mathbf{F}(\mathbf{x})]^T \mathbf{S}_\epsilon^{-1}[\mathbf{y} - \mathbf{F}(\mathbf{x})] + [\mathbf{x} - \mathbf{x}_a]^T \mathbf{S}_a^{-1}[\mathbf{x} - \mathbf{x}_a]}_{\text{Cost function}} + c_3. \qquad (881)$$

The first two terms on the right-hand side of Eq. (881) constitute the *cost functions*

$$\Phi(\mathbf{x}) = [\mathbf{y} - \mathbf{F}(\mathbf{x})]^T \mathbf{S}_\epsilon^{-1}[\mathbf{y} - \mathbf{F}(\mathbf{x})] + [\mathbf{x} - \mathbf{x}_a]^T \mathbf{S}_a^{-1}[\mathbf{x} - \mathbf{x}_a]. \qquad (882)$$

As in the linear case, the task in the nonlinear case is to find the best estimate $\hat{\mathbf{x}}$ and an error estimate. To find the maximum probability state $\hat{\mathbf{x}}$, we set the derivative of Eq. (882) to zero (using $\nabla_x(\mathbf{x}^T\mathbf{S}\mathbf{x}) = (\partial(\mathbf{x}^T\mathbf{S}\mathbf{x}))/(\partial\mathbf{x}) = 2\mathbf{S}\mathbf{x}$):

$$\nabla_x\{\Phi(\mathbf{x})\} = \frac{\partial\Phi(\mathbf{x})}{\partial\mathbf{x}} \equiv \mathbf{g}(\mathbf{x}) = 0 \tag{883}$$

which implies

$$\mathbf{g}(\mathbf{x}) = -[\nabla_x\mathbf{F}(\mathbf{x})]^T\mathbf{S}_\epsilon^{-1}[\mathbf{y} - \mathbf{F}(\mathbf{x})] + \mathbf{S}_a^{-1}[\mathbf{x} - \mathbf{x}_a] = 0. \tag{884}$$

Note that the gradient ∇_x of a vector-valued function is a matrix-valued function. Putting $\mathbf{J}(\mathbf{x}) = \nabla_x\mathbf{F}(\mathbf{x})$, we find the following **implicit** equation for $\hat{\mathbf{x}}$:

$$\mathbf{g}(\hat{\mathbf{x}}) = -\mathbf{J}^T(\hat{\mathbf{x}})\mathbf{S}_\epsilon^{-1}[\mathbf{y} - \mathbf{F}(\hat{\mathbf{x}})] + \mathbf{S}_a^{-1}[\hat{\mathbf{x}} - \mathbf{x}_a] = 0. \tag{885}$$

This equation must be solved numerically.

Newton and Newton–Gauss Methods

If the problem is not **too** nonlinear, we may use *Newton's method* to solve $\nabla_x\{\Phi(\mathbf{x})\} = (\partial\Phi(\mathbf{x}))/(\partial\mathbf{x}) \equiv \mathbf{g}(\mathbf{x}) = 0$. The iteration is analogous to Newton's method for the scalar case.

$$x_{i+1} = x_i - \frac{g(x_i)}{g'(x_i)} \quad \rightarrow \quad \mathbf{x}_{i+1} = \mathbf{x}_i - [\nabla_\mathbf{x}\mathbf{g}(\mathbf{x}_i)]^{-1}\mathbf{g}(\mathbf{x}_i) \tag{886}$$

where the inverse is a matrix inverse. Using Eq. (885) for $\mathbf{g}(\mathbf{x})$, we find

$$\begin{aligned}\nabla_\mathbf{x}\mathbf{g}(\mathbf{x}) &= \nabla_\mathbf{x}\Big\{ -\mathbf{J}^T(\mathbf{x})\mathbf{S}_\epsilon^{-1}[\mathbf{y} - \mathbf{F}(\mathbf{x})] + \mathbf{S}_a^{-1}[\mathbf{x} - \mathbf{x}_a]\Big\} \\ &= \mathbf{S}_a^{-1} + \mathbf{J}^T\mathbf{S}_\epsilon^{-1}\mathbf{J} - [\nabla_x\mathbf{J}^T]\mathbf{S}_\epsilon^{-1}[\mathbf{y} - \mathbf{F}(\mathbf{x})].\end{aligned} \tag{887}$$

Since $\mathbf{g}(\mathbf{x}) = \nabla_\mathbf{x}\{\Phi(\mathbf{x})\}$ is the derivative of the cost function Eq. (882) $\nabla_\mathbf{x}\mathbf{g}(\mathbf{x}) = \nabla_\mathbf{x}^2\{\Phi(\mathbf{x})\}$ is the second derivative known as the *Hessian.*

The Hessian involves the Jacobian $\mathbf{J}(\mathbf{x}) = \nabla_x\mathbf{F}(\mathbf{x})$, the **first** derivative of the forward model, and $\nabla_x\mathbf{J}^T$, the **second** derivative of the forward model. If we ignore the second derivative (i.e., assume $\nabla_x\mathbf{J}^T = 0$), then by substituting

$$\text{Eq. (885)} \rightarrow \mathbf{g}(\mathbf{x}) = -\mathbf{J}^T(\mathbf{x})\mathbf{S}_\epsilon^{-1}[\mathbf{y} - \mathbf{F}(\mathbf{x})] + \mathbf{S}_a^{-1}[\mathbf{x} - \mathbf{x}_a] = 0$$

$$\text{and Eq. (887)} \rightarrow \nabla_\mathbf{x}\mathbf{g}(\mathbf{x}) \approx \mathbf{S}_a^{-1} + \mathbf{J}^T\mathbf{S}_\epsilon^{-1}\mathbf{J} \equiv \nabla\mathbf{g}$$

$$\text{into Eq. (886)} \rightarrow \mathbf{x}_{i+1} = \mathbf{x}_i - [\nabla_\mathbf{x}\mathbf{g}(\mathbf{x}_i)]^{-1}\mathbf{g}(\mathbf{x}_i),$$

we obtain the *Gauss–Newton method.*

$$\mathbf{x}_{i+1} = \mathbf{x}_i + \nabla\mathbf{g}_i^{-1}\underbrace{[\mathbf{J}_i^T\mathbf{S}_\epsilon^{-1}(\mathbf{y} - \mathbf{F}(\mathbf{x}_i)) - \mathbf{S}_a^{-1}(\mathbf{x}_i - \mathbf{x}_a)]}_{-\mathbf{g}(\mathbf{x}_i)} \tag{888}$$

where $\mathbf{J}_i = \mathbf{J}(\mathbf{x}_i)$, and using the result above $\mathbf{S}_a^{-1} = \nabla \mathbf{g}_i - \mathbf{J}_i^T \mathbf{S}_\epsilon^{-1} \mathbf{J}_i$ in Eq. (888), we obtain

$$\begin{aligned}
\mathbf{x}_{i+1} &= \mathbf{x}_i + (\mathbf{S}_a^{-1} + \mathbf{J}_i^T \mathbf{S}_\epsilon^{-1} \mathbf{J}_i)^{-1} [\mathbf{J}_i^T \mathbf{S}_\epsilon^{-1} (\mathbf{y} - \mathbf{F}(\mathbf{x}_i)) - \mathbf{S}_a^{-1} (\mathbf{x}_i - \mathbf{x}_a)] \\
&= \mathbf{x}_i + \nabla \mathbf{g}_i^{-1} [\mathbf{J}_i^T \mathbf{S}_\epsilon^{-1} (\mathbf{y} - \mathbf{F}(\mathbf{x}_i)) - (\nabla \mathbf{g}_i - \mathbf{J}_i^T \mathbf{S}_\epsilon^{-1} \mathbf{J}_i)(\mathbf{x}_i - \mathbf{x}_a)] \\
&= \mathbf{x}_a + \nabla \mathbf{g}_i^{-1} \mathbf{J}_i^T \mathbf{S}_\epsilon^{-1} [\mathbf{y} - \mathbf{F}(\mathbf{x}_i) + \mathbf{J}_i(\mathbf{x}_i - \mathbf{x}_a)] \\
&= \mathbf{x}_a + \mathbf{S}_a \mathbf{J}_i^T (\mathbf{J}_i \mathbf{S}_a \mathbf{J}_i^T + \mathbf{S}_\epsilon)^{-1} [\mathbf{y} - \mathbf{F}(\mathbf{x}_i) + \mathbf{J}_i(\mathbf{x}_i - \mathbf{x}_a)] \\
&= \mathbf{x}_a + \mathbf{G}_i [\mathbf{y} - \mathbf{F}(\mathbf{x}_i) + \mathbf{J}_i(\mathbf{x}_i - \mathbf{x}_a)]
\end{aligned} \tag{889}$$

where [see Eqs. (830) and (831)]

$$\mathbf{G}_i = \nabla \mathbf{g}_i^{-1} \mathbf{J}_i^T \mathbf{S}_\epsilon^{-1} = (\mathbf{S}_a^{-1} + \mathbf{J}_i^T \mathbf{S}_\epsilon^{-1} \mathbf{J}_i)^{-1} \mathbf{J}_i^T \mathbf{S}_\epsilon^{-1} = \mathbf{S}_a \mathbf{J}^T (\mathbf{J} \mathbf{S}_a \mathbf{J}^T + \mathbf{S}_\epsilon)^{-1}.$$

5.9.2
Levenberg–Marquardt Method

Recall Tikhonov's zeroth-order regularization of the least-squares (LS) solution for the **linear** inversion problem [see §5.8.2, Eq. (870)]

$$\mathbf{x} = (\mathbf{J}^T \mathbf{J} + \alpha^2 \mathbf{1})^{-1} \mathbf{J}^T \mathbf{y}$$

where α^2 is a regularization parameter and $\mathbf{1}$ is a measure of smoothness. If we set $\gamma \equiv \alpha^2$, we obtain a *constrained linear LS inversion*

$$\mathbf{x} = (\mathbf{J}^T \mathbf{J} + \gamma \mathbf{1})^{-1} \mathbf{J}^T \mathbf{y}.$$

For the **nonlinear** LS problem, Levenberg proposed the iteration

$$\mathbf{x}_{i+1} = \mathbf{x}_i + (\mathbf{J}_i^T \mathbf{J}_i + \gamma_i \mathbf{1})^{-1} \mathbf{J}_i^T [\mathbf{y} - \mathbf{F}(\mathbf{x}_i)] \tag{890}$$

where $0 \le \gamma_i \le \infty$ is chosen at each iterative step to minimize the cost function [see Eq. (882)]. Applied to the *Gauss–Newton method* [see first line of Eq. (889)], we have

$$\mathbf{x}_{i+1} = \mathbf{x}_i + (\mathbf{S}_a^{-1} + \mathbf{J}_i^T \mathbf{S}_\epsilon^{-1} \mathbf{J}_i + \gamma_i \mathbf{1})^{-1} [\mathbf{J}_i^T \mathbf{S}_\epsilon^{-1} (\mathbf{y} - \mathbf{F}(\mathbf{x}_i)) - \mathbf{S}_a^{-1} (\mathbf{x}_i - \mathbf{x}_a)].$$

The computation needed for choosing γ_i is significant, because $F(\mathbf{x})$ must be evaluated for each γ_i tried. Marquardt simplified the choice of γ_i by starting a new iteration step as soon as a value was found that reduced the cost function. A simplified version of this strategy, known as the *Levenberg–Marquardt method*, is (see Press *et al.* [222])

- if χ^2 **increases** in a step, do not update $\mathbf{x}_i$, **increase** γ, and try again;
- if χ^2 **decreases** in a step, update $\mathbf{x}_i$, and **decrease** γ for the next step.

Since elements of the state vector may have different magnitudes, replace $\gamma \mathbf{I}$ by $\gamma \mathbf{D}$, where $\mathbf{D}$ is a diagonal scaling matrix. A strategy for updating γ is based on the ratio

$$R = \frac{\text{change in } \Phi(\mathbf{x}) \text{ computed properly using } F(\mathbf{x})}{\text{change in } \Phi(\mathbf{x}) \text{ computed using } F(\mathbf{x}) \approx \mathbf{J}\mathbf{x}}.$$

- $R \to 1$ if a linear approximation ($F(\mathbf{x}) \approx \mathbf{Jx}$) is satisfactory;
- $R < 0$ if χ^2 has increased instead of decreased;
- goal: find a value of γ which restricts the new value of $\mathbf{x}$ to the **trust region**:
- that is, to lie within the linear range of the previous estimate.

The strategy becomes

- If $R > 0.75 \;\to\;$ reduce γ. If $R < 0.25 \;\to\;$ increase γ.
- If $0.25 < R < 0.75$ do not change γ.
- If γ less than some critical value, use $\gamma = 0$ (Gauss–Newton).

The numbers 0.75 and 0.25 were found by experiment, and are not crucial. Replacing $\gamma\mathbf{1}$ by $\gamma\mathbf{D}$, in the Gauss–Newton method, we have

$$\mathbf{x}_{i+1} = \mathbf{x}_i + (\mathbf{S}_a^{-1} + \mathbf{J}_i^T\mathbf{S}_\epsilon^{-1}\mathbf{J}_i + \gamma_i\mathbf{D})^{-1}[\mathbf{J}_i^T\mathbf{S}_\epsilon^{-1}(\mathbf{y} - \mathbf{F}(\mathbf{x}_i)) - \mathbf{S}_a^{-1}(\mathbf{x}_i - \mathbf{x}_a)]. \quad (891)$$

The scaling matrix $\mathbf{D}$ does not have to be diagonal, but must be positive definite. The simplest choice is $\mathbf{D} = \mathbf{S}_a^{-1}$, so that Eq. (891) becomes

$$\mathbf{x}_{i+1} = \mathbf{x}_i + [(1 + \gamma_i)\mathbf{S}_a^{-1} + \mathbf{J}_i^T\mathbf{S}_\epsilon^{-1}\mathbf{J}_i]^{-1}\{\mathbf{J}_i^T\mathbf{S}_\epsilon^{-1}(\mathbf{y} - \mathbf{F}(\mathbf{x}_i)) - \mathbf{S}_a^{-1}(\mathbf{x}_i - \mathbf{x}_a)\}. \quad (892)$$

The computation required is similar to that in the *Gauss–Newton method* ($\gamma_i = 0$), but the number of steps may be larger because the problem may be more difficult.

Problems

5.1 Verify Eqs. (648)–(650).
5.2 Verify Eq. (657).
5.3 Derive the result shown in Eq. (716).
5.4 Derive Eqs. (720) and (723).

6
Applications

6.1
Principal Component (PC) Analysis

A *PC analysis* method was introduced by Natraj *et al.* [105] to overcome the limitations of the correlated-k and spectral mapping methods. As discussed in Appendix B, if a dataset consists of s IOPs in L atmospheric layers at N_λ wavenumbers, then a set of *empirical orthogonal functions (EOFs)* $\varphi_k, k = 1, \dots, s \times L$, can be constructed based on a certain *covariance matrix* representing the data. The PCs are the projections of the original dataset onto these EOFs:

$$P_{ki} = \sum_{\ell=1}^{sL} \frac{\varphi_{k\ell} F_{\ell i}}{\sqrt{\lambda_k}} \tag{893}$$

where $\varphi_{k\ell}$ is the ℓth component of the kth EOF and P_{ki} is the ith component of the kth PC.

An atmospheric radiative transfer (RT) problem is characterized by (i) the optical depth $d\tau$, (ii) the *single-scattering albedo* ϖ, (iii) the *scattering phase function*, and (iv) the *surface reflectance*. Assuming for simplicity that (iii) and (iv) are independent of wavenumber, one may assign the first L components of each EOF to the optical depth, and the remaining $L + 1$ to $2L$ components to the single-scattering albedo. As in correlated-k and spectral mapping methods, the goal is reduce the number of RT computations by grouping wavenumbers with similar IOPs.

To determine how to do the grouping, Natraj *et al.* [105] inspected the profiles of the layer optical depth and the single-scattering albedo versus wavenumber, and found that for the O_2 A band the maximum variability in the optical depth occurred in the lower half of the atmosphere, while the single-scattering albedo was fairly uniform. Based on this inspection, the following grouping criteria were obtained:

- $c_1 \le \ln(2\tau_2) \le c_2$, where τ_2 is the cumulative optical depth of the lower half of the atmosphere,
- $c_3 \le \varpi_1 \le c_4$, where ϖ_1 is the single-scattering albedo of the top layer.

A particular choice of the parameters $c_1, \dots, c_4$ defines a "case" consisting of a range of *IOPs* corresponding to a single EOF computation. The EOFs and the PCs can then be computed using the covariances and Eq. (893).

Radiative Transfer in Coupled Environmental Systems: An Introduction to Forward and Inverse Modeling,
First Edition. Knut Stamnes and Jakob J. Stamnes.

For each case, the optical depth profile at the associated wavenumbers can be reconstructed from the EOFs and the PCs as follows:

$$\ln d\tau_{\ell i} = \overline{\ln d\tau_{\ell i}} + \sum_{\ell=1}^{2L} P_{ki}\varphi_{k\ell} \tag{894}$$

where $d\tau_{\ell i}$ is the total optical depth of layer ℓ at the ith wavenumber. The over-bar denotes the mean over all wavenumbers. A similar procedure can be used to reconstruct the *single-scattering albedo.*

6.1.1
Application to the O_2 A Band

Natraj *et al.* [105] used two *multiple scattering* codes to generate the O_2 A band spectrum: (i) DISORT [161] and (ii) a fast two-stream code TWOSTR [223]. Line-by-line IOPs were used in both codes to generate reflectance spectra at the TOA. Figure 37 shows (a) the results obtained from DISORT, (b) the correlation between DISORT and scaled TWOSTR spectra, and (c) the difference between the two computations. The scaling of the TWOSTR spectrum was done by first performing a least-squares fit to find linear regression coefficients m (slope) and c (y-intercept) as follows:

$$\text{TWOSTR}_{\text{fitted}} = m\text{DISORT} + c \tag{895}$$

and then computing scaled TWOSTR reflectances using

$$\text{TWOSTR}_{\text{scaled}} = \frac{\text{TWOSTR} - c}{m}. \tag{896}$$

For each case, TOA reflectances were computed for the associated mean IOPs using DISORT and TWOSTR, and the difference was denoted by I_d. A similar computation was done for an IOP perturbation of magnitude 1 EOF, with $I_d^+(k)$ denoting the result of a positive perturbation and $I_d^-(k)$ the result of a negative perturbation, where k refers to a particular *EOF*. Then first and second differences with respect to the EOF were computed as

$$\delta I_k = \frac{I_d^+(k) - I_d^-(k)}{2} \tag{897}$$

$$\delta^2 I_k = I_d^+(k) - 2I_d + I_d^-(k) \tag{898}$$

and finally the total TOA reflectance for the ith wavenumber was computed as

$$I_i = I_i^{TS} + I_d + \sum_{k=1}^{4} \delta I_k P_{ki} + \frac{1}{2}\sum_{k=1}^{4} \delta^2 I_k P_{ki}^2 \tag{899}$$

where I_i^{TS} is the result obtained from the TWOSTR code. The sum indicates that only *four* EOFs are sufficient to reconstruct the spectrum.

Figure 38 shows the O_2 A band spectrum obtained by applying the *PC analysis* method. The corresponding residuals indicate that the correlation between

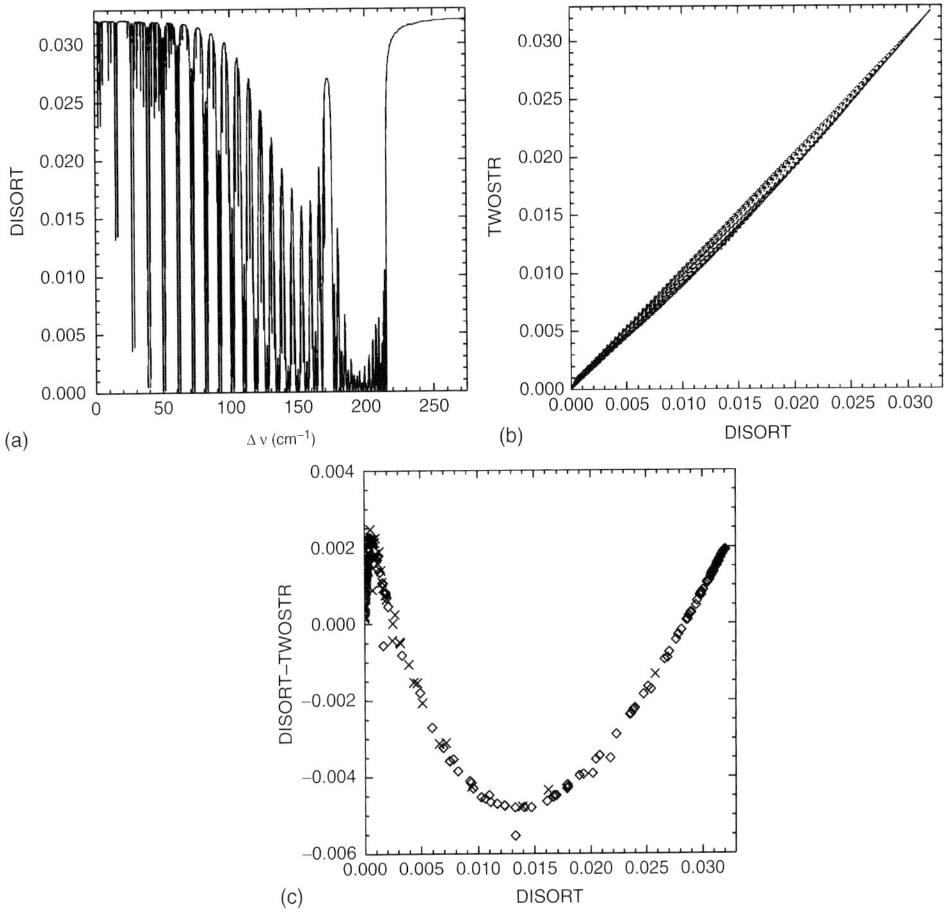

Figure 37 TOA reflectance spectra. (a) Obtained from DISORT, $\tilde{\nu}\ (\mathrm{cm}^{-1}) = \tilde{\nu}_0 + \Delta\tilde{\nu}$, $\tilde{\nu}_0 = 12950\ \mathrm{cm}^{-1}$. (b) Correlation between DISORT and TWOSTR reflectance spectra. (c) Difference between TWOSTR and DISORT reflectance spectra. The residuals are plotted as a function of the DISORT reflectance to show systematic deviations more clearly. After Natraj *et al.* [105] with permission.

the approximate two-stream and accurate multistream results can be used to reproduce the O_2 A band spectrum with an accuracy better than 1% with an order of magnitude increase in speed (see Natraj *et al.* [105] for details).

6.2 Simultaneous Retrieval of Total Ozone Column (TOC) Amount and Cloud Effects

Ultraviolet (UV) radiation received by the Earth occupies only 7% of the total solar energy, yet it plays a critical role in the biosphere. UV radiation is divided

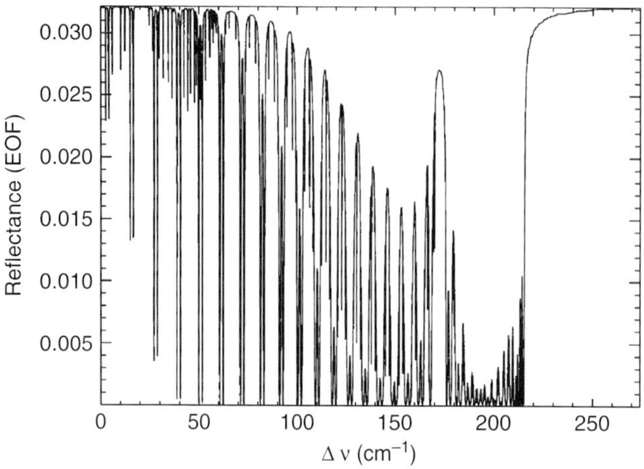

Figure 38 Same as Figure 37 (a), but reconstructed using PC analysis. After Natraj *et al.* [105] with permission.

into three parts: UVA (315–400 nm), UVB (280–315 nm), and UVC (100–280 nm). UVC radiation is absent at the surface of the Earth because absorption by diatomic oxygen and ozone prevents it from penetrating the atmosphere. UVB radiation is significantly attenuated before reaching the surface of the Earth by the *total ozone column (TOC)* in the atmosphere, while UVA radiation is little affected by the ozone layer. Since UVB radiation is harmful to human beings, animals, and plants, the TOC is a very important quantity to be monitored. Although ozone constitutes only a tiny fraction by mass ($\sim$ 0.0000007%) of the Earth's atmosphere, it is very important for life, because it shields the biosphere against harmful *UV radiation* [224]. The ozone layer, which contains about 90% of the total amount of ozone in the atmosphere, lies in the stratosphere between about 15 and 35 km altitude, and the TOC varies with location and season.

Several different kinds of instruments are available for measuring the UV irradiance, including spectroradiometers; multichannel, narrow, and moderate bandwidth filter instruments; and broadband filter instruments. Many studies have been conducted based on multichannel, moderate bandwidth filter instruments, including the ground-based UV (GUV) instrument (manufactured by Biospherical Instruments, USA) and the Norwegian Institute for Air Research UV (NILU-UV) instrument (manufactured by Innovation NILU, Norway) [225–229]. The *NILU-UV instrument* is a multichannel, moderate bandwidth filter instrument that measures the UV irradiance in five UV channels and one channel covering the visible spectral range (400–700 nm), which can be used to infer *TOC* values and quantify the influence of cloud cover on the amount of UV radiation at the surface of the Earth. In a recent study, comparisons were made between NILU-UV measurements of solar UV radiation at four sites on the Tibetan plateau at altitudes ranging from 2995 to 4510 m with very high UV exposure. In this study, good agreement was found between TOC values derived from NILU-UV measurements and *Ozone Monitoring Instrument (OMI)* measurements, the difference between average TOC values being less than 2.5% at all sites [230].

In this section, we compare two methods for analyzing data obtained by multichannel, moderate bandwidth filter instruments, such as the NILU-UV instrument. One method is based on *look-up tables (LUTs)* to infer TOC and a *radiation modification factor (RMF)* that accounts for the combined effects of clouds, aerosols, and surface albedo, while the other makes use of a *radial basis function neural network (RBF-NN)* to infer TOC and *cloud optical depth (COD)* values, where COD represents the combined radiative effect of clouds, aerosols, and surface albedo. In the LUT approach, TOC and COD values are retrieved *independently* by using irradiances in two UV channels, while in the *RBF-NN* approach these two parameters are retrieved *simultaneously* from irradiances in three UV channels. Thus, in the latter approach, data obtained by the NILU-UV instrument in the three channels centered at 305, 320, and 340 nm with a 10-nm spectral width (FWHM) are used, and an RT model is used to compute irradiances in these NILU-UV channels (output parameters) as a function of three input parameters consisting of (i) the solar zenith angle (20–70°), (ii) the TOC (200–500 *Dobson units (DU)*), and (iii) the *COD* (0–150). These computed irradiances can be used to infer the *TOC* and *RMF* values using a traditional LUT approach [231, 232], but they can also be used in an RBF-NN to create a relationship between the input and output parameters in terms of a set of coefficients. To retrieve the desired atmospheric parameters (TOC and COD values), these coefficients are then applied to data from the *NILU-UV instrument.*

6.2.1
NILU-UV Versus OMI

The NILU-UV instrument measures irradiances in one channel covering the visible spectral range (400–700 nm) and five UV channels with center wavelengths at 305, 312, 320, 340, and 380 nm, each with a 10-nm spectral width (FWHM). It has been demonstrated that three such instruments, deployed side by side in the New York area for a 3-year period, provided essentially identical TOC and RMF values [233]. Here, data from one of them will be used to compare the LUT and the RBF-NN approaches.

The *OMI* is a satellite instrument deployed on NASA's *AURA satellite.* AURA was launched in July 2004 and its main purpose is to study the Earth's atmosphere. AURA's swath is 2600 km and the nadir viewing foot print is 13×24 km^2. From OMI measurements, one can derive the TOC in the atmosphere as well as aerosol loading and *UV radiation.*

6.2.2 Atmospheric Radiative Transfer Model

When solar radiation passes through the ozone layer of the atmosphere, a portion of the UV radiation will be absorbed by ozone, while the portion that penetrates the ozone layer will be multiply scattered or absorbed by air molecules, *aerosols*, and *cloud particles* [18]. Therefore, a *radiative transfer model (RTM)* is needed to quantify how the UV radiation is affected by ozone, other molecules, and particles in the atmosphere, and to compute the fraction of the incoming solar radiation that reaches the Earth's surface. For this purpose, the *DISORT* (discrete-ordinate) RTM [161] (see Sections 3.4.4 and 4.3), with corrections for Earth curvature effects [53], was used to simulate the radiation measured by the NILU-UV instruments.

6.2.3 LUT Methodology

In the *look-up table (LUT) method*, the TOC is determined from the ratio N of irradiances in two different UV channels with spectral responses $R_i(\lambda)$ and $R_j(\lambda)$, one of which is quite sensitive to the TOC while the other is significantly less sensitive, that is

$$N(\theta_0, \text{TOC}) = \frac{\sum_{\lambda=0}^{\infty} R_i(\lambda)F(\theta_0, \lambda, \text{TOC})}{\sum_{\lambda=0}^{\infty} R_j(\lambda)F(\theta_0, \lambda, \text{TOC})} \tag{900}$$

where θ_0 is the solar zenith angle and $F(\theta_0, \lambda, \text{TOC})$ is the spectral irradiance [231].

The *RMF* is a simple way to quantify the combined effect of clouds, aerosols, and surface reflection on measured *downward hemispherical (DH) irradiances*. The RMF is defined as the ratio of the measured DH irradiance and the DH irradiance computed by an *RTM* under cloud-free sky conditions for a ground surface at sea level, which is assumed to be totally absorbing (black) [234], that is

$$\text{RMF} = \frac{F^m(\theta_0, \Delta\lambda)}{F^c(\theta_0, \Delta\lambda)} \times 100 \tag{901}$$

where θ_0 is the solar zenith angle, $F^m(\theta_0, \Delta\lambda)$ is the measured *DH irradiance*, and $F^c(\theta_0, \Delta\lambda)$ is the computed DH irradiance (using the RTM) under cloud-free sky conditions. To determine the RMF, we used channel 4 (centered at 340 nm, $\Delta\lambda = (340 \pm 5)$ nm) of the *NILU-UV instrument*.

6.2.4 Radial Basis Function Neural Network Methodology

The concept of a *neural network forward model* was discussed in Section 3.5.2. In a *radial basis function neural network (RBF-NN)*, the complete RBF-NN function is given by [see Eq. (482)]

$$p_i = \sum_{j=1}^{N} a_{ij} \exp[-b^2 \sum_{k=1}^{N_{\text{in}}} (c_{jk} - R_k)^2] + d_i \tag{902}$$

where N is the total number of neurons, and N_{in} is the number of input parameters. The purpose of the training of the RBF-NN is to determine the coefficients a_{ij}, b, c_{jk}, and d_i appearing in Eq. (902). The four input parameters R_k are the DH irradiances measured in the three NILU-UV channels centered at 305, 320, and 340 nm, as well as the solar zenith angle, and the two output parameters p_i are the desired retrieval (state) parameters, namely the *TOC* and the *COD*.

6.2.5
Training of the RBF-NN

Getting adequate data for the neural network training to calculate the coefficients is very important in order to get accurate final results from Eq. (902). Hence, in order to apply the *RBF-NN*, we need a set of *model data* for training. We used an RTM to generate model data. The three input parameters used in the RTM were θ_0 (ranging between 20 and 70°), and the two *retrieval parameters* TOC (ranging between 200 and 500 DU) and COD (ranging between 0 and 150). By means of a random number generator 20,000 different combinations of the three RTM input parameters were generated by random sampling of each of them within the ranges indicated above. These 20,000 combinations of θ_0, TOC, and COD values were used as input to the RTM to simulate the DH irradiances in the three UV bands centered at 305, 320, and 340 nm in order to construct a dataset consisting of 20,000 *synthetic measurements* of DH UV irradiances covering the ranges of θ_0, TOC, and COD densely enough to ensure adequate accuracy. To simulate the impact of clouds and other environmental factors on DH UV irradiances, the COD was used in the RTM as a proxy for the combined effect of clouds, aerosols, and *surface reflectance*. The COD is a measure of the attenuation due to absorption and scattering of sunlight by cloud particles (water droplets or ice crystals). For simplicity, the cloud was assumed to consist of water droplets, and cloud IOPs were calculated using the parameterization developed by Hu and Stamnes [120], as reviewed in Section 2.7.6. Figure 39 shows the relation between the RMF [Eq. (901)] and the COD obtained from the generated *model data*.

6.2.6
COD and TOC Values Inferred by the LUT and RBF-NN Methods

Applying Eq. (902) with the trained coefficients a_{ij}, b, c_{jk}, and d_i to the NILU-UV measurements adjusted by the instrument response function and instrument drift factors [233], we inferred the TOC and COD values. We used the solar zenith angle θ_0, the ratio of DH irradiances at 305 and 320 nm, and the DH irradiance at 340 nm obtained from the NILU-UV instrument as input to Eq. (902), from which the *TOC* and *COD* values were derived. The TOC values derived from *RBF-NN*, *LUT*,

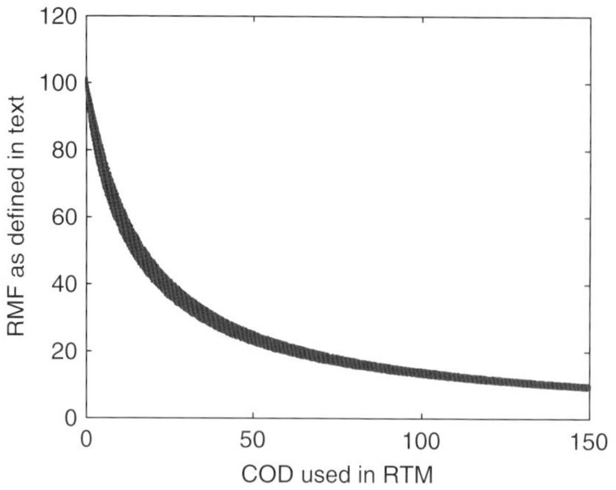

Figure 39 Relation between the radiation modification factor (RMF) defined in Eq. (901) and cloud optical depth (COD) based on simulated data obtained from the RTM (after Fan *et al.* [235] with permission).

and *OMI* were generally in good agreement, and the COD values derived from the RBF-NN and the RMF values derived from the LUT agreed well with the relation presented in Figure 39 as will be discussed in more detail below.

COD and TOC values were derived by using the RBF-NN method, and the corresponding RMF and TOC values were derived by the LUT method based on measurements recorded by the NILU-UV 29 instrument every day (minute by minute) from 08/05/2010 to 03/01/2013. Figure 40 (a) shows correlations between the TOC derived by the RBF-NN and LUT methods in 2012. The corresponding correlations for the other years (not shown here) are similar to those in Figure 40 (a). The correlations for the entire period from 2010 to 2013 vary little with time and are larger than 0.99, indicating a good match. Figure 40 (b) shows the relation between *COD* and *RMF* values, which is the similar to the model data (Figure 39)

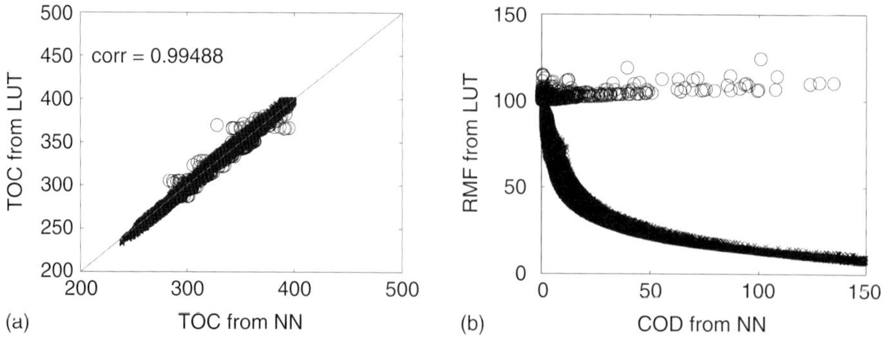

Figure 40 Correlations between TOC values derived by the RBF-NN and LUT methods (a), and relation between COD values derived using the RBF-NN method and RMF values derived using the LUT method (b) in 2012 (after Fan *et al.* [235] with permission).

except for some points marked by circles above the curve, which are caused by incorrect RMF values occurring for heavy cloud cover, broken clouds, or snow on the ground. Such circumstances will enhance the DH irradiance, leading to RMF values that are larger than 100. Also, on days with incorrect RMF values, the TOC values indicated by circles derived by the LUT are different from those derived by the RBF-NN method. COD values derived by the *RBF-NN* method are more reliable than the RMF values, which are incorrect for heavy cloud cover and broken cloud situations or snow-covered ground. *TOC* values derived by the RBF-NN method are more accurate for all weather conditions. Also, in the RBF-NN method, the cloud effect is accounted for in the retrieval method, while in the *LUT* method it is ignored.

6.2.7
TOC Inferred from NILU-UV (RBF-NN and LUT) and OMI

While *OMI* provides only a daily averaged value of the TOC, the NILU-UV instrument records data every minute. In order to compare the TOC values from OMI with the corresponding values obtained from NILU-UV measurements, we calculated the daily averaged TOC values by applying both the RBF-NN and LUT methods to the NILU-UV data. Our results show that RBF-NN-derived TOC values have better agreement with OMI TOC values than LUT-derived TOC values. The relative differences are smaller and the correlations higher between TOC values inferred from the RBF-NN method and OMI than between TOC values inferred from the LUT method and OMI. The OMI-inferred TOC value was used as a reference for calculating the relative differences. For the period 2010–2013, the mean relative differences and standard deviations for TOC values derived from OMI versus those derived from RBF-NN and LUT methods applied to NILU-UV data were smaller for OMI versus RBF-NN than for OMI versus LUT.

Clouds appear to have a smaller effect on the correlation between TOC values derived from OMI and those obtained by applying the RBF-NN method to NILU-UV data than they have on the correlation between TOC values derived from OMI and those obtained by applying the LUT method to NILU-UV data. Figure 41 shows how the COD value impacts the agreement between TOC values derived from OMI and by applying the RBF-NN method to NILU-UV data (a), and also the agreement between TOC values derived from OMI and by applying the LUT method to NILU-UV data (b). The left panel shows the distribution of the ratio of the TOC value from RBF-NN to that from OMI for different COD values in the period 2010–2013, while the right panel shows the distribution of the ratio of the TOC value from LUT to that from OMI for different RMF values in the same period. Shades of gray are used to indicate the number of a specific value of the ratio in a certain COD or RMF interval. For OMI versus RBF-NN, 80% of the data correspond to COD values less than 10, while for OMI versus LUT 80% of the data correspond to RMF values larger than 80. The larger the COD or the smaller the RMF, the more the ratio deviates from 1.0. When the ratio is larger (smaller)

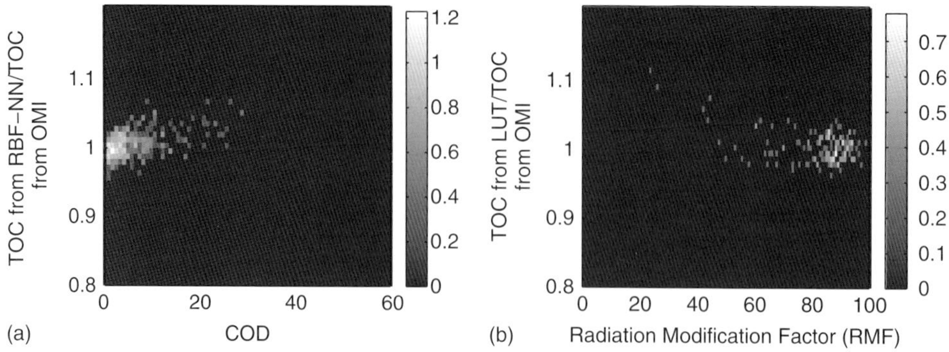

Figure 41 (a) COD impact on the ratio of the TOC values derived from the RBF-NN method and OMI (2010–2013). (b) RMF impact on the TOC values derived from the LUT method and OMI (2010–2013) (after Fan *et al.* [235] with permission).

than 1, the TOC from RBF-NN or LUT is larger (smaller) than the TOC from OMI. Thus, TOC from RBF-NN or LUT is overestimated compared with TOC from OMI on overcast days. There are many more values of the ratio close to 1 in the RBF-NN versus OMI panel than in the LUT versus OMI panel. Also, the ratio interval is larger in the LUT versus OMI panel than in the RBF-NN versus OMI panel.

6.2.8 Summary

Simultaneous derivation of *TOC* and *COD* from moderate-bandwidth multichannel instruments can be accomplished by the RBF-NN method by which TOC and COD values are derived directly from the measured irradiances in three UV channels. Application of the method to three years of data recorded by a NILU-UV irradiance meter using channels at 305, 320, and 340 nm showed that it generally yields results in close agreement with those derived from the traditional LUT method. However, the RBF-NN method is less influenced by environmental factors including clouds, aerosols, and *surface reflectance*, represented in this study as a "cloud effect", than the LUT method, and TOC values derived from the RBF-NN method are in better agreement with the corresponding results derived from OMI. Thus, compared to the LUT method, the RBF-NN method yields an increase of 0.03 in the correlation with OMI results. Furthermore, the RBF-NN method retrieves more valid results than the LUT method. One plausible reason is that both the RBF NN and the OMI methods take "cloud effects" into account, while the LUT method ignores "cloud effects" in the TOC retrieval. In essence, by performing a simultaneous retrieval of TOC and COD values from the NILU-UV measurements, the RBF-NN method leads to improved accuracy compared to the LUT method.

6.3 Coupled Atmosphere–Snow–Ice Systems

The *C-DISORT coupled RTM (CRTM)* can be used to compute the *BRDF*, defined in Eq. (315), for sea ice as described in [123]. To quantify the BRDF, one needs the backscattered radiance distribution as a function of the polar angles θ_0 and θ of incidence and observation, respectively, as well as the corresponding azimuth-difference angle $\Delta\phi = \phi_0 - \phi$. According to Eq. (359), the radiance distribution can be expressed as a *Fourier* cosine *series* in $\Delta\phi$, in which the *expansion coefficients* $I^m(\tau, \mu, \mu_0)$ depend on the polar angles θ_0 ($\mu_0 = \cos\theta_0$) and θ ($\mu = \cos\theta$) as well as on the sea ice IOPs. Each expansion coefficient satisfies Eq. (360), which is readily computed using the CRTM to provide $I^m(\tau, \mu, \mu_0)$, and Eq. (359) then yields the complete angular distribution of the radiance as a function of θ_0, θ, and $\Delta\phi$. With no atmosphere assumed to be present, so that $S_1^*(\tau, \mu, \mu_0, \phi) = 0$ [Eq. (1072)], the expansion coefficients $I^m(\tau, \mu, \mu_0)$ for a number of values of θ_0 (solar zenith angle), θ (observation angle), and sea ice IOPs were computed [123] and stored in a set of LUTs.

To create LUTs for the *sea ice BRDF* to be used for interpolation, Stamnes *et al.* [123] assumed the sea ice to float on water with a known albedo A_w, and the IOPs for a slab of sea ice to be characterized in terms of its optical thickness τ, its *single-scattering albedo* ϖ, and its asymmetry factor g. Then the CRTM was used to tabulate the expansion coefficients $I^m(\tau, \varpi, g, \mu_0, \mu, A_w)$, which determine the sea ice BRDF as a function of τ, ϖ, g, μ_0, μ, and A_w. Stamnes *et al.* [123] also created a tool [ISIOP] for computing sea ice IOPs (τ, ϖ, and g) for any desired wavelength from sea ice physical parameters: real and imaginary parts of the sea ice *refractive index*, brine pocket concentration and effective size, air bubble concentration and effective size, volume fraction and absorption coefficient of sea ice *impurities*, *asymmetry factors* for scattering by *brine pockets* and *air bubbles*, and sea ice thickness. This approach enabled a reliable computation of the wavelength-dependent BRDF as a function of sea ice IOPs. The BRDF for snow-covered sea ice was readily obtained by including snow as a "cloud on top of the sea ice".

A combination of the two different tools developed by Stamnes *et al.* [123], namely (i) *ISIOP* for computing *IOPs* for ice and snow, and (ii) *ISBRDF* for computing the *BRDF* of sea ice (with or without snow cover), can be used to quantify the BRDF of sea ice in a very efficient manner. An example is shown in Figure 42, where the computed albedos of clean snow are compared with laboratory measurements. This figure shows that flux reflectances generated from BRDF values obtained using the ISIOP and ISBRDF tools agree well with computations done independently [236] as well as with experimental values [237]. In [123], it was shown that sea ice spectral albedo values derived from the ISIOP/ISBRDF tools are consistent with independently computed [238] as well as observed values [239, 240] for a variety of ice types and thicknesses.

In the work discussed here, it was assumed that snow/ice particles have spherical shapes so that a *Mie code* could be used to compute their IOPs, although a

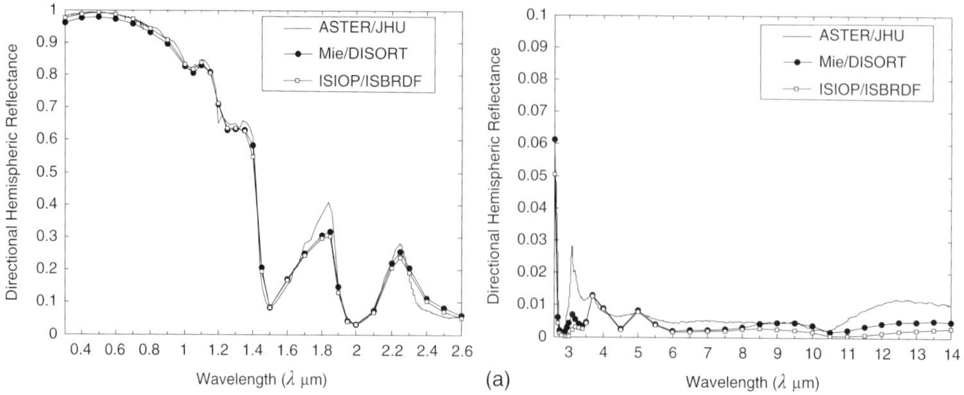

Figure 42 Directional hemispherical reflectance of clean snow for 24-μm-radius snow grain from 1) *ISIOP* computed IOPs and *ISBRDF*, 2) Mie computed IOPs and *DISORT* [236], and 3) ASTER spectral library observations [237] for 10° solar zenith angle for the visible and near-infrared (a) and infrared (b) spectral regions (adapted from Stamnes *et al.* [123]).

Henyey–Greenstein scattering phase function was used with the *asymmetry factor* provided by the Mie computation. However, it should be kept in mind that, for BRDF computations, the nonspherical shape of ice crystals may be important [241–243].

6.3.1 Retrieval of Snow/Ice Parameters from Satellite Data

It is well recognized that snow cover has a strong impact on the surface energy balance in any part of the world. Satellite remote sensing provides a very useful tool for estimating the spatial and temporal changes in snow cover and for retrieving snow optical characteristics. Data from numerous satellite sensors have been used to retrieve snow optical properties, including *Landsat Thematic Mapper (TM)* data [244, 245], *Airborne Visible/Infrared Imaging Spectrometer (AVIRIS)* data [246–248], and *Moderate Resolution Imaging Spectroradiometer (MODIS)* [249] data. The retrieval of snow *grain size* and impurity concentration is possible because snow reflectance depends primarily on the impurity concentration (assumed to be soot contamination) in the visible range, but also on the snow grain size [250] in the NIR, as shown in Figure 43.

The *GLobal Imager (GLI)* sensor was launched onboard Japan's *Advanced Earth Observing Satellite II (ADEOS-II)* on December 14, 2002. GLI was an optical sensor similar to MODIS, which observed solar radiation reflected from the Earth's atmosphere and surface including land, oceans, and clouds as well as terrestrial infrared radiation. In addition to atmospheric parameters, GLI, like MODIS, was designed to gather information about several other quantities including marine and land parameters such as *chlorophyll concentration, dissolved organic matter, surface temperature*, vegetation distribution and *biomass*, distribution of snow

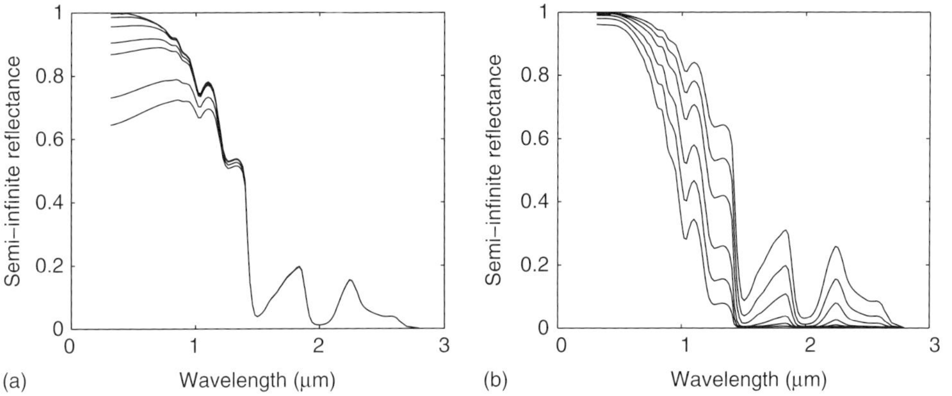

Figure 43 Spectral albedo of snow as a function of wavelength. (a) grain size 1000 μm, and impurity concentrations (in top→down order) 0, 0.01, 0.1, 1, 5, 10 ppmw. (b) Pure snow with grain size (in top→down order) 50, 100, 200, 500, 1000, and 2000 μm.

and ice, and albedo of snow and ice. The GLI sensor acquired data from April 2 to October 24, 2003. After that date, no useful data were retrieved from ADEOS-II due to a power failure. Because the GLI sensor was similar in many respects to the MODIS sensor that was launched prior to ADEOS-II, algorithms developed to retrieve information about the cryosphere from GLI data [251] were tested by the use of MODIS data [252], and are therefore applicable also to data obtained with the MODIS sensor [253].

Snow can be regarded as a mixture of pure ice, air, liquid water, and *impurities*. Pure ice is highly transparent in the visible spectral region, so that an increase in *snow grain size* has little effect on the reflectance. However, because ice is moderately absorptive in the NIR, the reflectance is sensitive to *grain size*, especially in the wavelength region 0.8–1.3 μm (see Figure 43). For satellite measurements, spectral channels should be selected to lie in wavelength regions where the effect of atmospheric scattering and absorption is small, so that when the radiance values are atmospherically corrected to yield surface reflectance, errors in the characterization of the atmosphere, particularly atmospheric *water vapor*, are minimized [254–256]. For these reasons, GLI channels 19 (0.86 μm), 24 (1.05 μm), 26 (1.24 μm), and 28 (1.64 μm) would be suitable for retrieval of *snow grain size*, because the impact of changes in snow grain size is large, whereas the effects of the atmosphere and *snow impurities* are relatively small [251].

Accurate estimates of *surface temperature* can provide an early signal of such change, particularly in the Arctic, which is known to be quite sensitive to *climate change*. The surface temperature in the polar regions controls sea ice growth, *snow melt*, and surface–atmosphere energy exchange. During the past decade, significant progress has been made in the estimation of sea surface temperature [257–259] and snow/ice surface temperature [260–263] from satellite thermal infrared data. Algorithms for surface temperature retrievals in the Arctic based on GLI measurements were developed [251] for retrieval of snow/ice surface temperature (IST), as well as for open-ocean *sea surface temperature (SST)*. The SST algorithm can be applied to areas consisting of a mixture of snow/ice and *melt ponds*. GLI channels 35 (10.8 μm) and 36 (12.0 μm) were used in conjunction with RT simulations and a multilinear regression to determine empirical coefficients in the expression for the surface temperature [251].

6.3.2 Cloud Mask and Surface Classification

In order to infer information about snow and ice properties from visible and IR satellite imagery a *cloud mask* is required to discriminate between clear and cloudy sky scenes. For scenes that are determined to be cloud-free, the next step is to do a *surface classification*. Thus, algorithms are needed to (i) determine whether a given field of view is obstructed by clouds, and (ii) distinguish bare sea ice from snow-covered *sea ice*. The snow/sea ice discriminator is designed to discriminate bare sea ice from snow-covered sea ice during the bright polar summer. The surface is classified into five possible types: snow, sea ice, cloud shadow, land (tundra), and open ocean. When sea ice is covered by snow (even only a few centimeters), the surface radiative characteristics will be similar to snow [28, 29, 39]. Thus, sea ice covered by snow will be classified as a snow surface, while only bare sea ice is classified as sea ice.

6.3.2.1 Snow Sea Ice Cover and Surface Temperature

Figure 44 shows the seasonal variations of the extents of snow and sea ice cover around the northern polar region derived from GLI data. For comparison, seasonal variations of MODIS land snow cover and of the *Advanced Microwave Scanning Radiometer (AMSR)* sea ice cover are also shown. For the period April 7–22, 2003, the snow and sea ice cover maps derived from GLI and MODIS + AMSR data are shown as well. The GLI and MODIS snow cover extents are consistent except for slight differences in the periods from June 10 to August 12, possibly due to differences in the cloud detection scheme and the ability to detect snow cover. The trend of the GLI sea ice cover also follows closely the variation of the AMSR sea ice cover except for slight negative biases, which become larger in the later 16-day periods. The bias is caused by the loss of valid sea ice pixels in the GLI results, partly due to persisting cloudiness over the Arctic Ocean during the 16-day averaging period, particularly for the July–October time frame, and partly due to the drift of the sea ice itself over the averaging period.

Figure 45 shows the 16-day average *snow surface temperature* around the northern polar region from April 7 to May 8, 2003. White areas indicate snow- or sea-ice-covered areas for which no snow physical parameters were determined because at least one of the four snow physical parameters retrieved was beyond the valid range of the analysis. Possible surface types of the white areas can be one of the following: bare ice, spatially inhomogeneous snow (e.g., snow cover contaminated by clouds or vegetation), invalid geometric conditions (e.g., too large solar or sensor zenith angles).

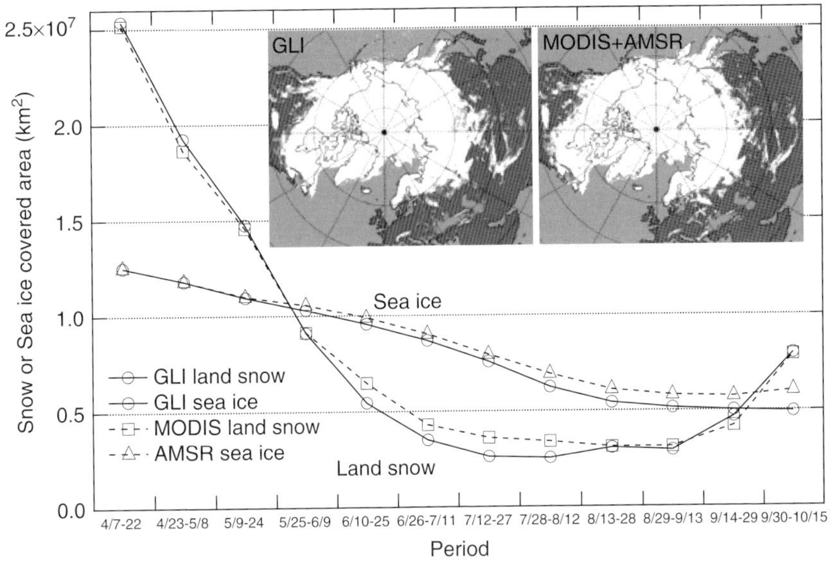

Figure 44 Comparison of the temporal variations (16-day averages from April 7 to October 15, 2003) of GLI-derived snow-covered land area and sea-ice-covered area with those derived from MODIS (land snow) and AMSR (sea ice). Images of the extents of snow and sea ice cover for the period of April 7–22 from GLI and MODIS + AMSR are also shown (after Hori *et al.* [253] with permission).

To assess the accuracy of the GLI-derived *surface temperature*, a comparison between MODIS and GLI snow surface temperatures is shown in Figure 46. The GLI surface temperatures are well correlated with the MODIS temperatures having slight negative biases of about −2.0, −1.0, and −0.5 at 250, 260, and 270 K, respectively. Comparisons between GLI-derived snow grain sizes and surface temperatures also indicate that the GLI-derived surface temperatures have about −0.5 K negative bias at around the melting point of ice (273 K), which is estimated from the temperature at which the retrieved snow grain size distribution shifts to a coarser mode due to melting of snow.

6.3.3
Snow Impurity Concentration and Grain Size

Hori *et al.* [253] showed that the regional dependence of the retrieved *snow impurity* is different from that of the snow surface temperature. The snow impurity values were found to be mostly less than 0.3 ppmw over the arctic sea ice, tundra, polar desert areas, and the Greenland ice sheet. In particular, impurity fractions at the Greenland ice sheet were found to be the lowest (mostly less than 0.05 ppmw) among the snow-covered areas in the Arctic during the 7-month observation period from April 7 to October 15, 2003. Although the retrieved impurity

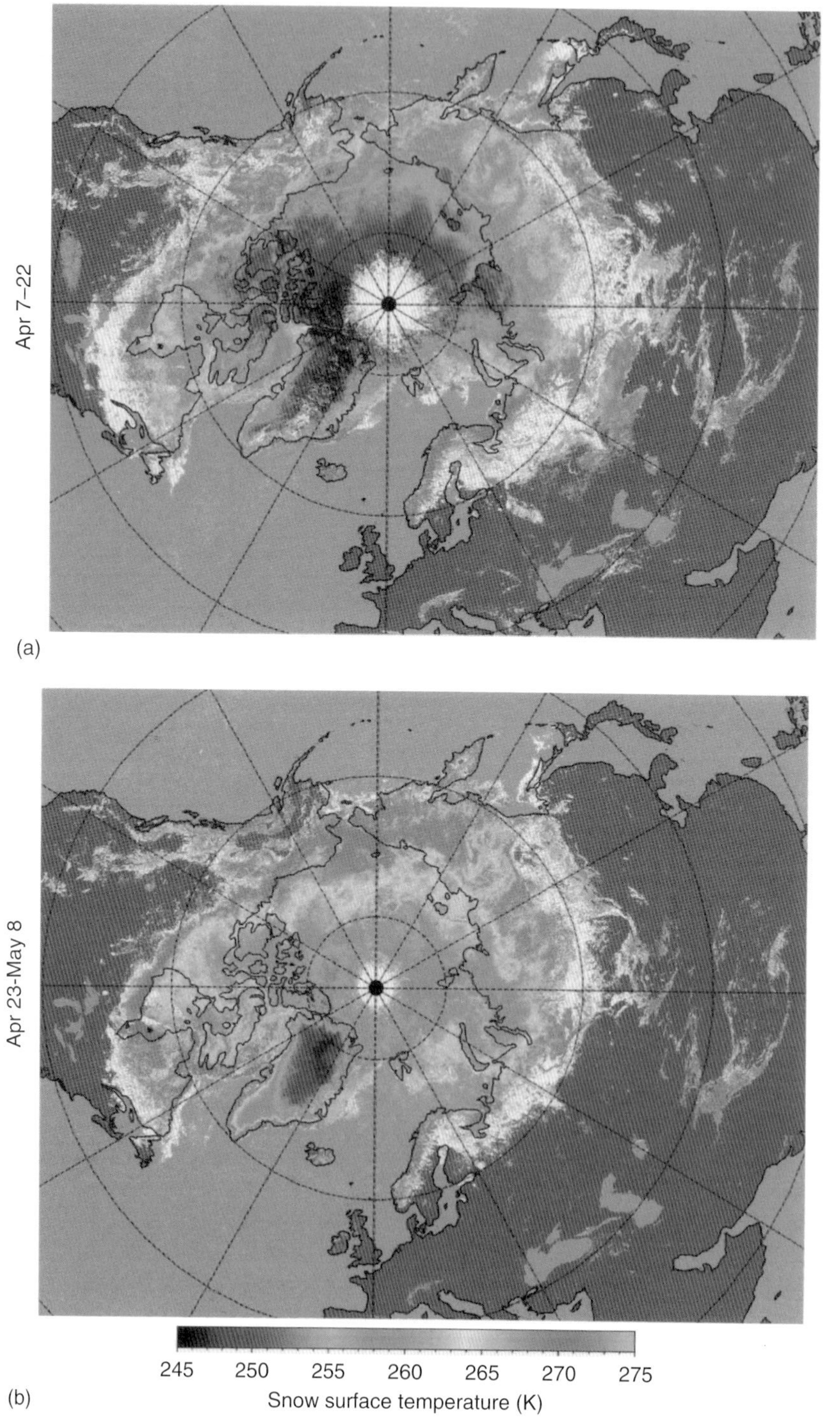

Figure 45 Sixteen-day average GLI snow surface temperature around the northern polar region from April 7 to May 8, 2003 (after Hori *et al.* [253] with permission).

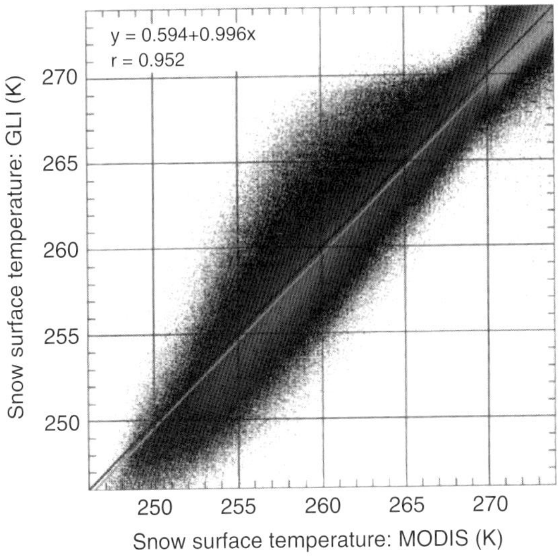

Figure 46 Scatter plot between snow surface temperatures from MODIS and GLI (after Hori *et al.* [253] with permission).

concentrations appear reasonable, their accuracy is uncertain, as discussed by Aoki *et al.* [252].

Figure 47 shows the 16-day average spatial distribution of snow grain size of the shallow layer (0–20 cm) retrieved from the GLI NIR channel at 0.875 μm ($R_{s0.9}$). The spatial distributions of the snow grain size exhibit not only a large-scale variation but also several regional patterns. The large-scale variation is the latitudinal dependence similar to those of the *snow surface temperature*, i.e., the higher the latitude, the smaller the *grain size*, and vice versa. The regional patterns are related to local weather or the thermal environment (e.g., relatively fine newly fallen snow in the mid-latitude area around the northern prairie in the United States seen in the April 7–22 period and coarse, probably melting snow over sea ice in the Arctic around Baffin Island in April).

The *snow grain size* of the top surface layer (0 – 2 cm) can be retrieved from the $\lambda = 1.64$ μm GLI channel ($R_{s1.6}$) [251–253]. When comparing the $R_{s0.9}$ distribution with the $R_{s1.6}$ distribution (not shown), one finds not only the spatial variability of $R_{s1.6}$ to be different from that of $R_{s0.9}$ but also the absolute values of $R_{s1.6}$ to be one order of magnitude smaller than those of $R_{s0.9}$. The ratio of $R_{s1.6}$ to $R_{s0.9}$ makes those features clearer, as shown for the April 7–22, 2003, period in Figure 48. The difference in the spatial distribution between $R_{s0.9}$ and $R_{s1.6}$ may be explained by a possible vertical inhomogeneity of the grain size in the upper several centimeters of the snow cover or by a depth variation in the snow cover taking into account the light penetration depth difference at $\lambda = 0.865$ and 1.64 μm.

The close relationship between *snow grain size* ($R_{s0.9}$) and *surface temperature* (T_s) seen in the melting season may be considered as an average feature of the

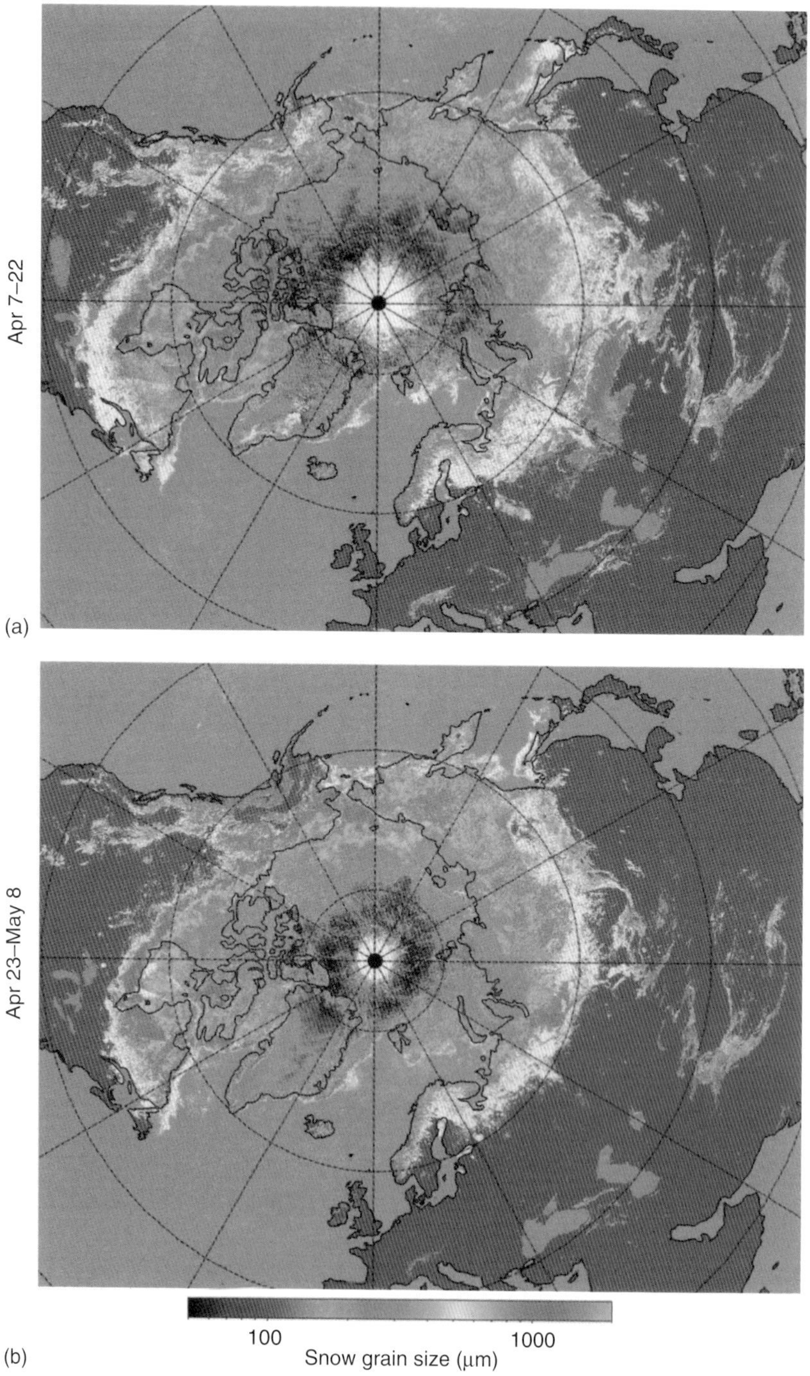

Figure 47 Sixteen-day average of the GLI-derived snow grain size of the shallow layer ($R_{s0.9}$) around the northern polar region from April 7 to May 8, 2003 (after Hori *et al.* [253] with permission).

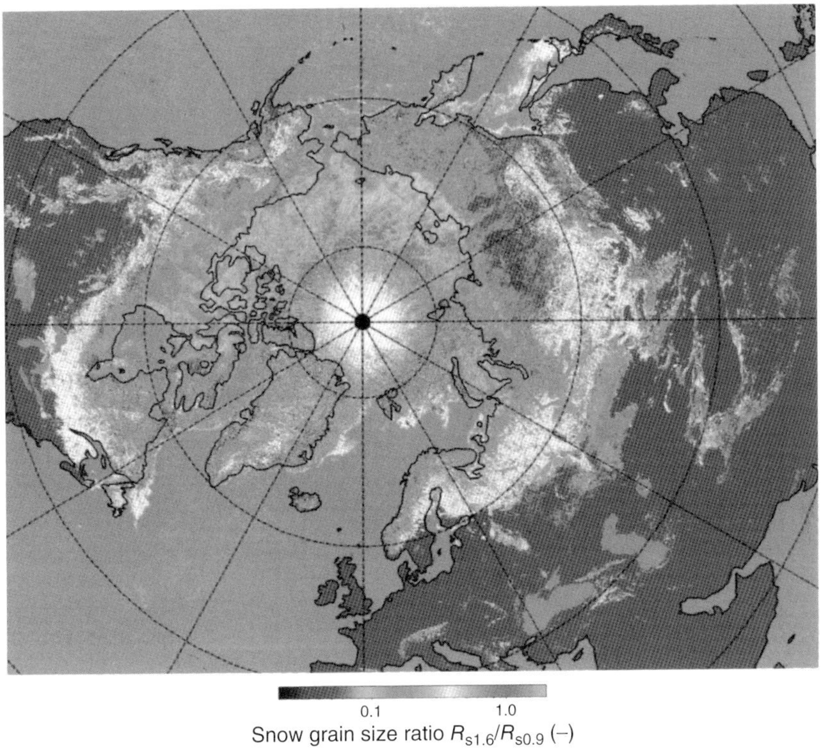

Figure 48 Ratio of the snow grain radius of the top surface ($R_{s1.6}$) to that of the shallow layer ($R_{s0.9}$) for the April 7–22 period. White colored areas are the same as in Figure 45 (after Hori *et al.* [253] with permission).

seasonal snow cover on a hemispheric scale. On a local scale, however, the snow cover can shift temporally to different states of the temperature–grain size relationship, for example, a state with small grain size under warm temperature or coarse grain sizes under cold temperatures depending on the recent history of the thermal environment to which snow grains were exposed after a snow fall. As an example, Figure 49 shows the spatial distribution of the same snow cover as in Figure 48 for the period of April 7–22, 2003, but color-coded using the two-dimensional temperature–grain size ($T_s - R_{s0.9}$) relationship. Warm (orange) color denotes small grains under high temperature indicating high potential for *metamorphosis* into larger grains, whereas cold (blue) color indicates coarse grains under low temperature with sizes that are likely to remain intact for a while. Thus, the map has information about the potential of snow grains to metamorphose in the near future. For example, the snow cover in each of the two elliptically inscribed areas in Figure 49, of which one is shown in orange color, implying high potential for *metamorphosis* and the other in blue with low potential, has similar grain size (around 200 μm) but exists in different temperature regimes

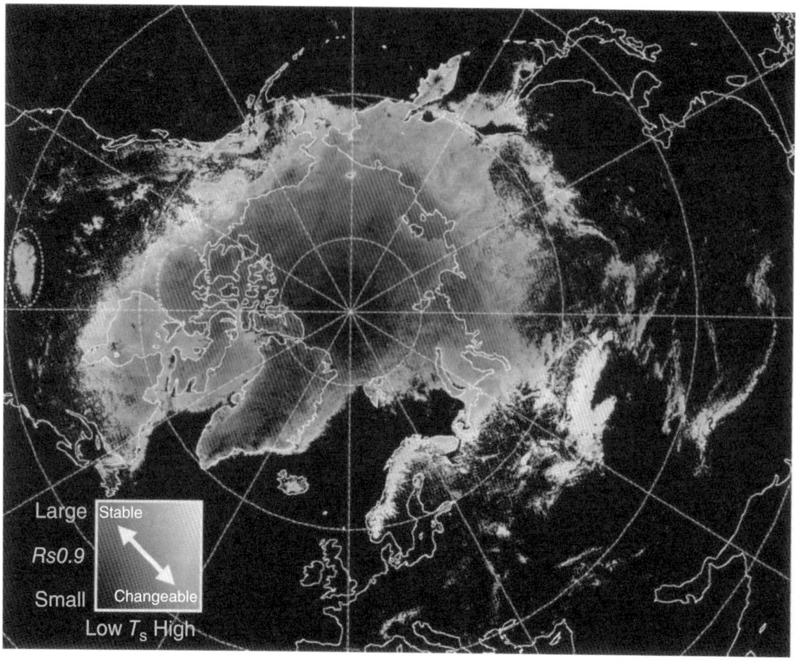

Figure 49 Map of snow *metamorphism* potential around the northern polar region for the period of April 7–22, 2003, determined from the relation between snow surface temperature (T_s) and snow grain size ($R_{s0.9}$). Warm (orange) color denotes small grains under high temperature, whereas cold (blue) color indicates coarse grains under low temperature (after Hori *et al.* [253] with permission).

(272–273 K in the left orange area and 253 K in the right blue area, see Figures 45 and 47). This information will be useful for the validation of *snow metamorphism models* such as CROCUS [264].

Another more practical application of the $T_s - R_{s0.9}$ relationship is for detection of the onset of *snow melt* at the hemispheric scale. Figure 50 shows a map of the date for the onset of snow melt over the Arctic derived from the daily maps of $R_{s0.9}$ and T_s. The *melt onset date* is defined as the average of the first three days (if determinable) when $R_{s0.9}$ becomes larger than 500 μm in a warm environment, that is, with T_s higher than 272 K. Black areas in Figure 50 indicate nonmelted regions (e.g., the central area of the Greenland ice sheet), or areas where the snow cover evolves from dry to wet under cloudy conditions so that the GLI observation cannot detect the transition in the $T_s - R_{s0.9}$ relationship (e.g., some parts of the arctic sea ice). The map clearly illustrates the development of the melt zones of the snow cover in the Northern Hemisphere, for example, Julian Day (JD) 90–150 for the snow cover over the continents, JD 110–180 for the sea ice zone of the Arctic Ocean, and JD 150–220 for the marginal Greenland ice sheet. Thus, because of their higher spatial resolution, *melt onset maps* derived from optical sensor data

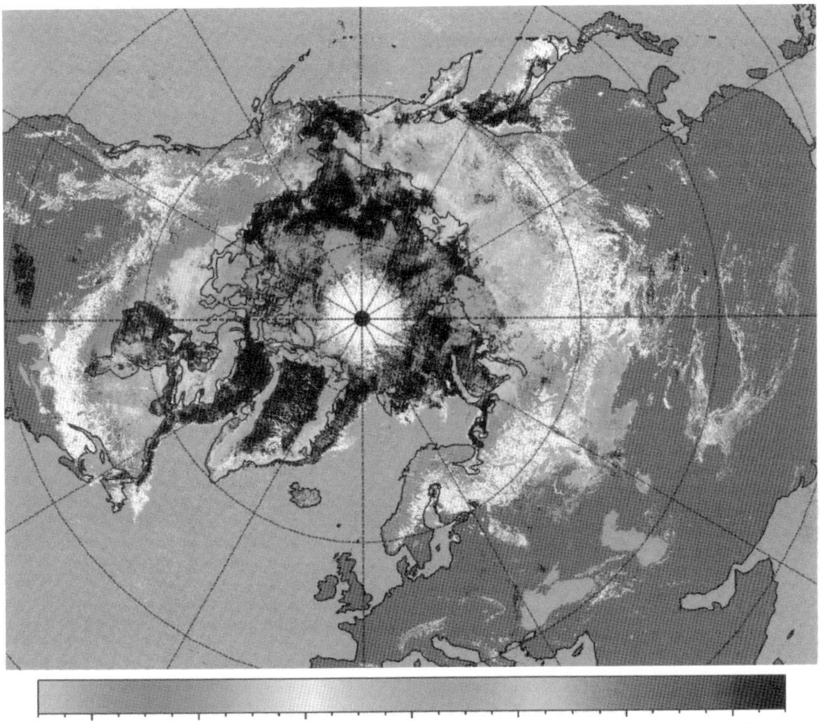

Figure 50 Spatial distribution of melt onset date around the northern polar region in 2003 determined from the relation between snow surface temperature (T_s) and snow grain size ($R_{s0.9}$). Date is indicated by Julian Day (after Hori *et al.* [253] with permission).

can be useful in the interpretation of similar maps derived from *microwave sensors (SSM/I, AMSR, NSCAT, etc.)*, which have coarser spatial resolution.

6.4 Coupled Atmosphere–Water Systems

Figure 51 illustrates the transfer of solar radiation in a coupled atmosphere–ocean system. For this kind of system, Mobley *et al.* [26] presented a comparison of underwater light fields computed by several different methods including *MC methods* [40, 66, 67, 164, 265–280], invariant imbedding [281], and the *discrete-ordinate method* [39], demonstrating similar results for a limited set of test cases. However, these comparisons were qualitative rather than quantitative because of the different ways in which the models treated the RT in the atmosphere, leading to a spread in the downwelling irradiance just above the water surface of 18%, which persisted throughout the water column.

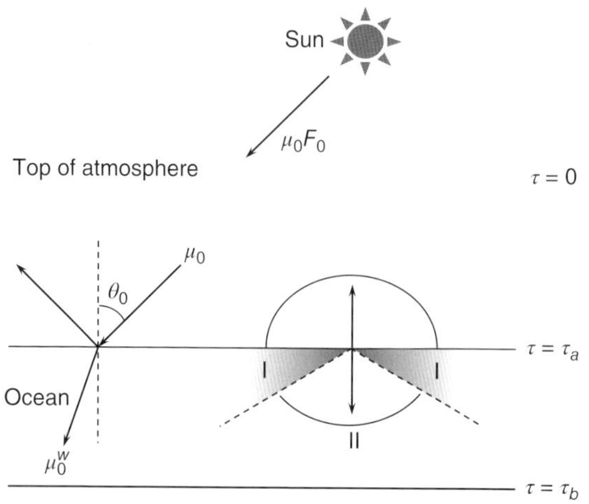

Figure 51 Schematic illustration of the atmosphere and ocean with incident solar irradiance $\mu_0 F_0$ and optical depth increasing downward from $\tau = 0$ at the top of the atmosphere. The incident polar angle is $\theta_0 = \cos^{-1}\mu_0$, which after refraction according to Snell's law changes into the angle $\cos^{-1}\mu_0^w$. Since the ocean has a larger refractive index than the atmosphere, radiation distributed over 2π sr in the atmosphere will be confined to a cone less than 2π sr in the ocean (region II). Upward radiation in the ocean with directions in region I will undergo total internal reflection at the ocean–air interface (adapted from Thomas and Stamnes [18]).

6.4.1 Comparisons of C-DISORT and C-MC Results

Gjerstad *et al.* [27] compared the irradiances obtained from an MC model for a coupled atmosphere–ocean system (C-MC) with those obtained from a discrete-ordinate method (C-DISORT). By treating the scattering and absorption processes in the two slabs in the same manner in both methods, they were able to provide a more detailed and quantitative comparison than those previously reported [26]. Figure 52 shows a comparison of direct and diffuse downward irradiances computed with the C-MC and C-DISORT codes, demonstrating that when precisely the same *IOPs* are used in the two models, computed irradiances agree to within 1% throughout the coupled atmosphere–ocean system.

6.4.2 Impact of Surface Roughness on Remotely Sensed Radiances

The *bidirectional reflectance distribution function (BRDF)* was defined in Eq. (315) as

$$\rho(\hat{\Omega}', \hat{\Omega}) = \frac{dI_r^+(\hat{\Omega})}{I^-(\hat{\Omega}')\cos\theta' d\omega'}. \tag{903}$$

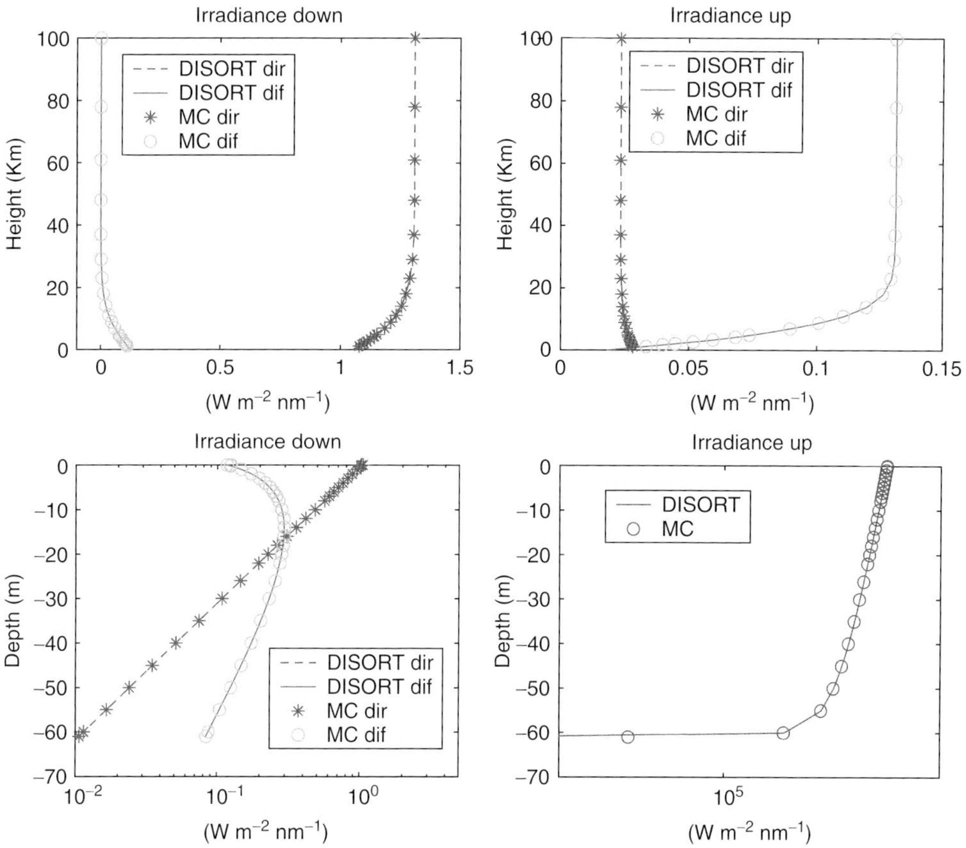

Figure 52 Comparison of irradiance results obtained with C-DISORT and a C-MC code for RT in a coupled atmosphere-ocean system. The simulations are for an atmosphere containing only molecular absorption and Rayleigh scattering and for an ocean having no Rayleigh scattering, only absorption and scattering from a chlorophyll concentration of 0.02 mg/m^3, uniformly distributed to a depth of 61 m, below which the albedo is zero (adapted from Gjerstad *et al.* [27] with permission).

Here, $I^-(\hat{\Omega}')\cos\theta' d\omega'$ is the radiant energy incident on a flat surface due to an angular beam of radiation with *radiance* $I^-(\hat{\Omega}')$ within a cone of *solid angle* $d\omega'$ around the direction $\hat{\Omega}'(\theta', \phi')$, whereas $dI_r^+(\hat{\Omega})$ is the radiance of reflected light leaving the surface within a cone of solid angle $d\omega$ around the direction $\hat{\Omega}(\theta, \phi)$. θ' is the polar angle between the incident beam direction $\hat{\Omega}'$ and the normal to the surface.

For simplicity, the interface between the two slabs is sometimes assumed to be flat, but natural surfaces are not flat. For example, if the ocean surface were flat as illustrated in Figure 51, a perfect image of the Sun's disk would be observed in the specular direction. The effect of surface roughness is to spread the specular reflection over a range of angles referred to as the *sunglint* region in the case of reflections from a wind-roughened water surface. As discussed in Section 2.11.4, if the surface is characterized by a Gaussian random height distribution $z = f(x, y)$ with mean height $\langle z \rangle = \langle f(x, y) \rangle = 0$, and the *tangent plane approximation* is invoked, according to which the radiation fields at any point on the surface are approximated by those that would be present at the tangent plane at that point [282], the *BRDF* in Eq. (315) for an isotropic rough water surface can be expressed as [see Eq. (307)]

$$R_{rs}(\theta_s, \phi_s; \theta_i, \phi_i) = \frac{1}{4\cos\theta_s \cos^4\theta_n} \frac{1}{\pi\sigma^2} \exp\left[-\frac{\tan^2\theta_n}{\sigma^2}\right] R_{\text{unpol}} \tag{904}$$

where R_{unpol} is the *Fresnel reflectance* of unpolarized light given by Eq. (260), θ_n is the tilt angle between the vertical and the normal to the tangent plane, and the factor

$$P(\theta_n, \sigma) = \frac{1}{\pi\sigma^2} \exp\left[-\frac{\tan^2\theta_n}{\sigma^2}\right] \tag{905}$$

describes the probability density of the distribution of tangent plane facet slopes with variance σ^2.

To analyze remotely sensed *radiances* obtained by instruments such as the *Sea-viewing Wide Field of view Sensor (SeaWiFS, on-board SeaStar),* the *MODerate-resolution Imaging Spectroradiometer (MODIS, deployed on both the Terra and Aqua spacecrafts),* and the *MEdium Resolution Imaging Spectrometer (MERIS, deployed onboard the European Space Agency (ESA)'s Envisat platform),* NASA has developed a comprehensive data analysis software package (*SeaWiFS Data Analysis System, SeaDAS*), which performs a number of tasks, including cloud screening and calibration, required to convert the raw satellite signals into calibrated top-of-the-atmosphere (TOA) *radiances.* In addition, the SeaDAS software package has tools for quantifying and removing the atmospheric contribution to the TOA radiance (*atmospheric correction*) as well as contributions from whitecaps and sunglint due to reflections from the ocean surface [35].

If one ignores the effects of shadowing and multiple reflections due to surface facets, the sunglint reflectance can be expressed by Eq. (904) with the distribution of surface slopes as given by Eq. (905), where $\sigma^2 = 0.003 + 0.00512\times$ ws, ws being the wind speed in meters per second [145, 283, 284].

6.4.3
The Directly Transmitted Radiance (DTR) Approach

The sunglint radiance at the TOA can be expressed as a function of the following variables:

$$I_{\text{glint}}^{\text{TOA}} \equiv I_{\text{glint}}^{\text{TOA}}(\mu_0, \mu, \Delta\phi, \text{ws}, \text{AM}, \tau_{\text{tot}}, \lambda)$$

where μ_0, μ, and $\Delta\phi$ define the *sun–satellite geometry,* ws is the wind speed, and λ is the wavelength. The atmosphere is characterized by its total optical depth τ_{tot} and the choice of an aerosol model (AM).

In the SeaDAS algorithm, a sunglint flag is activated for a given pixel when the reflectance or BRDF, as calculated from Eq. (904), with the slope distribution in Eq. (905), exceeds a certain threshold. If the reflectance for a given pixel is above the threshold, the signal is not processed. If the reflectance is below the threshold, a *directly transmitted radiance (DTR)* approach is used to calculate the TOA *sunglint radiance* in the *SeaDAS* algorithm. Thus, it is computed assuming that the direct beam and its reflected portion experience only exponential attenuation through the atmosphere [286], that is

$$I_{\text{glint}}^{\text{TOA}}(\mu_0, \mu, \Delta\phi) = F_0(\lambda)T_0(\lambda)T(\lambda)I_{GN} \tag{906}$$

$$T_0(\lambda)T(\lambda) = \exp\left\{ -[\tau_M(\lambda) + \tau_A(\lambda)]\left(\frac{1}{\mu_0} + \frac{1}{\mu}\right)\right\} \tag{907}$$

where the normalized sunglint radiance I_{GN} is the radiance that would result in the absence of the atmosphere if the incident solar irradiance were $F_0 = 1$, and where τ_M and τ_A ($\tau_{\text{tot}} = \tau_M + \tau_A$) are the Rayleigh (molecular) and aerosol optical thicknesses. Multiple scattering is ignored in the *DTR* approach, implying that photons removed from the direct beam path through scattering will not be accounted for at the TOA.

6.4.4
The Multiply Scattered Radiance (MSR) Approach

Wheras the DTR approach accounts only for the direct beam (Beam 2 in Figure 53), the *multiply scattered radiance (MSR)* approach is based on

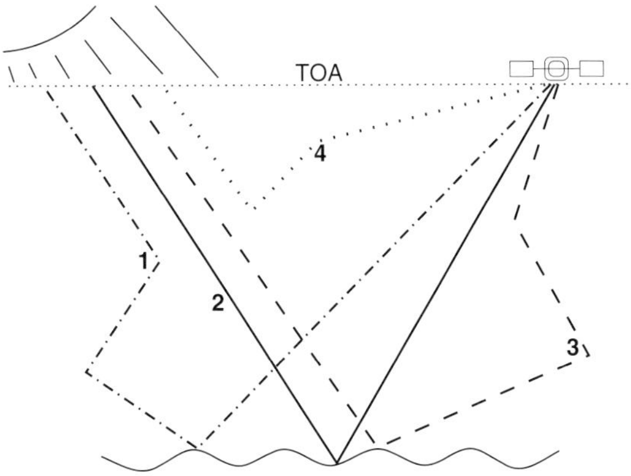

Figure 53 Schematic illustration of various contributions to the TOA radiance in the case of a wind-roughened ocean surface. (1) Diffuse downward component reflected from the ocean surface; (2) direct, ocean-surface reflected beam; (3) beam undergoing multiple scattering after ocean-surface reflection, and (4) (multiply) scattered beam reaching the TOA without hitting the ocean surface (adapted from Ottaviani *et al.* [286] with permission.)

computing the TOA radiance by solving Eq. (350) subject to the *boundary condition*

$$I(\tau_1,\mu,\phi) = \frac{\mu F_0}{\pi} e^{-\tau_1/\mu_0} \rho_{\rm glint}(-\mu_0,\phi_0;\mu,\phi) + \frac{1}{\pi}\int_0^{2\pi} d\phi' \int_0^1 d\mu' \rho_{\rm glint}(-\mu',\phi';\mu,\phi) I(\tau_1,\mu',\phi') \quad (908)$$

thereby allowing *multiple scattering* to be included in the computation. Here, τ_1 is the thickness of slab$_1$, and $\rho_{\rm glint}(-\mu_0,\phi_0;\mu,\phi) = \rho_s(-\mu_0,\mu,\Delta\phi)$ [see Eq. (904)] is the BRDF of the slab$_1$-slab$_2$ interface. For consistency with the definition of *sunglint*, radiation reflected from the surface after being scattered on its way down to the ocean surface (Beam 1 in Figure 53) is neglected by not allowing the presence of a downward diffuse term in Eq. (908). The complete solution of Eqs. (350) and (908) gives the total TOA radiance $I_{\rm TOA}^{\rm tot}(\mu_0,\mu,\Delta\phi)$, which includes light scattered into the observation direction without being reflected from the ocean surface (Beam 4 in Figure 53). This contribution is denoted by $I_{\rm TOA}^{\rm bs}(\mu_0,\mu,\Delta\phi)$, since it can be computed by considering a black or totally absorbing ocean surface, for which $\rho_{\rm glint} = 0$ in Eq. (908). To isolate the glint contribution, one must subtract this "black-surface" component from the complete radiation field:

$$I_{\rm TOA}^{\rm glint}(\mu_0,\mu,\Delta\phi) = I_{\rm TOA}^{\rm tot}(\mu_0,\mu,\Delta\phi) - I_{\rm TOA}^{\rm bs}(\mu_0,\mu,\Delta\phi). \quad (909)$$

Equation (909) includes multiply scattered reflected radiation, but ignores multiply scattered sky radiation undergoing ocean-surface reflection (Beam 1 in Figure 53). Thus, it guarantees that the difference between the TOA radiances obtained by the *DTR* and *MSR* approaches is due solely to that component of the TOA radiance, which is scattered along its path from the ocean surface to the TOA (Beam 3 in Figure 53). In order to quantify the error introduced by the DTR assumption, Ottaviani *et al.* [285] used a fully coupled atmosphere–ocean discrete-ordinate code with a Gaussian [Cox–Munk, Eq. (905)] surface slope distribution [54].

6.4.5
Comparison of DTR and MSR

To correct for the *sunglint* signal, Wang and Bailey [286] added a procedure to the SeaDAS algorithm based on the DTR assumption, which ignores *multiple scattering* in the path between the ocean surface and the TOA as well as in the path from the TOA to the ocean surface. To quantify the error introduced by the DTR assumption, Ottaviani *et al.* [285] neglected the effect of whitecaps as well as the wavelength dependence of the refractive index. Figure 54 shows a comparison of DTR and MSR results at 490 nm for several wind speeds and different aerosol types and loads. The incident solar irradiance was set to $F_0 = 1$ and in the computation of the *Fresnel reflectance* in Eq. (260) the imaginary part of the *refractive index* n_2' was assumed to be zero. A standard molecular atmospheric model (mid-latitude) with a uniform aerosol distribution below 2 km

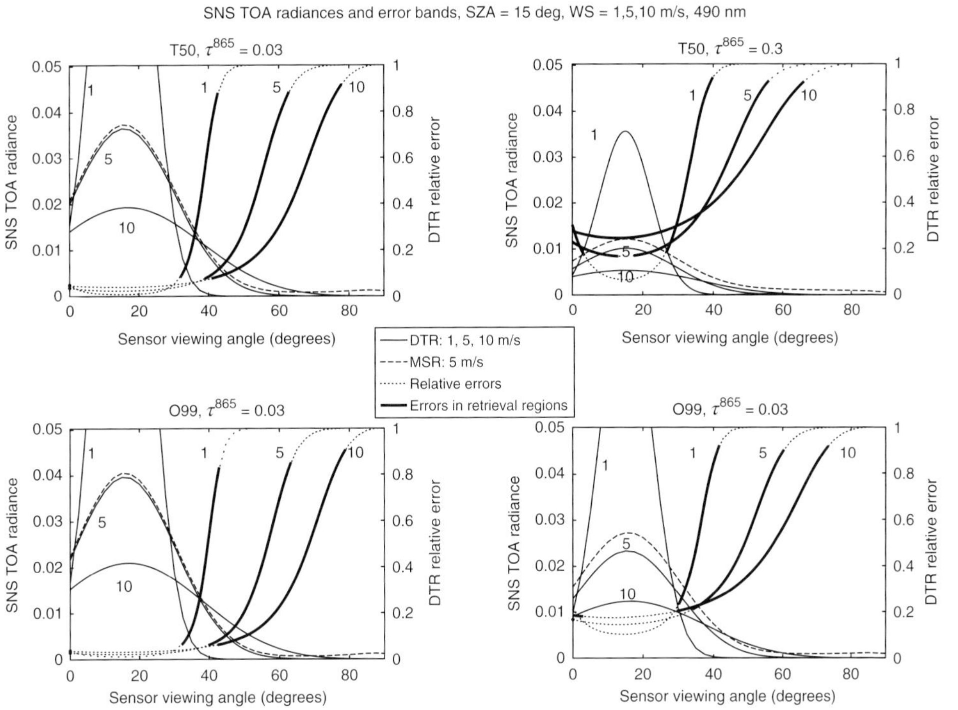

Figure 54 Sun-normalized *sunglint* TOA radiance (solid and thin curves) at 490 nm for an SZA of 15°, along the principal plane of reflection, and relative error incurred by ignoring *multiple scattering* along the path from the surface to the TOA (dotted curves). Each plot contains three representative wind speeds (1, 5, and 10 m/s). The upper row pertains to small aerosol particles in small amounts ($\tau = 0.03$, left panel) and larger amounts ($\tau = 0.3$, right panel). The bottom row is similar to the top one, but for large *aerosol particles*. The error curves have been thickened within the angular ranges in which retrievals are attempted (corresponding to $0.0001 \le I_{\rm TOA}^{\rm tot} \le 0.001$ in normalized radiance units) (adapted from Ottaviani *et al.* [285] with permission).

was used in the computations. Thus, below 2 km, the *aerosol optical thickness* due to scattering ($\tau_{A,\beta}$) and absorption ($\tau_{A,\alpha}$) was added to the molecular optical thickness τ_M:

$$\tau_{\rm tot} = \tau_M + \tau_A = (\tau_{M,\beta} + \tau_{M,\alpha}) + (\tau_{A,\beta} + \tau_{A,\alpha}). \tag{910}$$

The IOPs for aerosols were computed by a *Mie code*, and the IOPs of a multicomponent mixture were then obtained as a concentration-weighted average of the IOPs of each aerosol component [see Eqs. (85)–(87)].

The upper panels in Figure 54 pertain to small *aerosol particles* with optical depths of 0.03 and 0.3, while the lower panels are for large aerosol particles. The DTR curves are shown for wind speeds of 1, 5, and 10 m/s, while only one MSR curve at 5 m/s is shown for clarity. The errors incurred by ignoring multiple scattering in the path from the surface to the TOA are high, typically ranging from 10% to 90% at 490 nm (Figure 54). These error ranges are determined by the radiance threshold values that mark the retrieval region boundaries; the errors are smaller closer to the specular reflection peak (higher threshold). Surface roughness affects only the angular location and extent of the retrieval region where these errors occur. The minimum errors grow significantly in an atmosphere with a heavy *aerosol loading*, and asymmetries are found close to the horizon, especially in the presence of large (coarse-mode) particles.

Figure 54 pertains to the *principal plane* of *specular reflection*. Similar computations showed that the errors are azimuth-dependent [285]. Thus, in a typical maritime situation the errors tend to grow as the radiance decreases away from the specular direction, and the high directionality of the radiance peak at low wind speeds causes larger minimum errors away from the principal plane. Correcting for *sunglint contamination* including *multiple scattering* effects in future processing of ocean color satellite data is feasible, and would be desirable in view of the magnitude of the errors incurred by the DTR approach.

6.5
Simultaneous Retrieval of Aerosol and Aquatic Parameters

Traditional ocean color remote sensing algorithms start by the application of an *atmospheric correction* step to estimate the *aerosol optical thickness* at a single near-infrared channel (865 nm for SeaWiFS), for which the ocean is assumed to be nonscattering (the *black-pixel approximation*). Based on this atmospheric correction, *water-leaving radiances* are generated for visible channels [35, 287]. Next, marine constituents are estimated from two or three visible-channel water-leaving radiances, either through regression or by *LUT* matching based on a suitable *bio-optical model*. One shortcoming of this approach is that the black-pixel approximation may not be valid [288]. Also, the atmospheric correction step is based on the assumption that the radiation in the atmosphere can be decoupled from that in the ocean, which is potentially a large source of uncertainty, because the oceanic contribution to the total TOA radiance is typically less than 10%. Further, it is difficult, if not impossible, to quantify systematically error sources in such two-step *ad hoc* inversion procedures.

To remedy these shortcomings, Stamnes *et al.* [289] devised a one-step *iterative inversion* scheme, based on simulated radiances stored in LUTs, for simultaneous retrieval of two aerosol parameters and one ocean parameter (chlorophyll concentration). To minimize uncertainties caused by forward model assumptions, an accurate RT model for the coupled atmosphere–ocean system [39, 163] was used. Atmospheric correction was not treated separately, since any atmospheric effects were fully accounted for in the coupled RT model (CRTM).

The Stamnes *et al.* [289] one-step algorithm cannot easily be extended beyond the estimation of three (two aerosol and one ocean) parameters. Therefore, Li *et al.* [38] developed a new method that employs a linearized version of the CRTM [34] and simultaneously uses all available visible and near-infrared SeaWiFS measurements (eight channels at 412, 443, 490, 510, 555, 670, 765, and 865 nm). The *linearized CRTM* computes not only radiances but also *Jacobians* (radiance partial derivatives) that are required for inversion by standard methods such as the iterative fitting technique based on nonlinear least squares or *optimal estimation* (OE [37]). According to the new method [38], the *retrieval parameters* contained in the retrieval state vector includes both boundary-layer aerosol parameters and several marine parameters. At each iteration step, the *CRTM forward model* is linearized about the current estimate of the retrieval state vector, and used to generate both simulated radiances and *Jacobians* with respect to the state vector elements and other parameters that are required in the OE fitting.

6.5.1 Atmospheric IOPs

For altitudes up to 2 km, Li *et al.* [38] used a *bimodal aerosol model*, in which the IOPs are defined in terms of the optical depth τ_0 at 865 nm and the fractional weighting f between the two aerosol modes [see Eqs. (85)–(87)]:

$$\tau_{A,\gamma} \equiv \tau_0 \tilde{\gamma}_A = \tau_0 \left[(1-f)\tilde{\gamma}_1 + f\tilde{\gamma}_2\right] \tag{911}$$

$$\tau_{A,\beta} \equiv \tau_0[(1-f)\varpi_1\tilde{\gamma}_1 + f\varpi_2\tilde{\gamma}_2]; \quad \varpi_A = \frac{\tau_{A,\beta}}{\tau_{A,\gamma}} \tag{912}$$

$$\chi_{A,\ell} = \frac{(1-f)\varpi_1\tilde{\gamma}_1\chi_{1,\ell} + f\varpi_2\tilde{\gamma}_2\chi_{2,\ell}}{\tau_{A,\beta}}. \tag{913}$$

Here, $\tau_{A,\gamma}$ is the total extinction aerosol optical depth in the layers containing aerosols, $\tau_{A,\beta}$ is the total scattering aerosol optical depth, ϖ_A is the *single-scattering albedo* for aerosols, and $\chi_{A,\ell}$ are the total *Legendre polynomial* expansion coefficients for the *bimodal aerosol mixture*. The two atmospheric *retrieval parameters* are τ_0 and f. All other quantities in Eqs. (911)–(913) are assumed model parameterizations: $\tilde{\gamma}_1$, ϖ_1, and $\chi_{1,\ell}$ are, respectively, the extinction coefficient normalized to the value at 865 nm, the single-scattering albedo, and the *scattering phase function expansion coefficients* for aerosol type 1 ("fine-mode"), and $\tilde{\gamma}_2$, ϖ_2, and $\chi_{2,\ell}$ are the corresponding values for aerosol type 2 ("coarse-mode"). The fine mode (subscript 1) was assumed to be a tropospheric aerosol model with 70% humidity, while the coarse mode (subscript 2) was assumed to be a coastal aerosol model with 99% humidity. IOPs were calculated for the eight SeaWiFS channels using a *Mie code* for *spherical particles* with size and *refractive index* depending on humidity [113, 290]. A justification for adopting just one large and one small aerosol model (instead of several models of each type) can be found elsewhere [291]. To obtain the total IOPs in the marine boundary layer (MBL) containing aerosols, the *Rayleigh scattering* coefficient β_{Ray} and the molecular absorption coefficient α_{gas} are also needed. Rayleigh scattering cross-sections and depolarization ratios were taken from standard sources. Absorption by O_3 (visible channels), O_2 (A-band), and water vapor was included.

6.5.2
Aquatic IOPs

In the ocean, *IOPs* can be derived from simple wavelength-dependent parameterizations of (i) the phytoplankton absorption coefficient $\alpha_{\rm ph}(\lambda)$ in terms of the overall *chlorophyll concentration CHL* in [mg·m^{-3}], and (ii) the detrital and *colored dissolved material (CDM)* absorption coefficient $\alpha_{\rm dg}(\lambda)$ and the *backscattering coefficient* $b_{\rm bp}(\lambda)$ in terms of their respective values $\text{CDM} \equiv \alpha_{\rm dg}(\lambda_0)$ and $\text{BBP} \equiv b_{\rm bp}(\lambda_0)$ at some reference wavelength λ_0 [38]:

$$\alpha_{\rm ph}(\lambda) = a_1(\lambda)\text{CHL}^{a_2(\lambda)} \tag{914}$$

$$\alpha_{\rm dg}(\lambda) = \text{CDM}\exp\left[-S(\lambda-\lambda_0)\right] \tag{915}$$

$$b_{\rm bp}(\lambda) = \text{BBP}\left(\frac{\lambda}{\lambda_0}\right)^{-\eta}. \tag{916}$$

Thus, this *bio-optical model* is described by the three retrieval elements CHL, CDM, and BBP, and the four model parameters $a_1(\lambda), a_2(\lambda), S$, and η. For a_1 and a_2, wavelength-dependent coefficients are determined by fitting the power-law expression in Eq. (914) to field measurements of chlorophyll absorption. From spectral fittings of measurements to the expressions for $\alpha_{\rm dg}(\lambda)$ and $b_{\rm bp}(\lambda)$ in Eqs. (915) and (916), it was found [38] that $S = 0.012$ and $\eta = 1.0$. All coefficients are in units of [m^{-1}]. Together with the *pure water absorption* and *scattering coefficients* $\alpha_w(\lambda)$ and $\beta_w(\lambda)$ [116, 118] expressed in the same units, the layer total optical depth and total *single-scattering albedo*, and *Legendre polynomial expansion coefficients* IOPs for the marine medium become [see Eqs. (85)–(87)]:

$$\tau_{\rm tot} = \Delta z\left[\alpha_{\rm ph}(\lambda) + \alpha_{\rm dg}(\lambda) + \beta_p(\lambda) + \alpha_w(\lambda) + \beta_w(\lambda)\right] \tag{917}$$

$$\varpi = \Delta z\frac{\beta_p(\lambda) + \beta_w(\lambda)}{\tau_{\rm tot}} \tag{918}$$

$$\chi_\ell = \frac{\beta_p(\lambda)\chi_{\ell,FF} + \beta_w(\lambda)\chi_{\ell,\rm water}}{\beta_p(\lambda) + \beta_w(\lambda)} \tag{919}$$

where Δz is the layer thickness in [m].

As alluded to in Section 2.9.3 , the *particle size distribution (PSD)* function in oceanic water is frequently described by an inverse power law [*Junge distribution*, Eq. (246)]: $n(r) \propto r^{-\xi}$, where $n(r)$ is the number of particles per unit volume per unit bin width, r is the particle radius, and ξ is the slope of the distribution. The *PSD slope* ξ typically varies between 3.0 and 5.0 [76, 77]. Based on this PSD, Forand and Fournier [78] derived an analytic expression for the scattering phase function [see Eq. (27)]. Values of the relative refractive index $n = 1.069$ and $\xi = 3.38$ were chosen in the FF (Forand–Fournier) *scattering phase function* to give a *backscattering ratio* $b_{\rm FF} = 0.0067$, consistent with a certain mixture of living organisms and resuspended sediments [79]. A moment-fitting code [91] was used to generate

Legendre polynomial expansion coefficients for the FF scattering phase function. Linearized IOPs were obtained by differentiation in the study by Li *et al.* [38].

6.5.3
Inverse Modeling

In *optimal estimation (OE)*, the update of the retrieval state vector $\mathbf{x}_i$ at iteration step i is given by [see Eq. (892)]

$$\mathbf{x}_{i+1} = \mathbf{x}_i + \mathbf{G}_i\{\mathbf{J}_i^T\mathbf{S}_\epsilon^{-1}(\mathbf{y}_m - \mathbf{y}_i) - \mathbf{S}_a^{-1}(\mathbf{x}_i - \mathbf{x}_a)\} \tag{920}$$

where the gain matrix is given by

$$\mathbf{G}_i = [(1+\gamma_i)\mathbf{S}_a^{-1} + \mathbf{J}_i^T\mathbf{S}_\epsilon^{-1}\mathbf{J}_i]^{-1}. \tag{921}$$

The measurement vector $\mathbf{y}_m$ has covariance error matrix $\mathbf{S}_\epsilon$, $\mathbf{y}_i = \mathbf{F}(\mathbf{x}_i)$ are simulated radiances generated by the *forward model* $\mathbf{F}(\mathbf{x}_i)$, which is a (nonlinear) function of $\mathbf{x}_i$. $\mathbf{J}_i$ is the *Jacobian matrix* of simulated radiance partial derivatives with respect to $\mathbf{x}_i$. The *a priori* state vector is $\mathbf{x}_a$ with covariance $\mathbf{S}_a$. The *Levenberg–Marquardt parameter* γ_i is chosen in each iterative step of minimization of the *cost function*, as discussed in Section 5.9.2. When $\gamma_i \to 0$, the iterative step is executed in accordance with the *Gauss–Newton formula*, and when $\gamma_i \to \infty$, it is executed in accordance with the *steepest descent method*. One may start with $\gamma_i = 0.01$. The inverse process starts from an initial guess $\mathbf{x}_0$, often set to $\mathbf{x}_a$. Alternatively, one may use the previous pixel's retrieved values as the next pixel's initial values. At each step, a *convergence criterion* is employed to check the progress toward the solution $\mathbf{x}$ that minimizes the *cost function*. If the error decreases, one updates $\mathbf{x}_i$ and decreases γ_i for the next step. If the error increases, one increases γ_i, keeps $\mathbf{x}_i$ the same, and tries again.

To illustrate the application of the method, Li *et al.* [38] considered a SeaWiFS image over Santa Barbara Channel obtained on February 28, 2003. The forward and inverse models described above were used for *simultaneous retrieval* of the five-element state vector $\{\tau_{865}, f, \mathrm{CHL}, \mathrm{CDM}, \mathrm{BBP}\}$. Figure 55 shows the retrieved values of the four parameters $\{\tau_{865}, f, \mathrm{CDM}, \text{and BBP}\}$. The most probable value for the aerosol optical depth τ_{865} was about 0.04, with a range between 0.002 and 0.10. The *aerosol fraction* f ranged from about 0.2 (predominantly small particles) to 0.9. On average, there appeared to be equal amounts of small and large particles for this image. The right top panel in Figure 56 shows that the distribution of the *CDM absorption coefficient* [Eq. (915)] lay between 0.02 and 0.07 m^{-1} and had a peak at around 0.04 m^{-1}. Also, it shows that the distribution of the *backscattering coefficient (BBP)* lay between 0.001 and 0.005 m^{-1} with a peak at 0.002 m^{-1}.

The retrieved *chlorophyll concentration*, shown in Figure 56, ranged from near 0 to about 3.0 $\mathrm{mg{\cdot}m^{-3}}$. In contrast to the traditional two-step "atmospheric correction and regression" approach, the *simultaneous retrieval* described above produced a direct assessment of the error by examining sensor radiance residuals, summarized in the table inserted in Figure 56. For the nearly 35,000 pixels in this

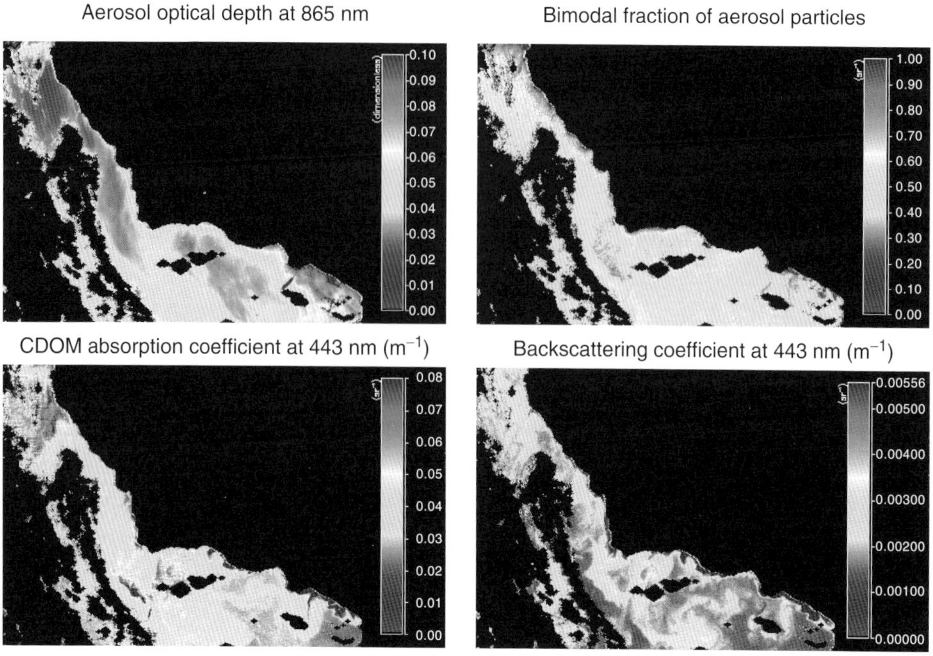

Figure 55 Retrieved values of four of the five parameters: aerosol optical depth, bimodal fraction of aerosols, CDM absorption coefficient at 443 nm, and backscattering coefficient at 443 nm (adapted from Stamnes *et al.* [292]).

SeaWiFS image, the residuals were less than 1% for seven of the eight SeaWiFS channels, and less than 2% for the remaining 765-nm (O_2 A-band) channel. It may be concluded that this simultaneous forward/inverse retrieval method yielded excellent retrieval capability, and that eight SeaWiFS channels were sufficient to retrieve two atmospheric and three marine parameters in *coastal waters*. In addition to well-calibrated SeaWiFS data, the good results are believed to be due to the availability of high-quality field data used to construct a reliable *bio-optical model*, and an adjustable *bimodal fraction* of large versus small *aerosol particles*.

The only drawback with the *OE* approach, as described above, is the slow speed. The most time-consuming step in the inversion process is the *CRTM forward model* computations. It is possible, however, to employ a fast forward model trained by a *radial-basis functions neural-network (RBF-NN)* to increase the speed considerably. As discussed in Section 3.5.2 [see Eq. (482)], the *RBF-NN* function can be written as

$$p_i = \sum_{j=1}^{N} a_{ij} \exp[-b^2 \sum_{k=1}^{N_{\rm in}} (c_{jk} - R_k)^2] + d_i \qquad (922)$$

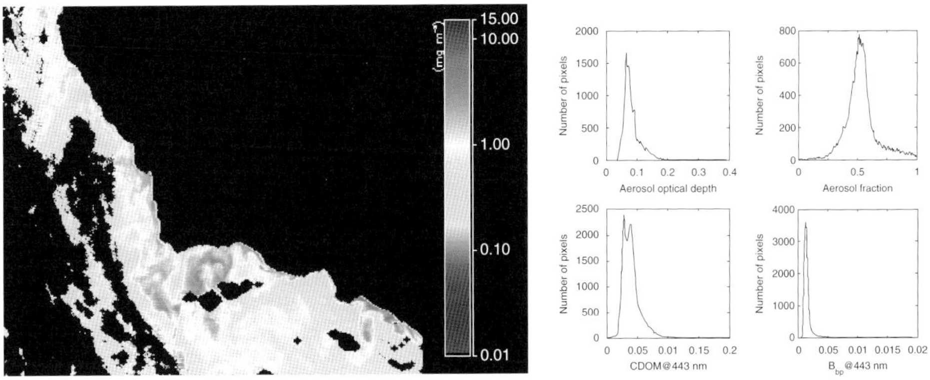

Left panel: Retrieved chlorophyll concentration (mg$\cdot$m^{-3}) from SeaWiFS image on Feb. 28, 2003 over the Santa Barbara Channel.

Right top panel: The distributions of the other retrieved parameters from the same image. (a) aerosol optical depth; (b) aerosol model fraction; (c) CDOM absorption coefficient at 443nm (m^{-1}) ; (d) backscattering at 443 nm (m^{-1})

Right bottom table: Radiance residuals at all SeaWiFS channels.

Wavelength	Average relative error (%)	Pixels with <2% relative error (%)
412 nm	±0.298	99.488
443 nm	±0.289	99.625
490 nm	±0.555	99.216
510 nm	±0.716	98.552
555 nm	±0.240	99.168
670 nm	±1.050	94.267
765 nm	±1.952	64.102
865 nm	±0.857	94.745

Figure 56 Retrieved chlorophyll concentration for the same SeaWiFS image as in Figure 55, showing distributions of the other for parameters and residuals (adapted from Stamnes *et al.* [292]).

where N is the total number of *neurons*, and $N_{\rm in}$ is the number of input parameters. Note that, since our goal is to use the RBF-NN to obtain the TOA radiances, the input parameters R_k should be the *state parameters* and the solar/viewing geometry, and the output parameters p_i should be the TOA radiances as discussed in Section 3.5.2. The elements of the Jacobian are obtained by calculating the partial derivatives with respect to the retrieval parameters R_k:

$$J(k) = \frac{\partial p_i}{\partial R_k} = -2b^2(c_{jk} - R_k)\sum_{j=1}^{N} a_{ij}\exp[-b^2\sum_{k=1}^{N_{\rm in}}(c_{jk} - R_k)^2]. \tag{923}$$

The purpose of training the RBF-NN is to determine the coefficients a_{ij}, b, c_{jk}, and d_i appearing in Eq. (922). It has been demonstrated that use of RBF-NN training leads to a computing efficiency enhancement of about 1500 [292].

6.6
Polarized RT in a Coupled Atmosphere–Ocean System

As discussed in the previous section, in order to solve the *inverse RT problem* it is important to have an accurate and efficient *forward RTM*. Accuracy is important in order to obtain reliable and robust retrievals, and efficiency is an issue because standard iterative solutions of the nonlinear inverse RT problem require running the forward RTM repeatedly to compute the radiation field and its partial derivatives (the elements of the *Jacobian*) with respect to the *retrieval parameters (RPs)* [37, 38, 292].

While solutions to the *scalar RTE,* which includes only the first component of the Stokes vector (the radiance or intensity), are well developed, modern *RT models* that solve the *vector RTE* are capable of accounting also for polarization effects described by the second, third, and fourth component of the Stokes vector. Even if one's interest lies primarily in the radiance, it is important to realize that solutions of the *scalar RTE,* which ignores polarization effects, introduce errors in the computed radiances [40–42].

As mentioned above, in scalar RT theory only the first component of the Stokes vector (the radiance or intensity) is considered. Numerical realizations using the *discrete-ordinate method,* developed by Chandrasekhar and others in the 1940s [11] for the scalar problem, were first explored in the early 1970s [172]. The scalar RT method was further developed in the 1980s [172–174], and eventually implemented numerically in a robust, well-tested, and freely available computer code (*DISORT*) in 1988 [161]. The *DISORT* code computes radiances and irradiances in a vertically inhomogeneous medium that is divided into homogenous layers to resolve the vertical variation in the IOPs. The method was extended to be applicable to two adjacent media with different refractive indices, such as a coupled atmosphere–water system [39], and the performance of the resulting *C-DISORT* code was tested against *Monte Carlo* (MC) computations [27], as discussed in Section 6.4.1.

For a layered medium with constant *refractive index,* solutions of the *vector RTE* [see Eq. (353)] were developed and implemented in the *VDISORT* code [176–179]. For a coupled system consisting of two media with different refractive indices, the coupled vector discrete-ordinate (*C-VDISORT*) RTM was developed and tested for pure *Rayleigh scattering* [87].

It may seem trivial to extend a *vector RT method* valid for *Rayleigh scattering* to scattering by large spherical particles (Mie scattering). However, it was found that this extension required a lot of attention to detail and also further development of both the coupled VDISORT (C-VDISORT) and Monte Carlo (C-PMC) codes in order to get them to produce correct results [84]. For a *Rayleigh scattering* medium, it was easy to distinguish between the direct and diffuse components of the Stokes vector in the C-PMC code. The direct component was simply separated in postprocessing [87]. For a medium consisting of strongly forward-scattering particles, it was necessary to add a counter to keep track of the scattering events in order to separate the diffuse component from the direct one. The quadrature of the *scattering phase matrix* integration also had to be increased, and the number of azimuthal and polar detector bins had to be significantly increased, to get accurate results from the C-PMC code [84]. For the C-VDISORT code, a new subroutine had to be used to handle a generalized scattering phase matrix, and the δ-M technique [89] was needed to get accurate results without using a large number of streams, which adversely affects the computational efficiency [84].

6.6.1 C-VDISORT and C-PMC Versus Benchmark – Aerosol Layer – Reflection

Here we show a comparison of the results produced by C-VDISORT and C-PMC with *benchmark computations* provided by Kokhanovsky *et al.* [181] for a homogeneous slab of aerosol particles. The aerosol particles considered in the benchmark computations were assumed to have a *log-normal size distribution* [see Eq. (102)] with $r_n = 0.3$ μm, and $\sigma = 0.92$, ($r_{\text{eff}} = r_n \exp[2.5\sigma^2] = 2.5$ μm) and the smallest and largest particle radii were selected to be $r_1 = 0.005$ μm and $r_2 = 30$ μm. The *refractive index* of the *aerosol particles* was set to $m = 1.385$, which yields a *single-scattering albedo* of 1.0, and an asymmetry factor $g = 0.79275$ [181] [see Eq. (10)].

The left four panels of Figure 57 show a comparison of C-VDISORT results (dotted curves) with benchmark results (solid curves) for a layer of aerosol particles as described above. In the simulations using the C-VDISORT and C-PMC RT models, we set the refractive index to 1.0 in both slabs, and put half of the aerosol particles (optical depth = 0.1631) in each slab. The C-VDISORT results are seen to be indistinguishable from the benchmark results; in fact, the dotted and solid curves lie on top of each other in the left panels of Figure 57. The right four panels of Figure 57 show comparisons between the *reflected polarized radiation components* computed by the C-VDISORT model (same results as in the four left panels) and the C-PMC RT model for the benchmark case [181]. Except for *statistical fluctuations* in the C-PMC results, the agreement is seen to be very good. A closer examination shows that the difference between the C-VDISORT results and the benchmark results is less than a fraction of a percent for all viewing directions [84].

6.6.2 C-VDISORT and C-PMC Versus Benchmark – Aerosol Layer – Transmission

The left four panels of Figure 58 show comparisons between C-VDISORT and benchmark results for the transmitted radiation for the same aerosol benchmark case [181] as in Figure 57. Except for artifacts in the I, Q, and U parameters around the forward direction ($\cos\theta = -0.5$), the agreement between C-VDISORT and the benchmark results [181] is excellent. These artifacts are most likely due to the treatment of the forward scattering peak in the C-VDISORT RT model, which uses the δ-M approximation [89] to truncate the forward scattering peak of the *scattering phase function*. A comparison between C-VDISORT and C-PMC results for *transmitted polarized radiation components* is shown in the right four panels of Figure 58. The C-VDISORT results are the same as those shown in the left four panels. The C-VDISORT and C-PMC RT models give very similar results except for small differences due to statistical fluctuations in the C-PMC results

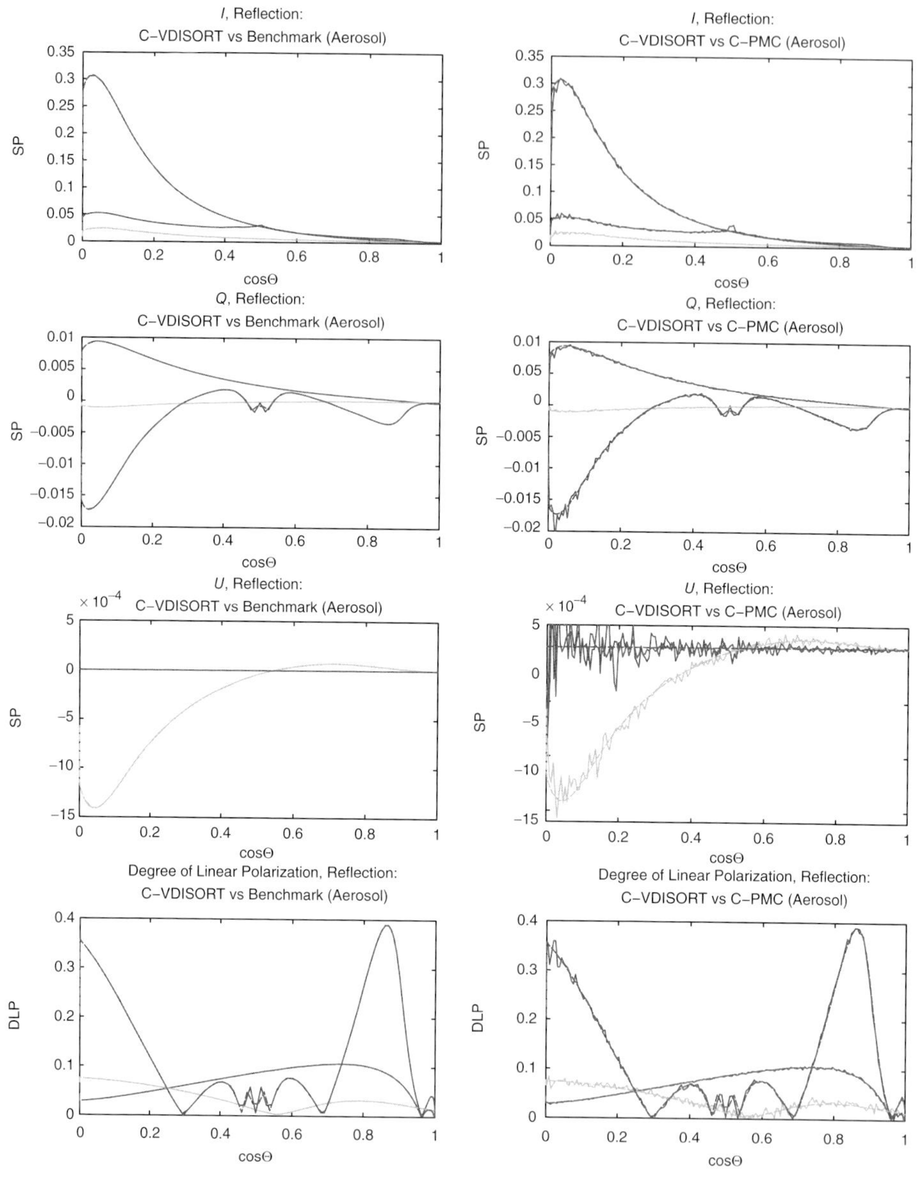

Figure 57 Left four panels: Comparison of *reflected polarized radiation components* computed by C-VDISORT (dashed curves) with benchmark results [181] (solid curves) for a homogeneous layer of nonabsorbing aerosol particles, simulated by putting half of the aerosol particles (optical depth = 0.1631) in each slab, and setting the refractive index to 1.0 in both slabs. From top to bottom: Stokes parameters I, Q, U, and the degree of linear polarization. Right four panels: Same as left four panels except that the comparison is between C-VDISORT (dashed curves) and C-PMC (solid curves). Red: $\Delta\phi = 0°$; green: $\Delta\phi = 90°$; blue: $\Delta\phi = 180°$. Two-hundred discrete-ordinate streams, 10^9 photons (after Cohen *et al.* [84]).

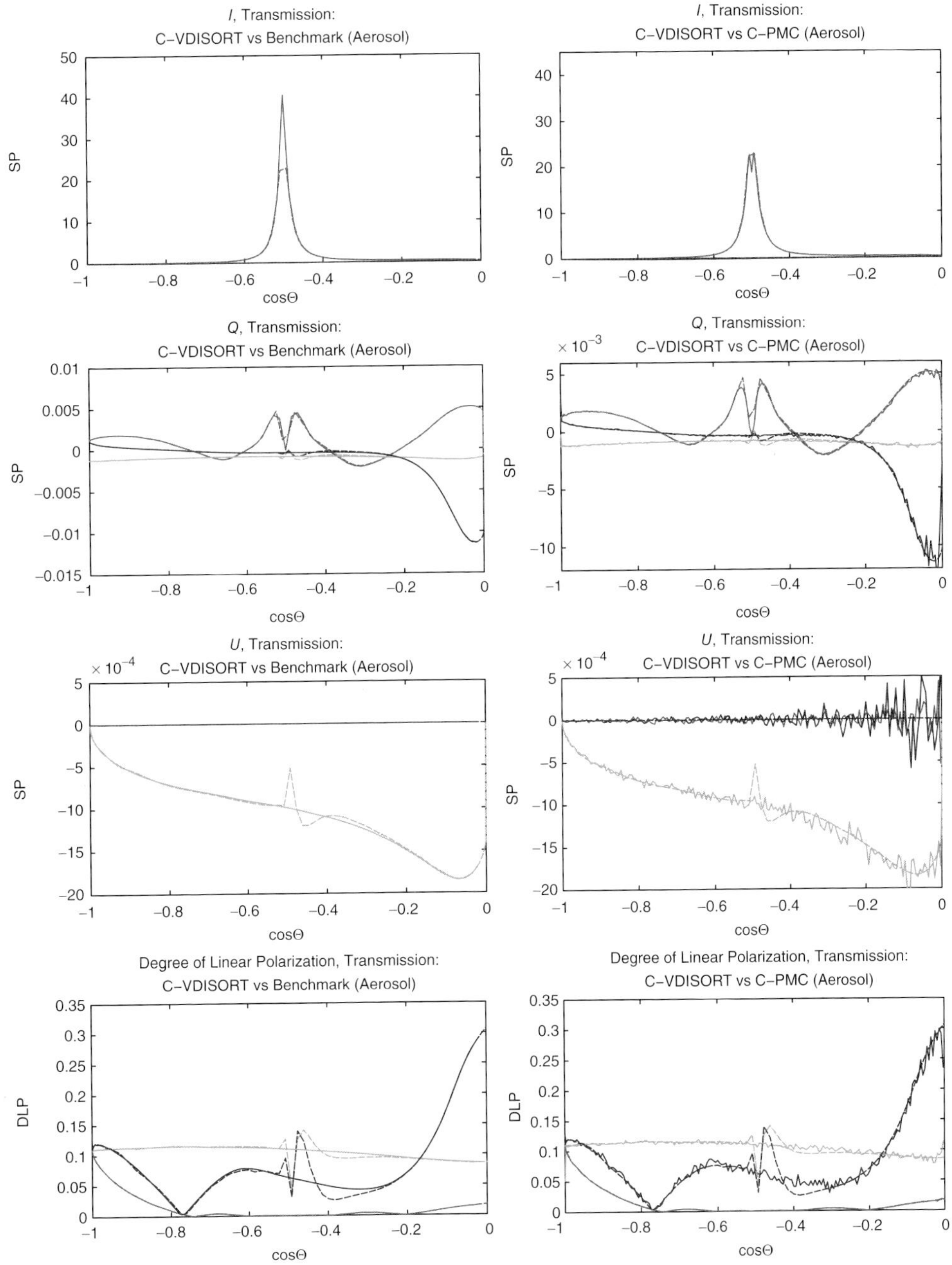

Figure 58 Similar results as in Figure 57 but for the *transmitted polarized radiation components* (after Cohen *et al.* [84]).

and artifacts around the forward scattering direction due to the use of the *δ-M approximation* in C-VDISORT.

6.6.3
C-VDISORT and C-PMC Versus Benchmark – Cloud Layer – Reflection

For cloud particles, Kokhanovsky *et al.* [181] used a *log-normal particle size distribution* with mode radius of $r_n = 5$ µm, and a standard deviation of $\sigma = 0.4$ ($r_{\mathrm{eff}} = r_n \exp[2.5\sigma^2] = 7.5$ µm), and the smallest and largest particle radii were selected to be $r_1 = 0.005$ µm and $r_2 = 100$ µm. The refractive index was set to $m = 1.339$, which yields a single-scattering albedo of 1.0, and an asymmetry factor $g = 0.86114$ for the cloud particles. In the simulations with C-VDISORT and C-PMC, we set the refractive index equal to 1.0 in both slabs, and put half of the cloud particles (optical depth = 2.5) in each slab. The left four panels of Figure 59 show comparisons of C-VDISORT and benchmark results [181] for the *reflected polarized radiation components* for this two-layer system of *cloud particles* with a total optical depth of 5.0. The agreement is seen to be very good, and a quantitative comparison shows that the differences between C-VIDORT and benchmark results are similar to those for the aerosol case [84]. The right four panels of Figure 59 show the comparisons between reflected polarized radiation components computed by the C-VDISORT and C-PMC RT models for the cloud benchmark case. The C-VDISORT and C-PMC RT models give almost indistinguishable results except for small differences due to statistical fluctuations in the C-PMC results and artifacts in the I and Q parameters for $\Delta\phi = 180°$ around the forward scattering direction due to the use of the *δ-M approximation*.

6.6.4
C-VDISORT and C-PMC Versus Benchmark – Cloud Layer – Transmission

The left four panels of Figure 60 show the comparisons between C-VDISORT and benchmark results for the transmitted radiation for the same cloud benchmark case as in Figure 59. The agreement between C-VDISORT and the benchmark results [181] is excellent. Comparisons between C-VDISORT and C-PMC results for the *transmitted polarized radiation components* are shown in the right four panels of Figure 60. Disregarding the small differences due to *statistical fluctuations* in the C-PMC results, we see that the agreement between the C-VDISORT and the C-PMC results is generally very good, except in the forward direction where the C-PMC result is smaller than the C-VDISORT result (which agrees with the benchmark), presumably due to a *sampling problem* in the *Monte Carlo* algorithm. For example, if the resolution of detector bins in polar or azimuth space is insufficiently sampled, then a beam scattered in the forward direction will not be detected at exactly the forward direction, but rather at angles close to the forward direction. Or, if the quadrature of the *scattering phase matrix* is too coarse, then

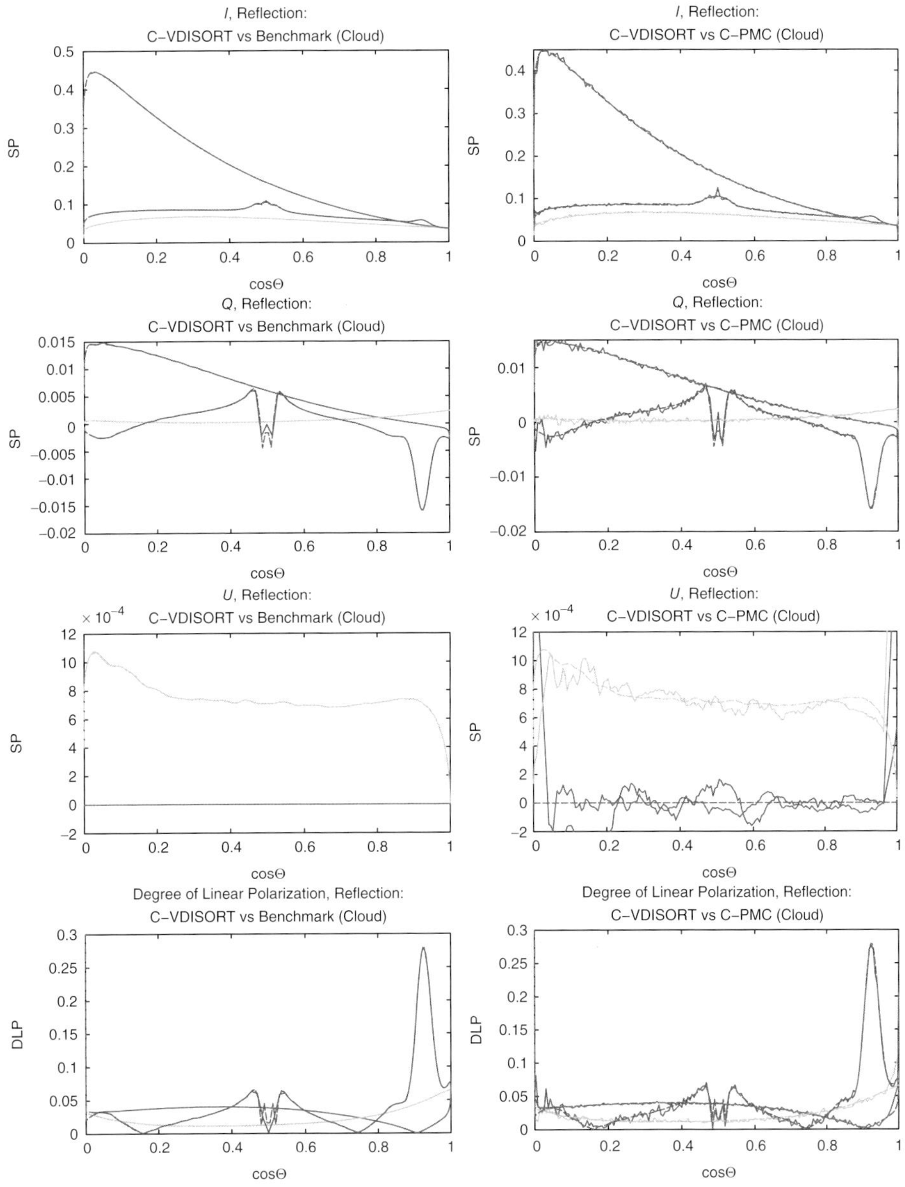

Figure 59 Left four panels: Comparisons of *reflected polarized radiation components* computed by C-VDISORT (dashed curves) with benchmark results [181] (solid curves) for a homogeneous layer of nonabsorbing cloud particles, simulated by setting the refractive index equal to 1.0 in both slabs, and putting half of the cloud particles (optical depth = 2.5) in each slab. From top to bottom: Stokes parameters I, Q, U, and the degree of linear polarization. Right four panels: Same as the left four panels except that the comparisons are between C-VDISORT (dashed curves) and C-PMC (solid curves). Red: $\Delta\phi = 0°$; green: $\Delta\phi = 90°$; blue: $\Delta\phi = 180°$. Two hundred *discrete-ordinate* streams, 10^9 photons (after Cohen *et al.* [84]).

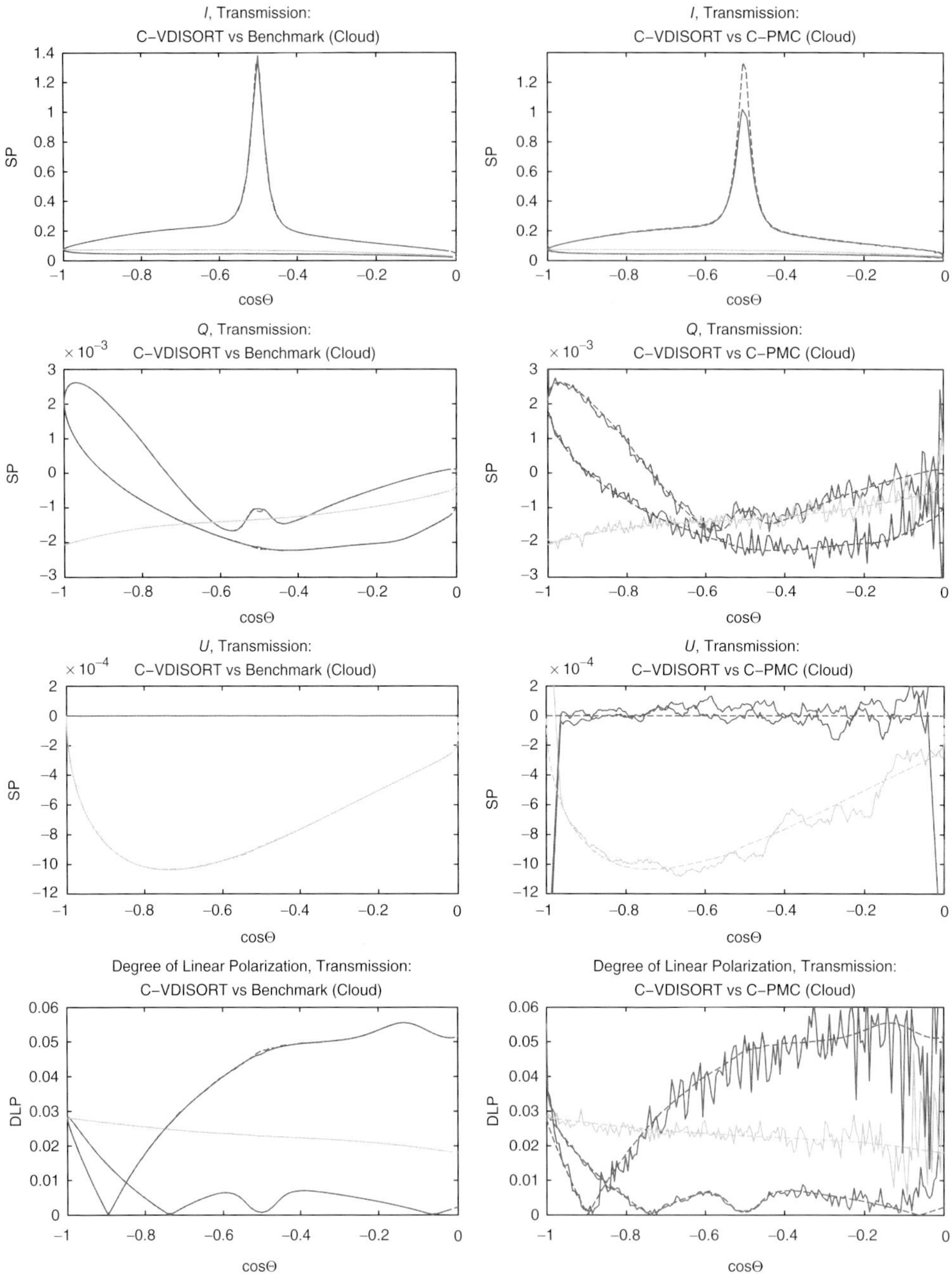

Figure 60 Results similar to those in Figure 59, but for the *transmitted polarized radiation components*. Two hundred discrete-ordinate streams, 10^9 photons (after Cohen *et al.* [84]).

a forward scattered beam will not be scattered precisely in the forward direction, but at an angle close to the forward direction. These problems could presumably be solved by increasing the resolution of the detector bins about the forward direction, and by increasing the resolution of the scattering phase matrix quadrature about the 0° scattering direction. A change in the resolution of the detector bins will also need to be accounted for when averaging over the bins to reduce noise. The C-VDISORT results are the same as those shown in the left four panels.

6.6.5
C-VDISORT Versus C-PMC – Aerosol Particles – Coupled Case

Next, we consider results for a case similar to that in Figures 57 and 58 (right panels), except that the refractive indices are different in the two slabs. To simulate the atmosphere–water system, we set the refractive index equal to 1.0 in the upper slab and equal to 1.338 in the lower slab, and as before we put half of the particles (optical depth = 0.1631) in each slab. Comparisons between *polarized radiation components* computed by the C-VDISORT and C-PMC RT models are shown in Figure 61 just above the interface (upper four panels), and just below the interface (lower four panels). The two codes give very similar results except for small discrepancies in the Q and U components for directions close to the horizon at $\Delta\phi = 45°$. The corresponding results at the TOA are displayed in Figure 63 (upper panel), showing generally very good agreement also at the TOA.

6.6.6
C-VDISORT Versus C-PMC – Aerosol/Cloud Particles – Coupled Case

Here we consider results for a case similar to that in Figures 59 and 60, except that we set the *refractive index* equal to 1.0 in the upper slab and equal to 1.338 in the lower slab, and we put *aerosol particles* in the upper slab (atmosphere, optical depth = 0.3262) and cloud particles in the lower slab of optical depth = 5.0 to mimic large particles in the ocean. This choice may seem artificial, but our purpose is simply to compare the performance of the two codes for identical inputs. Comparisons between polarized radiation components computed by the C-VDISORT and C-PMC RT models are shown in Figure 62 just above the interface and just below the interface, and the corresponding results at the TOA are shown in Figure 63 (lower panel). The two codes give very similar results. The increased differences compared to the corresponding comparisons with no coupling between the slabs are likely due to *statistical noise* and could be dealt with by smoothing using simple running mean averaging over the detector bins, as well as by using more photons. In *Monte Carlo simulations*, there is a trade-off between the number of detector bins one adopts (an increase in the number of bins increases both accuracy and noise) and the number of photons one uses (an increase in the number of photons tends to reduce noise but increase the computing time.)

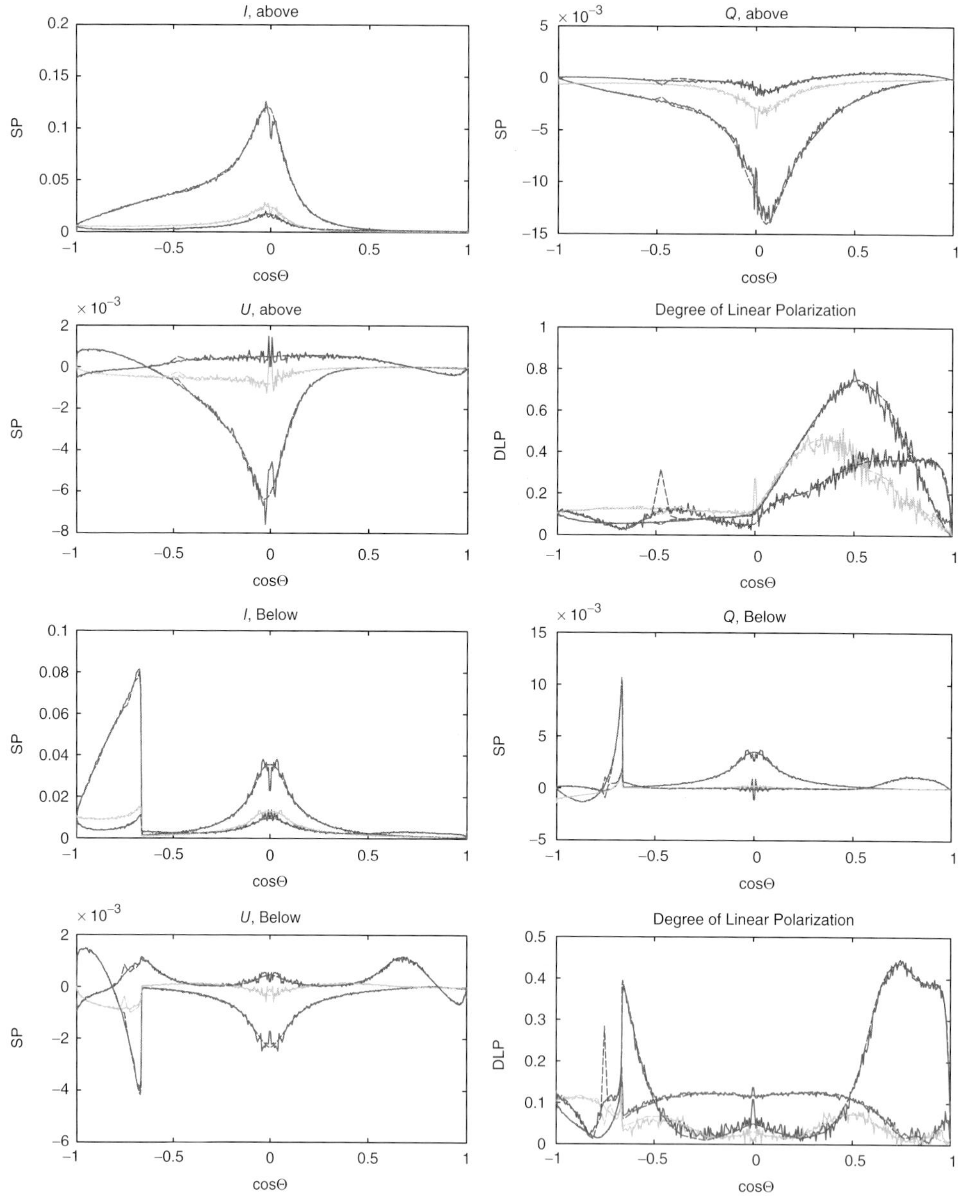

Figure 61 Comparisons between *polarized radiation components* computed with C-VDISORT (dashed curves) and C-PMC (solid curves) for the same physical situation as in Figure 57, except that the refractive index was set equal to 1.0 in the upper slab and equal to 1.338 in the lower slab. Upper four panels: just above interface. Lower four panels: just below the interface. Red: $\Delta\phi = 45°$; green: $\Delta\phi = 90°$; blue: $\Delta\phi = 135°$. Sixty-four discrete-ordinate streams in upper slab, 96 streams in lower slab, 10^9 photons (after Cohen *et al.* [84]).

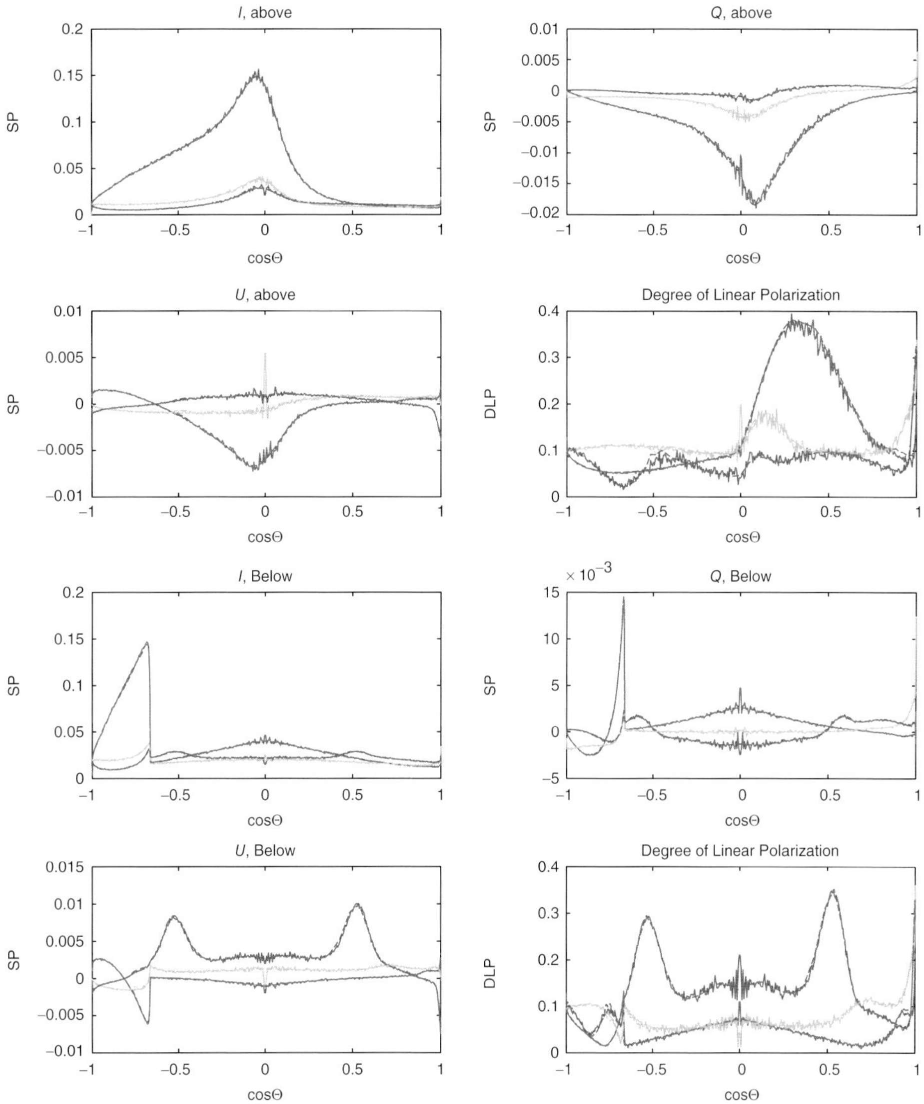

Figure 62 Similar to Figure 61 (aerosol particles with an optical depth of 0.1631 in each slab) except that we put nonabsorbing aerosol particles of optical depth 0.3262 in the upper slab, and cloud-like particles with single-scattering albedo of 0.9 and optical depth of 5.0 in the lower slab. 96 discrete-ordinate streams in upper slab. One-hundred and twenty-eight streams in lower slab, 10^9 photons (after Cohen *et al.* [84]).

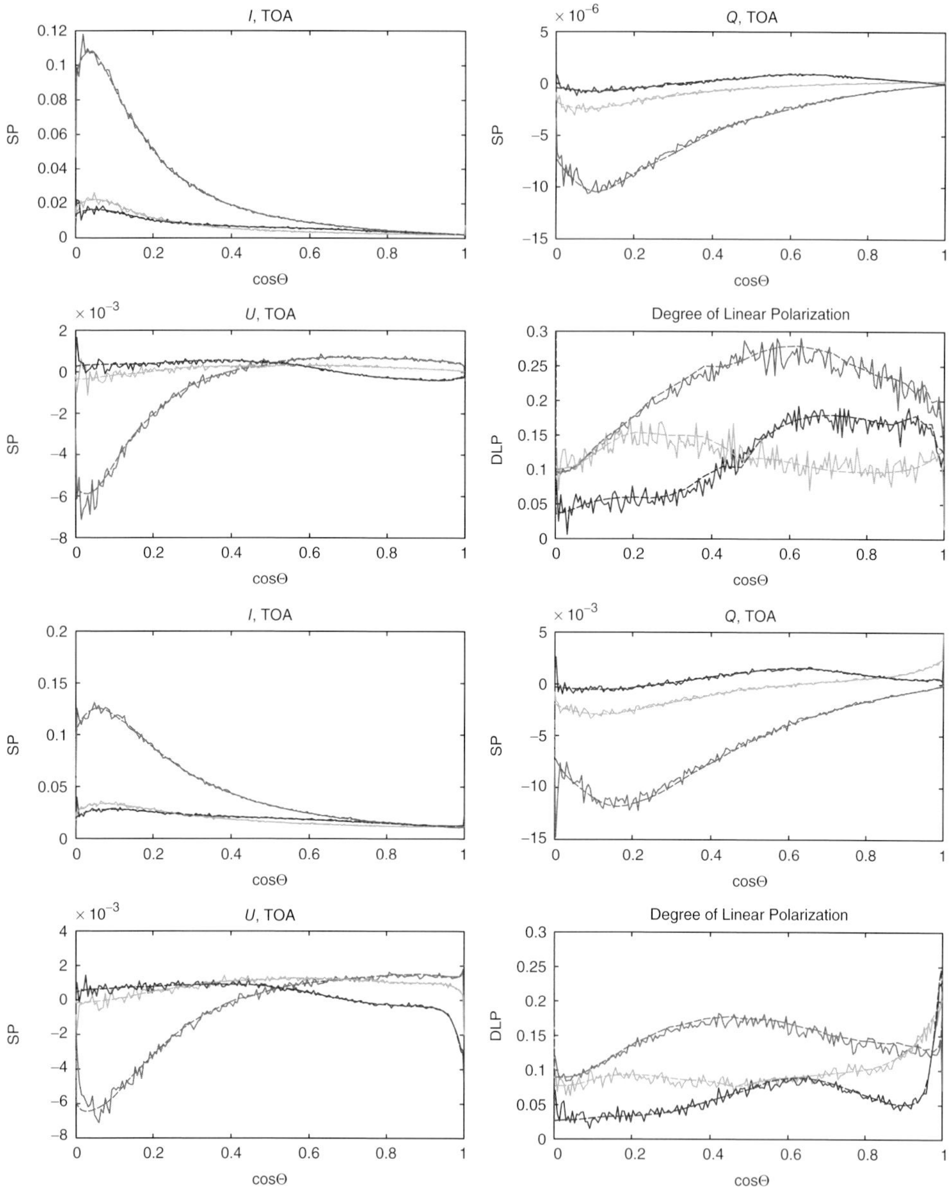

Figure 63 Upper four panels: Similar to Figure 61, but at TOA for nonabsorbing aerosol particles with an optical depth 0.1631 in each slab. Lower four panels: Similar to Figure 62, but at TOA for nonabsorbing aerosol particles with an optical depth of 0.3262 in the upper slab, and cloud particles with an optical depth of 5.0 and single-scattering albedo of 0.9 in the lower slab (after Cohen *et al.* [84]).

6.6.7
Summary

In this section, results from two *vector RT models* were compared for a coupled system consisting of two adjacent slabs with different refractive indices. The two RT models were based on the coupled vector discrete-ordinate (C-VDISORT) method and the coupled polarized Monte Carlo (C-PMC) method, which apply to particle scattering beyond the *Rayleigh limit*. Emphasis was placed on testing the RT models for scattering by *spherical particles* with sizes comparable to or larger than the wavelength of light (*Mie scattering*). If the refractive indices of the two adjacent slabs are different, they could describe a coupled atmosphere–water system. The comparisons were carried out for large spherical particles (Mie scattering), and for the same or different refractive indices in the two slabs with identical input *IOPs* to the two *RT models*. For a number of different cases, the two *RTMs* were found to give very similar results for identical input IOPs.

The results show that the two RT models give results that agree with benchmark results in the limit of no coupling or change in the refractive index between the two slabs. Also, good agreement was obtained between the two RT models for cases in which the refractive indices in the two slabs were different, indicating that the *scattering phase matrix* as well as the interface conditions were correctly implemented in the two RT models.

The C-VDISORT RT model solves the *integro-differential radiative transfer equation* for polarized radiation using a combination of analytic and numerical techniques based on linear algebra, while the C-PMC RT model is based on probabilistic and statistical concepts. Since these two radically different RT models give essentially the same results, we can confidently apply either RT model in various applications. The advantage of the C-VDISORT RT model is that it is very fast compared to the C-PMC RT model. The disadvantage of the current C-VDISORT RT model is that it is limited to a plane-parallel geometry, while the C-PMC RT model can readily be extended to apply to any desired geometry including 3-D RT simulations.

6.7
What if MODIS Could Measure Polarization?

6.7.1
Motivation

The *MODIS* is a scientific instrument launched into Earth orbit by NASA in 1999 on board the *Terra satellite* and in 2002 on board the *Aqua satellite*. *MODIS* instrument captures data in 36 spectral bands ranging in wavelength from 0.4 to 14.4 μm. The spatial resolution is: 2 bands at 250 m, 5 bands at 500 m, and 29 bands at 1 km. Together, the two MODIS instruments image the entire earth every 1 to 2 days. They are designed to provide measurements of large-scale global dynamics including changes in Earth's cloud cover, radiation budget, and processes occurring in the oceans, on land, and in the lower atmosphere. *MERIS* was one of the main instruments on board the European Space Agency's *Envisat platform*, which was in operation from 2002 until 2012. The MERIS instrument was equipped with spectrometers to measure reflected sunlight in several spectral bands between 390 and 1040 nm. It was specifically designed to study/monitor the health of water bodies including *open ocean* and *coastal waters*.

Since MODIS and MERIS provide radiance measurements only, we asked the question: what if MODIS could measure the Stokes parameter $Q = I_{\|} - I_{\perp}$ in addition to $I = I_{\|} + I_{\perp}$? As discussed in the preceding, compared with traditional two-step methods that rely on an *atmospheric correction* to yield *surface reflectance* (over land) or *water-leaving radiance* (over water), *simultaneous retrieval* of aerosol and surface properties using inverse techniques based on coupled atmosphere–surface radiative transfer and *optimal estimation* can yield considerable improvement in retrieval accuracy based on radiances measured by MERIS, MODIS, and similar instruments.

However, there are uniqueness problems associated with remote sensing measurements that ignore polarization effects and provide only the radiance. Use of *polarization measurements* is particularly important for retrieval of absorbing aerosol properties over coastal waters as well as over *bright targets* such as snow and ice, where it has proved difficult to retrieve the aerosol *single-scattering albedo* from radiance-only spectrometers such as *MERIS* and *MODIS*.

6.7.2 Goals of the Study

The goals of this study were to use a vector radiative transfer model for the coupled atmosphere–surface system in conjunction with optimal estimation to (i) quantify how polarization measurements can be used to overcome the uniqueness problems associated with radiance-only retrieval of aerosol parameters, and (ii) explore how future instruments, which unlike MERIS and MODIS would measure the *Stokes parameters* Q and U in addition to the radiance I, could be used to enhance our ability to retrieve accurate aerosol parameters over turbid coastal waters and bright targets such as snow and ice.

Using a *bimodal mixture* of aerosols with one nonabsorbing population of particles (sea-salt type) and another absorbing population (soot type), our aim was to investigate to what extent it is possible to retrieve the optical depth ($\tau_{\lambda_{ref}}$), the relative fraction of absorbing particles (f), and the vertical distribution of the aerosol components (Δz_i).

6.7.3 Study Design

To simplify the study, we created a synthetic dataset by randomly varying the following input parameters to our vector radiative transfer code (C-VDISORT):

θ_0, θ, $\Delta\phi$, τ_{ref}, f, Δz_i $(i = 1, 2)$, where θ_0 is the solar zenith angle (fixed at 30°), θ is the sensor polar viewing angle, varying in the range [30°, 60°], and $\Delta\phi$ is the azimuthal angle between the sun and the sensor, varying in the range [120°, 150°]. The set τ_{ref}, f, Δz_i $(i = 1, 2)$ represented our *retrieval parameters* (RPs), where τ_{ref} is the *aerosol optical depth* at λ_{ref}, varying in the range [0.001, 0.5], f is the bimodal aerosol fraction, varying in the range [0,1], and Δz_i $(i = 1, 2)$ is the location of absorbing aerosols, either between 0 and 2 km [Δz_1] or between 2 and 4 km [Δz_2].

We used a modified version of the SeaDAS aerosol models (see Figure 64), which are based on *AERONET* data [106], and selected a bimodal aerosol combination in which a sea salt particle represented a nonabsorbing aerosol type, and a **soot** particle represented an **absorbing** aerosol type. Specifically, our question was: from simulated "MODIS" data of I, Q and U, would we be able to infer the aerosol optical depth τ_{ref}, fraction f, and location Δz_i ($i = 1$ or 2) of absorbing aerosols?

We considered a bimodal aerosol mixture consisting of a total of N particles per unit volume in a layer of thickness Δz: $N = N_a + N_c$, where N_a and N_c are concentrations of absorbing and nonabsorbing particles, respectively. To compute aerosol IOPs, we defined $\beta_{n,j}$ = *scattering cross section*, $\alpha_{n,j}$ = *absorption cross section*, and $\gamma_{n,j} = \beta_{n,j} + \alpha_{n,j}$ = *extinction cross section*, where $j = a$ stands for "**absorbing**", and $j = c$ stands for "nonabsorbing". We used standard mixing rules to create IOPs for the *bimodal mixture* of aerosols [see Eqs. (85)–(87)]. Thus, weighting by number concentrations was used to combine the absorption and scattering cross sections, and the moments of the *scattering phase matrix* elements, so that the IOPs of the mixture were (subscript m stands for mixture)

$$\Delta\tau_m = \gamma_m \Delta z = \left[N_a\gamma_{n,a} + N_c\gamma_{n,c}\right]\Delta z = [\gamma_a + \gamma_c]\Delta z = \Delta\tau_a + \Delta\tau_c \tag{924}$$

$$N_a = fN, \qquad N_c = (1-f)N, \qquad N = N_a + N_c \tag{925}$$

$$\gamma_a = \gamma_{n,a}N_a, \qquad \gamma_c = \gamma_{n,c}N_c \qquad \gamma_m = \gamma_a + \gamma_c \tag{926}$$

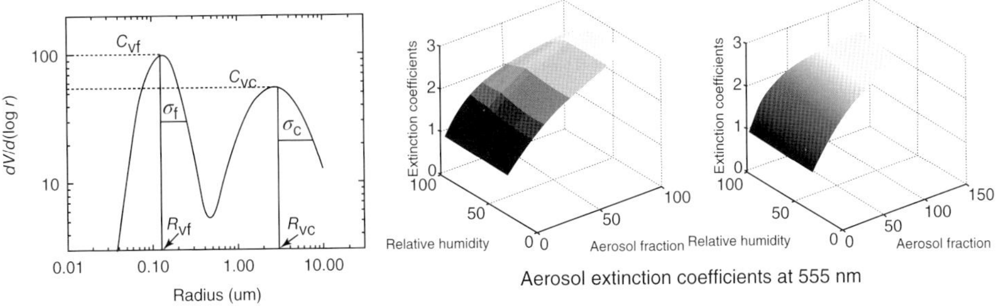

Figure 64 Left: Schematic illustration of the two (coarse and fine) aerosol modes. Middle: Ten different aerosol fractions: f = 0, 1, 2, 5, 10, 20, 30, 50, 80, 95%, and eight different relative humidities: RH = 30, 50, 70, 75, 80, 85, 90, 95%. Right: "Continuum" of models obtained by interpolation between the discrete ones.

$$\varpi_m = \frac{\varpi_a \gamma_a + \varpi_c \gamma_c}{\gamma_m} = \frac{f \varpi_a \gamma_{n,a} + (1-f)\varpi_c \gamma_{n,c}}{f \gamma_{n,a} + (1-f)\gamma_{n,c}} \quad (927)$$

$$\chi_{m,\ell} = \frac{f \varpi_a \gamma_{n,a} \chi_{a,\ell} + (1-f)\varpi_c \gamma_{n,c} \chi_{c,\ell}}{f \beta_{n,a} + (1-f)\beta_{n,c}} \quad (928)$$

where $\Delta\tau_m$ = *aerosol optical depth*; γ_m = *extinction coefficient*; ϖ_m = *single-scattering albedo*; $\chi_{m,\ell}$ = *scattering phase function expansion coefficient*; and f = the fraction of absorbing particles. A mixing rule similar to Eq. (928) was used for each element of the *scattering phase matrix*.

6.7.4
Forward Model

Since the most time-consuming step in the inversion process was the C-VDISORT forward model computations, we used C-VDISORT to train a *radial basis function neural network (RBF-NN)*, as discussed in Sections (3.5.2) and (6.5), to obtain a speed enhancement on the order of 1000. Thus, the C-VDISORT forward model (thousands of lines of code) was replaced with the following single RBF-NN equation:

$$p_i = \sum_{j=1}^{N} a_{ij} \exp[-b^2 \sum_{k=1}^{K} (c_{jk} - R_k)^2] + d_i. \quad (929)$$

Since our goal was to use the RBF-NN as a fast forward model, the input parameters R_k are the *state parameters* and the solar/viewing geometry, and the output parameters p_i are the desired TOA *Stokes parameters, I,* and Q at nine different wavelengths, in our case. A comparison of C-VDISORT versus RBF-NN results for the Stokes parameter I is given in Figure 65 and for the *Stokes parameter Q* in Figure 66. The performance of the *neural network* is seen to be quite good, with correlations greater than 0.9995 across all nine channels for both the I and Q Stokes parameters.

6.7.5
Optimal estimation/Inverse model

Our goal was to explore the retrieval feasibility using C-VDISORT/RBF-NN and *Optimal Estimation/Levenberg–Marquardt* (OE/LM) inversion with a state vector consisting of three aerosol parameters: the optical depth τ_{ref} at λ_{ref}, the bimodal fraction f of absorbing particles, and the location Δz_i of the absorbing aerosols.

Hence, the state vector was

$$\mathbf{x} = \{\tau_{\text{ref}}, f, \Delta z_i \ (i = 1 \text{ or } 2)\}. \quad (930)$$

For simplicity, the ocean was fixed to consist of pure sea water, although embedded *impurities* could be added: three marine parameters (the *chlorophyll concentration*

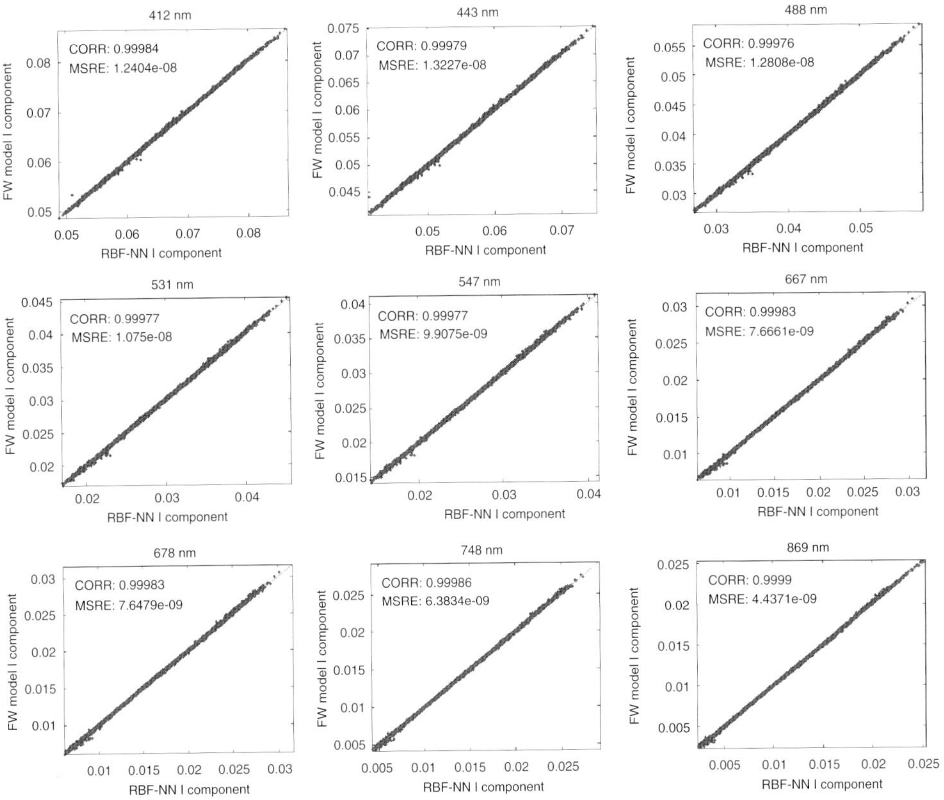

Figure 65 Comparison of RBF-NN forward model results for *I* with C-VDISORT results (after Stamnes *et al.* [293] with permission).

CHL, the *absorption coefficient CDM* at 443 nm due to detrital and dissolved material, and the *backscattering coefficient BBP* at 443 nm due to suspended particles).

To find the "best" answer from the simulated measurements of I, Q, and U, we employed OE/LM inversion, implying that in each iteration the next estimate of the state vector was given by

$$\mathbf{x}_{i+1} = \mathbf{x}_i + [(1+\gamma_i)\mathbf{S}_a^{-1} + \mathbf{J}_i^T \mathbf{S}_m^{-1} \mathbf{J}_i]^{-1} \times \{\mathbf{J}_i^T \mathbf{S}_m^{-1}(\mathbf{y}_m - \mathbf{y}_i) - \mathbf{S}_a^{-1}(\mathbf{x}_i - \mathbf{x}_a)\} \tag{931}$$

where $\mathbf{y}_m$ and $\mathbf{y}_i$ are the actual and simulated measurements, and $\mathbf{x}_a$ and $\mathbf{S}_a$ are the *a priori* state vector and covariance matrix, respectively. $\mathbf{S}_m$ is the measurement error covariance matrix, γ_i is the *Levenberg–Marquardt parameter*, which, as $\gamma_i \to 0$, implies *Gauss–Newton optimal estimation.*

Note that C-VDISORT/RBF-NN returned simulated *Stokes parameters* ($\mathbf{y}_i$) and *Jacobians* ($\mathbf{J}_i$) as required to update the state vector estimate ($\mathbf{x}_i$) according to Eq. (931).

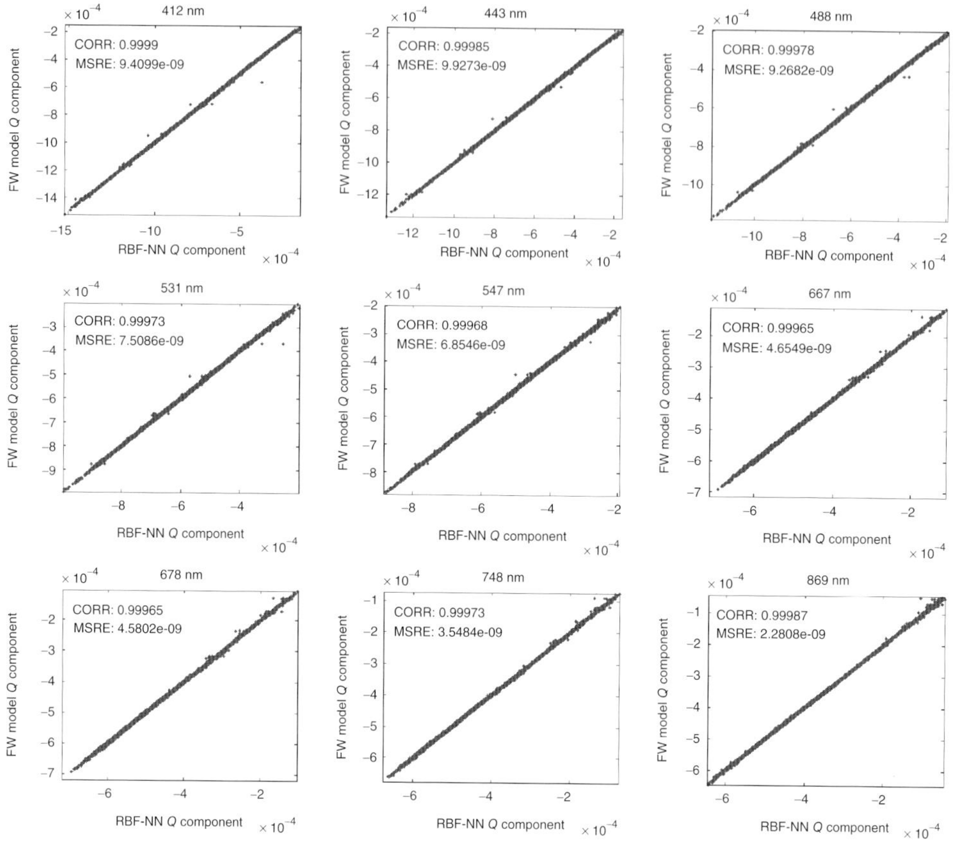

Figure 66 Comparison of RBF-NN forward model results for Q with C-VDISORT results (after Stamnes *et al.* [293] with permission).

6.7.6
Results

In order to explore the advantage of having access to polarization information, we created a *synthetic dataset* of simulated TOA Stokes parameters I, Q, and U for a range of aerosol optical depths (AODs), and fractions f of absorbing versus nonabsorbing aerosol particles embedded in a background molecular atmosphere using a fast and accurate RBF-NN forward model, as discussed above. We then used an *optimal estimation* scheme for retrieval of AOD, fraction f, and location of the absorbing aerosols to obtain the results shown in Figures 67–72.

Figure 67 shows a comparison of retrievals obtained from two independent implementations of the optimal estimation algorithm, one of which was programmed in FORTRAN, while the other was done in C++ using the Armadillo library for matrix operations [293]. Both implementations used the I as well as Q Stokes parameters and an intelligent first guess obtained by training a neural

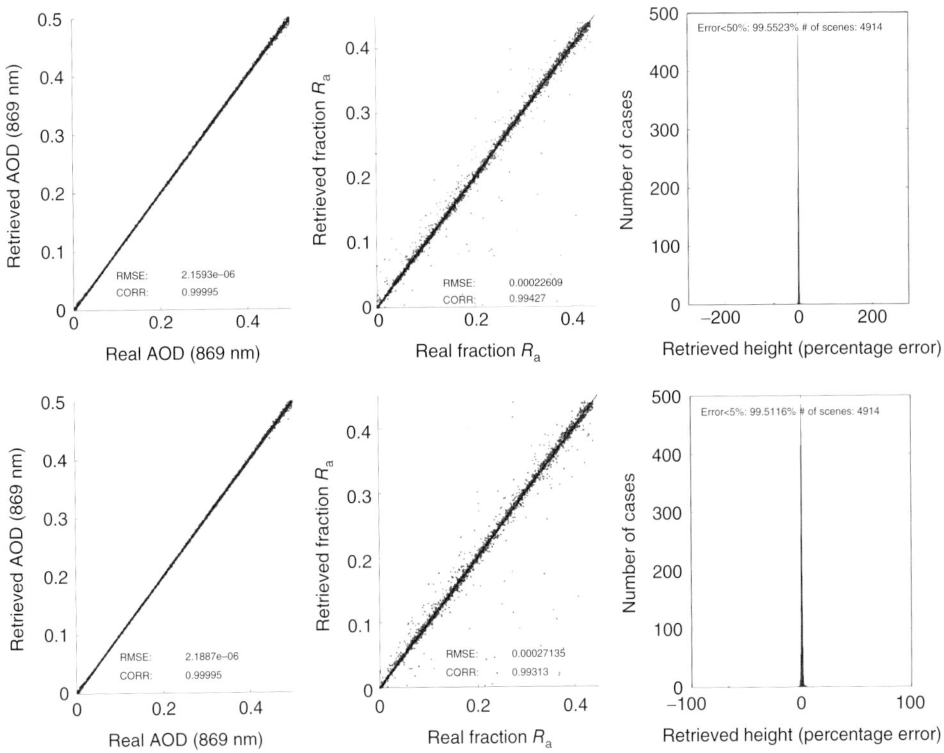

Figure 67 Comparison of two different implementations for a synthetic retrieval for MODIS assuming the instrument could measure the Stokes parameter Q in addition to I (after Stamnes *et al.* [293] with permission).

network to go directly from the measurements to the *retrieval parameters*. In both implementations, the prior $\mathbf{x}_a$ was set equal to this intelligent first guess. There is good agreement between the two implementations, and we use the bottom curves in Figure 67 as a **benchmark** in the following comparisons. With our intelligent first guess, we were able to retrieve optical depth, fraction, and location. Thus, the result in Figure 67 was considered to be our "best" result achievable, and is plotted in the bottom panel of the following figures as a reference. Note that in the following figures the prior is always set to be equal to the first guess (intelligent or not), and the *a priori covariance matrix* $\mathbf{S}_a$ is assumed to be a *Gaussian* [37] diagonal matrix with equal elements, each being a function of the *a priori* vector $\mathbf{x}_a$:

$$\mathbf{S}_a = (10\,\mathbf{x}_a)^2\,\mathbf{I}. \tag{932}$$

Thus, the *a priori covariance matrix* $\mathbf{S}_a$ has an assumed *variance* of $(10\,\mathbf{x}_a)^2$ with all nondiagonal elements set equal to 0.

Figure 68 is based on using both I and Q, as in Figure 67, but not an intelligent first guess. Instead, the first guess was assumed to be $\mathbf{x}_1 = (0.03, 0.2, 2.0)$, and the

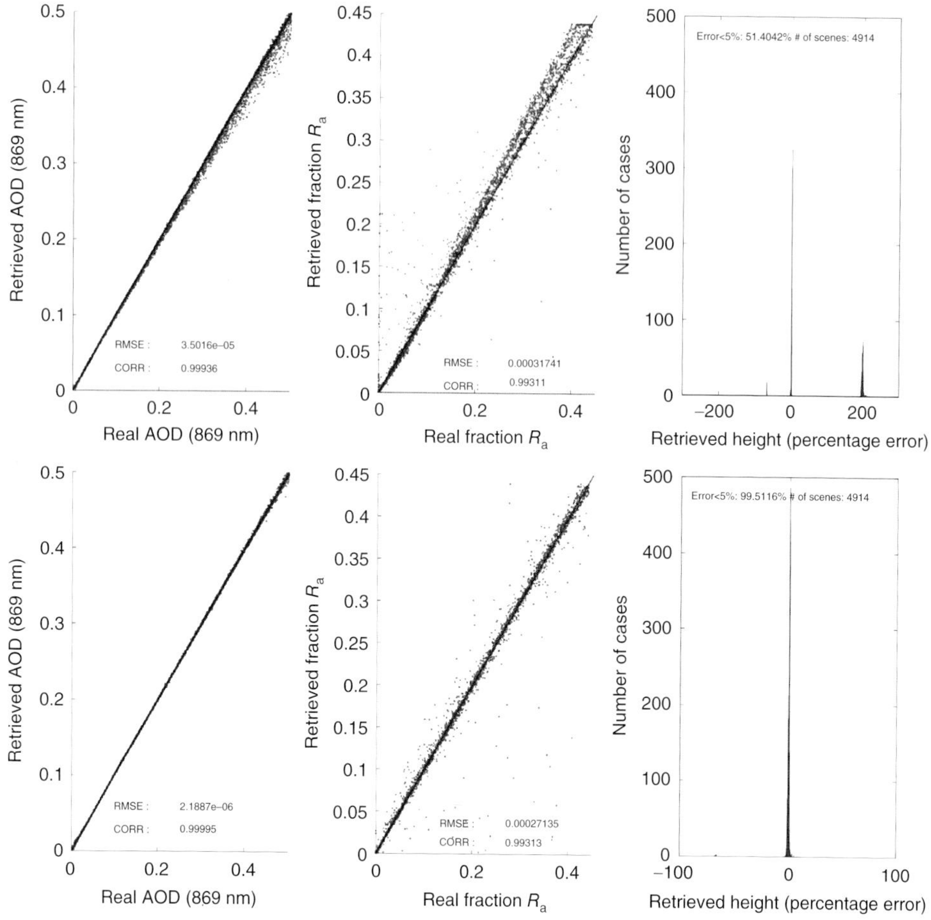

Figure 68 Top, same as Figure 67 **except with a prior assumed to be $\mathbf{x}_a = \mathbf{x}_1 =$** $(0.03, 0.2, 2.0)$ versus benchmark retrieval (bottom). A retrieval that makes use of both I and Q as well as a prior set equal to the assumed first guess appears to work reasonably well for retrieving optical depth and fraction, but not location (after Stamnes *et al.* [293] with permission).

prior $\mathbf{x}_a$ was set equal to the first guess. Also in this case the result is seen to be quite good. We were able to retrieve both the *aerosol optical depth* and the fraction, and, in most cases, also the location.

Figure 69 is based on performing the retrieval using only the total radiance I in the OE/LM optimization scheme, but on using the intelligent first guess obtained from training using both I and Q. The strong performance of the *optimal estimation* scheme without the use of Q may be surprising, but it makes sense because, when the starting point is close to the right answer, the intensity will converge. Hence, as long as the first guess provides a good estimate, the retrieval is expected to work reasonably well.

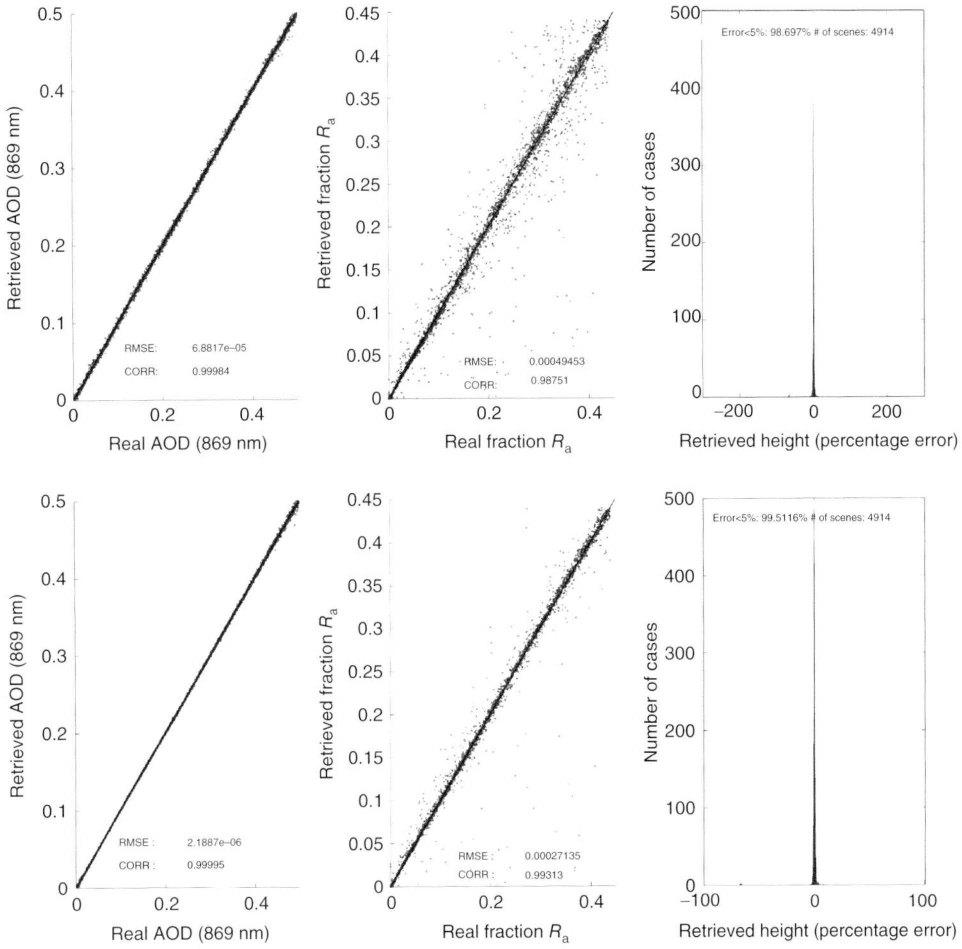

Figure 69 Comparison of *intensity-only* retrieval (top) versus benchmark retrieval (bottom). Identical to Figure 67 except that only *I* was used in the retrieval, not *Q*. **Basing our first guess on the use of both *I* and** *Q*, we were also able to retrieve optical depth, fraction, and location when using *intensity only* in the OE/LM retrieval step! (after Stamnes *et al.* [293] with permission).

Figure 70 shows what would happen in the absence of an intelligent first guess. Instead, an assumed first guess was used. In contrast with Figure 68, which also was based on the use of an assumed first guess, it is seen that the additional polarization information is a big help for cases in which the first guess is not very good. A comparison with Figure 69, which compares favorably with the benchmark (bottom), shows that an accurate first guess is very important if radiance alone is used in the retrieval step.

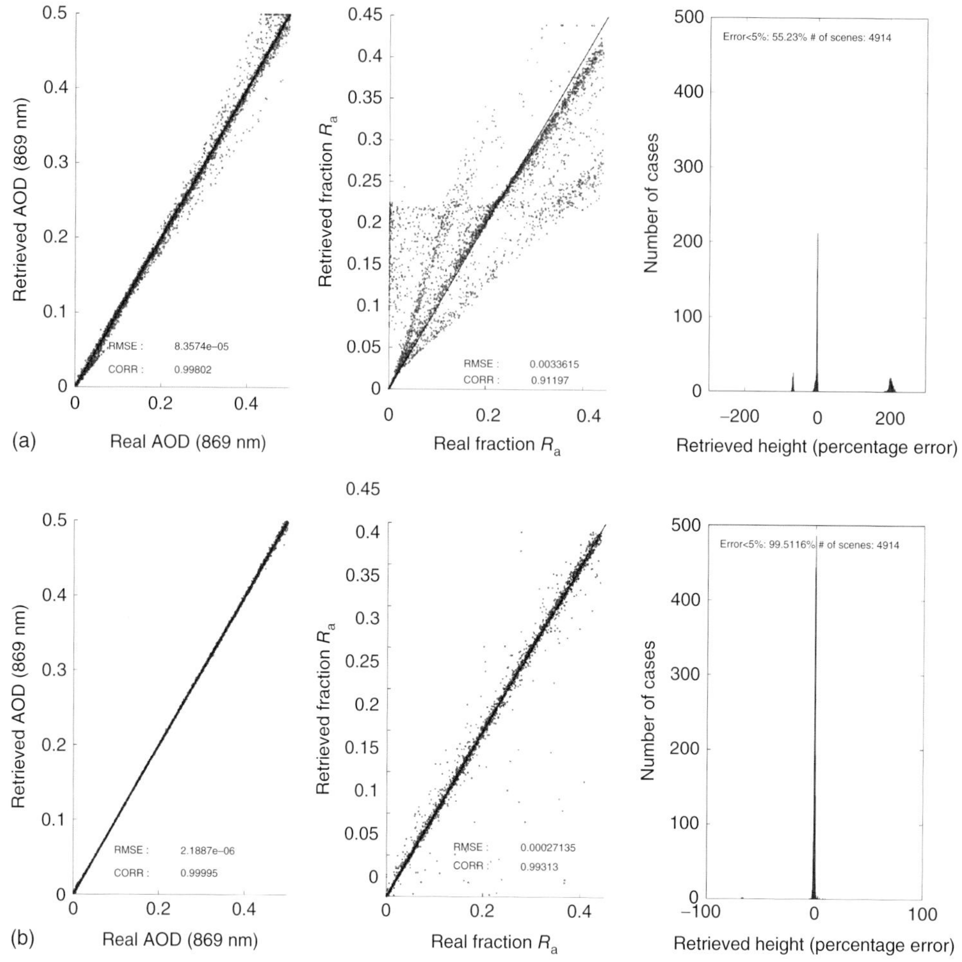

Figure 70 Comparison of *intensity-only* retrieval with an **assumed first guess** (a) versus benchmark retrieval (b). Identical to Figure 68 except only *I* was used in the retrieval, not *Q*. Identical to Figure 69 except that an assumed first guess was used (after Stamnes *et al.* [293] with permission).

Figure 71 demonstrates what would happen if the *neural network forward model* was used to obtain the intelligent first guess using only the total radiance *I*, while ignoring *Q*. In Figures 69 and 70, the *Q* component was used in the training but ignored in the OE/LM retrieval. Figure 71 shows that, when one trains the neural network forward model using the intensity alone produced by the C-VDISORT, the results obtained by ignoring the *Q* component in the *OE/LM* retrieval step seem comparable to those obtained when both *I* and *Q* are used in the OE/LM retrieval step. However, the correlation for the retrieved fraction of absorbing aerosols is slightly lower than when both *I* and *Q* are used.

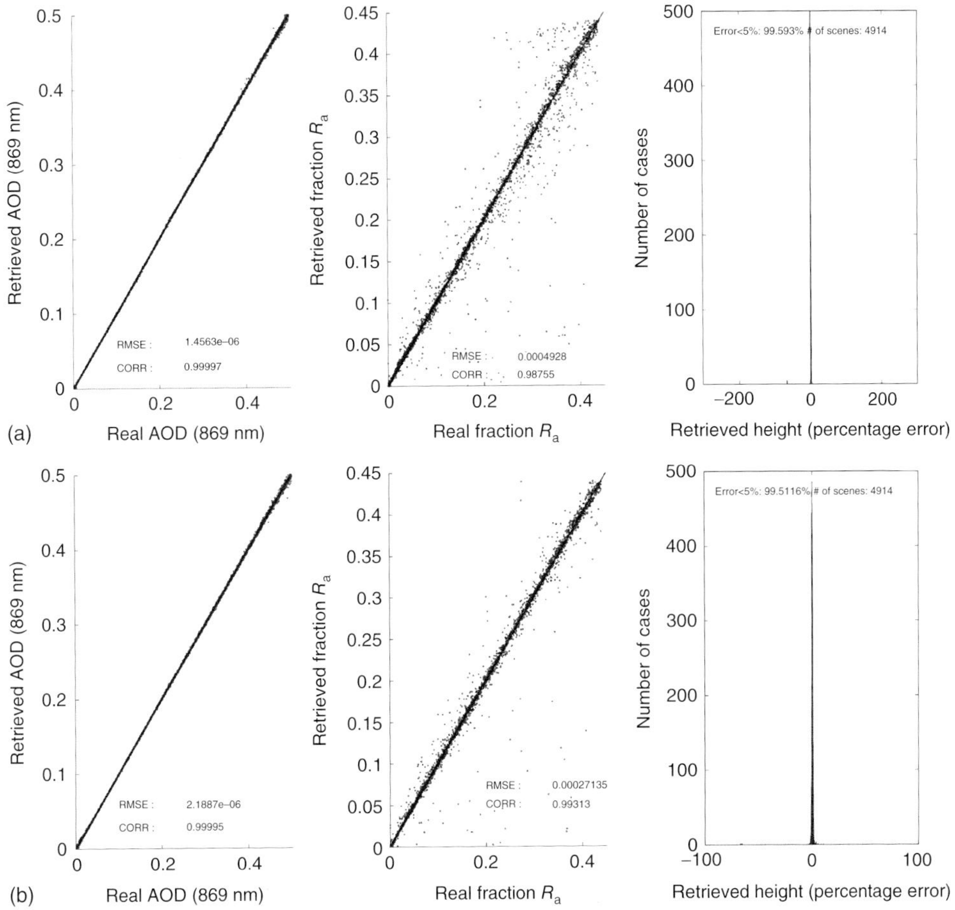

Figure 71 Comparison of *intensity-only* retrieval based on a *fast forward model* trained using only the *I* component of the Stokes vector computed by the C-VDISORT radiative transfer model (a) versus benchmark retrieval (b) (after Stamnes *et al.* [293] with permission).

A comparison of the results shown in Figure 71 with those in Figure 69 shows that the retrievals appear to be comparable whether or not the training of the neural network *fast forward model* is based on using both *I* and *Q* or just *I*. Figure 72 was based on using only the radiance computed by C-VDISORT as input to the neural network training of the fast forward model, but using an assumed first guess. The retrieval results in Figure 72 are seen to be quite poor, despite the fact that the fast forward model trained by using only the radiance (produced by C-VDISORT) produces retrievals that compare favorably to those produced by using both *I* and *Q* to train the fast forward model. Hence, the quality of the first guess appears to very important. Note that the input for producing Figure 72 was identical to that used for producing Figure 70 except that the former (top) was based on using only *I* to train the fast forward model.

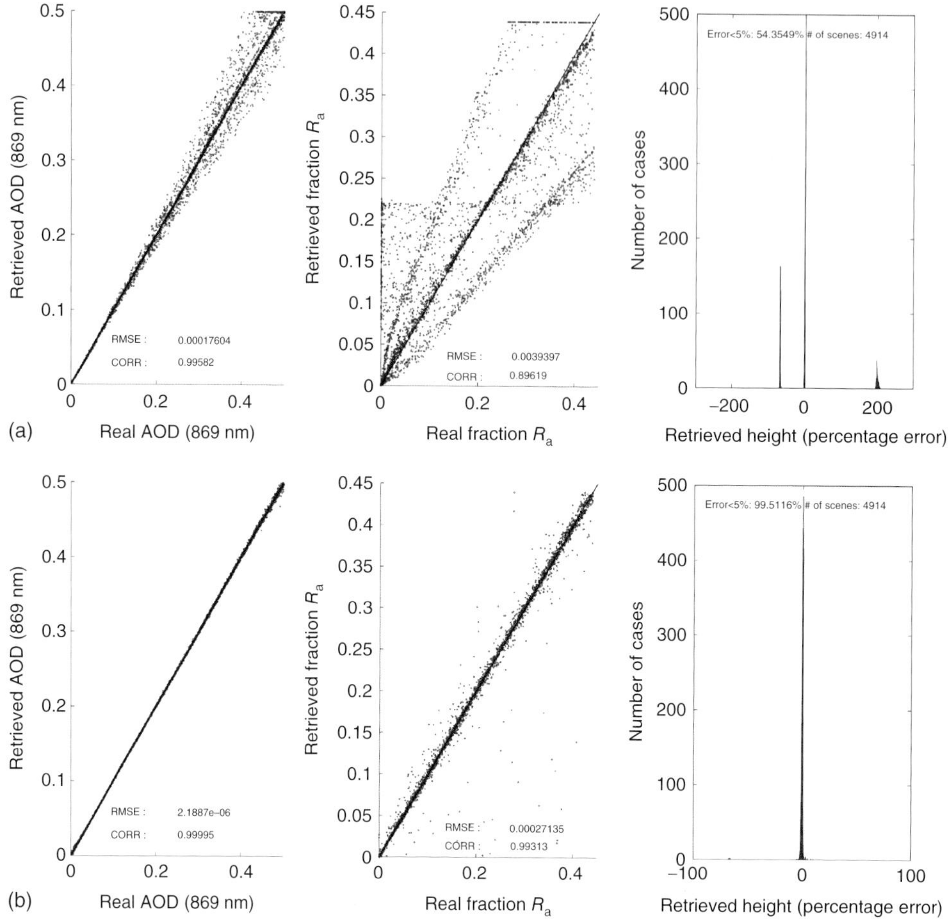

Figure 72 Comparison of *intensity-only* retrieval based on a fast forward model trained using only the *I* component of the Stokes vector computed by the C-VDISORT radiative transfer model and an **assumed first guess** (a) versus benchmark retrieval (b) (after Stamnes *et al.* [293] with permission).

6.7.7
Concluding Remarks

In summary, it appears that use of radiance alone (the *I* component) is generally insufficient to retrieve accurate values of the three *retrieval parameters*: the *aerosol optical depth*, the fraction of absorbing particles, and the location of the particles. However, use of an accurate first guess based on a direct inversion resulting from accurate forward model simulations that include polarization seems to help significantly in that retrievals based on radiance only (computed by a *vector radiative transfer model*) then yield adequate results. This finding suggests that use of accurate forward model simulations of the polarized radiation could improve retrievals based on existing *optimal estimation* schemes using radiance-only measurements with little modification, whereas use of Q (if available) in addition to I would lead to improved retrievals of all three aerosol parameters.

A
Scattering of Electromagnetic Waves

To study scattering of electromagnetic waves, one must start with Maxwell's equations, which in Gaussian units are

$$\nabla \times \mathbf{E} = -\frac{1}{c}\frac{\partial \mathbf{B}}{\partial t} \tag{933}$$

$$\nabla \times \mathbf{H} = \frac{1}{c}\frac{\partial \mathbf{D}}{\partial t} + \frac{4\pi}{c}\mathbf{J} = \frac{4\pi}{c}(\mathbf{J}_0 + \sigma\mathbf{E}) \tag{934}$$

$$\nabla \cdot \mathbf{D} = 4\pi\rho \tag{935}$$

$$\nabla \cdot \mathbf{B} = 0. \tag{936}$$

Taking the divergence of Eq. (934) and the temporal derivative of Eq. (935), and using that $\nabla \cdot (\nabla \times \mathbf{A}) = 0$ for an arbitrary vector $\mathbf{A}$, one obtains the continuity equation

$$\nabla \cdot (\mathbf{J}_0 + \sigma\mathbf{E}) + \frac{\partial \rho}{\partial t} = 0. \tag{937}$$

For an isotropic, nondispersive material, the constitutive relations are given by

$$\mathbf{D} = \varepsilon\mathbf{E} \;\; ; \;\; \mathbf{B} = \mu\mathbf{H} \tag{938}$$

and for a time-harmonic field

$$A(\mathbf{r}, t) = \Re\{A(\mathbf{r})e^{-i\omega t}\} \tag{939}$$

where A stands for ρ or any scalar component of $\mathbf{E}$, $\mathbf{D}$, $\mathbf{B}$, $\mathbf{H}$, and $\mathbf{J}$.

In a uniform, time-independent medium, ε, μ, and σ do not vary with position or time. Thus, in source-free regions of space, where $\mathbf{J}_0 = 0$, the time-harmonic form of Maxwell's equations in a uniform, isotropic, time-independent medium become

$$\nabla \times \mathbf{E} = \frac{i\omega}{c}\mu\,\mathbf{H} \tag{940}$$

$$\nabla \times \mathbf{H} = -\frac{i\omega}{c}\left(\varepsilon + 4\pi i\frac{\sigma}{\omega}\right)\mathbf{E} \tag{941}$$

$$\nabla \cdot \mathbf{E} = 0 \tag{942}$$

$$\nabla \cdot \mathbf{H} = 0. \tag{943}$$

Taking the curl of Eqs. (940) and (941), we obtain the vector wave equations

$$\nabla \times (\nabla \times \mathbf{E}) - k^2 = \nabla \times (\nabla \times \mathbf{H}) - k^2 = 0, \tag{944}$$

Radiative Transfer in Coupled Environmental Systems: An Introduction to Forward and Inverse Modeling,
First Edition. Knut Stamnes and Jakob J. Stamnes.

where

$$k^2 = \frac{\omega^2}{c^2}\mu\left(\varepsilon + 4\pi i\frac{\sigma}{\omega}\right). \tag{945}$$

By using Eqs. (942) and (943) and the vector identity

$$\nabla \times (\nabla \times \mathbf{A}) = \nabla(\nabla \cdot \mathbf{A}) - \nabla^2\mathbf{A} = 0 \tag{946}$$

each of the vector wave equations becomes equal to the Helmholtz equation:

$$(\nabla^2 + k^2)\mathbf{E} = (\nabla^2 + k^2)\mathbf{H} = 0. \tag{947}$$

Consider an electromagnetic field $(\mathbf{E}^i, \mathbf{H}^i)$ that is incident upon a finite-size particle and generates an internal field $(\mathbf{E}^{\text{int}}, \mathbf{H}^{\text{int}})$ inside the scatterer as well as a scattered field $(\mathbf{E}^{\text{sc}}, \mathbf{H}^{\text{sc}})$ outside the scatterer. The *scattering problem* may be formulated as follows:

1) The incident field $(\mathbf{E}^i, \mathbf{H}^i)$ is known and the internal field $(\mathbf{E}^{\text{int}}, \mathbf{H}^{\text{int}})$ and the scattered field $(\mathbf{E}^{\text{sc}}, \mathbf{H}^{\text{sc}})$ are to be determined.
2) The scattered field $(\mathbf{E}^{\text{sc}}, \mathbf{H}^{\text{sc}})$ must satisfy the vector wave equations Eqs. (947) with $k = k_2$.
3) The internal field $(\mathbf{E}^{\text{int}}, \mathbf{H}^{\text{int}})$ must satisfy the vector wave equations Eqs. (947) with $k = k_1$.
4) The scaterred field $(\mathbf{E}^{\text{sc}}, \mathbf{H}^{\text{sc}})$ must satisfy the radiation condition at infinity, that is, it must behave as an outgoing spherical wave.
5) The internal field $(\mathbf{E}^{\text{int}}, \mathbf{H}^{\text{int}})$ must be finite at the origin.
6) The electromagnetic field must satisfy certain boundary conditions at the surface of the scatterer:
   (a) If σ_1 is finite, then the tangential components of $\mathbf{E}$ and $\mathbf{H}$ must be continuous across the interface between the scatterer and the surrounding medium.
   (b) If $\sigma_1 = \infty$, the material of the scatterer is a perfect conductor, which is also a perfect reflector, and the tangential component of $\mathbf{E}$ must vanish at the surface of the scatterer.

A.1 Absorption and Scattering by a Particle of Arbitrary Shape

A.1.1 General Formulation

Consider a particle with material properties ϵ_1, μ_1, σ_1 that is embedded in a medium with material properties ϵ_2, μ_2, $\sigma_2 = 0$, and let a time-harmonic plane electromagnetic wave given by

$$\mathbf{E}^i = \Re\left\{\mathbf{E}_0^i e^{i(k_2 z - \omega t)}\right\} \;;\; \mathbf{H}^i = \Re\left\{\mathbf{H}_0^i e^{i(k_2 z - \omega t)}\right\} \;;\; k_2^2 = \frac{\omega_2^2}{c^2}\epsilon_2\mu_2 \tag{948}$$

be incident upon the particle and give rise to a scattered field ($\mathbf{E}^{sc}, \mathbf{H}^{sc}$) and a field ($\mathbf{E}^{int}, \mathbf{H}^{int}$) inside the particle. From Maxwell's equations

$$\nabla \cdot \mathbf{E} = 0 \quad ; \quad \nabla \cdot \mathbf{H} = 0 \tag{949}$$

$$\nabla \times \mathbf{E} = i\frac{\omega}{c}\mu\mathbf{H} \quad ; \quad \nabla \times \mathbf{H} = -i\frac{\omega}{c}\epsilon\mathbf{E} \tag{950}$$

it follows that $\mathbf{E}$ and $\mathbf{H}$ must satisfy the vector wave equation

$$\left(\nabla^2 + k^2\right)\mathbf{E} = \left(\nabla^2 + k^2\right)\mathbf{H} = 0, \tag{951}$$

where $k^2 = k_2^2 = (\omega^2/c^2)\epsilon_2\mu_2$ outside the particle and $k^2 = k_1^2 = (\omega^2/c^2)\mu_1 \left(\varepsilon_1 + 4\pi i(\sigma_1/\omega)\right)$ inside the particle. At the surface of the particle, the boundary conditions are

$$\left(\mathbf{E}^i + \mathbf{E}^{sc} - \mathbf{E}^{int}\right) \times \hat{\mathbf{n}} = \left(\mathbf{H}^i + \mathbf{H}^{sc} - \mathbf{H}^{int}\right) \times \hat{\mathbf{n}} = 0, \tag{952}$$

where $\hat{\mathbf{n}}$ is a unit vector along the outward surface normal, and the electric field in the medium surrounding the particle is the sum of the incident electric field $\mathbf{E}^i$ and the scattered electric field $\mathbf{E}^{sc}$.

A.1.2
Amplitude Scattering Matrix

One may decompose the electric field into components parallel and perpendicular to the *scattering plane*, which is spanned by the direction of the incident plane wave (along $\hat{\mathbf{e}}_z$) and the observation direction (along $\hat{\mathbf{e}}_r$). Thus, for linear polarization, the electric field amplitude of the incident plane wave becomes

$$\mathbf{E}_0^i = \left(E_{0\parallel}\hat{\mathbf{e}}_{\parallel i} + E_{0\perp}\hat{\mathbf{e}}_{\perp i}\right) = E_\parallel^i \hat{\mathbf{e}}_{\parallel i} + E_\perp^i \hat{\mathbf{e}}_{\perp i}, \tag{953}$$

where $k_2 = (2\pi/\lambda)n_2$, with λ and $n_2 = \sqrt{\varepsilon_2\mu_2}$ being, respectively, the wavelength in vacuum and the refractive index in the surrounding medium.

Let the projection of $\hat{\mathbf{e}}_r$ onto the plane $z = 0$ make an angle ϕ with the x-axis. Since this projection is along $\hat{\mathbf{e}}_{\parallel i}$, it follows that

$$\hat{\mathbf{e}}_{\parallel i} = \cos\phi\hat{\mathbf{e}}_x + \sin\phi\hat{\mathbf{e}}_y \; ; \; \hat{\mathbf{e}}_{\perp i} = \sin\phi\hat{\mathbf{e}}_x - \cos\phi\hat{\mathbf{e}}_y \; ; \; \hat{\mathbf{e}}_{\perp i} \times \hat{\mathbf{e}}_{\parallel i} = \hat{\mathbf{e}}_z \tag{954}$$

$$\hat{\mathbf{e}}_{\parallel i} = \cos\theta\hat{\mathbf{e}}_\theta + \sin\theta\hat{\mathbf{e}}_r \; ; \; \hat{\mathbf{e}}_{\perp i} = -\hat{\mathbf{e}}_\phi \tag{955}$$

$$E_\parallel^i = E_x^i(\hat{\mathbf{e}}_{\parallel i} \cdot \hat{\mathbf{e}}_x) + E_y^i(\hat{\mathbf{e}}_{\parallel i} \cdot \hat{\mathbf{e}}_y) = E_x^i \cos\phi + E_y^i \sin\phi \tag{956}$$

$$E_\perp^i = E_x^i(\hat{\mathbf{e}}_{\perp i} \cdot \hat{\mathbf{e}}_x) + E_y^i(\hat{\mathbf{e}}_{\perp i} \cdot \hat{\mathbf{e}}_y) = E_x^i \sin\phi - E_y^i \cos\phi. \tag{957}$$

In the *far zone*, where $k_2 r >> 1$, the expression for the scattered electric field becomes

$$\mathbf{E}^{sc} \sim i\frac{e^{ik_2 r}}{k_2 r}\mathbf{X} \quad ; \quad \mathbf{X} \cdot \hat{\mathbf{e}}_r = 0 \quad ; \quad \mathbf{H}^{sc} = \frac{c}{\omega\mu_2}\mathbf{k}^{sc} \times \mathbf{E}^{sc} \sim i\frac{ck_2}{\omega\mu_2}\frac{e^{ik_2 r}}{k_2 r}\hat{\mathbf{e}}_r \times \mathbf{X}, \tag{958}$$

where

$$\mathbf{E}^{sc} = E_\parallel^{sc}\hat{\mathbf{e}}_{\parallel s} + E_\perp^{sc}\hat{\mathbf{e}}_{\perp s} \; ; \; \hat{\mathbf{e}}_{\parallel s} = \hat{\mathbf{e}}_\theta \; ; \; \hat{\mathbf{e}}_{\perp s} = -\hat{\mathbf{e}}_\phi \; ; \; \hat{\mathbf{e}}_{\perp s} \times \hat{\mathbf{e}}_{\parallel s} = \hat{\mathbf{e}}_r. \tag{959}$$

Because Maxwell's equations (949)–(950) and the boundary conditions in (952) are linear, the scattered electric field components $E_{\parallel}^{\rm sc}$ and $E_{\perp}^{\rm sc}$ must be linearly related to the incident electric field components $E_{\parallel}^{i}$ and $E_{\perp}^{i}$, that is,

$$\begin{pmatrix} E_{\parallel}^{\rm sc} \\ E_{\perp}^{\rm sc} \end{pmatrix} = i\frac{e^{ik_2 r}}{k_2 r}\begin{pmatrix} S_2 & S_3 \\ S_4 & S_1 \end{pmatrix}\begin{pmatrix} E_{\parallel}^{i} \\ E_{\perp}^{i} \end{pmatrix}, \tag{960}$$

where the elements S_j ($j = 1, 2, 3, 4$) of the *amplitude scattering matrix* depend on the *scattering angle* Θ and the *azimuth angle* ϕ. From (958) and (960), it follows that

$$\mathbf{X} = (S_2 E_{\parallel}^{i} + S_3 E_{\perp}^{i})\hat{\mathbf{e}}_{\parallel s} + (S_4 E_{\parallel}^{i} + S_1 E_{\perp}^{i})\hat{\mathbf{e}}_{\perp s}. \tag{961}$$

A.1.3
Scattering Matrix

The time-averaged Poynting vector of the field in the medium surrounding the particle is

$$\mathbf{S}_2 = \frac{c}{8\pi}\Re\{\mathbf{E}_2 \times \mathbf{H}_2^*\} = \mathbf{S}^{\rm i} + \mathbf{S}^{\rm sc} + \mathbf{S}^{\rm ext}, \tag{962}$$

where

$$\mathbf{E}_2 = \mathbf{E}^i + \mathbf{E}^{\rm sc} \quad ; \quad \mathbf{H}_2 = \mathbf{H}^i + \mathbf{H}^{\rm sc} \tag{963}$$

so that

$$\mathbf{S}^i = \frac{c}{8\pi}\Re\{\mathbf{E}^i \times (\mathbf{H}^i)^*\} \tag{964}$$

$$\mathbf{S}^{\rm sc} = \frac{c}{8\pi}\Re\{\mathbf{E}^{\rm sc} \times (\mathbf{H}^{\rm sc})^*\} \tag{965}$$

$$\mathbf{S}^{\rm ext} = \frac{c}{8\pi}\Re\{\mathbf{E}^i \times (\mathbf{H}^{\rm sc})^* + \mathbf{E}^{\rm sc} \times (\mathbf{H}^i)^*\} \tag{966}$$

where

- $\mathbf{S}^i$ = Poynting vector of the incident field,
- $\mathbf{S}^{\rm sc}$ = Poynting vector of the scattered field, and
- $\mathbf{S}^{\rm ext}$ = Poynting vector of the field that is due to the interaction between the incident and scattered fields.

For a scattering arrangement in which the detector "sees" only the scattered field, the recorded signal is proportional to

$$\mathbf{S}^{\rm sc} \cdot \hat{\mathbf{e}}_r \Delta A, \tag{967}$$

where ΔA is the detector area. From Eqs. (958) and (965), it follows that

$$\begin{aligned} \mathbf{S}^{\rm sc} \cdot \hat{\mathbf{e}}_r \Delta A &= \frac{c}{8\pi}\Re\left(\mathbf{E}^{\rm sc} \times (\mathbf{H}^{\rm sc})^*\right) \cdot \hat{\mathbf{e}}_r \Delta A \\ &= \frac{c}{8\pi}\Re\left(i\frac{e^{ik_2 r}}{k_2 r}\mathbf{X} \times (-i)\frac{ck_2}{\omega\mu_2}\frac{e^{-ik_2 r}}{k_2 r}\hat{\mathbf{e}}_r \times \mathbf{X}^*\right) \cdot \hat{\mathbf{e}}_r \Delta A \\ &= \frac{c}{8\pi}\frac{ck_2}{\omega\mu_2}|C|^2|\mathbf{X}|^2\Delta\Omega \ ; \ C = i\frac{e^{ik_2 r}}{k_2 r}, \end{aligned} \tag{968}$$

where $\Delta\Omega = (\Delta A/r^2)$ is the solid angle subtended by the detector.

For the time-harmonic incident electromagnetic plane wave given in Eq. (948) with $\mathbf{H}^i = (c/(\omega\mu_2))\mathbf{k}_2 \times \mathbf{E}^i$, the Poynting vector becomes

$$\mathbf{S}^i = \frac{c}{8\pi}\Re\left(\mathbf{E}^i \times (\mathbf{H}^i)^*\right) = \frac{c}{8\pi}\Re\left(\mathbf{E}^i \times \frac{c}{\omega\mu_2}\mathbf{k}_2 \times (\mathbf{E}^i)^*\right) = S^i\hat{\mathbf{s}}, \tag{969}$$

where $\hat{\mathbf{s}}$ is a unit vector in the direction of the Poynting vector and S^i is the magnitude of the Poynting vector of the incident plane wave given by

$$S^i = \frac{c}{8\pi}\frac{ck_2}{\omega\mu_2}|\mathbf{E}^i|^2 = \frac{c}{8\pi}\frac{ck_2}{\omega\mu_2}|E|^2. \tag{970}$$

Letting $\mathbf{X} = E\mathbf{X}'$, where E is the electric field amplitude of the incident plane wave, one obtains from Eq. (968)

$$\mathbf{S}^{\mathrm{sc}} \cdot \hat{\mathbf{e}}_r = S^i|C|^2|\mathbf{X}'|^2 \;\; ; \;\; C = i\frac{e^{ik_2 r}}{k_2 r}. \tag{971}$$

Thus, by recording the detector response at various positions on a hemisphere surrounding the particle, one can determine the angular variation of $|\mathbf{X}|^2$.

The Stokes parameters of the light scattered by the particle are

$$I^{\mathrm{sc}} = E_{\parallel}^{\mathrm{sc}}(E_{\parallel}^{\mathrm{sc}})^* + E_{\perp}^{\mathrm{sc}}(E_{\perp}^{\mathrm{sc}})^* \tag{972}$$

$$Q^{\mathrm{sc}} = E_{\parallel}^{\mathrm{sc}}(E_{\parallel}^{\mathrm{sc}})^* - E_{\perp}^{\mathrm{sc}}(E_{\perp}^{\mathrm{sc}})^* \tag{973}$$

$$U^{\mathrm{sc}} = E_{\parallel}^{\mathrm{sc}}(E_{\perp}^{\mathrm{sc}})^* + E_{\perp}^{\mathrm{sc}}(E_{\parallel}^{\mathrm{sc}})^* \tag{974}$$

$$V^{\mathrm{sc}} = i[E_{\parallel}^{\mathrm{sc}}(E_{\perp}^{\mathrm{sc}})^* - E_{\perp}^{\mathrm{sc}}(E_{\parallel}^{\mathrm{sc}})^*] \tag{975}$$

consistent with the representation $\mathbf{I}_S = [I, Q, U, V]^T$, with $I = I_{\parallel} + I_{\perp}$ and $Q = I_{\parallel} - I_{\perp}$, where $I_{\parallel}$, $I_{\perp}$, U, and V are given in Eq. (29). The *scattering matrix*, which by definition is the *Mueller* matrix for scattering by a single particle, follows from Eqs. (960) and (972)–(975):

$$\begin{pmatrix} I^{\mathrm{sc}} \\ Q^{\mathrm{sc}} \\ U^{\mathrm{sc}} \\ V^{\mathrm{sc}} \end{pmatrix} = \begin{pmatrix} S_{11} & S_{12} & S_{13} & S_{14} \\ S_{21} & S_{22} & S_{23} & S_{24} \\ S_{31} & S_{32} & S_{33} & S_{34} \\ S_{41} & S_{42} & S_{43} & S_{44} \end{pmatrix} \begin{pmatrix} I^i \\ Q^i \\ U^i \\ V^i \end{pmatrix}, \tag{976}$$

where

$$S_{11} = \frac{1}{2}\left(|\,S_1\,|^2 + |\,S_2\,|^2 + |\,S_3\,|^2 + |\,S_4\,|^2\right) \tag{977}$$

$$S_{12} = \frac{1}{2}\left(|\,S_2\,|^2 - |\,S_1\,|^2 + |\,S_4\,|^2 - |\,S_3\,|^2\right) \tag{978}$$

$$S_{13} = \Re\left(S_2 S_3^* + S_1 S_4^*\right) \tag{979}$$
$$S_{14} = \Im\left(S_2 S_3^* - S_1 S_4^*\right) \tag{980}$$
$$S_{21} = \frac{1}{2}\left(|S_2|^2 - |S_1|^2 - |S_4|^2 + |S_3|^2\right) \tag{981}$$
$$S_{22} = \frac{1}{2}\left(|S_2|^2 + |S_1|^2 - |S_4|^2 - |S_3|^2\right) \tag{982}$$
$$S_{23} = \Re\left(S_2 S_3^* - S_1 S_4^*\right) \tag{983}$$
$$S_{24} = \Im\left(S_2 S_3^* + S_1 S_4^*\right) \tag{984}$$
$$S_{31} = \Re\left(S_2 S_4^* + S_1 S_3^*\right) \tag{985}$$
$$S_{32} = \Re\left(S_2 S_4^* - S_1 S_3^*\right) \tag{986}$$
$$S_{33} = \Re\left(S_1 S_2^* + S_3 S_4^*\right) \tag{987}$$
$$S_{34} = \Im\left(S_2 S_1^* + S_4 S_3^*\right) \tag{988}$$
$$S_{41} = \Im\left(S_2^* S_4 + S_3^* S_1\right) \tag{989}$$
$$S_{42} = \Im\left(S_2^* S_4 - S_3^* S_1\right) \tag{990}$$
$$S_{43} = \Im\left(S_1 S_2^* - S_3 S_4^*\right) \tag{991}$$
$$S_{44} = \Re\left(S_1 S_2^* - S_3 S_4^*\right). \tag{992}$$

See Problem 2.8 for derivations of the relations given by Eqs. (977)–(992).

It can be shown that only 7 of the 16 elements S_{ij} are independent. Each of the Stokes parameters for scattering by a collection of randomly oriented particles is the sum of the corresponding Stokes parameters for light scattered by the individual particles. Hence, the scattering matrix for such a collection is the sum of the scattering matrices for the individual particles. In general, there will be 16 nonzero independent matrix elements for such a collection, but this number may be reduced because of symmetry, as discussed in Section 2.3.1.

A.1.4
Extinction, Scattering, and Absorption

Consider one or more particles that are placed in a beam of electromagnetic radiation. Then the electromagnetic energy measured by a detector situated in the forward scattering direction would be reduced compared to that measured in the absence of the particles. Thus, the presence of the particles would result in *extinction* of the incident beam. If the medium in which the particles are placed is nonabsorbing, the extinction must be due to *absorption* in the particles and *scattering* by the particles.

Consider now the extinction by a single particle of arbitrary shape that is embedded in a nonabsorbing medium and illuminated by a plane wave. The net rate at which electromagnetic energy crosses the surface A of a sphere of radius r around the particle is

$$W^{\text{ab}} = -\int_A \mathbf{S}_2 \cdot \hat{\mathbf{e}}_r dA, \tag{993}$$

where the Poynting vector $\mathbf{S}_2$ of the fields in the medium surrounding the particle is given by Eq. (962). If $W^{\mathrm{ab}} > 0$, energy is absorbed within the sphere of radius r, and if the medium surrounding the particle is nonabsorbing, the energy must be absorbed by the particle.

Using Eq. (962), one may write W^{ab} as a sum of three terms:

$$W^{\mathrm{ab}} = W^{i} - W^{\mathrm{sc}} + W^{\mathrm{ext}}, \tag{994}$$

where

$$W^{i} = -\int_A \mathbf{S}^{i} \cdot \hat{\mathbf{e}}_r dA \;\; ; \;\; W^{\mathrm{sc}} = \int_A \mathbf{S}^{\mathrm{sc}} \cdot \hat{\mathbf{e}}_r dA \;\; ; \;\; W^{\mathrm{ext}} = -\int_A \mathbf{S}^{\mathrm{ext}} \cdot \hat{\mathbf{e}}_r dA \tag{995}$$

but since $W^{i} = 0$ for a nonabsorbing medium, Eq. (994) yields

$$W^{\mathrm{ext}} = W^{\mathrm{ab}} + W^{\mathrm{sc}}. \tag{996}$$

For an x-polarized incident plane wave with electric field amplitude given by Eq. (953), the scattered electric field in the far zone is [cf. Eq. (958)]

$$\mathbf{E}^{\mathrm{sc}} \sim i\frac{e^{ik_2 r}}{k_2 r}\mathbf{X} \;\; ; \;\; \mathbf{X}\cdot\hat{\mathbf{e}}_r = 0 \;\; ; \;\; \mathbf{H}^{\mathrm{sc}} = \frac{c}{\omega\mu_2}\mathbf{k}^{\mathrm{sc}} \times \mathbf{E}_s \sim i\frac{c}{\omega\mu_2}\frac{e^{ik_2 r}}{r}\hat{\mathbf{e}}_r \times \mathbf{X}, \tag{997}$$

where $\mathbf{X}$ is the *vector scattering amplitude* given in Eq. (961) with $E_y^i = 0$ and $E_x^i = E$:

$$\mathbf{X} = E\left[\left(S_2\cos\phi + S_3\sin\phi\right)\hat{\mathbf{e}}_{\parallel s} + \left(S_4\cos\phi + S_1\sin\phi\right)\hat{\mathbf{e}}_{\perp s}\right] = \mathbf{X}'E. \tag{998}$$

For the time-harmonic incident electromagnetic plane wave given by Eq. (953), the Poynting vector is given by [see Eqs. (969) and (970)]

$$\mathbf{S}^{i} = S^{i}\hat{\mathbf{s}} \;\; ; \;\; S^{i} = \frac{c}{8\pi}\frac{ck_2}{\omega\mu_2}E^2, \tag{999}$$

where $\hat{\mathbf{s}}$ is a unit vector in the direction of the Poynting vector, S^i is the magnitude of the Poynting vector, and E is the electric field amplitude of the incident plane wave.

Consider now $\mathbf{S}^{\mathrm{ext}}$ given by [see Eq. (966)]

$$\mathbf{S}^{\mathrm{ext}} = \frac{c}{8\pi}\Re\{\mathbf{E}^{i} \times (\mathbf{H}^{\mathrm{sc}})^* + \mathbf{E}^{\mathrm{sc}} \times (\mathbf{H}^{i})^*\}, \tag{1000}$$

where $(\mathbf{E}^i, \mathbf{H}^i)$ denotes the time-harmonic plane wave given by Eq. (953), which for x polarization becomes (omitting the time dependence $\exp(-i\omega t)$)

$$\mathbf{E}^{i} = Ee^{ik_2 z}\hat{\mathbf{e}}_x \;\; ; \;\; \mathbf{H}^{i} = \frac{c}{\omega\mu_2}\mathbf{k}^{i} \times \mathbf{E}^{i} = \frac{ck_2}{\omega\mu_2}Ee^{ik_2 z}\hat{\mathbf{e}}_z \times \hat{\mathbf{e}}_x = \frac{ck_2}{\omega\mu_2}Ee^{ik_2 z}\hat{\mathbf{e}}_y \tag{1001}$$

and where $(\mathbf{E}^{\mathrm{sc}}, \mathbf{H}^{\mathrm{sc}})$ is given by Eq. (997).

Substituting Eqs. (997) and (1001) into Eq. (1000) and using $\mathbf{X}\cdot\hat{\mathbf{e}}_r = 0$, one obtains

$$\begin{aligned}\mathbf{S}^{\mathrm{ext}} \cdot \hat{\mathbf{e}}_r = -S^{i}\Re\Big\{ & - C^*\left[f_1(\cos\Theta)e^{-ik_2 r\cos\Theta}\right]^* \\ & - Cf_2(\cos\Theta)e^{-ik_2 r\cos\Theta} \\ & + C\cos\phi f_3(\cos\Theta)e^{-ik_2 r\cos\Theta} \Big\},\end{aligned} \tag{1002}$$

where $S^i = (c/(8\pi))((ck_2)/(\omega\mu_2))|E|^2$, $C = i((e^{ik_2 r})/(k_2 r))$, and

$$f_1(\cos\Theta) = \hat{\mathbf{e}}_x \cdot \mathbf{X}' \; ; \; f_2(\cos\Theta) = \cos\Theta f_1(\cos\Theta) \; ; \; f_3(\cos\Theta) = \sqrt{1-\cos^2\Theta}\,\hat{\mathbf{e}}_z \cdot \mathbf{X}'. \tag{1003}$$

Through integration by parts with respect to Θ in the expression for W^{ext} in Eq. (995), one obtains (with $j = 1, 2, 3$)

$$\begin{aligned}\int_0^\pi e^{-ik_2 r\cos\Theta} f_j(\cos\Theta)\sin\Theta d\Theta &= \int_{-1}^{1} e^{-ik_2 r\mu} f_j(\mu)d\mu \\ &= \frac{e^{ik_2 r} f_j(-1) - e^{-ik_2 r} f_j(+1)}{ik_2 r} + O\left(\frac{1}{(k_2 r)^2}\right) \\ &= -[Cf_j(-1) + C^* f_j(+1)] + O\left(\frac{1}{(k_2 r)^2}\right).\end{aligned} \tag{1004}$$

Since $f_3(\pm 1) = 0$, the third term in Eq. (1002) gives no contribution to the integral in Eq. (1004). Also, the ϕ integration of the factor $\cos\phi$ in the third term in Eq. (1002) gives zero. For the other two terms in Eq. (1002), the ϕ integration gives a factor of 2π, and the Θ integration associated with the first and second terms in Eq. (1003) gives

$$\begin{aligned}W^{\text{ext}} &= -\int_A \mathbf{S}^{\text{ext}} \cdot \hat{\mathbf{e}}_r dA = -r^2 \int_0^{2\pi} d\phi \int_0^\pi \mathbf{S}^{\text{ext}} \cdot \hat{\mathbf{e}}_r \sin\Theta d\Theta \\ &= 2\pi r^2 S^i \Re\left\{C^*[Cf_1(-1) + C^* f_1(+1)]^* + C[Cf_2(-1) + C^* f_2(+1)]\right\} \\ &= 2\pi r^2 S^i \Re\left\{|C|^2 [f_1^*(+1) + f_1(+1)] + (C^*)^2 f_1^*(-1) - (C)^2 f_1(-1)\right\} \\ &= 2\pi r^2 S^i \Re\left\{2|C|^2 \Re[f_1(+1)] - i\Im\left[C^2 f_1(-1)\right]\right\},\end{aligned} \tag{1005}$$

where we have used $f_2(\pm 1) = \pm f_1(\pm 1)$. Thus, one obtains the following result:

$$W^{\text{ext}} = \frac{4\pi}{k_2^2} S^i \Re\left(\hat{\mathbf{e}}_x \cdot \mathbf{X}'\right)_{\Theta=0}. \tag{1006}$$

The ratio of W^{ext} to S^i is denoted by C^{ext} and has the dimension of area. It is called the *extinction cross section*:

$$C^{\text{ext}} = \frac{W^{\text{ext}}}{S^i} = \frac{4\pi}{k_2^2} \Re\left(\hat{\mathbf{e}}_x \cdot \mathbf{X}'\right)_{\Theta=0}. \tag{1007}$$

This result is known as the *optical theorem* and is common to all kinds of scattering phenomena involving, for example, acoustic waves, electromagnetic waves, and elementary particles.

It follows from Eq. (996) that the extinction cross section is the sum of the *absorption cross section* C^{ab} and the *scattering cross section* C^{sc}:

$$C^{\text{ext}} = C^{\text{ab}} + C^{\text{sc}} \quad ; \quad C^{\text{ab}} = \frac{W^{\text{ab}}}{S^i} \quad ; \quad C^{\text{sc}} = \frac{W^{\text{sc}}}{S^i}. \tag{1008}$$

If one defines the differential scattering cross section as $\beta_n(\Theta) \equiv \frac{|\mathbf{X}'|^2}{k_2^2}$, then it follows from Eqs. (971) and (995) that

$$\beta_n \equiv C^{\rm sc} = \frac{W^{\rm sc}}{S^i} = \int_0^{2\pi} \int_0^{\pi} \beta_n(\Theta) \sin\Theta d\Theta d\phi.$$

If the scattering potential is spherically symmetric, the integration over azimuth gives 2π, so that

$$\beta_n = 2\pi \int_0^{\pi} \beta_n(\Theta) \sin\Theta d\Theta = 2\pi \int_{-1}^{1} \beta_n(x) dx$$

and we may define the scattering phase function as

$$p(\Theta) = \frac{\beta_n(\Theta)}{\frac{1}{4\pi}\int_{4\pi} \beta_n(\Theta) d\omega}; \qquad \frac{1}{4\pi}\int_{4\pi} p(\Theta) d\omega = 1.$$

The average cosine of the scattering angle or the *asymmetry parameter* g is given by [see Eq. (10)]

$$g = \langle \cos\Theta \rangle = \int_{4\pi} p \cos\Theta d\omega = 2\pi \int_{-1}^{1} p(x) x dx \tag{1009}$$

and has the following properties:

- $g = 0$ for *isotropic* scattering (which is the same in all directions) and for scattering that is symmetric about a scattering angle of 90°.
- $g > 0$ if the particle scatters more light in the forward direction ($\Theta = 0°$).
- $g < 0$ if the particle scatters more light in the backward direction ($\Theta = 180°$).

The scattering *efficiencies* may be defined as

$$Q^{\rm ext} = \frac{C^{\rm ext}}{G} \quad ; \quad Q^{\rm sc} = \frac{C^{\rm sc}}{G} \quad ; \quad Q^{\rm ab} = \frac{C^{\rm ab}}{G}, \tag{1010}$$

where G is the cross-sectional area of the particle projected onto a plane perpendicular to the incident beam direction. The theory discussed so far applies to a particle of arbitrary shape. In the remainder of this Appendix, we will focus on particles of spherical shape.

A.2
Absorption and Scattering by a Sphere – Mie Theory

In a linear, isotropic, and homogeneous medium, a time-harmonic electromagnetic field must satisfy the vector wave equation

$$(\nabla^2 + k^2)\mathbf{E} = (\nabla^2 + k^2)\mathbf{H} = 0 \quad ; \quad k^2 = \frac{\omega^2}{c^2}\varepsilon\mu \tag{1011}$$

and, in source-free regions of space, $\mathbf{E}$ and $\mathbf{H}$ must be divergence-free:

$$\nabla \cdot \mathbf{E} = \nabla \cdot \mathbf{H} = 0. \tag{1012}$$

In addition, **E** and **H** are related through Maxwell's curl equations

$$\nabla \times \mathbf{E} = i\frac{\omega}{c}\mu\mathbf{H} \quad ; \quad \nabla \times \mathbf{H} = -i\frac{\omega}{c}\varepsilon\mathbf{E}. \tag{1013}$$

A.2.1
Solutions of Vector Wave Equations in Spherical Polar Coordinates

Solutions of Eqs. (1011)–(1013) may be constructed by considering a scalar function ψ and a vector **c**, which either may be a constant vector or have zero curl, that is, $\nabla \times \mathbf{c} = 0$. From ψ and **c**, one may construct a divergence-free vector function **M**:

$$\mathbf{M} = \nabla \times (\mathbf{c}\psi) = \psi\nabla \times \mathbf{c} - \mathbf{c} \times \nabla\psi = -\mathbf{c} \times \nabla\psi. \tag{1014}$$

which fulfills $\nabla \cdot \mathbf{M} = 0$, since $\nabla \cdot (\nabla \times \mathbf{A}) = 0$ for any arbitrary vector function **A**.

Applying the vector identity $\nabla \times (\nabla \times \mathbf{A}) = \nabla(\nabla \cdot \mathbf{A}) - \nabla^2\mathbf{A}$ to Eq. (1014), one obtains

$$\nabla \times \mathbf{M} = \nabla \times [\nabla \times (\mathbf{c}\psi)] = \nabla[\nabla \cdot (\mathbf{c}\psi)] - \nabla^2(\mathbf{c}\psi) \tag{1015}$$

$$\nabla \times (\nabla \times \mathbf{M}) = \nabla(\nabla \cdot \mathbf{M}) - \nabla^2\mathbf{M} = \nabla \times \{\nabla[\nabla \cdot (\mathbf{c}\psi)]\} - \nabla \times [\nabla^2(\mathbf{c}\psi)]. \tag{1016}$$

Next, one may use Eq. (1014) and the vector identity $\nabla \times (\nabla\varphi) = 0$, which is valid for any arbitrary scalar function φ, and assume that $\nabla^2\mathbf{c} = 0$ to obtain

$$(\nabla^2 + k^2)\mathbf{M} = \nabla \times [\nabla^2(\mathbf{c}\psi)] + k^2\nabla \times (\mathbf{c}\psi) = \nabla \times \{\mathbf{c}[\nabla^2\psi + k^2\psi]\}. \tag{1017}$$

Thus, **M** satisfies the *vector* wave equation, provided ψ is a solution of the *scalar* wave equation

$$(\nabla^2 + k^2)\psi = 0. \tag{1018}$$

From $\mathbf{M} = -\mathbf{c} \times \nabla\psi$ [cf. Eq. (1014)], it follows that $\mathbf{c} \cdot \mathbf{M} = 0$. Next, one may construct another vector function **N** given by

$$\mathbf{N} = \frac{1}{k}\nabla \times \mathbf{M} \tag{1019}$$

which clearly has zero divergence: ($\nabla \cdot \mathbf{N} = \frac{1}{k}\nabla \cdot (\nabla \times \mathbf{M}) = 0$), and which satisfies

$$\nabla \times \mathbf{N} = (1/k)\nabla \times (\nabla \times \mathbf{M}) = -\frac{1}{k}\nabla^2\mathbf{M} = k\mathbf{M}, \tag{1020}$$

where we have used $(\nabla^2 + k^2)\mathbf{M} = 0$. Therefore, the vector functions **M** and **N** have all the required properties of an electromagnetic field:

- they satisfy the vector wave equation,
- they have zero divergence, and
- as required by Maxwell's curl equations [Eqs. (1013)], the curl of **M** is proportional to **N**, and the curl of **N** is proportional to **M**.

Thus, one may start by finding solutions ψ of the scalar wave equation. The scalar function ψ is called a *generating function* for the *vector harmonics* **M** and **N**.

Since one is interested in scattering by a sphere, one may choose functions ψ that are solutions of the scalar wave equation in spherical polar coordinates r, θ, and ϕ. By choosing $\mathbf{c} = \mathbf{r} = x\hat{\mathbf{e}}_x + y\hat{\mathbf{e}}_y + z\hat{\mathbf{e}}_z = r\hat{\mathbf{e}}_r$, which is in accordance with the previous assumptions that $\nabla \times \mathbf{c} = 0$ and $\nabla^2\mathbf{c} = 0$, one obtains

$$\mathbf{M} = \nabla \times (\mathbf{r}\psi) = -\mathbf{r} \times \nabla\psi \quad ; \quad \mathbf{r} \cdot \mathbf{M} = 0, \tag{1021}$$

where **r** is the radius vector and **M** is a solution of the vector wave equation in spherical polar coordinates. Since $\mathbf{r} \cdot \mathbf{M} = 0$, **M** is tangential to any sphere $|\mathbf{r}| =$ constant.

The scalar wave equation in spherical polar coordinates is

$$\frac{1}{r^2}\frac{\partial}{\partial r}\left(r^2\frac{\partial\psi}{\partial r}\right) + \frac{1}{r^2\sin\theta}\frac{\partial}{\partial\theta}\left(\sin\theta\frac{\partial\psi}{\partial\theta}\right) + \frac{1}{r^2\sin^2\theta}\frac{\partial^2\psi}{\partial\phi^2} + k^2\psi = 0. \tag{1022}$$

By multiplying Eq. (1022) by r^2, writing $\psi(r, \theta, \phi) = R(r)F(\theta, \phi)$, and dividing through by ψ, one obtains

$$\frac{1}{R(r)}\frac{\partial}{\partial r}\left(r^2\frac{\partial R(r)}{\partial r}\right) + k^2r^2 = -\frac{1}{F(\theta, \phi)}\Lambda(\theta, \phi)F(\theta, \phi) = c, \tag{1023}$$

where c is a separation constant and $\Lambda(\theta, \phi)$ is a differential operator given by

$$\Lambda(\theta, \phi) = \frac{1}{\sin^2\theta}\left[\sin\theta\frac{\partial}{\partial\theta}\left(\sin\theta\frac{\partial}{\partial\theta}\right) + \frac{\partial^2}{\partial\phi^2}\right]. \tag{1024}$$

For the radial part, the result is

$$\frac{d}{dr}\left(r^2\frac{dR(r)}{dr}\right) + [k^2r^2 - c]R(r) = 0. \tag{1025}$$

Writing $F(\theta, \phi) = \Theta(\theta)\Phi(\phi)$, the angular part becomes

$$\frac{1}{\Theta(\theta)}\sin\theta\frac{\partial}{\partial\theta}\left(\sin\theta\frac{\partial}{\partial\theta}\Theta(\theta)\right) + c\sin^2\theta = -\frac{1}{\Phi(\phi)}\frac{\partial^2}{\partial\phi^2}\Phi(\phi) = m^2, \tag{1026}$$

where m^2 is another separation constant. Thus, one obtains

$$\left(\frac{d^2}{d\phi^2} + m^2\right)\Phi(\phi) = 0 \tag{1027}$$

$$\left\{\frac{1}{\sin\theta}\frac{d}{d\theta}\left(\sin\theta\frac{d}{d\theta}\right) + c - \frac{m^2}{\sin^2\theta}\right\}\Theta(\theta) = 0. \tag{1028}$$

The solutions of Eq. (1027) are

$$\Phi_e = \cos m\phi \quad ; \quad \Phi_o = \sin m\phi, \tag{1029}$$

where the subscripts e and o denote even and odd. From the requirement that ψ must be *periodic* with period 2π, it follows that m must be an integer or zero. Negative integers do not provide solutions that are linearly independent from those with positive integers, so it suffices to consider positive integers and zero.

The solutions of Eq. (1028) that are finite at $\theta = 0$ and $\theta = \pi$ are the *associated Legendre functions* of the first kind $P_n^{(m)}$ of degree n and order m, where the separation constant c is given by $c = n(n+1)$, with $n = m, m+1, m+2, \ldots$. The $P_n^{(m)}$ functions are orthogonal:

$$\int_{-1}^{1} P_n^{(m)}(\mu)P_{n'}^{(m)}(\mu)d\mu = \delta_{n,n'}\frac{2}{2n+1}\frac{(n+m)!}{(n-m)!}, \tag{1030}$$

where $\mu = \cos\theta$ and $\delta_{n,n'}$ is the Kronecker delta, which is 1 if $n = n'$ and zero otherwise. When $m = 0$, then $P_n^{(0)} = P_n$, where P_n is a Legendre polynomial.

With $z = kr$, the radial equation in Eq. (1025) with $c = n(n+1)$ becomes

$$\left\{ z^2\frac{d^2}{dz^2} + 2z\frac{d}{dz} + z^2 - n(n+1) \right\} z_n(z) = 0 \tag{1031}$$

which has the following solutions for $z_n(z)$:

- Spherical Bessel functions of the first kind: $j_n(z) = \sqrt{(\pi/2z)}J_{n+(1/2)}(z)$
- Spherical Bessel functions of the second kind: $y_n(z) = \sqrt{(\pi/2z)}Y_{n+(1/2)}(z)$
- Spherical Bessel functions of the third kind: $h_n^{(1)}(z) = j_n(z) + iy_n(z) = \sqrt{(\pi/2z)}H_{n+(1/2)}^{(1)}(z)$ and $h_n^{(2)}(z) = j_n(z) - iy_n(z) = \sqrt{(\pi/2z)}H_{n+(1/2)}^{(2)}(z)$.

The pairs $j_n(z), y_n(z)$ and $h_n^{(1)}(z), h_n^{(1)}(z)$ are linearly independent solutions for every n, and the spherical Bessel functions satisfy the recurrence relations

$$z_{n-1}(z) + z_{n+1}(z) = \frac{2n+1}{z}z_n(z) \tag{1032}$$

$$nz_{n-1}(z) - (n+1)z_{n+1}(z) = (2n+1)\frac{d}{dz}z_n(z) \tag{1033}$$

$$\frac{n+1}{z}z_n(z) + \frac{d}{dz}z_n(z) = z_{n-1}(z) \tag{1034}$$

$$\frac{n}{z}z_n(z) - \frac{d}{dz}z_n(z) = z_{n+1}(z) \tag{1035}$$

implying that from the first two orders one can generate higher order functions by recurrence. The first three orders are given by

$$\begin{aligned} j_0(z) &= \frac{\sin z}{z}\ ;\ j_1(z) = \frac{\sin z}{z^2} - \frac{\cos z}{z}\ ; \\ j_2(z) &= \left(\frac{3}{z^3} - \frac{1}{z}\right)\sin z - \frac{3}{z^2}\cos z \end{aligned} \tag{1036}$$

$$\begin{aligned} y_0(z) &= -\frac{\cos z}{z}\ ;\ y_1(z) = -\frac{\cos z}{z^2} - \frac{\sin z}{z}\ ; \\ y_2(z) &= \left(-\frac{3}{z^3} + \frac{1}{z}\right)\cos z - \frac{3}{z^2}\sin z. \end{aligned} \tag{1037}$$

From the solutions of the scalar wave equation in spherical polar coordinates, one may construct even and odd generating functions:

$$\psi_{emn} = \cos m\phi P_n^{(m)}(\cos\theta)z_n(kr)\ ;\ \psi_{omn} = \sin m\phi P_n^{(m)}(\cos\theta)z_n(kr). \tag{1038}$$

The vector spherical harmonics generated by ψ_{emn} and ψ_{omn} are

$$\mathbf{M}_{emn} = \nabla \times (\mathbf{r}\psi_{emn}) \quad ; \quad \mathbf{M}_{omn} = \nabla \times (\mathbf{r}\psi_{omn}) \tag{1039}$$

$$\mathbf{N}_{emn} = \frac{1}{k}\nabla \times \mathbf{M}_{emn} \quad ; \quad \mathbf{N}_{omn} = \frac{1}{k}\nabla \times \mathbf{M}_{omn} \tag{1040}$$

which in component form become

$$\mathbf{M}_{emn} = -\frac{m}{\sin\theta}\sin m\phi P_n^{(m)}(\cos\theta) z_n(z)\hat{\mathbf{e}}_\theta - \cos m\phi \frac{dP_n^{(m)}(\cos\theta)}{d\theta} z_n(z)\hat{\mathbf{e}}_\phi \tag{1041}$$

$$\mathbf{M}_{omn} = \frac{m}{\sin\theta}\cos m\phi P_n^{(m)}(\cos\theta) z_n(z)\hat{\mathbf{e}}_\theta - \sin m\phi \frac{dP_n^{(m)}(\cos\theta)}{d\theta} z_n(z)\hat{\mathbf{e}}_\phi \tag{1042}$$

$$\begin{aligned}\mathbf{N}_{emn} = {} & \frac{z_n(z)}{z}\cos m\phi\, n(n+1)P_n^{(m)}(\cos\theta)\hat{\mathbf{e}}_r \\ & + \cos m\phi \frac{dP_n^{(m)}(\cos\theta)}{d\theta}\frac{1}{z}\frac{d}{dz}\left[z z_n(z)\right]\hat{\mathbf{e}}_\theta \\ & - m\sin m\phi \frac{P_n^{(m)}(\cos\theta)}{\sin\theta}\frac{1}{z}\frac{d}{dz}\left[z z_n(z)\right]\hat{\mathbf{e}}_\phi \end{aligned} \tag{1043}$$

$$\begin{aligned}\mathbf{N}_{omn} = {} & \frac{z_n(z)}{z}\sin m\phi\, n(n+1)P_n^{(m)}(\cos\theta)\hat{\mathbf{e}}_r \\ & + \sin m\phi \frac{dP_n^{(m)}(\cos\theta)}{d\theta}\frac{1}{z}\frac{d}{dz}\left[z z_n(z)\right]\hat{\mathbf{e}}_\theta \\ & + m\cos m\phi \frac{P_n^{(m)}(\cos\theta)}{\sin\theta}\frac{1}{z}\frac{d}{dz}\left[z z_n(z)\right]\hat{\mathbf{e}}_\phi, \end{aligned} \tag{1044}$$

where the r components of $\mathbf{N}_{emn}$ and $\mathbf{N}_{omn}$ have been simplified by using the fact that $P_n^{(m)}$ satisfies Eq. (1028) with $c = n(n+1)$.

A.2.2
Expansion of Incident Plane Wave in Vector Spherical Harmonics

Let the incident field be an x-polarized plane wave propagating in the z direction, so that in spherical polar coordinates

$$\mathbf{E}^i = Ee^{ikr\cos\theta}\hat{\mathbf{e}}_x, \tag{1045}$$

where

$$\hat{\mathbf{e}}_x = \sin\theta\cos\phi\hat{\mathbf{e}}_r + \cos\theta\cos\phi\hat{\mathbf{e}}_\theta - \sin\phi\hat{\mathbf{e}}_\phi. \tag{1046}$$

Note that, since the propagation direction of the incident plane wave is along the z-axis, the scattering angle Θ is equal to the polar angle θ. It follows from Eqs. (956) and (957) with $E_y^i = 0$ and $E_x^i = E$ that

$$E_\parallel^i = E\cos\phi \quad ; \quad E_\perp^i = E\sin\phi. \tag{1047}$$

Expanding the incident field in vector spherical harmonics, one obtains

$$\mathbf{E}^i = \sum_{m=0}^{\infty}\sum_{n=m}^{\infty}\{B_{emn}\mathbf{M}_{emn} + B_{omn}\mathbf{M}_{omn} + A_{emn}\mathbf{N}_{emn} + A_{omn}\mathbf{N}_{omn}\}, \tag{1048}$$

where the coefficients B_{emn}, B_{omn}, A_{emn}, and A_{omn} are to be determined. All vector spherical harmonic functions $\mathbf{M}_{pmn}$ and $\mathbf{N}_{pmn}$ $(p = o, e)$ in Eq. (1048) are orthogonal such that for all $m' \neq m$, all $n' \neq n$, and all combinations of $p = o, e$, $q = o, e$, $\mathbf{X} = \mathbf{M}, \mathbf{N}$, and $\mathbf{Y} = \mathbf{M}, \mathbf{N}$:

$$\int_0^{2\pi}\int_0^{\pi} \mathbf{X}_{pm'n'} \cdot \mathbf{Y}_{qmn} \sin\theta d\theta d\phi = 0. \tag{1049}$$

Therefore, to obtain the expansion coefficients B_{pmn} or A_{pmn} $(p = o, e)$, one may first multiply both sides of Eq. (1048) by $\mathbf{M}_{pm'n'}$ or $\mathbf{N}_{pm'n'}$ and then integrate on both sides over the full 4π angular range. Because of the orthogonality expressed in Eq. (1049), one obtains

$$B_{pmn} = \frac{\int_0^{2\pi}\int_0^{\pi} \mathbf{E}^i \cdot \mathbf{M}_{pmn} \sin\theta d\theta d\phi}{\int_0^{2\pi}\int_0^{\pi} |\mathbf{M}_{pmn}|^2 \sin\theta d\theta d\phi} \tag{1050}$$

$$A_{pmn} = \frac{\int_0^{2\pi}\int_0^{\pi} \mathbf{E}^i \cdot \mathbf{N}_{pmn} \sin\theta d\theta d\phi}{\int_0^{2\pi}\int_0^{\pi} |\mathbf{N}_{pmn}|^2 \sin\theta d\theta d\phi}. \tag{1051}$$

It follows from the expressions for $\mathbf{M}_{emn}$ and $\mathbf{N}_{omn}$ in Eqs. (1041) and (1044), the expression for $\hat{\mathbf{e}}_x$ in Eq. (1046), and the orthogonality of sine and cosine that $B_{emn} = A_{omn} = 0$, and that all remaining coefficients vanish unless $m = 1$.

Further, the appropriate radial function to be used in the vector spherical harmonics is $j_n(kr)$ since it is finite at the origin. Labeling such functions $\mathbf{M}_{pmn}$ and $\mathbf{N}_{pmn}$ $(p = o, e)$ by the superscript (1), one obtains from Eq. (1048)

$$\mathbf{E}^i = \sum_{n=1}^{\infty}\{B_{o1n}\mathbf{M}^{(1)}_{o1n} + A_{e1n}\mathbf{N}^{(1)}_{e1n}\}. \tag{1052}$$

A long and tedious calculation shows that

$$B_{o1n} = -A_{e1n} = i^n E\frac{2n+1}{n(n+1)} \tag{1053}$$

so that Eq. (1052) gives

$$\mathbf{E}^i = \sum_{n=1}^{\infty} E_n(\mathbf{M}^{(1)}_{o1n} - i\mathbf{N}^{(1)}_{e1n}) \quad ; \quad E_n = i^n E\frac{2n+1}{n(n+1)}. \tag{1054}$$

The corresponding magnetic field follows from Maxwell's equation:

$$\nabla \times \mathbf{E}^i = -\frac{\mu_2}{c}\frac{\partial}{\partial t}\mathbf{H}^i = \frac{i\omega\mu_2}{c}\mathbf{H}^i \tag{1055}$$

which in combination with $\nabla \times \mathbf{M}^{(1)}_{pmn} = k_2\mathbf{N}^{(1)}_{pmn}$ and $\nabla \times \mathbf{N}^{(1)}_{pmn} = k_2\mathbf{M}^{(1)}_{pmn}$ $(p = o, e)$ gives

$$\mathbf{H}^i = -i\frac{c}{\omega\mu_2}\nabla \times \mathbf{E}^i = -\frac{ck_2}{\omega\mu_2}\sum_{n=1}^{\infty} E_n(\mathbf{M}^{(1)}_{e1n} + i\mathbf{N}^{(1)}_{o1n}). \tag{1056}$$

A.2.3
Internal and Scattered Fields

The boundary conditions to be satisfied at the surface of the sphere, where $r = a$, are

$$(\mathbf{E}^i + \mathbf{E}^{\rm sc} - \mathbf{E}^{\rm int}) \times \hat{\mathbf{e}}_r = 0 \quad ; \quad (\mathbf{H}^i + \mathbf{H}^{\rm sc} - \mathbf{H}^{\rm int}) \times \hat{\mathbf{e}}_r = 0, \tag{1057}$$

where $(\mathbf{E}^{\rm sc}, \mathbf{H}^{\rm sc})$ is the scattered field and $(\mathbf{E}^{\rm int}, \mathbf{H}^{\rm int})$ is the field inside the sphere. The boundary conditions, the orthogonality of the vector spherical harmonics, and the form of the expansion for the incident field dictate the form of the expansions for the scattered field and the field inside the sphere:

$$\mathbf{E}^{\rm int} = \sum_{n=1}^{\infty} E_n \left(c_n \mathbf{M}^{(1)}_{o1n} - i d_n \mathbf{N}^{(1)}_{e1n} \right) \tag{1058}$$

$$\mathbf{H}^{\rm int} = -\frac{ck_1}{\omega\mu_1} \sum_{n=1}^{\infty} E_n \left(d_n \mathbf{M}^{(1)}_{e1n} + i c_n \mathbf{N}^{(1)}_{o1n} \right) \tag{1059}$$

$$\mathbf{E}^{\rm sc} = \sum_{n=1}^{\infty} E_n \left(i a_n \mathbf{N}^{(3)}_{e1n} - b_n \mathbf{M}^{(3)}_{o1n} \right) \tag{1060}$$

$$\mathbf{H}^{\rm sc} = \frac{ck_2}{\omega\mu_2} \sum_{n=1}^{\infty} E_n \left(i b_n \mathbf{N}^{(3)}_{o1n} + a_n \mathbf{M}^{(3)}_{e1n} \right), \tag{1061}$$

where the superscript (3) has been used to denote vector spherical harmonics with radial dependence given by $h_n^{(1)}(k_2 r)$, which for each order n behaves as an outgoing spherical wave when $k_2 r \gg 1$. Here, the coefficients c_n and d_n of the field inside the sphere and the scattering coefficients a_n and b_n are to be determined.

It is convenient to define the following angular functions:

$$\pi_n = \frac{P_n^{(1)}}{\sin\theta} \quad ; \quad \tau_n = \frac{dP_n^{(1)}}{d\theta} \tag{1062}$$

which have the property that their sum and difference are orthogonal functions:

$$\int_0^{\pi} (\tau_n + \pi_n)(\tau_m + \pi_m) \sin\theta d\theta = \int_0^{\pi} (\tau_n - \pi_n)(\tau_m - \pi_m) \sin\theta d\theta = 0 \quad (m \neq n). \tag{1063}$$

The vector spherical harmonics in Eqs. (1041)–(1044) with $m = 1$ now become:

$$\mathbf{M}_{e1n} = -\sin\phi\ \pi_n(\cos\theta) z_n(z) \hat{\mathbf{e}}_\theta - \cos\phi\ \tau_n(\cos\theta) z_n(z) \hat{\mathbf{e}}_\phi \tag{1064}$$

$$\mathbf{M}_{o1n} = \cos\phi\ \pi_n(\cos\theta) z_n(z) \hat{\mathbf{e}}_\theta - \sin\phi \tau_n(\cos\theta) z_n(z) \hat{\mathbf{e}}_\phi \tag{1065}$$

$$\begin{aligned}\mathbf{N}_{e1n} &= \cos\phi\ n(n+1) \sin\theta\ \pi_n(\cos\theta) \frac{z_n(z)}{z} \hat{\mathbf{e}}_r \\ &\quad + \cos\phi\ \tau_n(\cos\theta) \frac{[z z_n(z)]'}{z} \hat{\mathbf{e}}_\theta \\ &\quad - \sin\phi\ \pi_n(\cos\theta) \frac{[z z_n(z)]'}{z} \hat{\mathbf{e}}_\phi \end{aligned} \tag{1066}$$

$$\mathbf{N}_{o1n} = \sin\phi\, n(n+1)\sin\theta\, \pi_n(\cos\theta)\frac{z_n(z)}{z}\hat{\mathbf{e}}_r + \sin\phi\, \tau_n(\cos\theta)\frac{[zz_n(z)]'}{z}\hat{\mathbf{e}}_\theta + \cos\phi\, \pi_n(\cos\theta)\frac{[zz_n(z)]'}{z}\hat{\mathbf{e}}_\phi. \tag{1067}$$

To determine the elements S_j ($j = 1, 2, 3, 4$) of the amplitude scattering matrix [see Eq. (960)], one may use Eqs. (1065) and (1066) to express the scattered electric field in Eq. (1060) in terms of its components parallel and perpendicular to the scattering plane and in terms of the corresponding components given in Eq. (1047) for the incident electric field. Substitution of Eqs. (1065) and (1066) into Eq. (1060) gives

$$E_{\parallel}^{\rm sc} = E_{\theta}^{\rm sc} = E_{\parallel}^{i}\sum_{n=1}^{\infty} i^n \frac{2n+1}{n(n+1)}[ia_n\tau_n(\cos\theta)f_n(z) - b_n\pi_n(\cos\theta)z_n(z)] \tag{1068}$$

$$E_{\perp}^{\rm sc} = -E_{\phi}^{\rm sc} = -E_{\perp}^{i}\sum_{n=1}^{\infty} i^n \frac{2n+1}{n(n+1)}[-ia_n\pi_n(\cos\theta)f_n(z) + b_n\tau_n(\cos\theta)z_n(z)], \tag{1069}$$

where we have made use of Eqs. (1047) and (1054), and where

$$z_n(z) = h_n^{(1)}(z) \;\; ; \;\; f_n(z) = \frac{[zz_n(z)]'}{z} \;\; ; \;\; z = k_z r. \tag{1070}$$

Comparison of Eqs. (1068) and (1069) with Eq. (960) shows that $S_3 = S_4 = 0$ and that ($C = i((e^{ik_2 r})/(k_2 r))$)

$$CS_2 = \sum_{n=1}^{\infty} i^n \frac{2n+1}{n(n+1)}[ia_n\tau_n(\cos\theta)f_n(z) - b_n\pi_n(\cos\theta)z_n(z)] \tag{1071}$$

$$CS_1 = -\sum_{n=1}^{\infty} i^n \frac{2n+1}{n(n+1)}[-ia_n\pi_n(\cos\theta)f_n(z) + b_n\tau_n(\cos\theta)z_n(z)]. \tag{1072}$$

For large values of $z = k_2 r$, one may use the asymptotic forms of the expressions for $z_n(z)$ and $f_n(z)$, given by

$$z_n(z) = h_n^{(1)}(z) = \sqrt{\frac{\pi}{2z}}H_{n+1/2}^{(1)}(z) \sim \frac{1}{z}(-i)^{n+1}e^{iz} \tag{1073}$$

$$f_n(z) = \frac{[zz_n(z)]'}{z} \sim \frac{1}{z}(-i)^{n+1}ie^{iz}. \tag{1074}$$

Substitution of these asymptotic results in Eqs. (1071) and (1072) leads to

$$S_2 = \sum_{n=1}^{\infty} \frac{2n+1}{n(n+1)}[a_n\tau_n(\cos\theta) + b_n\pi_n(\cos\theta)] \tag{1075}$$

$$S_1 = -\sum_{n=1}^{\infty} \frac{2n+1}{n(n+1)}[a_n\pi_n(\cos\theta) + b_n\tau_n(\cos\theta)]. \tag{1076}$$

Extinction Cross Section

To determine the *extinction cross section* given by [Eq. (1007)]

$$C^{\text{ext}} = \frac{4\pi}{k_2^2}\Re\left(\hat{\mathbf{e}}_x \cdot \mathbf{X}'\right)_{\Theta=0} \tag{1077}$$

one may use Eq. (998) with $S_3 = S_4 = 0$, from which it follows that

$$\mathbf{X}' = S_2 \cos\phi\hat{\mathbf{e}}_{\parallel s} + S_1 \sin\phi\hat{\mathbf{e}}_{\perp s} \tag{1078}$$

and hence

$$\mathbf{X}' \cdot \hat{\mathbf{e}}_x = S_2 \cos\phi\; \hat{\mathbf{e}}_{\parallel s} \cdot \hat{\mathbf{e}}_x + S_1 \sin\phi\; \hat{\mathbf{e}}_{\perp s} \cdot \hat{\mathbf{e}}_x. \tag{1079}$$

Using $\hat{\mathbf{e}}_{\parallel s} \cdot \hat{\mathbf{e}}_x = -\hat{\mathbf{e}}_\phi \cdot \hat{\mathbf{e}}_x = \cos\phi$ and $\hat{\mathbf{e}}_{\perp s} \cdot \hat{\mathbf{e}}_x = \hat{\mathbf{e}}_\theta \cdot \hat{\mathbf{e}}_x = \cos\theta\cos\phi$, one finds (on noting that $\Theta = \theta$)

$$C^{\text{ext}} = \frac{4\pi}{k_2^2}\Re\left(\hat{\mathbf{e}}_x \cdot \mathbf{X}\right)_{\theta=0,\phi=0} = \frac{4\pi}{k_2^2}\Re[S_2(\theta=0,\phi=0)]. \tag{1080}$$

From the definitions of the angular functions [see Eqs. (1062)]

$$\pi_n = \frac{P_n^{(1)}(\cos\theta)}{\sin\theta} \quad ; \quad \tau_n = \frac{dP_n^{(1)}(\cos\theta)}{d\theta} \tag{1081}$$

and the expansion (with $x = \cos\theta$)

$$P_n^{(1)}(x) = \frac{1}{2}\frac{(n+1)!}{(n-1)!}(1-x)^{1/2}\left[1 + c_1\frac{1-x}{2} + c_2\left(\frac{1-x}{2}\right)^2\right], \tag{1082}$$

where the coefficients c_j ($j = 1, 2, \ldots$) do not depend on x, it follows that

$$\pi_n(\theta=0) = \frac{1}{2}n(n+1) \quad ; \quad \tau_n(\theta=0) = \frac{1}{2}n(n+1) \tag{1083}$$

and hence Eq. (1075) gives

$$S_2(\theta=0,\phi=0) = \frac{1}{2}\sum_{n=1}^{\infty}(2n+1)[a_n + b_n] \tag{1084}$$

so that Eq. (1077) becomes

$$C^{\text{ext}} = \frac{4\pi}{k_2^2}\Re[S_2(\theta=0,\phi=0)] = \frac{2\pi}{k_2^2}\Re\sum_{n=1}^{\infty}(2n+1)[a_n + b_n]. \tag{1085}$$

Scattering Cross Section

The scattering cross section is given by [see Eqs. (965), (995), and (999)]

$$C^{\text{sc}} = \frac{W^{\text{sc}}}{S^i} \quad ; \quad W^{\text{sc}} = \int_0^{2\pi}\int_0^{\pi} \mathbf{S}^{\text{ext}} \cdot \hat{\mathbf{e}}_r r^2 \sin\theta d\theta d\phi, \tag{1086}$$

where

$$\mathbf{S}^{\text{sc}} = \frac{c}{8\pi}\Re\{\mathbf{E}^{\text{sc}} \times (\mathbf{H}^{\text{sc}})^*\} \tag{1087}$$

$$S^i = \frac{c}{8\pi}\frac{ck_2}{\omega\mu_2}E^2 \tag{1088}$$

with [see Eqs. (1054), (1060), and (1061)]

$$\mathbf{E}^{\text{sc}} = \sum_{n=1}^{\infty} E_n \left(ia_n \mathbf{N}_{e1n}^{(3)} - b_n \mathbf{M}_{o1n}^{(3)} \right) \quad ; \quad E_n = i^n E \frac{2n+1}{n(n+1)} \tag{1089}$$

$$\mathbf{H}^{\text{sc}} = \frac{ck_2}{\omega\mu_2} \sum_{n=1}^{\infty} E_n \left(ib_n \mathbf{N}_{o1n}^{(3)} + a_n \mathbf{M}_{e1n}^{(3)} \right). \tag{1090}$$

From Eqs. (1064)–(1067), it follows that

$$\left[\mathbf{N}_{e1n}^{(3)} \times (\mathbf{N}_{o1n}^{(3)})^* \right] \cdot \hat{\mathbf{e}}_r = f_n f_m^* [\cos^2\phi\ \tau_n \pi_m + \sin^2\phi\ \tau_m \pi_n], \tag{1091}$$

where

$$f_n(z) = \frac{[z z_n(z)]'}{z} \quad ; \quad z_n(z) = h_n^{(1)}(z) \quad ; \quad z = k_z r. \tag{1092}$$

Since the integration of $\cos^2\phi$ and $\sin^2\phi$ gives 1/2, the contribution W_1^{sc} to W^{sc} in Eq. (1086) becomes

$$W_1^{\text{sc}} = r^2 \frac{c}{8\pi} \frac{ck_2}{\omega\mu_2} \Re \left\{ \sum_{m=1}^{\infty} \sum_{n=1}^{\infty} \frac{1}{2} E_n E_m^* i a_n (i b_m)^* f_n f_m^* \int_0^{\pi} [\tau_n \pi_m + \tau_m \pi_n] \sin\theta d\theta \right\}. \tag{1093}$$

Similarly, it follows from Eqs. (1064)–(1067)

$$\left[\mathbf{M}_{o1n}^{(3)} \times (\mathbf{M}_{e1n}^{(3)})^* \right] \cdot \hat{\mathbf{e}}_r = -z_n z_m^* [\cos^2\phi\ \tau_m \pi_n + \sin^2\phi\ \tau_n \pi_m] \tag{1094}$$

which gives a contribution W_2^{sc} to W^{sc} in Eq. (1086):

$$W_2^{\text{sc}} = -r^2 \frac{c}{8\pi} \frac{ck_2}{\omega\mu_2} \Re \left\{ \sum_{m=1}^{\infty} \sum_{n=1}^{\infty} \frac{1}{2} z_n z_m^* E_n E_m^* b_n a_m^* \int_0^{\pi} [\tau_n \pi_m + \tau_m \pi_n] \sin\theta d\theta \right\}. \tag{1095}$$

Since

$$(\tau_n + \pi_n)(\tau_m + \pi_m) + (\tau_n - \pi_n)(\tau_m - \pi_m) = 2(\tau_n \pi_m + \pi_n \tau_n) \tag{1096}$$

it follows from Eq. (1063) that $W_1^{\text{sc}} = W_2^{\text{sc}} = 0$.

Next, it follows from Eqs. (1064)–(1067)

$$\left[\mathbf{N}_{e1n}^{(3)} \times (\mathbf{M}_{e1n}^{(3)})^* \right] \cdot \hat{\mathbf{e}}_r = -f_n z_m^* [\cos^2\phi\ \tau_m \tau_n + \sin^2\phi\ \pi_n \pi_m] \tag{1097}$$

which gives a contribution W_3^{sc} to W^{sc} in Eq. (1086):

$$W_3^{\text{sc}} = -r^2 \frac{c}{8\pi} \frac{ck_2}{\omega\mu_2} \Re \left\{ \sum_{m=1}^{\infty} \sum_{n=1}^{\infty} \frac{1}{2} f_n z_m^* E_n E_m^* i a_n a_n^* \int_0^{\pi} [\tau_n \tau_m + \pi_m \pi_n] \sin\theta d\theta \right\}. \tag{1098}$$

To evaluate this integral, one may use the relations

$$\int_0^{2\pi} \int_0^{\pi} \mathbf{M}_{e1n}^{(3)} \cdot (\mathbf{M}_{e1m}^{(3)})^* \sin\theta d\theta d\phi = 0 \tag{1099}$$

$$\int_0^{2\pi}\int_0^{\pi} \mathbf{M}^{(3)}_{e1n}\cdot(\mathbf{M}^{(3)}_{e1n})^*\sin\theta d\theta d\phi = \pi|z_n|^2\frac{2n(n+1)(n+1)!}{(2n+1)(n-1)!}. \tag{1100}$$

Substitution from Eq. (1064) in Eqs. (1099) and (1100) gives

$$\int_0^{\pi}(\pi_n\pi_m + \tau_n\tau_m)\sin\theta d\theta = 0 \tag{1101}$$

$$\int_0^{\pi}(\pi_n^2 + \tau_n^2)\sin\theta d\theta = 2\pi\frac{2n(n+1)(n+1)!}{(2n+1)(n-1)!} = 2\pi\frac{2n^2(n+1)^2}{2n+1}. \tag{1102}$$

Substitution of these results in (1098) yields

$$W_3^{\mathrm{sc}} = 2\pi r^2\ S^i\ \Re\left\{\sum_{n=1}^{\infty} if_n z_n^*(2n+1)|a_n|^2\right\}. \tag{1103}$$

From the asymptotic results in (1073) and (1074), it follows that

$$if_n z_n^* \sim -\frac{1}{(k_2 r)^2} \tag{1104}$$

so that Eq. (1103) becomes

$$W_3^{\mathrm{sc}} = 2\pi\frac{S^i}{k_2^2}\sum_{n=1}^{\infty}(2n+1)|a_n|^2. \tag{1105}$$

Finally, it follows from Eqs. (1064)–(1067)

$$\left[\mathbf{M}^{(3)}_{o1n}\times\left(\mathbf{N}^{(3)}_{o1n}\right)^*\right]\cdot\hat{\mathbf{e}}_r = z_n f_m^*[\cos^2\phi\ \tau_m\tau_n + \sin^2\phi\ \pi_n\pi_m] \tag{1106}$$

which gives a contribution W_4^{sc} to W^{sc} in Eq. (1086):

$$W_4^{\mathrm{sc}} = -r^2\frac{c}{8\pi}\frac{ck_2}{\omega\mu_2}\Re\left\{\sum_{m=1}^{\infty}\sum_{n=1}^{\infty}\frac{1}{2}z_n f_m^* E_n E_m^* b_n(ib_m)^*\int_0^{\pi}[\tau_n\tau_m + \pi_m\pi_n]\sin\theta d\theta\right\} \tag{1107}$$

which, in a similar way as above, leads to

$$W_4^{\mathrm{sc}} = 2\pi\frac{S^i}{k_2^2}\sum_{n=1}^{\infty}(2n+1)|b_n|^2. \tag{1108}$$

Thus, the result for the sacttering cross section becomes

$$C^{\mathrm{sc}} = \frac{W^{\mathrm{sc}}}{S^i} = \frac{W_3^{\mathrm{sc}} + W_4^{\mathrm{sc}}}{S^i} = \frac{2\pi}{k_2^2}\sum_{n=1}^{\infty}(2n+1)\left(|a_n|^2 + |b_n|^2\right). \tag{1109}$$

Mueller Matrix

From Eqs. (976)–(992) it follows that, for scattering by a spherical particle, for which $S_3 = S_4 = 0$, the Mueller matrix becomes

$$\begin{pmatrix} I^{sc} \\ Q^{sc} \\ U^{sc} \\ V^{sc} \end{pmatrix} = \begin{pmatrix} S_{11} & S_{12} & 0 & 0 \\ S_{12} & S_{11} & 0 & 0 \\ 0 & 0 & S_{33} & S_{34} \\ 0 & 0 & -S_{34} & S_{33} \end{pmatrix} \begin{pmatrix} I^i \\ Q^i \\ U^i \\ V^i \end{pmatrix}, \tag{1110}$$

where

$$S_{11} = \frac{1}{2}\left(|S_1|^2 + |S_2|^2\right) \tag{1111}$$

$$S_{12} = \frac{1}{2}\left(|S_2|^2 - |S_1|^2\right) \tag{1112}$$

$$S_{33} = \Re\left(S_1 S_2^*\right) \tag{1113}$$

$$S_{34} = \Im\left(S_2 S_1^*\right) \tag{1114}$$

which, as expected, is in a form similar to that of the Stokes scattering matrix in Eq. (36) with $a_1 = a_2$ and $a_3 = a_4$.

Determination of Expansion Coefficients

The boundary conditions in Eq. (1057) at the surface of the sphere $r = a$ are as follows in component form:

$$E_\theta^i + E_\theta^s - E_\theta^{\text{int}} = 0 \quad ; \quad E_\phi^i + E_\phi^s - E_\phi^{\text{int}} = 0 \tag{1115}$$

$$H_\theta^i + H_\theta^s - H_\theta^{\text{int}} = 0 \quad ; \quad H_\phi^i + H_\phi^s - H_\phi^{\text{int}} = 0. \tag{1116}$$

For a given value of n, there are four unknown coefficients a_n, b_n, c_n, and d_n. Thus, one needs four independent equations, which are obtained by substituting the field expansions into the four boundary condition equations above. Using the orthogonality of $\sin\phi$ and $\cos\phi$ and the orthogonality relation for the sum and difference of π_n and τ_n in Eq. (1063), one obtains the following four equations:

$$j_n(mx)c_n + h_n^{(1)}(x)b_n = j_n(x) \tag{1117}$$

$$\mu_2[mxj_n(mx)]'c_n + \mu_1[xh_n^{(1)}(x)]'b_n = \mu_1[xj_n(x)]' \tag{1118}$$

$$\mu_2 m j_n(mx)d_n + \mu_1 h_n^{(1)}(x)a_n = \mu_1 j_n(x) \tag{1119}$$

$$[mxj_n(mx)]'d_n + m\left[xh_n^{(1)}(x)\right]' a_n = m[xj_n(x)]', \tag{1120}$$

where the prime indicates differentiation with respect to the argument in the parentheses, and the *size parameter* x and the *relative refractive index* m are given by

$$x = k_2 a = \frac{2\pi n_2 a}{\lambda} \quad ; \quad m = \frac{k_1}{k_2} = \frac{n_1}{n_2} \tag{1121}$$

with $n_1 = \sqrt{(\varepsilon_1 + i((4\pi\sigma_1)/\omega))\mu_2}$ and $n_2 = \sqrt{\varepsilon_2\mu_2}$ being the refractive indices of the medium inside the particle and the surrounding medium, respectively.

Solving the four equations (1117)–(1120), one obtains the following expressions for the scattering coefficients:

$$a_n = \frac{\mu_2 m^2 j_n(mx)[xj_n(x)]' - \mu_1 j_n(x)[mxj_n(mx)]'}{\mu_2 m^2 j_n(mx)\left[xh_n^{(1)}(x)\right]' - \mu_1 h_n^{(1)}(x)[mxj_n(mx)]'} \tag{1122}$$

$$b_n = \frac{\mu_1 j_n(mx)[xj_n(x)]' - \mu_2 j_n(x)[mxj_n(mx)]'}{\mu_1 j_n(mx)\left[xh_n^{(1)}(x)\right]' - \mu_2 h_n^{(1)}(x)[mxj_n(mx)]'} \tag{1123}$$

and for the coefficients of the internal field

$$c_n = \frac{\mu_1 j_n(x)\left[xh_n^{(1)}(x)\right]' - \mu_1 h_n^{(1)}(x)[xj_n(x)]'}{\mu_1 j_n(mx)\left[xh_n^{(1)}(x)\right]' - \mu_2 h_n^{(1)}(x)[mxj_n(mx)]'} \tag{1124}$$

$$d_n = \frac{\mu_1 m j_n(x)\left[xh_n^{(1)}(x)\right]' - \mu_1 m h_n^{(1)}(x)[xj_n(x)]'}{\mu_2 m^2 j_n(mx)\left[xh_n^{(1)}(x)\right]' - \mu_1 h_n^{(1)}(x)[mxj_n(mx)]'}. \tag{1125}$$

Note that the denominators of c_n and b_n are identical and so are those of a_n and d_n. If for a particular n the frequency (or radius) is such that one of these denominators is very small, the corresponding normal mode will dominate the scattered field. The a_n mode is dominant if the condition

$$\frac{\left[xh_n^{(1)}(x)\right]'}{h_n^{(1)}(x)} = \frac{\mu_1[mxj_n(mx)]'}{\mu_2 m^2 j_n(mx)} \tag{1126}$$

is approximately fulfilled, and the b_n mode is dominant if the condition

$$\frac{\left[xh_n^{(1)}(x)\right]'}{h_n^{(1)}(x)} = \frac{\mu_2[mxj_n(mx)]'}{\mu_1 j_n(mx)} \tag{1127}$$

is approximately fulfilled. The frequencies for which both of these conditions are exactly satisfied are called the *natural* frequencies of the sphere. The natural frequencies are *complex*, and if their imaginary parts are small compared with the real parts, the latter correspond approximately to the real frequencies of incident electromagnetic waves that will excite the various electromagnetic modes.

By introducing the *Ricatti–Bessel functions*

$$\psi_n(z) = zj_n(z) \quad ; \quad \xi_n(z) = zh_n^{(1)}(z) \tag{1128}$$

and letting $\mu_2 = \mu_1$, one can simplify the expressions for the scattering coefficients:

$$a_n = \frac{m\psi_n(mx)\psi_n'(x) - \psi_n(x)\psi_n'(mx)}{m\psi_n(mx)\xi_n'(x) - \xi_n(x)\psi_n'(mx)} \tag{1129}$$

$$b_n = \frac{\psi_n(mx)\psi_n'(x) - m\psi_n(x)\psi_n'(mx)}{\psi_n(mx)\xi_n'(x) - m\xi_n(x)\psi_n'(mx)}. \tag{1130}$$

Note that both a_n and b_n vanish as $m \to 1$, as expected.

Rayleigh Scattering

When the size of the sphere becomes very small compared to the wavelength of the incident light, so that $a \ll \lambda$, the results will reduce to those known as Rayleigh scattering. In this limit, the size parameter $x = k_2 a = ((2\pi n_2 a)/\lambda) \ll 1$, and the scattering coefficients in Eqs. (1129)–(1130) simplify considerably. From the expansion

$$j_n(x) = \frac{\sqrt{\pi}}{2}\left(\frac{x}{2}\right)^n \sum_{k=0}^{\infty} \frac{(-\frac{1}{4}x^2)^k}{k!\Gamma(n+3/2+k)} \tag{1131}$$

it follows that for small values of x

$$j_n(x) \approx \frac{\sqrt{\pi}}{2}\left(\frac{x}{2}\right)^n \left[\frac{1}{\Gamma(n+3/2)} - \frac{x^2}{4\Gamma(n+5/2)}\right]. \tag{1132}$$

Using the formula

$$\Gamma\left(\frac{1}{2}+n\right) = \frac{1\cdot 3\cdot 5\cdot \ldots \cdot (2n-1)}{2^n}\sqrt{\pi} = \frac{(2n-1)!!}{2^n}\sqrt{\pi} \tag{1133}$$

one obtains

$$\Gamma\left(n+\frac{3}{2}\right) = \Gamma\left(\frac{1}{2}+(n+1)\right) = \frac{(2n+1)!!}{2^{n+1}}\sqrt{\pi} \tag{1134}$$

$$\Gamma(n+\frac{5}{2}) = \Gamma(\frac{1}{2}+(n+2)) = \frac{(2n+3)!!}{2^{n+2}}\sqrt{\pi} \tag{1135}$$

so that Eq. (1136) gives

$$j_n(x) \approx \frac{x^n}{(2n+1)!!}\left[1 - \frac{x^2}{2(2n+3)}\right] \tag{1136}$$

and hence

$$\psi_n(x) = x j_n(x) \approx \frac{x^{n+1}}{(2n+1)!!}\left[1 - \frac{x^2}{2(2n+3)}\right] \tag{1137}$$

$$\psi_n'(x) \approx \frac{(n+1)x^n}{(2n+1)!!}\left[1 - \frac{x^2}{2(2n+3)}\right]. \tag{1138}$$

Next, one may use the formula

$$h_n^{(1)}(x) = j_n(x) - i(-1)^n j_{-(n+1)}(x) \tag{1139}$$

and Eq. (1132) to obtain

$$j_{-(n+1)}(x) \approx \frac{\sqrt{\pi}}{2}\left(\frac{x}{2}\right)^{-(n+1)} \left[\frac{1}{\Gamma(-(n+1)+3/2)} - \frac{x^2}{4\Gamma(-(n+1)+5/2)}\right]. \tag{1140}$$

Using the formula

$$\Gamma\left(\frac{1}{2}-n\right) = \frac{(-1)^n 2^n}{(2n-1)!!}\sqrt{\pi} \tag{1141}$$

one obtains

$$\Gamma(-(n+1)+3/2) = \Gamma\left(\frac{1}{2}-n\right) = \frac{(-1)^n 2^n}{(2n-1)!!}\sqrt{\pi} \tag{1142}$$

$$\Gamma(-(n+1)+5/2) = \Gamma\left(\frac{1}{2}-(n-1)\right) = \frac{(-1)^{n-1}2^{n-1}}{(2n-3)!!}\sqrt{\pi} \tag{1143}$$

so that Eq. (1140) gives

$$j_{-(n+1)}(x) \approx (-1)^n (2n-1)!!\, x^{-(n+1)} \left[1+\frac{x^2}{2(2n-1)}\right]. \tag{1144}$$

Substitution from Eqs. (1136) and (1144) in Eq. (1139) yields

$$h_n^{(1)}(x) = \frac{x^n}{(2n+1)!!}\left[1-\frac{x^2}{2(2n+3)}\right] - i(2n-1)!!\, x^{-(n+1)}\left[1+\frac{x^2}{2(2n-1)}\right] \tag{1145}$$

and hence

$$\xi_n = x h_n^{(1)}(x) = \frac{x^{n+1}}{(2n+1)!!}\left[1-\frac{x^2}{2(2n+3)}\right] - i(2n-1)!!\, x^{-n}\left[1+\frac{x^2}{2(2n-1)}\right] \tag{1146}$$

$$\xi_n' = \left[x h_n^{(1)}(x)\right]' = \frac{(n+1)x^n}{(2n+1)!!}\left[1-\frac{x^2}{2(2n+3)}\right] + in(2n-1)!!\, x^{-(n+1)}\left[1+\frac{x^2}{2(2n-1)}\right]. \tag{1147}$$

From Eq. (1129) it follows that

$$a_n = \frac{N_a}{D_a}, \tag{1148}$$

where

$$N_a = m\psi_n(mx)\psi_n'(x) - \psi_n(x)\psi_n'(mx) \tag{1149}$$

$$D_a = m\psi_n(mx)\xi_n'(x) - \xi_n(x)\psi_n'(mx) \tag{1150}$$

Substitution from Eqs. (1137) and (1138) in Eq. (1149) and from Eqs. (1137), (1138), (1146), and (1147) in Eq. (1150) yields to the lowest order

$$N_a \approx \frac{x^{2n+1}(n+1)m^n}{[(2n+1)!!]^2}(m^2-1) \tag{1151}$$

$$D_a \approx \frac{x^{2n+1}(n+1)m^n}{[(2n+1)!!]^2}(m^2-1) + im^n\frac{n}{2n+1}(m^2+n+1). \tag{1152}$$

Thus, from Eq. (1148) one obtains to the lowest order

$$a_n = \frac{N_a}{D_a} \approx -i\frac{(2n+1)(n+1)}{n[(2n+1)!!]^2}x^{2n+1}\,\frac{m^2-1}{m^2+\frac{n+1}{n}}. \tag{1153}$$

From Eq. (1130)

$$b_n = \frac{\psi_n(mx)\psi_n'(x) - m\psi_n(x)\psi_n'(mx)}{\psi_n(mx)\xi_n'(x) - m\xi_n(x)\psi_n'(mx)} = \frac{N_b}{D_b}, \tag{1154}$$

it follows by substitution from Eqs. (1137) and (1138) in the expression for the numerator in Eq. (1154) that, to the lowest order, $N_b = b_n = 0$. Since $x \ll 1$, only the first term a_1 in each of Eqs. (1075) and (1076) contributes, that is

$$S_2 = \frac{3}{2} a_1 \tau_1(\cos\theta) \tag{1155}$$

$$S_1 = -\frac{3}{2} a_1 \pi_1(\cos\theta) \tag{1156}$$

where from Eq. (1153)

$$a_1 \approx -i\frac{2}{3}x^3 \frac{m^2 - 1}{m^2 + 2}. \tag{1157}$$

With $\tau_1(\cos\theta) = 1$ and $\pi_1(\cos\theta) = \cos\theta$, the expressions for S_1 and S_2 become

$$S_2 = -i\frac{m^2 - 1}{m^2 + 2} x^3 \cos\theta \; ; \; S_2 = -i\frac{m^2 - 1}{m^2 + 2} x^3 \; ; \; x = k_2 a = \frac{2\pi n_2 a}{\lambda}. \tag{1158}$$

Substituting these expressions for S_2 and S_1 into Eq. (960) with $S_4 = S_3 = 0$, one obtains for Rayleigh scattering (recalling that $\theta = \Theta$)

$$\begin{pmatrix} E_{\parallel}^{sc} \\ E_{\perp}^{sc} \end{pmatrix} = \frac{m^2 - 1}{m^2 + 2}(k_2 a)^3 \frac{e^{ik_2 r}}{k_2 r} \begin{pmatrix} \cos\Theta & 0 \\ 0 & 1 \end{pmatrix} \begin{pmatrix} E_{\parallel}^{i} \\ E_{\perp}^{i} \end{pmatrix}. \tag{1159}$$

Since the scattered intensity is proportional to the electric field amplitude squared, it follows from this result that for Rayleigh scattering the intensity of the scattered light is proportional to $k_2^4 = \left((2\pi n_2)/\lambda\right)^4 = \left((\omega n_2)/c\right)^4$.

Using Eqs. (1110)–(1114), one obtains the Mueller matrix for Rayleigh scattering:

$$\begin{pmatrix} I^{sc} \\ Q^{sc} \\ U^{sc} \\ V^{sc} \end{pmatrix} = C \begin{pmatrix} 1 + \cos^2\Theta & -\sin^2\Theta & 0 & 0 \\ -\sin^2\Theta & 1 + \cos^2\Theta & 0 & 0 \\ 0 & 0 & 2\cos\Theta & 0 \\ 0 & 0 & 0 & 2\cos\Theta \end{pmatrix} \begin{pmatrix} I^{i} \\ Q^{i} \\ U^{i} \\ V^{i} \end{pmatrix}, \tag{1160}$$

where

$$C = \frac{1}{2}(k_2 a)^6 \left| \frac{m^2 - 1}{m^2 + 2} \right|^2. \tag{1161}$$

As expected, the result in Eq. (1160) is of the same form as in Eq. (59) with $f = ((1 - \rho)/(1 + \rho)) = 1$, since the depolarization factor $\rho = 0$ for scattering by a homogeneous spherical particle.

B
Spectral Sampling Strategies

The absorption properties of a molecular gas depend very strongly upon wavelength because (i) the line strengths can vary drastically over a given band, and (ii) the absorption coefficient may change by many orders of magnitude over a small wavelength range. Also, the radiation field itself will generally depend strongly on wavelength. For example, at low spectral resolution, the solar radiation field approximates a continuous spectrum, but at higher resolution it reveals a rich structure due to the physical conditions in the solar atmosphere.

The terrestrial IR radiation field has these same general characteristics, consisting of (i) a near-blackbody component emitted by a land surface or a water body, and (ii) a more complicated component arising from the atmospheric emission due to the complex absorption properties of the medium through *Kirchoff's Law*. Thermal emission is high at the centers of strong absorption lines and low in the transparent spectral windows. Even though there is an abundance of photons emitted at a line center, their mean free path can be very short. On the other hand, very few photons are emitted within the spectral window regions, yet they can be transmitted over a long distance in a medium. Consequently, it is not obvious which wavelengths contribute significantly to a given quantity, implying that care must be taken not to truncate a spectral line wing too close to a line center, since an important part of the energy in the wings could be missed.

Here we provide a brief review of the basic concepts based in part on the material presented in Chapter 10 of Thomas and Stamnes [18].

Transmission in an Isolated Spectral Line

The spectral beam transmittance and absorptance at wavenumber $\tilde{\nu} = 1/\lambda$ are defined as

$$T_b(\tilde{\nu}) = \exp\left[-\tau(\tilde{\nu})\right]; \qquad A_b(\tilde{\nu}) = 1 - \mathcal{T}_b(\tilde{\nu}). \tag{1162}$$

For a single spectral line, the optical thickness along the path $0 \to s$ is given by

$$\tau(\tilde{\nu}, s) = \int_0^s ds' \, S n(s') \Phi(\tilde{\nu}) \tag{1163}$$

where $n(s')$ is the number density in units of m^{-3}, S is the (frequency-integrated) line strength in units of $\mathrm{m}^2 \cdot \mathrm{s}^{-1}$, and $\Phi(\tilde{\nu})$ is the line profile function in units of s^{+1}. Both S and $\Phi(\tilde{\nu})$ will, in general, depend on the pressure and temperature (and

Radiative Transfer in Coupled Environmental Systems: An Introduction to Forward and Inverse Modeling,
First Edition. Knut Stamnes and Jakob J. Stamnes.

thus on the position s'), but assuming a homogeneous (e.g., a horizontal) optical path, we have

$$\tau(\tilde{\nu}) = S\mathcal{N}\Phi(\tilde{\nu}), \tag{1164}$$

where $\mathcal{N}$ is the column number (m^{-2}) of absorbing molecules. If we describe the path in terms of the **column mass** u, the line strength is defined **per unit mass**, and

$$\tau(\tilde{\nu}) = Su\Phi(\tilde{\nu}).$$

The **mean beam absorptance** and **mean beam transmittance** for the spectral line are defined as

$$\langle A_b \rangle \equiv 1 - \langle T_b \rangle \equiv \frac{1}{\Delta\tilde{\nu}} \int_{\Delta\tilde{\nu}} d\tilde{\nu} A_b(\tilde{\nu}) = \frac{1}{\Delta\tilde{\nu}} \int_{\Delta\tilde{\nu}} d\tilde{\nu}[1 - e^{-\tau(\tilde{\nu})}] \equiv \frac{W}{\Delta\tilde{\nu}}. \tag{1165}$$

The product $W = \langle A_b \rangle \Delta\tilde{\nu}$ is called the **equivalent width**, which is *equal to the area of a rectangle with complete absorption inside and zero absorption outside.*

Real absorption spectra reveal that line strengths are distributed over a wide range of values and line separations are far from constant. If we average over an interval that contains a large number of nonoverlapping lines of varying strengths, it is possible to define a continuous distribution function of line strengths, $p(S)$, such that the number of lines with strengths between S and $S + dS$ is $p(S)dS$. In the limit of an infinite number of lines, the mean beam absorptance becomes

$$\langle A_b(u) \rangle = \int_0^\infty dSp(S) \int_{\Delta\tilde{\nu}} \frac{d\tilde{\nu}}{\Delta\tilde{\nu}} \{1 - \exp[-Su\Phi(\tilde{\nu})]\}. \tag{1166}$$

Several analytic line-strength distributions are in use, including the Malkmus distribution defined as

$$p(S) = \left(\frac{1}{S}\right) \exp\left(-\frac{S}{\overline{S}}\right) \tag{1167}$$

which is normalized so that $\int_0^\infty dSp(S) = 1$. The average line strength is defined as

$$\overline{S} = \int_0^\infty dSSp(S). \tag{1168}$$

The Random Band Model

In some irregular spectral bands, such as those due to absorption by water vapor, the line positions appear to vary randomly over the spectrum. Thus, in the **statistical**, **random**, or **Goody–Meyer band model**, one assumes that the band interval of width $\Delta\tilde{\nu}$ consists of n lines of average separation δ, so that $\Delta\tilde{\nu} = n\delta$, and that the line positions are **uncorrelated**. Then the transmission of the band can be written in terms of the products of the individual line transmittances

$$\langle T_b \rangle = \langle T_1 \rangle \langle T_2 \rangle \cdots \langle T_n \rangle = \left[\frac{1}{\Delta\tilde{\nu}} \int_{\Delta\tilde{\nu}} d\tilde{\nu} \int_0^\infty dSp(S) e^{-Su\Phi(\tilde{\nu})}\right]^n = \left[1 - \frac{\langle A_b \rangle}{n}\right]^n. \tag{1169}$$

Taking the limit $n \to \infty$, and noting that $(1 - x/n)^n \to e^{-x}$, one finds

$$\langle \mathcal{T}_b(u) \rangle = e^{-\langle A_b(u) \rangle} \tag{1170}$$

which says that *the beam transmittance of randomly placed lines is equal to the exponential attenuation due to the single-line beam absorptance.*

B.1
The MODTRAN Band Model

MODTRAN [100] is a moderate-resolution atmospheric transmission, radiance, and irradiance model developed as the industry standard for arbitrary (0.1 cm^{-1} to spectrally broad) bandwidth radiative transfer applications. The model is unclassified and used by academia and corporations in the United States, as well as by international scientific communities. MODTRAN's spectral range covers 0–50,000 cm^{-1}, which spans from ultraviolet (UV) to far-infrared (IR) wavelengths (from 0.2 to greater than 40 μm). MODTRAN is based on molecular band model techniques and provides rapid calculations of atmospheric absorption, extinction, and emission.

Continuum molecular absorption features (e.g., ozone in the UV and visible, plus chlorofluorocarbons [CFCs] in the IR) are equally well represented in MODTRAN. The transmission accuracy over the entire spectral range is typically better than 1% at all spectral resolutions when compared to line-by-line (LBL) calculations. Solar irradiance at the top of the atmosphere (TOA) is defined for the same binning as the band models, such that correlations in telluric and solar line structures can be correctly convolved. MODTRAN also incorporates the inherent optical properties (IOPs) of two important polarizing agents, that is, aerosols and clouds.

MODTRAN includes multiple scattering options (based on DISORT [18, 161]) and allows for vertical specification of aerosol types, amounts, and IOPs. Such atmospheric profiles not only help in determining the extinction due to water vapor and other absorbing molecules but can also be used to calculate the optical slant path as a function of wavelength or a user-specified path or viewing geometry associated with atmospheric refraction for altitudes up to 100 km.

Three band model parameters are used in MODTRAN: (i) an effective absorption coefficient S/d, (ii) a line density, and (iii) a line width d. For a given wavenumber interval $\Delta\tilde{\nu}_i$, the effective absorption coefficient is a measure of the total strength of the lines, the line density is the average number of lines, and the line width d_i is the average wavenumber bin width:

$$\frac{S_i}{d_i} \equiv \frac{1}{\Delta\tilde{\nu}} \sum_{J=1}^{N} S_J^i(T); \qquad \frac{1}{d_i} \equiv \frac{\left[\sum S_J^i\right]^2}{\Delta\tilde{\nu} \sum (S_J^i)^2}, \tag{1171}$$

where $S_J^i \propto \exp((1/T) - (1/T_o))$ is the line strength of the Jth line of a specific molecule in wavenumber bin i at temperature T, and T_o is a reference temperature.

The molecular transmittance is based on the random band model in Eq. (1169) for a finite number of lines within a spectral interval $\Delta\tilde{\nu}$:

$$\langle T_b \rangle = \left[1 - \frac{\langle A_b \rangle}{\langle n \rangle}\right]^{\langle n \rangle}, \tag{1172}$$

where $\langle A_b \rangle \Delta\tilde{\nu}$ is the single-line equivalent width and $\langle n \rangle$ is the path-averaged effective number of lines in the bin $\langle n \rangle = \Delta\tilde{\nu}\langle 1/d \rangle$, where $\langle 1/d \rangle$ is the path-averaged line spacing or line density. For mixed Lorentz–Doppler absorption, the single-line equivalent width $\langle A_b \rangle \Delta\tilde{\nu}$ is computed from

$$\langle A_b \rangle \Delta\tilde{\nu} = \Delta\tilde{\nu}[1 - \langle T_b \rangle] = \int_{\Delta\tilde{\nu}} d\tilde{\nu} \left[1 - e^{-Su\Phi_V(\tilde{\nu})}\right], \tag{1173}$$

where $\Phi_V(\tilde{\nu})$ is the Voigt profile.

The contributions from lines located in the wing interval outside of a given bin but within $\pm$ 25 cm^{-1} are computed separately from line wing band model parameters determined by integrating the Lorentz profile over the wing interval. Wing contributions beyond $\pm$ 25 cm^{-1} are ignored.

Finally, it should be noted that MODTRAN, which has a "fixed-wavenumber" sampling of 0.1 cm^{-1} (and a nominal resolution of 0.2 cm^{-1}), is assumed to be **quasi-monochromatic** and therefore automatically **compatible with multiple scattering treatments.**

B.2
The *k*-Distribution Method

The *k*-**distribution** method and its associated **correlated** *k*-**distribution** method [18, 101] provide simplified computation with adequate accuracy for many applications such as computation of warming/cooling rates, and thus require much less computer time than line-by-line (LBL) methods. They can also accommodate multiple scattering in a straightforward manner. An LBL calculation is needed to derive some parameters, but since the simplified calculations are repeated many times, the cost savings can be substantial. Consider a spectral interval $\Delta\tilde{\nu} = \tilde{\nu}_2 - \tilde{\nu}_1$, which is large enough to contain a significant number (say >20) of spectral lines but small enough that the Planck function is essentially constant over $\Delta\tilde{\nu}$. The beam transmittance over a homogeneous mass path u is

$$\langle T_b(u) \rangle = \frac{1}{\Delta\tilde{\nu}} \int_{\tilde{\nu}_1}^{\tilde{\nu}_2} d\tilde{\nu}\, e^{-k(\tilde{\nu})u}, \tag{1174}$$

where $k(\tilde{\nu})$ denotes the mass extinction coefficient, which is equal to the mass absorption coefficient for a purely absorbing medium.

Accurate computation of $\langle T_b(u) \rangle$ using Eq. (1174) would require a division of the spectral interval into subintervals $\delta\tilde{\nu}$ small enough so that $k(\tilde{\nu})$ is essentially constant. Such a division requires $\delta\tilde{\nu}$ to be $\approx 10^{-3} - 10^{-5}$ cm^{-1}, leading to a total of $\Delta\tilde{\nu}/\delta\tilde{\nu} \approx 10^4 - 10^6$ quadrature points for a small part (say 10 cm^{-1}) of the spectrum, and this kind of division must be repeated over the entire band, for all absorption bands and over the full range of u-values. Also, it must be repeated for the range of pressures and temperatures encountered in the atmosphere.

Since such an approach would be computationally prohibitive, a more efficient procedure is desired. A transformation of Eq. (1174) which recognizes that **the same value of k is encountered repeatedly in a spectral interval** is a key to speeding up the computation.

By combining all values of k into groups and by ordering the groups into monotonically increasing values of k, one obtains a much more "orderly" function $f(k)$ than the wildly varying $k(\tilde{\nu})$. Choosing an interval Δk that is suitably small, the k-**distribution** can be formally defined by the following grouping algorithm [18, 101]:

$$f(k) \equiv \frac{1}{\Delta\tilde{\nu}} \sum_{l=1}^{M} \left|\frac{d\tilde{\nu}}{dk}\right| W_l(k), \tag{1175}$$

where $W_l(k)$ is a "window" function, equal to unity when $k^l_{\min} \leq k \leq k^l_{\max}$, and zero otherwise. In the lth subinterval, the absorption coefficient varies from $k^l_{\min}$ to $k^l_{\max}$, and M is the number of subintervals. Since k is considered to be a continuous variable, Eq. (1174) can be rewritten as a finite sum

$$\langle T_b(u)\rangle \approx \sum_{j=1}^{N} \Delta k_j f(k_j) e^{-k_j u}, \tag{1176}$$

where N is the total number of monotonic subintervals over the entire range of k-values. In the limit as $\Delta k \to 0$ (assuming that the number of lines within Δk is always large), the sum above becomes an integral

$$\langle T_b(u)\rangle = \int_{k_{\min}}^{k_{\max}} dk\, f(k) e^{-ku}, \tag{1177}$$

where $k_{\min}$ and $k_{\max}$ are the minimum and maximum values of k over the entire spectral interval $\Delta\tilde{\nu}$.

By summing $f(k)$ over all binned values of k, one should get unity:

$$\sum_{j=1}^{N} f(k_j)\Delta k_j = 1, \quad \text{or for } \Delta k_j \to 0, \quad \int_{k_{\min}}^{k_{\max}} dk\, f(k) = 1. \tag{1178}$$

By summing the distribution up to some value of $k_n < k_{\max}$, one obtains the **cumulative k-distribution** as

$$g(k_n) \equiv \sum_{j=1}^{n} f(k_j)\Delta k_j, \quad \text{or for } \Delta k_j \to 0, \quad g(k) = \int_0^k dk' f(k'). \tag{1179}$$

One can now write Eqs. (1176) and (1177) as

$$\langle T_b(u)\rangle \approx \sum_{j=1}^{N} e^{-k_j u}\Delta g_j, \quad \text{or for } \Delta g_j \to 0, \quad \langle T_b(u)\rangle = \int_0^1 dg e^{-k(g)u}. \tag{1180}$$

Note that the upper limit of unity is consistent with g being a cumulative k-distribution, that is the total number of k-values **smaller than** k.

Writing the sum in Eq. (1180) term by term

$$\langle \mathcal{T}_b(u) \rangle \approx \Delta g_1 e^{-k_1 u} + \Delta g_2 e^{-k_2 u} + \cdots + \Delta g_N e^{-k_N u} \tag{1181}$$

one obtains an approximation to the transmittance known as the **exponential-sum fit transmittance** (ESFT) approximation [102]. Clearly, the nongray problem has been reduced to a finite number of gray problems for which we have many computational tools. If we knew the transmittance measured in the laboratory under low spectral resolution, or had access to LBL calculations of $\langle T_b(u) \rangle$, we could in principle perform a nonlinear least-squares fit of Eq. (1181) to the "data" to yield the 'coefficients' of the fit $(g_1, g_2, \cdots; k_1, k_2, \cdots)$ at any desired accuracy. Unfortunately, such a nonlinear least-squares fit is mathematically ill-posed, and special analysis techniques must be applied for this method to be practical [102].

Fortunately, the availability of accurate synthetic absorption spectra, generated by LBL codes, means that we can compute the k-distribution directly, and the coefficients of the EFST approximation may be determined by numerical quadrature instead of least-squares fitting. It is a simple computational task to construct sorted tables of absorption coefficients to derive $f(k)$ and $g(k)$ from, say, a spectroscopic database (such as HITRAN). The **inverse k-distribution** $k(g)$ is also needed in order to perform spectral mapping. An "inverse table" of k versus g is easily constructed by computer methods.

Inhomogeneous Media – The Correlated-k Method

Unless the beam direction is horizontal, one must deal with the inhomogeneous nature of the atmosphere. When the beam makes an angle θ with the vertical, the mean beam transmittance in a plane-parallel medium over an inhomogeneous path becomes

$$\langle T_b(u, \theta) \rangle = \frac{1}{\Delta\tilde{\nu}} \int_{\Delta\tilde{\nu}} d\tilde{\nu} \left\{ \exp\left[-\int_0^u du' S(u')\Phi(u', \tilde{\nu}) \sec\theta \right] \right\}. \tag{1182}$$

The average transmittance in Eq. (1182) may be written in k-distribution form for an inhomogeneous (vertical) path, in analogy to Eq. (1177), as

$$\langle T_b(u) \rangle = \int_{k_{\min}}^{k_{\max}} dk\, f^*(k) \exp\left[-\int_0^u du' k(u') \right] \tag{1183}$$

or, in the finite-difference form as

$$\langle T_b(u_{l_1}, u_{l_2}) \rangle = \sum_{j=1}^{N} f^*(k_j) \exp\left[-\sum_{l=l_1}^{l_2} k_l \Delta u_l \right], \tag{1184}$$

where the sum extends from the center of the layer identified by u_{l_1} to the center of the layer u_{l_2}, and the function f^* denotes the distribution of the absorption coefficients *along the inhomogeneous path*. The absorption coefficients k_j are appropriate to the jth layer, because each layer is assumed to be thin enough to be considered homogeneous. In terms of the cumulative distribution variable $g^*(k^*) = \int_0^{k^*} dk' f^*(k')$, one can write Eq. (1184) as

$$\langle T_b(u)\rangle = \int_0^1 dg^* \exp\left[-\int_0^u du' k(g^*, u')\right]. \tag{1185}$$

Note the difference between Eq. (1185) and Eq. (1180): $\langle \mathcal{T}_b(u)\rangle = \int_0^1 dg e^{-k(g)u}$ applying to a homogeneous path:

- $k(g^*, u')$ refers to the distribution appropriate to the particular level u', and
- the distribution g^* is the cumulative distribution of k-values *for the inhomogeneous line of sight* associated with the beam angle θ.

The finite-difference form of Eq. (1185) is

$$\langle T_b(u)\rangle = \sum_{j=1}^{N} \Delta g_j^* \exp\left[-\sum_{l=l_1}^{l_2} k_l(g_j^*)\Delta u_l\right]. \tag{1186}$$

The **correlated**-k (c-k) method consists of replacing Eqs. (1185) and (1186) with

$$\langle T_b(u)\rangle = \int_0^1 dg \exp\left[-\int_0^u du' k(g, u')\right] \tag{1187}$$

$$\langle T_b(u)\rangle \approx \sum_{j=1}^{N} \Delta g_j \exp\left[-\sum_{l=l_1}^{l_2} k_l(g_j)\Delta u_l\right]. \tag{1188}$$

The replacement of the variable g^* with g implies that the single variable g maps into the distribution functions at **all** levels u'. Since g is effectively a wavenumber variable, this replacement implies that there is a one-to-one correspondence, or mapping, of wavenumbers from one level to another. Thus, the optical depth at a specific wavenumber g is given by the integral of $k(g, u')$ over the appropriate range of u', with g fixed.

For an isolated line, the monotonic ordering by strength of absorption coefficients retains the relative spectral alignment of absorption lines between different levels in the atmosphere so that there is a perfect **spectral correlation** at different pressure levels. Unfortunately, this one-to-one uniqueness does *not* work for a general molecular band, except in the weak-line and strong-line limits. The fundamental assumption of the c-k method is that a single mapping will produce a monotonically increasing k-spectrum in every layer. However, a mapping that produces a monotonically increasing k-spectrum for one layer may produce a k-spectrum for other layers in which neighboring k-values fluctuate by large amounts.

B.3
Spectral Mapping Methods

Like the c-k distribution method, spectral mapping methods identify spectral intervals with similar IOPs without making any assumptions about the spectral correlation along the optical path [103, 104]. Instead, a level-by-level comparison of monochromatic IOPs is performed, and only spectral regions that remain correlated at all points along the inhomogeneous path are combined. Intervals with similar IOPs are gathered into bins, and a single monochromatic multiple scattering computation may be done for each bin. Compared to the c-k method, this approach is less efficient because it requires a fine spectral binning.

B.4
Principal Component (PC) Analysis

A principal component (PC) analysis method was introduced by Natraj *et al.* [105] to overcome the limitations of the c-k and spectral mapping methods. Suppose a dataset consists of s IOPs in L layers at N_λ wavenumbers, and let this dataset be denoted by $F_{\ell i}$, where $\ell = 1, \ldots, s \times L$ and $i = 1, \ldots, s \times N_\lambda$. Then a set of empirical orthogonal functions (EOFs) $\varphi_k, k = 1, \ldots, s \times L$, which are eigenvectors of the algebraic eigenvalue problem

$$\mathbf{C}\vec{\varphi} = \lambda\vec{\varphi}, \tag{1189}$$

where $\mathbf{C}$ is the covariance matrix given by

$$C_{\ell j} = \overline{(F_{\ell i} - \overline{F}_\ell)(F_{ji} - \overline{F}_j)} \tag{1190}$$

and the overbar denotes an average over all wavenumbers, that is, $\overline{F}_\alpha = \sum_i F_{\alpha i}$; $\alpha = \ell$ or j. If λ_k is an eigenvalue corresponding to the eigenvector φ_k, then the PCs are the projections of the original dataset onto the EOFs (scaled by the square root of the eigenvalue λ_k)

$$P_{ki} = \sum_{\ell=1}^{sL} \frac{\varphi_{k\ell} F_{\ell i}}{\sqrt{\lambda_k}}, \tag{1191}$$

where $\varphi_{k\ell}$ is the ℓth component of the kth EOF and P_{ki} is the ith component of the kth PC.

In Chapter 6, an example is provided to illustrate how this PC analysis technique can be used to reduce the computational burden by exploiting the redundancy of the IOPs occurring within a given spectral interval.

B.5
Optimal Spectral Sampling

Optimal spectral sampling (OSS), developed by Moncet *et al.* [294], is a computationally rapid and accurate method for modeling sensor-band transmittances and radiances. It is currently used for operational algorithms for the Cross-track Infrared Sounder (CrIS) and the Advanced Technology Microwave Sounder (ATMS). The band-averaged beam transmittance over a homogeneous mass path u is given by Eq. (1174)

$$\langle T_b(u) \rangle = \frac{1}{\Delta\tilde{\nu}} \int_{\tilde{\nu}_1}^{\tilde{\nu}_2} d\tilde{\nu}\, e^{-k(\tilde{\nu})u}, \tag{1192}$$

where $k(\tilde{\nu})$ denotes the mass extinction coefficient, and $\Delta\tilde{\nu}$ includes several absorption lines. The discrete version of Eq. (1192) is given by Eq. (1180), which we rewrite as (setting $\Delta g_j = w_j$)

$$\langle T_b(u) \rangle \approx \sum_{j=1}^{N} e^{-k_j u} w_j. \tag{1193}$$

As discussed above, Wiscombe and Evans [102] used a nonlinear fitting technique to determine the values of k_j and the associated weights w_j, while in the k-distribution approach k_j and w_j are obtained by numerical integration over the cumulative distribution $g(k)$ within the spectral interval $\Delta\tilde{\nu}$. To avoid the problems discussed above with the c-k approach, which relies on the assumption that absorption coefficients in different altitude regimes are perfectly correlated, the OSS approach is based on replacing Eq. (1193) with

$$\langle T_b(u) \rangle \approx \sum_{j=1}^{N} e^{-k(\tilde{\nu}_j)u} w_j. \tag{1194}$$

Hence, rather than selecting k_j, the spectral points $\tilde{\nu}_j$ are selected, so that the extension from a single homogeneous layer to the inhomogeneous case consisting of multiple layers and mixtures of gases implies that Eq. (1194) becomes

$$\langle T_b \rangle \approx \sum_{j=1}^{N} e^{-\sum_\ell \sum_m k_{\ell m}(\tilde{\nu}_j) u_{\ell m}} w_j, \tag{1195}$$

where the subscripts ℓ and m refer to layers and molecular species, respectively. In Eq. (1195), the subscript j refers to a monochromatic computation at wavenumber $\tilde{\nu}_j$, called an OSS "node". The absorption coefficients $k_{\ell m}(\tilde{\nu}_j)$ can be derived directly from an LBL model avoiding any assumptions about the dependence of k on pressure, temperature, and gas mixtures used in the c-k approach.

For sounding applications, it is desirable to determine the OSS nodes and weights by directly fitting radiances rather than transmittances. The OSS approximation to the average radiance $\overline{R}$ in a channel

$$\overline{R} = \int_{\nu} \varphi(\tilde{\nu}) R(\tilde{\nu}) d\tilde{\nu} \Big/ \int_{\tilde{\nu}} \varphi(\tilde{\nu}) d\tilde{\nu} \approx \sum_{j=1}^{N} R(\tilde{\nu}_j) w_j, \tag{1196}$$

where φ is the sensor channel response at wavenumber $\tilde{\nu}$, $\tilde{\nu}_j$ is one of the N optimally selected wavenumbers, or "nodes", and w_j is an OSS weighting coefficient. The $R(\tilde{\nu})$ are the monochromatic radiances computed at the selected OSS nodes.

The gain in computational speed of the OSS method, relative to spectral integration of radiances, comes in part from the reduction of the sensor response integral to a summation over a small number of nodes, and in part from the use of prestored look-up-tables (LUTs) for the absorption coefficients at the nodes, from which radiances are calculated. One advantage of the OSS method is that it is capable of fitting reference-model radiance calculations however closely they may be required for any particular application.

The LBL model LBLRTM [295, 296] is used as the absorption coefficient reference for computing $R(\tilde{\nu})$ in the OSS training and for producing the pre-stored LUTs. LBLRTM is a state-of-the-art model that is widely used internationally as a reference standard for training various kinds of faster radiative transfer models.

C
Rough Surface Scattering and Transmission

C.1
Scattering and Emission by Random Rough Surfaces

Two basic analytic approaches have been used to study scattering from randomly rough surfaces [282, 297]:

- The Kirchhoff approach (KA) is based on the tangent plane approximation, according to which
  - the field at any point on the surface is approximated by the fields that would be present there if the actual surface were replaced by the tangent plane at that point;
  - a large radius of curvature relative to the wavelength of the incident radiation is required at every point on the surface.
- The small perturbation method (SPM), which is based on the assumption that
  - the variations in surface height are much smaller than the wavelength of the incident radiation;
  - the slopes of the rough surface are relatively small.

Kirchhoff Approach

As illustrated in Figure 15 we consider a plane wave incident on a random rough surface:

$$\mathbf{E}_i = \hat{e}_i E_0 e^{i\mathbf{k}_i \cdot \mathbf{r}}, \tag{1197}$$

where $\mathbf{k}_i$ denotes the incident wave vector and $\hat{e}_i$ is the polarization (i.e., the direction of the electric field vector). In the far-field region, the scattered field $\mathbf{E}_s(\mathbf{r})$ in direction $\hat{k}_s$ (medium 1) and the transmitted field $\mathbf{E}_t(\mathbf{r})$ in direction $\hat{k}_t$ (medium 2) can be expressed as integrals of the surface fields as follows:

$$\begin{aligned}\mathbf{E}_s(\mathbf{r}) = {} & \frac{ik_1 e^{ik_1 r}}{4\pi r}(\tilde{I} - \hat{k}_s\hat{k}_s) \\ & \times \int_{S'} dS' \left\{ \hat{k}_s \times [\hat{n} \times \mathbf{E}(\mathbf{r}')] + \eta_1[\hat{n} \times \mathbf{H}(\mathbf{r}')] \right\} e^{-i\mathbf{k}_{1s} \cdot \mathbf{r}'}\end{aligned} \tag{1198}$$

Radiative Transfer in Coupled Environmental Systems: An Introduction to Forward and Inverse Modeling,
First Edition. Knut Stamnes and Jakob J. Stamnes.

$$\mathbf{E}_t(\mathbf{r}) = \frac{ik_2 e^{ik_2 r}}{4\pi r}(\tilde{I} - \hat{k}_t\hat{k}_t)$$
$$\times \int_{S'} dS' \left\{ \hat{k}_t \times [\hat{n}_d \times \mathbf{E}(\mathbf{r}')] + \eta_2[\hat{n}_d \times \mathbf{H}(\mathbf{r}')] \right\} e^{-i\mathbf{k}_{2t}\cdot\mathbf{r}'}, \tag{1199}$$

where $\tilde{I}$ is the unit dyad, and $\eta_1 = \sqrt{\mu_1/\epsilon_1}$ and $\eta_2 = \sqrt{\mu_2/\epsilon_2}$ are the wave impedances in media 1 and 2, respectively.

C.1.1 Tangent Plane Approximation

The fields at any point $\mathbf{r}'$ on the surface are approximated by the fields that would be present at the tangent plane at that point [282]. We form an orthonormal coordinate system $(\hat{p}_i, \hat{q}_i, \hat{k}_i)$ at the point $\mathbf{r}'$ with

$$\hat{q}_i = \frac{\hat{k}_i \times \hat{n}}{|\hat{k}_i \times \hat{n}|} \qquad \hat{p}_i = \hat{q}_i \times \hat{k}_i, \tag{1200}$$

where $\hat{n} \equiv \hat{n}(\mathbf{r}')$ is the normal to the tangent plane. Thus, in the tangent plane approximation, the vectors $\hat{n}$, $\hat{k}_i$, and $\hat{k}_r$ define the local plane of incidence, and by definition $\hat{q}_i$ is $\perp$ to and $\hat{p}_i$ $\|$ to the local plane of incidence.

Decomposing the incident field into locally $\perp$ and $\|$ components, we find that the $\perp$ (or horizontal) component of the **local** incident electric field becomes

$$\mathbf{E}_{li\perp} = (\hat{e}_i \cdot \hat{q}_i)\hat{q}_i E_0 e^{i\mathbf{k}_i\cdot\mathbf{r}'}. \tag{1201}$$

Thus, the **local** reflected field is

$$\mathbf{E}_{lr\perp} = (\hat{e}_i \cdot \hat{q}_i)\hat{q}_i E_0 r_\perp e^{i\mathbf{k}_i\cdot\mathbf{r}'}, \tag{1202}$$

where $r_\perp$ is the local Fresnel reflection coefficient for the perpendicular or horizontal polarization

$$r_\perp = \frac{n_1 \cos\theta_{li} - n_2\sqrt{1 - \sin^2\theta_{li}/m_{rel}^2}}{n_1 \cos\theta_{li} + n_2\sqrt{1 - \sin^2\theta_{li}/m_{rel}^2}} \tag{1203}$$

with $m_{rel} = n_2/n_1 = \eta_1/\eta_2 = \sqrt{\varepsilon_2/\varepsilon_1}$, and θ_{li} is the local angle of incidence at the point $\mathbf{r}'$: $\cos\theta_{li} = -\hat{n}\cdot\hat{k}_i$.

The magnetic fields associated with the incident and reflected fields lie in the local plane of incidence. Thus, they are given by:

$$\mathbf{H}_{li\|} = \frac{1}{\eta}(\hat{k}_i \times \mathbf{E}_{li\perp}) = \frac{1}{\eta}(\hat{k}_i \times \hat{q}_i)(\hat{e}_i \cdot \hat{q}_i)E_0 e^{i\mathbf{k}_i\cdot\mathbf{r}'}$$
$$\mathbf{H}_{lr\|} = \frac{1}{\eta}(\hat{k}_r \times \mathbf{E}_{lr\perp}) = \frac{1}{\eta}(\hat{k}_r \times \hat{q}_i)(\hat{e}_i \cdot \hat{q}_i)E_0 r_\perp e^{i\mathbf{k}_i\cdot\mathbf{r}'}, \tag{1204}$$

where $\hat{k}_r$ is the local reflected direction, related to the incident direction by

$$\hat{k}_r = \hat{k}_i - 2\hat{n}(\hat{n}\cdot\hat{k}_i). \tag{1205}$$

The tangential electric field of the total horizontal component becomes

$$\hat{n} \times \mathbf{E}_{l\perp} = \hat{n} \times (\mathbf{E}_{li\perp} + \mathbf{E}_{lr\perp}) = (\hat{n} \times \hat{k}_i)(\hat{e}_i \cdot \hat{q}_i)(1 + r_\perp)E_0 e^{i\mathbf{k}_i \cdot \mathbf{r}'} \tag{1206}$$

and the associated magnetic field is

$$\begin{aligned} \hat{n} \times \mathbf{H}_{l\|} &= \hat{n} \times (\mathbf{H}_{li\|} + \mathbf{H}_{lr\|}) \\ &= \frac{1}{\eta}(\hat{e}_i \cdot \hat{q}_i)\hat{n} \times [(\hat{k}_i \times \hat{q}_i) + r_\perp(\hat{k}_r \times \hat{q}_i)]E_0 e^{i\mathbf{k}_i \cdot \mathbf{r}'} \\ &= -(1 - r_\perp)(\hat{n} \cdot \hat{k}_i)\frac{(\hat{e}_i \cdot \hat{h}_i)}{\eta}\hat{q}_i E_0 e^{i\mathbf{k}_i \cdot \mathbf{r}'}. \end{aligned} \tag{1207}$$

Here we have used the relations $\hat{n} \times (\hat{k}_i \times \hat{q}_i) = \hat{k}_i(\hat{n} \cdot \hat{q}_i) - \hat{q}_i(\hat{n} \cdot \hat{k}_i) = -\hat{q}_i(\hat{n} \cdot \hat{k}_i)$ and $\hat{n} \times (\hat{k}_r \times \hat{q}_i) = \hat{k}_r(\hat{n} \cdot \hat{q}_i) - \hat{q}_i(\hat{n} \cdot \hat{k}_r) = \hat{q}_i(\hat{n} \cdot \hat{k}_i)$ since $\hat{n} \cdot \hat{q}_i = 0$ and $\hat{n} \cdot \hat{k}_r = -\hat{n} \cdot \hat{k}_i$.

We may now repeat these **calculations for the local parallel** or vertical **component** of the electric field $\mathbf{E}_{l\|}$ with local reflection coefficient

$$r_\| = \frac{m_{rel}^2 n_1 \cos\theta_{li} - n_2\sqrt{1 - \sin^2\theta_{li}/m_{rel}^2}}{m_{rel}^2 n_1 \cos\theta_{li} + n_2\sqrt{1 - \sin^2\theta_{li}/m_{rel}^2}}. \tag{1208}$$

$r_\|$ yields its own contribution to the tangential component of the electric field at point $\mathbf{r}'$ and the associated magnetic field required to evaluate the integrals in Eqs. (1198) and (1199). Adding the contributions from the local parallel and perpendicular components, we obtain

$$\begin{aligned} \hat{n} \times \mathbf{E}_l &= \hat{n} \times (\mathbf{E}_{l\perp} + \mathbf{E}_{l\|}) = E_0 e^{i\mathbf{k}_i \cdot \mathbf{r}'} \\ &\times \left\{ (\hat{e}_i \cdot \hat{q}_i)(\hat{n} \times \hat{q}_i)(1 + r_\perp) + (\hat{e}_i \cdot \hat{p}_i)(\hat{n} \cdot \hat{k}_i)\hat{q}_i(1 - r_\|) \right\} \end{aligned} \tag{1209}$$

$$\begin{aligned} \hat{n} \times \mathbf{H}_l &= \hat{n} \times (\mathbf{H}_{l\perp} + \mathbf{H}_{l\|}) = \frac{E_0}{\eta} e^{i\mathbf{k}_i \cdot \mathbf{r}'} \\ &\times \left\{ -(\hat{e}_i \cdot \hat{q}_i)(\hat{n} \cdot \hat{k}_i)\hat{q}_i(1 - r_\perp) + (\hat{e}_i \cdot \hat{p}_i)(\hat{n} \times \hat{q}_i)(1 + r_\|) \right\}. \end{aligned} \tag{1210}$$

The local angle of incidence can be obtained from

$$\cos\theta_{li} = -\hat{n} \cdot \hat{k}_i \tag{1211}$$

and the normal vector at point $\mathbf{r}'$ is given by

$$\hat{n}(\mathbf{r}') = \frac{-\alpha\hat{x} - \beta\hat{y} + \hat{z}}{\sqrt{1 + \alpha^2 + \beta^2}}, \tag{1212}$$

where α and β are the local slopes in the x and y directions:

$$\alpha = \frac{\partial f(x', y')}{\partial x'} \qquad \beta = \frac{\partial f(x', y')}{\partial y'}. \tag{1213}$$

By substituting Eqs. (1209) and (1210) into Eqs. (1198) and (1199), one can show that the **far-field scattered and transmitted electric fields** become

$$\mathbf{E}_s(\mathbf{r}) = \frac{ik_1 e^{ik_1 r}}{4\pi r} E_0(\tilde{I} - \hat{k}_s\hat{k}_s) \int_{A_0} d\mathbf{r}'_\perp \mathbf{F}(\alpha,\beta) e^{i(\mathbf{k}_i - \mathbf{k}_{1s})\cdot\mathbf{r}'} \tag{1214}$$

$$\mathbf{E}_t(\mathbf{r}) = -\frac{ik_2 e^{ik_2 r}}{4\pi r} E_0(\tilde{I} - \hat{k}_t\hat{k}_t) \int_{A_0} d\mathbf{r}'_\perp \mathbf{N}(\alpha,\beta) e^{i(\mathbf{k}_i - \mathbf{k}_{2t})\cdot\mathbf{r}'}. \tag{1215}$$

Here, A_0 is the area of the rough surface S' projected onto the xy plane, and

$$\begin{aligned}\mathbf{F}(\alpha,\beta) = \sqrt{1+\alpha^2+\beta^2}\Big\{ &-(\hat{e}_i\cdot\hat{q}_i)(\hat{n}\cdot\hat{k}_i)\hat{q}_i(1-r_\perp) + (\hat{e}_i\cdot\hat{p}_i)(\hat{n}\times\hat{q}_i)(1+r_\parallel) \\ &+(\hat{e}_i\cdot\hat{q}_i)[\hat{k}_s\times(\hat{n}\times\hat{q}_i)](1+r_\perp) + (\hat{e}_i\cdot\hat{p}_i)(\hat{n}\cdot\hat{k}_i)(\hat{k}_s\times\hat{q}_i)(1-r_\parallel)\Big\}\end{aligned} \tag{1216}$$

$$\begin{aligned}\mathbf{N}(\alpha,\beta) = \sqrt{1+\alpha^2+\beta^2}\Big\{ &-\frac{\eta_2}{\eta_1}(\hat{e}_i\cdot\hat{q}_i)(\hat{n}\cdot\hat{k}_i)\hat{q}_i(1-r_\perp) + \frac{\eta_2}{\eta_1}(\hat{e}_i\cdot\hat{p}_i)(\hat{n}\times\hat{q}_i)(1+r_\parallel) \\ &+(\hat{e}_i\cdot\hat{q}_i)[\hat{k}_t\times(\hat{n}\times\hat{q}_i)](1+r_\perp) + (\hat{e}_i\cdot\hat{p}_i)(\hat{n}\cdot\hat{k}_i)(\hat{k}_t\times\hat{q}_i)(1-r_\parallel)\Big\}.\end{aligned} \tag{1217}$$

We note that

- the expressions in the integrands of Eqs. (1214) and (1215) are not explicit functions of $\mathbf{r}'$, except for the phase factors, but
- they depend explicitly on the slopes α and β which depend on $\mathbf{r}'$;
- Eqs. (1214) and (1215) do not take into account the effects of shadowing and multiple scattering.

Orthonormal coordinate systems for the incident, scattered, and transmitted fields may be defined as $(\hat{v}_i, \hat{h}_i, \hat{k}_i)$, $(\hat{v}_s, \hat{h}_s, \hat{k}_s)$, and $(\hat{v}_t, \hat{h}_t, \hat{k}_t)$, respectively, with

$$\begin{aligned}\hat{k}_i &= \hat{x}\sin\theta_i\cos\phi_i + \hat{y}\sin\theta_i\sin\phi_i - \hat{z}\cos\theta_i \\ \hat{h}_i &= -\hat{x}\sin\phi_i + \hat{y}\cos\phi_i \\ \hat{v}_i &= -\hat{x}\cos\theta_i\cos\phi_i - \hat{y}\cos\theta_i\sin\phi_i - \hat{z}\sin\theta_i\end{aligned} \tag{1218}$$

$$\begin{aligned}\hat{k}_s &= \hat{x}\sin\theta_s\cos\phi_s + \hat{y}\sin\theta_s\sin\phi_s + \hat{z}\cos\theta_s \\ \hat{h}_s &= -\hat{x}\sin\phi_s + \hat{y}\cos\phi_s \\ \hat{v}_s &= \hat{x}\cos\theta_s\cos\phi_s + \hat{y}\cos\theta_s\sin\phi_s - \hat{z}\sin\theta_s\end{aligned} \tag{1219}$$

$$\begin{aligned}\hat{k}_t &= \hat{x}\sin\theta_t\cos\phi_t + \hat{y}\sin\theta_t\sin\phi_t - \hat{z}\cos\theta_t \\ \hat{h}_t &= -\hat{x}\sin\phi_t + \hat{y}\cos\phi_t \\ \hat{v}_t &= -\hat{x}\cos\theta_t\cos\phi_t - \hat{y}\cos\theta_t\sin\phi_t - \hat{z}\sin\theta_t.\end{aligned} \tag{1220}$$

C.1.2
Geometrical Optics Solution

In the geometrical optics limit, as $k \to \infty$, an asymptotic solution to the integrals in Eqs. (1214) and (1215) can be derived using the method of stationary phase.

The coherent component of the scattered fields will vanish in this limit; only the incoherent component will remain.

C.1.2.1 Stationary-Phase Method

Consider first the reflected fields. The phase factor in Eq. (1214) is

$$\psi = (\mathbf{k}_i - \mathbf{k}_{1s}) \cdot \mathbf{r}' = \mathbf{k}_{1d} \cdot \mathbf{r}' = k_{1dx}x' + k_{1dy}y' + k_{1dz}f(x', y'). \tag{1221}$$

To determine the stationary-phase point, we set

$$\begin{aligned} \frac{\partial \psi}{\partial x'} &= 0 = k_{1dx} + k_{1dz}\alpha_0; & \alpha_0 &\equiv \frac{\partial f(x', y')}{\partial x'} \\ \frac{\partial \psi}{\partial y'} &= 0 = k_{1dy} + k_{1dz}\beta_0; & \beta_0 &\equiv \frac{\partial f(x', y')}{\partial y'}. \end{aligned} \tag{1222}$$

Thus, at the stationary point

$$\alpha_0 = -\frac{k_{1dx}}{k_{1dz}} \quad \text{and} \quad \beta_0 = -\frac{k_{1dy}}{k_{1dz}}. \tag{1223}$$

The slopes α_0 and β_0 are such that the incident and scattered wave directions form a specular reflection, which can be seen from the fact that from Eq. (1212), we have

$$\begin{aligned} \hat{n}(\alpha_0, \beta_0) &= \frac{-\alpha_0\hat{x} - \beta_0\hat{y} + \hat{z}}{\sqrt{1 + \alpha_0^2 + \beta_0^2}} = \frac{-k_{1dx}\hat{x} - k_{1dy}\hat{y} + k_{1dz}\hat{z}}{\sqrt{k_{1dx}^2 + k_{1dy}^2 + k_{1dz}^2}} \\ &= -\frac{(\mathbf{k}_i - \mathbf{k}_{1s})}{|\mathbf{k}_{1d}|} = -\frac{\mathbf{k}_{1d}}{|\mathbf{k}_{1d}|}. \end{aligned} \tag{1224}$$

Replacing the slopes α and β in Eq. (1214) with α_0 and β_0, we have

$$\mathbf{E}_s(\mathbf{r}) = \frac{ik_1e^{ik_1r}}{4\pi r}E_0(\tilde{I} - \hat{k}_s\hat{k}_s)\mathbf{F}(\alpha_0, \beta_0)I, \tag{1225}$$

where

$$I \equiv \int_{A_0} d\mathbf{r}'_\perp e^{i(\mathbf{k}_i - \mathbf{k}_{1s})\cdot\mathbf{r}'} = \int_{A_0} d\mathbf{r}'_\perp e^{i\mathbf{k}_{1d\perp}\cdot\mathbf{r}'_\perp}\, e^{ik_{1dz}f(\mathbf{r}'_\perp)} \tag{1226}$$

and

$$\mathbf{k}_{1d} = (\mathbf{k}_i - \mathbf{k}_{1s}) = k_{1dx}\hat{x} + k_{1dy}\hat{y} + k_{1dz}\hat{z} = \mathbf{k}_{1d\perp} + k_{1dz}\hat{z}. \tag{1227}$$

If we separate the fields into mean and fluctuating parts, that is

$$\mathbf{E}_s(\mathbf{r}) = \mathbf{E}_{sm}(\mathbf{r}) + \mathcal{E}_s(\mathbf{r}) \quad \text{with} \quad \langle \mathcal{E}_s(\mathbf{r}) \rangle = 0 \tag{1228}$$

the total scattered radiative energy is the sum of coherent and incoherent parts, given by

$$\langle |\mathbf{E}_s(\mathbf{r})|^2 \rangle = |\mathbf{E}_{sm}(\mathbf{r})|^2 + \langle |\mathcal{E}_s(\mathbf{r})|^2 \rangle. \tag{1229}$$

In view of Eq. (1225), we have

$$|\mathbf{E}_{sm}(\mathbf{r})|^2 = \frac{k_1^2|E_0|^2}{16\pi^2 r^2}\left\{|\hat{v}_s\cdot\mathbf{F}(\alpha_0,\beta_0)|^2 + |\hat{h}_s\cdot\mathbf{F}(\alpha_0,\beta_0)|^2\right\}|\langle I\rangle|^2$$

$$\langle|\mathcal{E}_s(\mathbf{r})|^2\rangle = \frac{k_1^2|E_0|^2}{16\pi^2 r^2}\left\{|\hat{v}_s\cdot\mathbf{F}(\alpha_0,\beta_0)|^2 + |\hat{h}_s\cdot\mathbf{F}(\alpha_0,\beta_0)|^2\right\}D_I \tag{1230}$$

where

$$D_I = \langle|I|^2\rangle - |\langle I\rangle|^2 = \langle II^*\rangle - |\langle I\rangle|^2. \tag{1231}$$

Since the coherent part is zero in the geometrical optics limit ($|\langle I\rangle|^2 = 0$), we have

$$\langle|\mathbf{E}_s(\mathbf{r})|^2\rangle = \frac{k_1^2|E_0|^2}{16\pi^2 r^2}\left|(\tilde{I} - \hat{k}_s\hat{k}_s)\cdot\mathbf{F}(\alpha_0,\beta_0)\right|^2\langle II^*\rangle, \tag{1232}$$

where the ensemble average of II^* in view of Eq. (1226) is given by

$$\langle II^*\rangle = \left\langle\int_{A_0} d\mathbf{r}_\perp \int_{A_0} d\mathbf{r}'_\perp e^{i\mathbf{k}_{1d\perp}\cdot(\mathbf{r}_\perp-\mathbf{r}'_\perp)}\, e^{ik_{1dz}[f(\mathbf{r}_\perp)-f(\mathbf{r}'_\perp)]}\right\rangle. \tag{1233}$$

The above integral can be solved by the method of asymptotic expansion, and it can be shown that for large k

$$\langle II^*\rangle = \frac{4\pi^2 A_0}{k_{dz}^2}p\left(-\frac{k_{1dx}}{k_{1dz}}, -\frac{k_{1dy}}{k_{1dz}}\right), \tag{1234}$$

where $p(\alpha,\beta)$ is the probability density function for the slopes at the surface. For a Gaussian rough surface

$$p(\alpha,\beta) = \frac{1}{2\pi\sigma^2}\exp\left[-\frac{\alpha^2+\beta^2}{2\sigma^2}\right], \tag{1235}$$

where σ^2 is the mean square surface slope. Using this probability density function in Eq. (1234), we find

$$\langle II^*\rangle = \frac{2\pi A_0}{k_{1dz}^2\sigma^2}\exp\left[-\frac{k_{1dx}^2+k_{1dy}^2}{2\sigma^2 k_{1dz}^2}\right]. \tag{1236}$$

For an incident field with polarization $\hat{b}$, the bistatic scattering coefficients for the **reflected** radiation with polarization $\hat{a}$ are defined as

$$\begin{aligned}\rho_{ab}^r(\hat{k}_i,\hat{k}_s) = \rho_{ab}(\theta_i,\phi_i;\theta_s,\phi_s) &= \lim_{r\to\infty}\frac{4\pi r^2|E_{as}|^2}{|E_{bi}|^2 A\cos\theta_i}\\ &= \frac{k_1^2}{4\pi A_0\cos\theta_i}\left|\hat{a}\cdot\mathbf{F}_b(\alpha_0,\beta_0)\right|^2\langle II^*\rangle\end{aligned} \tag{1237}$$

because the scattered radiative energy for polarization $\hat{a}$ is given by

$$\langle|E_{as}|^2\rangle = \frac{k_1^2|E_0|^2}{16\pi^2 r^2}\left|\hat{a}\cdot\mathbf{F}_b(\alpha_0,\beta_0)\right|^2\langle II^*\rangle, \tag{1238}$$

where $\mathbf{F}_b(\alpha_0, \beta_0) = \mathbf{F}(\alpha_0, \beta_0)|_{\hat{e}_i=\hat{b}}$. Using Eq. (1216) we find that

$$|\hat{a} \cdot \mathbf{F}_b(\alpha_0, \beta_0)|^2 = \frac{|\mathbf{k}_{1d}|^4}{k_1^2 |\hat{k}_i \times \hat{k}_s|^4 k_{1dz}^2} f_{ba}, \tag{1239}$$

where

$$\begin{aligned} f_{\|\|} &= \left|(\hat{h}_s \cdot \hat{k}_i)(\hat{h}_i \cdot \hat{k}_s) r_\perp + (\hat{v}_s \cdot \hat{k}_i)(\hat{v}_i \cdot \hat{k}_s) r_\|\right|^2 \\ f_{\perp\|} &= \left|(\hat{v}_s \cdot \hat{k}_i)(\hat{h}_i \cdot \hat{k}_s) r_\perp - (\hat{h}_s \cdot \hat{k}_i)(\hat{v}_i \cdot \hat{k}_s) r_\|\right|^2 \\ f_{\|\perp} &= \left|(\hat{h}_s \cdot \hat{k}_i)(\hat{v}_i \cdot \hat{k}_s) r_\perp - (\hat{v}_s \cdot \hat{k}_i)(\hat{h}_i \cdot \hat{k}_s) r_\|\right|^2 \\ f_{\perp\perp} &= \left|(\hat{v}_s \cdot \hat{k}_i)(\hat{v}_i \cdot \hat{k}_s) r_\perp + (\hat{h}_s \cdot \hat{k}_i)(\hat{h}_i \cdot \hat{k}_s) r_\|\right|^2. \end{aligned} \tag{1240}$$

Here, $r_\perp$ and $r_\|$ are evaluated at

$$\hat{n} = \frac{k_{1dx}\hat{x} + k_{1dy}\hat{y} + k_{1dz}\hat{z}}{\sqrt{k_{1dx}^2 + k_{1dy}^2 + k_{1dz}^2}}. \tag{1241}$$

Substituting Eqs. (1236) and (1239) in Eq. (1237), we find that the bistatic scattering coefficients for the reflected radiation become

$$\rho_{ab}^r(\hat{k}_i, \hat{k}_s) = \frac{|\mathbf{k}_{1d}|^4}{\cos\theta_i |\hat{k}_i \times \hat{k}_s|^4 k_{1dz}^4} \frac{1}{2\sigma^2} \exp\left[-\frac{k_{1dx}^2 + k_{1dy}^2}{2\sigma^2 k_{1dz}^2}\right] f_{ab}. \tag{1242}$$

In the backscattering direction $\hat{k}_s = -\hat{k}_i$

$$\beta_{ab}(\hat{k}_i) = \cos\theta_i \rho_{ab}^r(\hat{k}_i, -\hat{k}_i) \tag{1243}$$

which yields

$$\beta_{\perp\perp}^r(\theta_i) = \beta_{\|\|}^r(\theta_i) = \frac{|r_n|^2}{\cos^4\theta_i 2\sigma^2} \exp\left[-\frac{\tan^2\theta_i}{2\sigma^2}\right] \tag{1244}$$

$$\beta_{\|\perp}^r(\theta_i) = \beta_{\perp\|}^r(\theta_i) = 0, \tag{1245}$$

where r_n is the Fresnel reflection coefficient for normal incidence. From Eq. (1245) we see that there is **no depolarization in the backscattering direction**.

Transmitted Radiation Field

For an incident field with polarization $\hat{b}$, the bistatic scattering coefficients for the **transmitted** radiation can be derived in a similar manner. The stationary-phase points are given by

$$\alpha_0 = -\frac{k_{2dx}}{k_{2dz}} \qquad \beta_0 = -\frac{k_{tdy}}{k_{2dz}}$$

where

$$\mathbf{k}_{2d} = \mathbf{k}_i - \mathbf{k}_{2t} = k_{2dx}\hat{x} + k_{2dy}\hat{y} + k_{2dz}\hat{z}$$

At the stationary point

$$\hat{n} = \frac{-\alpha_0\hat{x} - \beta_0\hat{y} + \hat{z}}{\sqrt{1+\alpha_0^2+\beta_0^2}} = \frac{-k_{2dx}\hat{x} - k_{2dy}\hat{y} + k_{2dz}\hat{z}}{\sqrt{k_{2dx}^2 + k_{2dy}^2 + k_{2dz}^2}} = \frac{(\mathbf{k}_{2t} - \mathbf{k}_i)}{|\mathbf{k}_{2d}|} = -\frac{\mathbf{k}_{2d}}{|\mathbf{k}_{2d}|}$$

and it can be shown that the tangential components of $\mathbf{k}_i$ and $\mathbf{k}_{2t}$ are equal:

$$\mathbf{k}_i - (\mathbf{k}_i \cdot \hat{n})\hat{n} = \mathbf{k}_t - (\mathbf{k}_t \cdot \hat{n})\hat{n} \qquad \longleftarrow \qquad \text{Snell's law.}$$

The bistatic scattering coefficients for the transmitted radiation become

$$\rho^t_{ab}(\hat{k}_i, \hat{k}_t) = \frac{2|\mathbf{k}_{2d}|^2(\hat{n}\cdot\hat{k}_t)^2}{\cos\theta_i|\hat{k}_i \times \hat{k}_s|^4 k_{2dz}^4}\frac{\eta_1}{\eta_2}\frac{1}{\sigma^2}\exp\left[-\frac{k_{2dx}^2 + k_{2dy}^2}{2\sigma^2 k_{2dz}^2}\right] W_{ab}, \tag{1246}$$

where W_{ab} is given by

$$W_{\|\|} = \left|(\hat{h}_t\cdot\hat{k}_i)(\hat{h}_i\cdot\hat{k}_t)(1+r_\perp) + (\hat{v}_t\cdot\hat{k}_i)(\hat{v}_i\cdot\hat{k}_t)\frac{\eta_1}{\eta_2}(1+r_\|)\right|^2$$

$$W_{\perp\|} = \left|-(\hat{v}_t\cdot\hat{k}_i)(\hat{h}_i\cdot\hat{k}_t)(1+r_\perp) + (\hat{h}_t\cdot\hat{k}_i)(\hat{v}_i\cdot\hat{k}_t)\frac{\eta_1}{\eta_2}(1+r_\|)\right|^2$$

$$W_{\|\perp} = \left|(\hat{h}_t\cdot\hat{k}_i)(\hat{v}_i\cdot\hat{k}_t)(1+r_\perp) - (\hat{v}_t\cdot\hat{k}_i)(\hat{h}_i\cdot\hat{k}_t)\frac{\eta_1}{\eta_2}(1+r_\|)\right|^2$$

$$W_{\perp\perp} = \left|(\hat{v}_t\cdot\hat{k}_i)(\hat{v}_i\cdot\hat{k}_t)(1+r_\perp) + (\hat{h}_t\cdot\hat{k}_i)(\hat{h}_i\cdot\hat{k}_t)\frac{\eta_1}{\eta_2}(1+r_\|)\right|^2. \tag{1247}$$

The reflection coefficients should be evaluated at the stationary phase point $\alpha_0 = -(k_{2dx}/k_{2dz}) \quad \beta_0 = -(k_{2dy}/k_{2dz})$, so that

$$r_\perp = \frac{k_1(\hat{n}\cdot\hat{k}_i) - k_2(\hat{n}\cdot\hat{k}_t)}{k_1(\hat{n}\cdot\hat{k}_i) + k_2(\hat{n}\cdot\hat{k}_t)} = \frac{n_1\cos\theta_i - n_2\cos\theta_t}{n_1\cos\theta_i + n_2\cos\theta_t} \tag{1248}$$

$$r_\| = \frac{k_2(\hat{n}\cdot\hat{k}_i) - k_1(\hat{n}\cdot\hat{k}_t)}{k_2(\hat{n}\cdot\hat{k}_i) + k_1(\hat{n}\cdot\hat{k}_t)} = \frac{n_2\cos\theta_i - n_1\cos\theta_t}{n_2\cos\theta_i + n_1\cos\theta_t}. \tag{1249}$$

Flat Surface Limit

In the limit $\sigma^2 \to 0$, which corresponds to the vanishing of the slope variance, a specular (flat, smooth) surface is obtained. In this limit Goodman [298]

$$\lim_{\sigma^2\to 0}\frac{1}{2\sigma^2}\exp\left[-\frac{k_{dx}^2 + k_{dy}^2}{2\sigma^2 k_{dz}^2}\right] = \pi\delta\left(\frac{k_{dx}}{k_{dz}}, \frac{k_{dy}}{k_{dz}}\right). \tag{1250}$$

The δ functions can be expressed in terms of angular variables, which implies that the scattering is nonzero only at $\theta_s = \theta_i$ and $\phi_s = \phi_i$. In this case, the bistatic scattering coefficients given by Eq. (1242) simplify to

$$\rho^r_{ab}(\hat{k}_i, \hat{k}_s) = \frac{4\pi}{\sin\theta_{1i}}|r_{a0}|^2\delta(\theta_s - \theta_i)\delta(\phi_s - \phi_i)\delta_{ab}, \tag{1251}$$

where $r_{\perp 0}$ and $r_{\|0}$ are the Fresnel reflection coefficients of a flat surface. The corresponding expression for the transmittance coefficients are

$$\rho^t_{ab}(\hat{k}_i, \hat{k}_t) = \frac{4\pi}{\sin\theta_{2i}}|1 - r_{a0}|^2\delta(\theta_t - \theta_i)\delta(\phi_t - \phi_i)\delta_{ab}, \tag{1252}$$

where

$$\theta_{2i} = \sin^{-1}\left(\frac{k_1}{k_2}\sin\theta_{1i}\right). \tag{1253}$$

Assumptions

In the derivations above, we have assumed that the probability for the height distribution $f(\mathbf{r}_\perp)$ is (i) **independent of the position $\mathbf{r}_\perp$** on the rough surface and (ii) **Gaussian**:

$$p(f(\mathbf{r}_\perp)) = \frac{1}{\sqrt{2\pi}\sigma_h} e^{-f^2/\sigma_h^2}, \tag{1254}$$

where σ_h is the standard deviation of the surface height. For two points on the surface, $\mathbf{r}_{\perp 1}$ and $\mathbf{r}_{\perp 2}$, the joint probability density is

$$p(f_1(\mathbf{r}_{\perp 1}))p(f_2(\mathbf{r}_{\perp 2})) = \frac{1}{2\pi\sigma_h^2\sqrt{1-C^2}} \exp\left[-\frac{f_1^2 - 2Cf_1f_2 + f_2^2}{2\sigma_h^2(1-C^2)}\right], \tag{1255}$$

where C is the correlation coefficient between the two points, which is a function of $\mathbf{r}_{\perp 1}$ and $\mathbf{r}_{\perp 2}$. For a statistically isotropic surface, it is only a function of $\rho = \sqrt{(x_1 - x_2)^2 + (y_1 - y_2)^2}$:

$$\langle f_1(\mathbf{r}_{\perp 1}) f_2(\mathbf{r}_{\perp 2})\rangle = \sigma_h^2 C(\rho); \quad C(0) = 1; \quad C(\infty) = 0. \tag{1256}$$

If the correlation function $C(\rho)$ is assumed to have a Gaussian form

$$C(\rho) = e^{-\rho^2/l^2}, \tag{1257}$$

where l is the correlation length for the random variable $f(\mathbf{r}_\perp)$ in the transverse plane, then for a Gaussian random rough surface

$$p(\alpha, \beta) = \frac{1}{2\pi\sigma^2} \exp\left[-\frac{\alpha^2 + \beta^2}{2\sigma^2}\right] \tag{1258}$$

and the mean square surface slope is given by

$$\sigma^2 = \sigma_h^2 |C''(0)| = 2\frac{\sigma_h^2}{l^2}. \tag{1259}$$

Here, σ_h is the standard deviation of the rough surface height, and $C''(0) = -2/l^2$ is the double derivative of the correlation function $C(\rho)$ evaluated at $\rho = 0$.

Planar Dielectric Interface

In Figure 73, a plane wave incident from medium 1 onto medium 2 in direction $\hat{k}_i$ generates a reflected wave in direction $\hat{k}_r$ and a transmitted wave in direction $\hat{k}_t$. We have three orthogonal systems $(\hat{v}_i, \hat{h}_i, \hat{k}_i)$, $(\hat{v}_r, \hat{h}_r, \hat{k}_r)$, and $(\hat{v}_t, \hat{h}_t, \hat{k}_t)$, with $\hat{h} = \hat{z} \times \hat{k}/|\hat{z} \times \hat{k}|$. For the incident wave we have

$$\mathbf{E}_i = (\hat{v}_i E_{\parallel i} + \hat{h}_i E_{\perp i}) e^{i(k_x x - k_{1z} z)} \tag{1260}$$

$$\mathbf{H}_i = \frac{1}{\eta_1}(\hat{v}_i E_{\parallel i} - \hat{h}_i E_{\perp i}) e^{i(k_x x - k_{1z} z)} \tag{1261}$$

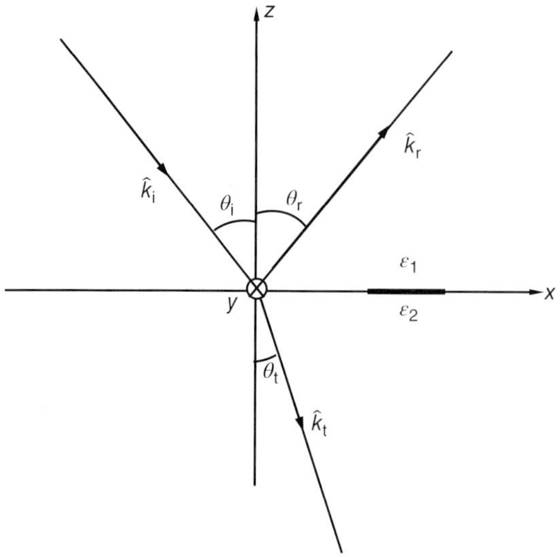

Figure 73 Geometric configuration of a plane interface between two dielectric media with permittivities ε_1 and ε_2, respectively.

where $k_x = k_1 \sin\theta_i$, $k_{1z} = k_1 \cos\theta_i$. The reflected and transmitted waves are

$$\mathbf{E}_r = (\hat{v}_r R_\parallel E_{\parallel i} + \hat{h}_r R_\perp E_{\perp i}) e^{i(k_x x + k_{1z} z)} \tag{1262}$$

$$\mathbf{H}_r = \frac{1}{\eta_1}(\hat{h}_r R_\parallel E_{\parallel i} - \hat{v}_r R_\perp E_{\perp i}) e^{i(k_x x + k_{1z} z)} \tag{1263}$$

$$\mathbf{E}_t = \left(\hat{v}_t \frac{\eta_2}{\eta_1} T_\parallel E_{\parallel i} + \hat{h}_t T_\perp E_{\perp i}\right) e^{i(k_x x - k_{2z} z)} \tag{1264}$$

$$\mathbf{H}_t = \left(\hat{h}_t \frac{1}{\eta_1} T_\parallel E_{\parallel i} - \hat{v}_t \frac{1}{\eta_2} T_\perp E_{\perp i}\right) e^{i(k_x x - k_{2z} z)} \tag{1265}$$

where $R_\perp$ and $R_\parallel$ are the reflection coefficients and $T_\perp$ and $T_\parallel$ are the transmission coefficients for horizontally and vertically polarized waves, respectively:

$$R_\perp = \frac{k_{1z} - k_{2z}}{k_{1z} + k_{2z}} = T_\perp - 1 \tag{1266}$$

$$R_\parallel = \frac{\epsilon_2 k_{1z} - \epsilon_1 k_{2z}}{\epsilon_2 k_{1z} + \epsilon_1 k_{2z}} = T_\parallel - 1. \tag{1267}$$

We also define reflectivity and transmissivity r_β and t_β ($\beta = \parallel, \perp$), as

$$r_\perp = |R_\perp|^2 = 1 - t_\perp \tag{1268}$$

$$r_\parallel = |R_\parallel|^2 = 1 - t_\parallel. \tag{1269}$$

From Eqs. (1260) and (1261), we know that the Poynting vector is $\tilde{S}_{pi} = \tilde{S}_{\parallel i} + \tilde{S}_{\perp i}$, where $\tilde{S}_{\parallel i}$ and $\tilde{S}_{\perp i}$ are the vertical and horizontal polarization components, with

$\tilde{S}_{\parallel i} = |E_{\parallel i}|^2/\eta_1$, $\tilde{S}_{\perp i} = |E_{\perp i}|^2/\eta_1$, and for the reflected wave, $\tilde{S}_{\parallel r} = r_\parallel \tilde{S}_{\parallel i}$ and $\tilde{S}_{\perp r} = r_\perp \tilde{S}_{\perp i}$. For the transmitted wave, we have from Eqs. (1264) and (1265)

$$\tilde{S}_{\parallel t} = \left|\frac{\eta_2}{\eta_1} T_\parallel\right|^2 \frac{\eta_1}{\eta_2} \tilde{S}_{\parallel i} \tag{1270}$$

$$\tilde{S}_{\perp t} = |T_\perp|^2 \frac{\eta_1}{\eta_2} \tilde{S}_{\perp i} \tag{1271}$$

for $\theta_i < \theta_{crit}$. Specific intensity (radiance) is defined as power per unit area per unit solid angle: $I_{\beta\ell} = \tilde{S}_{\beta\ell}/dAd\omega$, where $d\omega$ is differential solid angle, $\beta = \parallel, \perp$ and $\ell = i, t$. Since $d\omega_r = d\omega_i$, $\phi_t = \phi_i$, and (Snell's law)

$$\sqrt{\epsilon_2} \sin\theta_t = \sqrt{\epsilon_1} \sin\theta_i \tag{1272}$$

we find by differentiating Eq. (1272) and multiplying the result by Eq. (1272), we get

$$\epsilon_2 \cos\theta_t d\omega_t = \epsilon_1 \cos\theta_i d\omega_i . \tag{1273}$$

Using Eq. (1273) in Eqs. (1270) and (1271), we find ($I_{\beta\ell} = \tilde{S}_{\beta\ell}/dAd\omega$)

$$I_{\beta t} = \frac{\epsilon_2}{\epsilon_1} t_\beta I_{\beta i} \qquad \beta = \parallel, \perp . \tag{1274}$$

Equation (1274) relates the first two components of the Stokes vector for the incident and transmitted radiances at a plane interface. Note that there is a divergence of beam factor ϵ_2/ϵ_1 for the transmitted radiance.

The relation between the incident and reflected radiances are $I_{\parallel r} = r_\parallel I_{\parallel i}$ and $I_{\perp r} = r_\perp I_{\perp i}$. The final relation between the incident and reflected Stokes vectors become

$$\mathbf{I}_r = \mathbf{R}_F(\theta_i)\mathbf{I}_i, \tag{1275}$$

where

$$\mathbf{R}_F(\theta_i) = \begin{pmatrix} r_\parallel(\theta_i) & 0 & 0 & 0 \\ 0 & r_\perp(\theta_i) & 0 & 0 \\ 0 & 0 & \mathrm{Re}(R_\parallel R_\perp^*) & -\mathrm{Im}(R_\parallel R_\perp^*) \\ 0 & 0 & \mathrm{Im}(R_\parallel R_\perp^*) & \mathrm{Re}(R_\parallel R_\perp^*) \end{pmatrix}.$$

Similarly, the final relation between the incident and transmitted Stokes vectors become

$$\mathbf{I}_t = \mathbf{T}_F(\theta_i)\mathbf{I}_i, \tag{1276}$$

where ($\mu_t = \cos\theta_t$ and $\mu_i = \cos\theta_i$)

$$\mathbf{T}_F(\theta_i) = \frac{\epsilon_2}{\epsilon_1} \begin{pmatrix} t_\parallel(\theta_i) & 0 & 0 & 0 \\ 0 & t_\perp(\theta_i) & 0 & 0 \\ 0 & 0 & \frac{\mu_t}{\mu_i}\mathrm{Re}(T_\parallel T_\perp^*) & -\frac{\mu_t}{\mu_i}\mathrm{Im}(T_\parallel T_\perp^*) \\ 0 & 0 & \frac{\mu_t}{\mu_i}\mathrm{Im}(T_\parallel T_\perp^*) & \frac{\mu_t}{\mu_i}\mathrm{Re}(T_\parallel T_\perp^*) \end{pmatrix},$$

where $\theta_i < \theta_{crit}$ ($\mathbf{T}_F(\theta_i) = 0$ for $\theta_i > \theta_{crit}$).

Rough Dielectric Interface

We use the scattered and transmitted fields derived by a combination of the Kirchhoff approximation and geometrical optics. Note that

- unlike the planar case, where the coupling at the interface is to only the specular reflection and transmission directions, the incident radiance is coupled to all the reflected and transmitted directions;
- shadowing effects can be incorporated by modifying the coupling matrices;
- since only the single scattering solution is used, the reflected and transmitted radiances are always underestimated.

In Figure 15, a plane wave is incident from medium 1 onto medium 2 in direction $\hat{k}_i$ with electric field

$$\mathbf{E}_i = \hat{e}_i E_0 e^{i\mathbf{k}_i \cdot \mathbf{r}}, \tag{1277}$$

where $\mathbf{k}_i$ denotes the incident wave vector and $\hat{e}_i$ the polarization of the electric field vector. The incident field will generate reflected and transmitted fields in media 1 and 2, and the Kirchhoff-approximated diffraction integrals are calculated as discussed above [see Eqs. (1214) and (1215)].

The stationary-phase points for the reflected fields are given by

$$\alpha_0 = -\frac{k_{1dx}}{k_{1dz}} \qquad \beta_0 = -\frac{k_{1dy}}{k_{1dz}} \tag{1278}$$

and for the transmitted fields

$$\alpha_0' = -\frac{k_{2dx}}{k_{2dz}} \qquad \beta_0' = -\frac{k_{2dy}}{k_{2dz}}, \tag{1279}$$

where

$$\mathbf{k}_{1d} = \mathbf{k}_i - \mathbf{k}_{1r} = \hat{x}k_{1dx} + \hat{y}k_{1dy} + \hat{z}k_{1dz} \tag{1280}$$

$$\mathbf{k}_{2d} = \mathbf{k}_i - \mathbf{k}_{2t} = \hat{x}k_{2dx} + \hat{y}k_{2dy} + \hat{z}k_{2dz}. \tag{1281}$$

The reflection and transmission conditions at a rough interface are

$$\tilde{\mathbf{I}}_1(\hat{k}_s) = \int_0^{2\pi} d\phi_i \int_0^{\pi/2} d\theta_i \sin\theta_i \mathbf{R}_{rs}(\theta_s, \phi_s; \theta_i, \phi_i)\bar{\mathbf{I}}_1(\hat{k}_i) \tag{1282}$$

$$\tilde{\mathbf{I}}_2(\hat{k}_t) = \int_0^{2\pi} d\phi_i \int_0^{\pi/2} d\theta_i \sin\theta_i \mathbf{T}_{rs}(\theta_t, \phi_t; \theta_i, \phi_i)\bar{\mathbf{I}}_1(\hat{k}_i) \tag{1283}$$

where $\tilde{\mathbf{I}}_1$ and $\tilde{\mathbf{I}}_2$ are the Stokes vectors

$$\bar{\mathbf{I}}_p \equiv \begin{pmatrix} I_{\|p} \\ I_{\perp p} \\ U_p \\ V_p \end{pmatrix} \qquad p = 1, 2.$$

The reflected and transmitted fields at the directions $\hat{k}_s$ and $\hat{k}_t$ are given by integration of all the scattered fields which are coupled to that direction from the incident fields.

Explicit expressions for the reflectance matrix $\mathbf{R}_{rs}$ and the transmittance matrix $\mathbf{T}_{rs}$ at the rough surface interface are given by

$$\mathbf{R}_{rs}(\theta_s, \phi_s; \theta_i, \phi_i) = \frac{1}{\cos\theta_s} \frac{|\mathbf{k}_{1d}|^4}{4|\hat{k}_i \times \hat{k}_s|^4 k_{1dz}^4} \times \frac{1}{2\pi\sigma^2} \exp\left(-\frac{k_{1dx}^2 + k_{1dy}^2}{2k_{1dz}^2\sigma^2}\right) \mathbf{C}_{rs}^r(\theta_s, \phi_s; \theta_i, \phi_i) \tag{1284}$$

$$\mathbf{T}_{rs}(\theta_t, \phi_t; \theta_i, \phi_i) = \frac{1}{\cos\theta_t} \frac{k_2^2|\mathbf{k}_{2d}|^2(\hat{n}\cdot\hat{k}_t)^2}{|\hat{k}_i \times \hat{k}_s|^4 k_{2dz}^2} \frac{\eta_1}{\eta_2} \times \frac{1}{2\pi\sigma^2} \exp\left(-\frac{k_{2dx}^2 + k_{2dy}^2}{2k_{2dz}^2\sigma^2}\right) \mathbf{C}_{rs}^t(\theta_t, \phi_t; \theta_i, \phi_i) \tag{1285}$$

where σ^2 is the mean square surface slope,

$$\mathbf{C}_{rs}^{\alpha} = \begin{pmatrix} \langle |f_{\|\|}|^2\rangle & \langle |f_{\|\perp}|^2\rangle & \mathrm{Re}\langle f_{\|\perp}^{\alpha^*} f_{\|\|}^{\alpha}\rangle & -\mathrm{Im}\langle f_{\|\perp}^{\alpha^*} f_{\|\|}^{\alpha}\rangle \\ \langle |f_{\perp\|}|^2\rangle & \langle |f_{\perp\perp}|^2\rangle & \mathrm{Re}\langle f_{\perp\perp}^{\alpha^*} f_{\|\|}^{\alpha}\rangle & -\mathrm{Im}\langle f_{\perp\|}^{\alpha} f_{\perp\perp}^{\alpha}\rangle \\ 2\mathrm{Re}\langle f_{\|\|}^{\alpha} f_{\perp\|}^{\alpha^*}\rangle & 2\mathrm{Re}\langle f_{\|\perp}^{\alpha} f_{\perp\perp}^{\alpha^*}\rangle & \mathrm{Re}\langle C(3,3)^+\rangle & -\mathrm{Im}\langle C(3,3)^-\rangle \\ 2\mathrm{Im}\langle f_{\|\|}^{\alpha} f_{\perp\|}^{\alpha^*}\rangle & 2\mathrm{Im}\langle f_{\|\perp}^{\alpha} f_{\perp\perp}^{\alpha^*}\rangle & \mathrm{Im}\langle C(3,3)^+\rangle & \mathrm{Re}\langle C(3,3)^-\rangle \end{pmatrix}$$

$C(3,3)^+ = (f_{\|\|}^{\alpha} f_{\perp\perp}^{\alpha^*} + f_{\|\perp}^{\alpha} f_{\perp\|}^{\alpha^*})$, $C(3,3)^- = (f_{\|\|}^{\alpha} f_{\perp\perp}^{\alpha^*} - f_{\|\perp}^{\alpha} f_{\perp\|}^{\alpha^*})$, with $\alpha = r, t$ and

$$\begin{aligned} f_{\|\|}^r &= (\hat{h}_s\cdot\hat{k}_i)(\hat{h}_i\cdot\hat{k}_s)R_\perp + (\hat{v}_s\cdot\hat{k}_i)(\hat{v}_i\cdot\hat{k}_s)R_\| \\ f_{\perp\|}^r &= (\hat{v}_s\cdot\hat{k}_i)(\hat{h}_i\cdot\hat{k}_s)R_\perp - (\hat{h}_s\cdot\hat{k}_i)(\hat{v}_i\cdot\hat{k}_s)R_\| \\ f_{\|\perp}^r &= (\hat{h}_s\cdot\hat{k}_i)(\hat{v}_i\cdot\hat{k}_s)R_\perp - (\hat{v}_s\cdot\hat{k}_i)(\hat{h}_i\cdot\hat{k}_s)R_\| \\ f_{\perp\perp}^r &= (\hat{v}_s\cdot\hat{k}_i)(\hat{v}_i\cdot\hat{k}_s)R_\perp + (\hat{h}_s\cdot\hat{k}_i)(\hat{h}_i\cdot\hat{k}_s)R_\|. \end{aligned} \tag{1286}$$

and

$$\begin{aligned} f_{\|\|}^t &= (\hat{h}_t\cdot\hat{k}_i)(\hat{h}_i\cdot\hat{k}_t)T_\perp + (\hat{v}_t\cdot\hat{k}_i)(\hat{v}_i\cdot\hat{k}_t)T_\| \\ f_{\perp\|}^t &= -(\hat{v}_t\cdot\hat{k}_i)(\hat{h}_i\cdot\hat{k}_t)T_\perp + (\hat{h}_t\cdot\hat{k}_i)(\hat{v}_i\cdot\hat{k}_t)T_\| \\ f_{\|\perp}^t &= (\hat{h}_t\cdot\hat{k}_i)(\hat{v}_i\cdot\hat{k}_t)T_\perp - (\hat{v}_t\cdot\hat{k}_i)(\hat{h}_i\cdot\hat{k}_t)T_\| \\ f_{\perp\perp}^t &= (\hat{v}_t\cdot\hat{k}_i)(\hat{v}_i\cdot\hat{k}_t)T_\perp + (\hat{h}_t\cdot\hat{k}_i)(\hat{h}_i\cdot\hat{k}_t)T_\|. \end{aligned} \tag{1287}$$

Here, $T_\perp = 1 + R'_\perp$ and $T_\| = (\eta_2/\eta_1)(1 + R'_\|)$, and $R_\|$ and $R_\perp$ and $R'_\|$ and $R'_\perp$ are the local reflection coefficients for the vertical and horizontal polarizations evaluated at the stationary phase points (α_1, β_1) and (α_2, β_2), respectively.

As mentioned above, the geometrical optics results used to derive the conditions for a rough dielectric interface satisfies the principle of reciprocity, but violates the principle of energy conservation due to the neglect of multiple scattering and shadowing effects.

Shadowing Effects

When the angle of incidence is not normal to the xy plane, some points on the rough surface will not be illuminated directly, as illustrated in Figure 74. For some points (e.g., point 1 in Figure 74), the local angle of incidence θ_{li} is not defined because

$$\cos\theta_i = -\hat{n}\cdot\hat{k}_i < 0 \tag{1288}$$

and some other points (e.g., point 2 in Figure 74) are not directly illuminated even though the local angle of incidence is well defined because of the height at that point relative to the heights of the surrounding points.

To take shadowing into account when calculating the reflected field given by Eq. (1214), we may introduce an **illumination function** $L(\hat{k}_i, \hat{k}_s, \mathbf{r}')$ as follows

$$\mathbf{E}_s(\mathbf{r}) = \frac{ik_1 e^{ik_1 r}}{4\pi r} E_0(\tilde{I} - \hat{k}_s\hat{k}_s) \int_{A_0} d\mathbf{r}'_\perp \mathbf{F}(\alpha,\beta) L(\hat{k}_i, \hat{k}_s, \mathbf{r}') e^{i\mathbf{k}_{1d}\cdot\mathbf{r}'}. \tag{1289}$$

- $L(\hat{k}_i, \hat{k}_s, \mathbf{r}') = 1$ if a ray in direction $\hat{k}_i$ is not intersected by the surface and illuminates the point $\mathbf{r}'$ and if the line drawn from the point $\mathbf{r}'$ in the direction $\hat{k}_s$ does not strike the surface [Eq. (1289) → Eq. (1214)];
- $L(\hat{k}_i, \hat{k}_s, \mathbf{r}') = 0$ otherwise [299].

The integral in Eq. (1289) may be evaluated by the method of stationary phase. Using Eqs. (1232) and (1233), we find

$$\langle |\mathbf{E}_s(\mathbf{r})|^2 \rangle = \frac{k_1^2 |E_0|^2}{16\pi^2 r^2} \left| (\tilde{I} - \hat{k}_s\hat{k}_s) \cdot \mathbf{F}(\alpha_0, \beta_0) \right|^2 \langle II^* \rangle, \tag{1290}$$

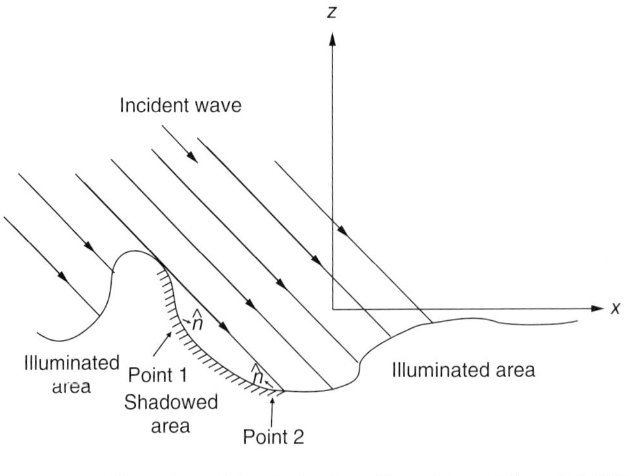

Figure 74 Illustration of the shadowing effect. Adapted from Ref. [300].

where the ensemble average of II^* in view of Eqs. (1233) and (1289) is given by

$$\langle II^* \rangle = \left\langle \int_{A_0} d\mathbf{r}_\perp \int_{A_0} d\mathbf{r}'_\perp e^{i\mathbf{k}_{1d\perp}\cdot(\mathbf{r}_\perp - \mathbf{r}'_\perp)} \right.$$
$$\left. \times L(\hat{k}_i, \hat{k}_s, \mathbf{r}_\perp) L(\hat{k}_i, \hat{k}_s, \mathbf{r}'_\perp) e^{ik_{1dz}[f(\mathbf{r}_\perp) - f(\mathbf{r}'_\perp)]} \right\rangle. \tag{1291}$$

For large k, the above integral can be solved by the method of asymptotics to yield an expression similar to Eq. (1234) except that $p(\alpha, \beta)$ is replaced by $p(\alpha, \beta, L)$:

$$\langle II^* \rangle = \frac{4\pi^2 A_0}{k_{dz}^2} p(\alpha, \beta, L), \tag{1292}$$

where $p(\alpha, \beta, L)$ is the **joint probability density function** for the slopes at the surface α and β and the illumination function L. We may express $p(\alpha, \beta, L)$ in terms of the conditional probability density

$$p(\alpha, \beta, L) = p(\alpha, \beta) p(L|\alpha, \beta) \tag{1293}$$

where

$$p(L|\alpha, \beta) = P_L(\hat{k}_i, \hat{k}_s|\alpha, \beta)\delta(L-1) + [1 - P_L(\hat{k}_i, \hat{k}_s|\alpha, \beta)]\delta(L) \tag{1294}$$

and $P_L(\hat{k}_i, \hat{k}_s|\alpha, \beta)$ is the **probability that given the slopes α and β at a point it will be illuminated with a ray in direction $\hat{k}_i$ that will be scattered in direction $\hat{k}_s$.**

Thus, in view of Eqs. (1292) and (1235), we have

$$\langle II^* \rangle = \frac{2\pi A_0}{k_{1dz}^2} \frac{1}{s^2} \exp\left[-\frac{k_{1dx}^2 + k_{1dy}^2}{2s^2 k_{1dz}^2} \right] \times P_L\left(\hat{k}_i, \hat{k}_s | -\frac{k_{1dx}}{k_{1dz}}, -\frac{k_{1dy}}{k_{1dz}} \right) \tag{1295}$$

and the bistatic reflection coefficients $\rho^r_{ab}(\hat{k}_i, \hat{k}_s)$ given by Eq. (1242) are modified to

$$\rho^{mr}_{ab}(\hat{k}_i, \hat{k}_s) = \rho^r_{ab}(\hat{k}_i, \hat{k}_s) P_L\left(\hat{k}_i, \hat{k}_s | -\frac{k_{1dx}}{k_{1dz}}, -\frac{k_{1dy}}{k_{1dz}} \right). \tag{1296}$$

The shadowing function [299, 301] becomes

$$P_L\left(\hat{k}_i, \hat{k}_s | -\frac{k_{1dx}}{k_{1dz}}, -\frac{k_{1dy}}{k_{1dz}} \right) \equiv S(\theta_i, \theta_s), \tag{1297}$$

where ($\mu = \cot\theta$)

$$S(\theta_i, \theta_s) = \begin{cases} \frac{1}{1+\Lambda(\mu_s)} & \phi_s = \phi_s + \pi, \theta_s \geq \theta_i \\ \frac{1}{1+\Lambda(\mu_i)} & \phi_s = \phi_s + \pi, \theta_i \geq \theta_s \\ \frac{1}{1+\Lambda(\mu_i)+\Lambda(\mu_s)} & \text{otherwise} \end{cases} \tag{1298}$$

$$\Lambda(\mu) = \frac{1}{2}\left[\sqrt{\frac{2}{\pi}} \frac{\sigma}{\mu} e^{-\mu^2/2\sigma^2} - \mathrm{erfc}\left(\frac{\mu}{\sqrt{2}\sigma} \right) \right], \tag{1299}$$

where σ^2 is the mean square surface slope given by Eq. (1259) and erfc is the complementary error function.

D
Boundary Conditions

D.1
The Combined Boundary Condition System

In order to get an overview of the entire system of linear algebraic equations that has to be solved to determine the unknown coefficients, we start by presenting the results derived below in condensed form:

1) Upper slab – top boundary [see Eq. 1311) below]:

$$\mathbf{A}_0\mathbf{c}_1 = -\mathbf{b}_0,$$

where the dimensions are $4N_1 \times 8N_1$ for $\mathbf{A}_0$, $8N_1 \times 1$ for $\mathbf{c}_1$ and $4N_1 \times 1$ for $\mathbf{b}_0$.

2) Layer interfaces in upper slab [Eq. (1329)]

$$\mathbf{A}'_{L_1}\mathbf{c}_{L_1} = \mathbf{b}_{L_1}$$

where the dimensions are $8N_1(L_1 - 1) \times 8N_1L_1$ for $\mathbf{A}'_{L_1}$, $8N_1(L_1 - 1) \times 1$ for $\mathbf{b}_{L_1}$, and $8N_1(L_1) \times 1$ for $\mathbf{c}_{L_1}$.

3) Noncritical region of the atmosphere–water interface [Eq. (1343)]:

$$\begin{pmatrix} -\mathbf{A}_t^- & \mathbf{S}^- - \mathbf{S}_r^+ \\ \mathbf{A}^+ - \mathbf{A}_r^- & -\mathbf{S}_t^+ \end{pmatrix} \times \begin{pmatrix} \mathbf{c}_{L_1}^- \\ \mathbf{c}_{L_1}^+ \end{pmatrix} = \begin{pmatrix} \mathbf{b}_{L_1}^t - \mathbf{b}_{L_1+1} + \mathbf{b}_{L_1+1}^r \\ \mathbf{b}_{L_1}^r - \mathbf{b}_{L_1} + \mathbf{b}_{L_1+1}^t \end{pmatrix},$$

where the dimensions are $4N_1 \times 8N_1$ for all $\mathbf{A}$ matrices, $4N_1 \times 8N_2$ for all $\mathbf{S}$ matrices, and $4N_1 \times 1$ for all vectors.

4) Critical region of the atmosphere–water interface [see Eq. 1347) below]:

$$(\mathbf{S}_c^- - \mathbf{S}_c^+)\mathbf{c}_{L_1}^+ = \mathbf{b}_{L_1+1,c}^+ - \mathbf{b}_{L_1+1,c}^-,$$

where the dimensions are $4N_c \times 8N_2$ $(N_c = N_2 - N_1)$ for the $\mathbf{S}_c^\pm$ matrices, $4N_c \times 1$ for the vectors $\mathbf{b}_{L_1+1,c}^\pm$, and $8N_2 \times 1$ for the vector $\mathbf{c}_{L_1}^+$.

5) Layer interfaces – lower slab [Eq. (1349)]:

$$\mathbf{S}'_{L_2}\mathbf{c}_{L_2} = \mathbf{b}_{L_2},$$

where the dimensions are $L_2 8N_2 \times (L_2 - 1)8N_2$ for $\mathbf{S}'_{L_2}$, $8N_2(L_2 - 1) \times 1$ for $\mathbf{b}_{L_2}$, and $8N_2L_2 \times 1$ for $\mathbf{c}_{L_2}$.

Radiative Transfer in Coupled Environmental Systems: An Introduction to Forward and Inverse Modeling,
First Edition. Knut Stamnes and Jakob J. Stamnes.

6) Lower slab – bottom boundary [Eq. (1369)]:

$$\mathbf{S}_b \times \mathbf{c}_L = \mathbf{b}_b,$$

where the dimensions are $4N_2 \times 8N_2$ for the matrix $\mathbf{S}_b$, $4N_2 \times 1$ for the vector $\mathbf{b}_b$, and $8N_2 \times 1$ for the vector $\mathbf{c}_L$.

Combining all results (1)–(6) from the top of the upper slab [see Eq. (1311)] to the bottom of the lower slab [see Eq. (1369)], we obtain the following band matrix

$$\mathbf{A}_{\text{tot}}\mathbf{c}_{\text{tot}} = \mathbf{b}_{\text{tot}}, \tag{1300}$$

where the band matrix $\mathbf{A}_{\text{tot}}$ is given by

$$\mathbf{A}_{\text{tot}} = \begin{array}{c} 4N_1 \\ \left.\begin{array}{c} \\ 8N_1\cdot \\ (L_1-1) \\ \\ \end{array}\right\} \\ 4N_1 \\ 4N_1 \\ 4N_2-4N_1 \\ \left.\begin{array}{c} \\ 8N_2\cdot \\ (L_2-1) \\ \\ \end{array}\right\} \\ 4N_2 \end{array} \overbrace{\hphantom{xxxxxxxxxxxxxxxxxxxx}}^{8N_1\cdot L_1}\overbrace{\hphantom{xxxxxxxxxxxxxxxx}}^{8N_2\cdot L_2} \left[\begin{array}{ccccccccccc}
\mathbf{A}_0 & 0 & 0 & \cdots & 0 & 0 & & & & & 0 \\
0 & -\mathbf{A}'_2 & 0 & & 0 & 0 & & & & & 0 \\
0 & 0 & -\mathbf{A}'_3 & & 0 & 0 & & & & & 0 \\
\vdots & & & \ddots & & \vdots & & & & & \vdots \\
0 & & & & \mathbf{A}'_{L_1-1} & 0 & 0 & & & & 0 \\
0 & & & & 0 & -\mathbf{A}_t^- & \mathbf{S}^- - \mathbf{S}_r^+ & 0 & & & 0 \\
0 & & & & 0 & \mathbf{A}^+ - \mathbf{A}_r^- & -\mathbf{S}_t^+ & 0 & & & 0 \\
0 & & & & & 0 & \mathbf{S}_c^- - \mathbf{S}_c^+ & 0 & & & 0 \\
0 & & & & & 0 & \mathbf{S}'_1 & 0 & 0 & & 0 \\
0 & & & & & & 0 & \mathbf{S}'_2 & 0 & & 0 \\
\vdots & & & & & & & & \ddots & & \vdots \\
0 & & & & & & & & 0 & \mathbf{S}'_{L_2-1} & 0 \\
0 & & & & & & & & 0 & 0 & \mathbf{S}_b
\end{array}\right] \tag{1301}$$

and the vectors $\mathbf{c}_{\text{tot}}$ and $\mathbf{b}_{\text{tot}}$ are

$$\mathbf{c}_{\text{tot}} = \begin{bmatrix} \mathbf{C}_1 \\ \mathbf{C}_2 \\ \vdots \\ \mathbf{C}_{L_1} \\ \mathbf{C}_{L_1+1} \\ \vdots \\ \mathbf{C}_{L-1} \\ \mathbf{C}_L \end{bmatrix}, \qquad \mathbf{b}_{\text{tot}} = \begin{bmatrix} \mathbf{b}_0 \\ \mathbf{b}_2 - \mathbf{b}_1 \\ \mathbf{b}_3 - \mathbf{b}_2 \\ \vdots \\ \mathbf{b}_{L_1} - \mathbf{b}_{L_1-1} \\ \mathbf{b}_{L_1}^t - \mathbf{b}_{L_1+1} + \mathbf{b}_{L_1+1}^r \\ \mathbf{b}_{L_1}^r - \mathbf{b}_{L_1} + \mathbf{b}_{L_1+1}^t \\ \mathbf{b}_{L_1+1,c}^+ - \mathbf{b}_{L_1+1,c}^- \\ \mathbf{b}_{L_1+2} - \mathbf{b}_{L_1+1} \\ \mathbf{b}_{L_1+3} - \mathbf{b}_{L_1+2} \\ \vdots \\ \mathbf{b}_{L-2} - \mathbf{b}_{L-1} \\ \mathbf{b}_b \end{bmatrix} \tag{1302}$$

This system of linear algebraic equations has a bandwidth of $8N_2 + 4N_2 - 1$, which quantifies how far from the diagonal the farthest nonzero element is. This system of equations can be solved by specialized numerical solvers for banded matrices such as LINPACK's SGBSL routine.

D.2
Top of Upper Slab

At the top of the upper slab, where $\ell = 1$ and $\tau = \tau_0$, the radiation field is given by

$$\mathbf{I}_{\alpha 1}(\tau_0, -\mu, \phi) = \mathbf{I}_i(\tau_0) + \mathbf{I}_t(\tau_0), \tag{1303}$$

where we have allowed for an isotropic illumination component $\mathbf{I}_i$ as well as a thermal component $\mathbf{I}_t$.[3] Since both of these boundary sources are isotropic, only the $m = 0$ Fourier component of the cosine modes will contribute. Assuming that the incident illumination is natural light, both components will also be unpolarized, implying that

$$\mathbf{I}_i = [I_i/2, I_i/2, 0, 0]^T \quad ; \quad \mathbf{I}_t = [I_t/2, I_t/2, 0, 0]^T. \tag{1304}$$

In our $8N$ notation scheme, the general solution in the upper slab (atmosphere) is given by Eq. (585) ($\alpha = c, s$):

$$\begin{aligned}\tilde{i}^m_{\alpha\ell}(\tau, i) = &\sum_{j=1}^{8N_1} \tilde{C}^m_{\alpha j\ell} g^m_{\alpha j\ell}(i) e^{-k^m_{\alpha j\ell}\tau} + z^m_{\alpha\ell,0}(i) e^{-\tau/\mu_0} + z^m_{\alpha\ell,1}(i) e^{\tau/\mu_0} \\ &+ \delta_{0m}\delta_{\alpha c}[x^t_{0,\ell}(i) + x^t_{1,\ell}(i)\tau], \quad \ell = 1, \ldots, L_1.\end{aligned} \tag{1305}$$

Hence, the boundary conditions at the top of the upper slab becomes [setting $I_{\rm top} \equiv (I_i(\tau_0) + I_t(\tau_0))/2)$]

$$\tilde{i}^m_{c1}(\tau_0, i) = \begin{cases} \delta_{0m} I_{\rm top} & i = 1, ..., N_1 \\ \delta_{0m} I_{\rm top} & i = N_1 + 1, ..., 2N_1 \\ 0 & i = 2N_1 + 1, ..., 3N_1 \\ 0 & i = 3N_1 + 1, ..., 4N_1 \\ 0 & i = 4N_1 + 1, ..., 5N_1 \\ 0 & i = 5N_1 + 1, ..., 6N_1 \\ 0 & i = 6N_1 + 1, ..., 7N_1 \\ 0 & i = 7N_1 + 1, ..., 8N_1 \end{cases} \tag{1306}$$

$$\tilde{i}^m_{s1}(\tau_0, i) = \mathbf{0} \quad \forall \quad i. \tag{1307}$$

3) Note that for a collimated beam source, such as the Sun, we have formulated the problem in terms of the diffuse radiation, so that the boundary condition at the top of the medium is zero, but there is an internal source, see Section 3.1.1.

Since the $8N$ Stokes vector is defined as

$$\tilde{\mathbf{I}}_c^m \equiv \begin{pmatrix} I_{\|c}^m(\tau,-\mu_i^a) \\ I_{\perp c}^m(\tau,-\mu_i^a) \\ U_s^m(\tau,-\mu_i^a) \\ V_s^m(\tau,-\mu_i^a) \\ I_{\|c}^m(\tau,+\mu_i^a) \\ I_{\perp c}^m(\tau,+\mu_i^a) \\ U_s^m(\tau,+\mu_i^a) \\ V_s^m(\tau,+\mu_i^a) \end{pmatrix}_{8N_1\times 1}$$

only the downward components $I_{\|c}^m(\tau,-\mu_i^a)$ and $I_{\perp c}^m(\tau,-\mu_i^a)$ are nonzero at the top boundary of the upper slab.

Inserting $\tilde{i}_{\alpha\ell}^m(\tau_0,i)$ from the general solution in Eq. (1305) with $\ell = 1$,

$$\tilde{i}_{\alpha 1}^m(\tau_0,i) = \sum_{j=1}^{8N_1} C_{\alpha j1}^m g_{\alpha j1}^m(i)e^{-k_{\alpha j1}^m\tau_0} + z_{\alpha 1,0}^m(i)e^{-\tau_0/\mu_0}$$
$$+ z_{\alpha 1,1}^m(i)e^{\tau_0/\mu_0} + \delta_{\alpha c}\delta_{0m}[x_{0,1}^t(i) + x_{1,1}^t(i)\tau_0] \qquad (1308)$$

into Eq. (1306), we obtain

$$\sum_{j=1}^{8N_1} C_{\alpha j1}^m g_{\alpha j1}^m(i)e^{-k_{\alpha j1}^m\tau_0} = \Big[\delta_{0m}[I_{\text{top}} - x_{0,1}^t(i) - x_{1,1}^t(i)\tau_0]$$
$$-z_{\alpha 1,0}^m(i)e^{-\tau_0/\mu_0} - z_{\alpha 1,1}^m(i)e^{\tau_0/\mu_0}\Big] \qquad i = 1,\dots,2N_1 \qquad \alpha = c,s$$
$$= \Big[-z_{c1,0}^m(i)e^{-\tau_0/\mu_0} - z_{c1,1}^m(i)e^{\tau_0/\mu_0} - \delta_{\alpha c}\delta_{0m}[x_{0,1}(i) + x_{1,1}(i)\tau_0]\Big]$$
$$i = 2N_1+1,\dots,4N_1. \qquad (1309)$$

We have dropped the $i = 4N_1+1,\dots,8N_1$ components because these values of i correspond to radiation in upward directions. To simplify the notation, we will now drop the index m for the Fourier mode, and the layer index $\ell = 1$, since the following expressions will be valid for all m, and apply only at the top of the upper slab (atmosphere) where $\tau = \tau_0$. Then, if we define $z_{\alpha,i} \equiv z_{\alpha i}$, $i = 0,1$, we may rewrite Eqs. (1309) in matrix form as follows:

$$\begin{pmatrix} g_{\alpha 1}(1)e^{-k_{\alpha 1}\tau_0} & \cdots & g_{k8N_1}(1)e^{-k_{\alpha 8N_1}\tau_0} \\ g_{\alpha 1}(2)e^{-k_{\alpha 1}\tau_0} & \cdots & g_{k8N_1}(2)e^{-k_{\alpha 8N_1}\tau_0} \\ \vdots & & \vdots \\ g_{\alpha 1}(4N_1)e^{-k_{\alpha 1}\tau_0} & \cdots & g_{\alpha 8N_1}(4N_1)e^{-k_{\alpha 8N_1}\tau_0} \end{pmatrix}_{4N_1\times 8N_1} \times \begin{pmatrix} C_{\alpha 1} \\ C_{\alpha 2} \\ \vdots \\ C_{\alpha 8N_1} \end{pmatrix}_{8N_1\times 1}$$
$$= \begin{pmatrix} \delta_{0m}\delta\alpha c[I_{\text{top}} - x_0^t(1) - x_1^t(1)\tau_0] - z_{\alpha 0}(1)e^{-\tau_0/\mu_0} - z_{\alpha 1}(1)e^{\tau_0/\mu_0} \\ \delta_{0m}\delta\alpha c[I_{\text{top}} - x_0^t(2) - x_1^t(2)\tau_0] - z_{\alpha 0}(2)e^{-\tau_0/\mu_0} - z_{\alpha 1}(2)e^{\tau_0/\mu_0} \\ \vdots \\ -\delta_{0m}\delta\alpha c[x_0^t(4N_1) + x_1^t(4N_1)\tau_0] - z_{\alpha 0}(4N_1)e^{-\tau_0/\mu_0} - z_{\alpha 1}(4N_1)e^{\tau_0/\mu_0} \end{pmatrix}_{4N_1\times 1}$$
$$(1310)$$

which may be rewritten in the compact matrix notation as

$$\mathbf{A}_0\mathbf{c}_1 = \mathbf{b}_0, \tag{1311}$$

where ($\alpha = c, s$)

$$\mathbf{A}_0 \equiv \begin{pmatrix} g_{\alpha 1}(1)e^{-k_{\alpha 1}\tau_0} & \cdots & g_{\alpha 8N_1}(1)e^{-k_{\alpha 8N_1}\tau_0} \\ g_{\alpha 1}(2)e^{-k_{\alpha 1}\tau_0} & \cdots & g_{\alpha 8N_1}(2)e^{-k_{\alpha 8N_1}\tau_0} \\ \vdots & & \vdots \\ g_{\alpha 1}(4N_1)e^{-k_{\alpha 1}\tau_0} & \cdots & g_{\alpha 8N_1}(4N_1)e^{-k_{\alpha 8N_1}\tau_0} \end{pmatrix}_{4N_1\times 8N_1}$$

and

$$\mathbf{c}_1 \equiv \begin{pmatrix} C_{\alpha 1} \\ C_{\alpha 2} \\ \vdots \\ C_{\alpha 8N_1} \end{pmatrix}_{8N_1\times 1}$$

$$\mathbf{b}_0 = \begin{pmatrix} \delta_{0m}\delta_{\alpha c}[I_{top} - x_0^t(1) - x_1^t(1)\tau_0] - z_{\alpha 0}(1)e^{-\tau_0/\mu_0} - z_{\alpha 1}(1)e^{+\tau_0/\mu_0} \\ \delta_{0m}\delta_{\alpha c}[I_{top} - x_0^t(2) - x_1^t(2)\tau_0] - z_{\alpha 0}(2)e^{-\tau_0/\mu_0} - z_{\alpha 1}(2)e^{+\tau_0/\mu_0} \\ \vdots \\ -\delta_{0m}\delta_{\alpha c}[x_0^t(4N_1) + x_1^t(4N_1)\tau_0] - z_{\alpha 0}(4N_1)e^{-\tau_0/\mu_0} - z_{\alpha 1}(4N_1)e^{+\tau_0/\mu_0} \end{pmatrix}_{4N_1\times 1}.$$

D.3
Layer Interface Conditions in the Upper Slab

Since the radiation field must be continuous across layer interfaces, we require

$$\tilde{i}^m_{\alpha\ell}(\tau_\ell, i) = \tilde{i}^m_{\alpha\ell+1}(\tau_\ell, i), \qquad i = 1, \ldots, 8N_1, \qquad \ell = 1, \ldots, L_1 - 1. \tag{1312}$$

Inserting the general solution Eq. (585) into Eq. (1312), we obtain

$$\begin{aligned} &\sum_{j=1}^{8N_1} \tilde{C}^m_{\alpha j\ell} g^m_{\alpha j\ell}(i)e^{-k^m_{\alpha j\ell}\tau_\ell} + z^m_{\alpha\ell,0}(i)e^{-\tau_\ell/\mu_0} + z^m_{\alpha\ell,1}(i)e^{\tau_\ell/\mu_0} \\ &\quad + \delta_{0m}\delta_{\alpha c}[x^t_{0,\ell}(i) + x^t_{1,\ell}(i)\tau_\ell] \\ &\quad = \sum_{j=1}^{8N_1} \tilde{C}^m_{\alpha j\ell+1} g^m_{\alpha j\ell+1}(i)e^{-k^m_{\alpha j\ell+1}\tau_\ell} + z^m_{\alpha\ell+1,0}(i)e^{-\tau_\ell/\mu_0} + z^m_{\alpha\ell+1,1}(i)e^{\tau_\ell/\mu_0} \\ &\quad + \delta_{0m}\delta_{\alpha c}[x^t_{0,\ell+1}(i) + x^t_{1,\ell+1}(i)\tau_\ell]. \end{aligned} \tag{1313}$$

We rearrange Eq. (1313) so that the sums involving the unknown coefficients appear on the left side as follows:

$$\begin{aligned} &\sum_{j=1}^{8N_1} [\tilde{C}^m_{\alpha j\ell} g^m_{\alpha j\ell}(i)e^{-k^m_{\alpha j\ell}\tau_p} - \tilde{\tilde{C}}^m_{\alpha j\ell+1} g^m_{\alpha j\ell+1}(i)e^{-k^m_{\alpha j\ell+1}\tau_\ell}] \\ &\quad = z^m_{\alpha\ell+1,0}(i)e^{-\tau_\ell/\mu_0} + z^m_{\alpha\ell+1,1}(i)e^{\tau_\ell/\mu_0} + \delta_{0m}\delta_{\alpha c}[x^t_{0,\ell+1}(i) + x^t_{1,\ell+1}(i)\tau_\ell] \\ &\qquad - z^m_{\alpha\ell,0}(i)e^{-\tau_\ell/\mu_0} - z^m_{\alpha\ell,1}(i)e^{\tau_\ell/\mu_0} - \delta_{0m}\delta_{\alpha c}[x^t_{0,\ell}(i) + x^t_{1,\ell}(i)\tau_\ell]. \end{aligned} \tag{1314}$$

There will be L_1 layers in the upper slab, L_2 layers in the lower slab, and hence a total of $L = L_1 + L_2$ layers in both slabs. To simplify the exposition, we introduce the following notation. In the upper slab ($p = 1$), the layer index will be $\ell = 1, ..., L_1$, and in the lower slab ($p = 2$) we will use the same index $\ell = L_1 + 1, \dots, L$. Thus, in the upper slab, where $\ell = 1, ..., L_1$, we define

$$C_{j\ell} \equiv \tilde{C}^m_{\alpha j\ell} \tag{1315}$$

$$A_{j\ell}(i) \equiv g^m_{\alpha j\ell}(i)e^{-k^m_{\alpha j\ell}\tau_\ell} \tag{1316}$$

$$A_{j\ell+1}(i) \equiv g^m_{\alpha j\ell+1}(i)e^{-k^m_{\alpha j\ell+1}\tau_\ell} \tag{1317}$$

$$\begin{aligned} b_\ell(i) \equiv{}& z^m_{\alpha\ell,0}(i)e^{-\tau_\ell/\mu_0} + z^m_{\alpha\ell,1}(i)e^{\tau_\ell/\mu_0} \\ &+ \delta_{0m}\delta_{\alpha c}[x^t_{0,\ell}(i) + x^t_{1,\ell}(i)\tau_\ell] \end{aligned} \tag{1318}$$

$$\begin{aligned} b_{\ell+1}(i) \equiv{}& z^m_{\alpha\ell+1,0}(i)e^{-\tau_\ell/\mu_0} + z^m_{\alpha\ell+1,1}(i)e^{\tau_\ell/\mu_0} \\ &+ \delta_{0m}\delta_{\alpha c}[x^t_{0,\ell+1}(i) + x^t_{1,\ell+1}(i)\tau_\ell]. \end{aligned} \tag{1319}$$

Similarly, in the lower slab, where $\ell = L_1 + 1, \dots, L$, and the general solution given by Eq. (587) applies, we define

$$C_{j\ell} \equiv \tilde{C}^m_{\alpha j\ell} \tag{1320}$$

$$S_{j\ell}(i) \equiv g^m_{\alpha j\ell}(i)e^{-k^m_{\alpha j\ell}\tau_\ell} \tag{1321}$$

$$S_{j\ell+1}(i) \equiv g^m_{\alpha jp+1}(i)e^{-k^m_{\alpha j\ell+1}\tau_\ell} \tag{1322}$$

$$b_\ell(i) \equiv z^m_{\alpha\ell,0}(i)e^{-\tau_\ell/\mu^w_0} + \delta_{0m}\delta_{\alpha c}[x^t_{0,\ell}(i) + x^t_{1,\ell}(i)\tau_\ell] \tag{1323}$$

$$b_{\ell+1}(i) \equiv z^m_{\alpha\ell+1,0}(i)e^{-\tau_\ell/\mu^w_0} + \delta_{0m}\delta_{\alpha c}[x^t_{0,\ell+1}(i) + x^t_{1,\ell+1}(i)\tau_\ell]. \tag{1324}$$

This simplification is convenient because the layer continuity conditions in both slabs are the same for all Fourier modes m and for cosine ($\alpha = c$) and sine ($\alpha = s$) modes, although one should keep in mind that the particular source terms [$b_\ell(i)$ and $b_{\ell+1}(i)$] differ somewhat between the two slabs.

Using this notation, we may proceed to write the system of linear algebraic equations [Eqs. (1314)] in matrix form. Defining the $8N_1 \times 8N_1$ matrices

$$\mathbf{A}_\ell = \begin{pmatrix} A_{1\ell}(1) & \cdots & A_{8N_1\ell}(1) \\ A_{1\ell}(2) & \cdots & A_{8N_1\ell}(2) \\ \vdots & & \\ A_{1\ell}(8N_1) & \cdots & A_{8N_1\ell}(8N_1) \end{pmatrix} \tag{1325}$$

$$\mathbf{A}_{\ell+1} = \begin{pmatrix} A_{1\ell+1}(1) & \cdots & A_{8N_1\ell+1}(1) \\ A_{1\ell+1}(2) & \cdots & A_{8N_1\ell+1}(2) \\ \vdots & & \\ A_{1\ell+1}(8N_1) & \cdots & A_{8N_1\ell+1}(8N_1) \end{pmatrix} \tag{1326}$$

and the vectors

$$\mathbf{c}_\ell = \begin{pmatrix} C_{1\ell} \\ \vdots \\ C_{8N_1\ell} \\ C_{1\ell+1} \\ \vdots \\ C_{8N_1\ell+1} \end{pmatrix}_{2\cdot 8N_1\times 1} ; \mathbf{b}_\ell = \begin{pmatrix} b_\ell(1) \\ \vdots \\ b_\ell(8N_1) \end{pmatrix}_{8N_1\times 1} ; \mathbf{b}_{\ell+1} = \begin{pmatrix} b_{\ell+1}(1) \\ \vdots \\ b_{\ell+1}(8N_1) \end{pmatrix}_{8N_1\times 1} \quad (1327)$$

we may rewrite Eqs. (1314) in matrix form for each layer interface in the upper slab (atmosphere) as

$$(\mathbf{A}_\ell - \mathbf{A}_{\ell+1})\mathbf{c}_\ell = \mathbf{b}_{\ell+1} - \mathbf{b}_\ell, \qquad \ell = 1, \ldots, L_1 - 1. \quad (1328)$$

Defining $\mathbf{A}'_\ell \equiv \mathbf{A}_\ell - \mathbf{A}_{\ell+1}$, and applying Eq. (1328) to every layer interface, we obtain the following system of equations for the entire upper slab:

$$\mathbf{A}'_{L_1}\mathbf{c}_{L_1} = \mathbf{b}_{L_1}, \quad (1329)$$

where

$$\mathbf{A}'_{L_1} = \begin{pmatrix} \mathbf{A}'_1 & 0 & \cdots & 0 \\ 0 & \mathbf{A}'_2 & 0 & \vdots \\ \vdots & & & \vdots \\ 0 & \cdots & \mathbf{A}'_{L_1-2} & 0 \\ 0 & \cdots & 0 & \mathbf{A}'_{L_1-1} \end{pmatrix}_{8N_1\cdot L_1\times 8N_1\cdot(L_1-1)} \quad (1330)$$

and

$$\mathbf{c}_{L_1} = \begin{pmatrix} \mathbf{c}_1 \\ \vdots \\ \mathbf{c}_{L_1} \end{pmatrix}_{8N_1\cdot L_1\times 1}, \quad \mathbf{b}_{L_1} = \begin{pmatrix} \mathbf{b}_2 - \mathbf{b}_1 \\ \vdots \\ \mathbf{b}_{L_1} - \mathbf{b}_{L_1-1} \end{pmatrix}_{8N_1\cdot(L_1-1)\times 1}. \quad (1331)$$

D.3.1
Interface Between the Two Slabs (Atmosphere–Water System)

For an atmosphere–water interface, we have to consider the quadrature angles in the atmosphere, which according to Snell's law are refracted into the water (the noncritical region) and the additional quadrature angles required in the total reflection region (the critical region). It is important to keep in mind that since we use an $8N$ scheme, the quadrature angles in the critical region will be scattered at equal intervals throughout the matrix equations. These facts require some special book-keeping. The structure of the quadrature points is illustrated below for the lower slab:

$$
\mathbf{I}_{\rm wat} = \begin{pmatrix}
I_{\|,\rm wat}(\tau,-\mu_1^w) & \\
\vdots & \Leftarrow \text{critical region} \\
I_{\|,\rm wat}(\tau,-\mu_{N_2-N_1}^w) & \\
I_{\|,\rm wat}(\tau,-\mu_{N_2-N_1+1}^w) & \\
\vdots & \Leftarrow \text{noncritical region} \\
I_{\|,\rm wat}(\tau,-\mu_{N_2}^w) & \\
I_{\perp,\rm wat}(\tau,-\mu_1^w) & \\
\vdots & \Leftarrow \text{critical region} \\
I_{\perp,\rm wat}(\tau,-\mu_{N_2-N_1}^w) & \\
I_{\perp,\rm wat}(\tau,-\mu_{N_2-N_1+1}^w) & \\
\vdots & \Leftarrow \text{noncritical region} \\
I_{\perp,\rm wat}(\tau,-\mu_{N_2}^w) & \\
U_{\rm wat}(\tau,-\mu_1^w) & \\
\vdots & \Leftarrow \text{same for U components} \\
U_{\rm wat}(\tau,-\mu_{N_2}^w) & \\
V_{\rm wat}(\tau,-\mu_1^w) & \\
\vdots & \Leftarrow \text{same for V components} \\
V_{\rm wat}(\tau,-\mu_{N_2}^w) & \\
I_{\|,\rm wat}(\tau,\mu_1^w) & \\
\vdots & \Leftarrow \text{critical region} \\
I_{\|,\rm wat}(\tau,\mu_{N_2-N_1}^w) & \\
I_{\|,\rm wat}(\tau,\mu_{N_2-N_1+1}^w) & \\
\vdots & \Leftarrow \text{noncritical region} \\
I_{\|,\rm wat}(\tau,\mu_{N_2}^w) & \\
\vdots & \text{same for upper } I_\perp, \text{ U and V components}
\end{pmatrix}
$$

Since we use the cosine of the quadrature angles, the angles in the critical region will be

$$
\theta > \theta_{\rm crit} \quad \Rightarrow \quad \mu < \mu_{\rm crit} = \sqrt{1 - \frac{1}{m_{\rm rel}^2}} \qquad 0 \le \theta \le 90,
$$

where the subscript crit stands for critical. Furthermore, we use the convention $\mu_i < \mu_{i+1}$, so that the μ's are in ascending order.

To implement the boundary conditions, we use the component form of the Stokes vector $\tilde{i}^m_{\alpha\ell}(\tau, i)$, [see Eqs. (585) and (587)], where i indicates the quadrature angle. There are $8N_1$ i-values ($i = 1, \ldots, 8N_1$) in the upper slab (atmosphere), but $8N_2$ i-values ($i = 1, \ldots, 8N_2$) in the lower slab. To prevent confusion, we denote the i's in the upper slab by i^a, and those in the lower slab by i^w.

To relate the quadrature angles in the upper slab (atmosphere) to those in the lower slab (water), we define:

$$i_-^a = 1, \ldots, 4N_1 \quad \text{(corresponds to } i_-^w \text{ in the water)}$$
$$i_+^a = 4N_1 + 1, \ldots, 8N_1 \quad \text{(corresponds to } i_+^w \text{ in the water)}$$
$$i_-^w = N_2 - N_1 + 1, \ldots, N_2;\ 2N_2 - N_1 + 1, \ldots, 2N_2;$$
$$3N_2 - N_1 + 1, \ldots, 3N_2; 4N_2 - N_1 + 1, \ldots, 4N_2$$
$$i_+^w = 5N_2 - N_1 + 1, \ldots, 5N_2;\ 6N_2 - N_1 + 1, \ldots, 6N_2;$$
$$7N_2 - N_1 + 1, \ldots, 7N_2;\ 8N_2 - N_1 + 1, \ldots, 8N_2,$$

where all the i's have $4N_1$ entries linking the $4N_1$ Stokes components in the upper slab (atmosphere) with the $4N_1$ Stokes components in the lower slab (water). The additional $N_c = N_2 - N_1$ quadrature points in the critical region are then

$$i_-^c = 1, \ldots, N_2 - N_1; N_2 + 1, \ldots, 2N_2 - N_1; 2N_2 + 1, \ldots, 3N_2 - N_1,$$
$$3N_2 + 1, \ldots, 4N_2 - N_1$$
$$i_+^c = 4N_2 + 1, \ldots, 5N_2 - N_1; 5N_2 + 1, \ldots, 6N_2 - N_1; 6N_2 + 1, \ldots, 7N_2 - N_1;$$
$$7N_2 + 1, \ldots, 8N_2 - N_1.$$

Example: $N_1 = 4; N_2 = 6$

$i_-^a = 1, \ldots, 16.$

$i_+^a = 17, \ldots, 32.$

$i_-^w = 3, \ldots, 6;\ 9, \ldots, 12;\ 15, \ldots, 18;\ 21, \ldots, 24.$

$i_+^w = 27, \ldots, 30;\ 33, \ldots, 36;\ 39, \ldots, 42;\ 45, \ldots, 48.$

$i_-^c = 1, \ldots, 2; 7, \ldots, 8; 13, \ldots, 14; 19, \ldots, 20.$

$i_+^c = 25, \ldots, 26; 31, \ldots, 32; 37, \ldots, 38; 43, \ldots, 44.$

Noncritical Region, $\mu_i^w > \mu_{\text{crit}}$

In the noncritical region, reflection and refraction occur at the interface according to the Fresnel formulas, and there is a one-to-one correspondence between each downward angle in the upper slab and the lower slab. The following conditions apply:

$$\tilde{i}^m_{\alpha L_1+1}(\tau_{L_1}, i_-^w) = R(\mu^w_{i_+^w}, m_{\text{rel}}^{-1})\tilde{i}^m_{\alpha L_1+1}(\tau_{L_1}, i_+^w) + m_{\text{rel}}^2 T(\mu^a_{i_-^a}, m_{\text{rel}})\tilde{i}^m_{\alpha L_1}(\tau_{L_1}, i_-^w) \tag{1332}$$

$$\tilde{i}^m_{\alpha L_1}(\tau_{L_1}, i_+^a) = R(\mu^a_{i_-^a}, m_{\text{rel}})\tilde{i}^m_{\alpha L_1}(\tau_{L_1}, i_+^a) + \frac{1}{m_{\text{rel}}^2} T(\mu^a_{i_+^w}, m_{\text{rel}}^{-1})\tilde{i}^m_{\alpha L_1+1}(\tau_{L_1}, i_+^w). \tag{1333}$$

Equation (1332) states that the downward radiation field at the interface in the lower slab consists of the reflected component originating in the lower slab plus the transmitted component originating in the upper slab. Similarly, Eq. (1333) states that the upward radiation field at the interface in the upper slab consists of the specularly reflected downward (atmospheric) radiation plus the transmitted upward radiation from the lower slab (water).

Inserting the general solutions for the upper and lower slabs [Eqs. (585) and (587)] into Eq. (1332), we obtain

$$\sum_{j=1}^{8N_2} \tilde{C}^m_{\alpha jL+1} g^m_{\alpha jL+1}(i^w_-) e^{-k^m_{\alpha jL+1}\tau_{L_1}} + z^m_{\alpha L_1+1,2}(i^w_-) e^{-\tau_{L_1}/\mu_0^w}$$
$$+ \delta_{0m}\delta_{\alpha c}[x^t_{0,L_1+1}(i^w_-) + x^t_{1,L_1+1}(i^w_-)\tau_{L_1}]$$
$$= R(\mu^w_{i^w_+}, m^{-1}_{\mathrm{rel}}) \left[\sum_{j=1}^{8N_2} \tilde{C}^m_{\alpha jL+1} g^m_{\alpha jL+1}(i^w_+) e^{-k^m_{\alpha jL+1}\tau_{L_1}} + z^m_{\alpha jL+1}(i^w_+) e^{-\tau_{L_1}/\mu_0^w} \right.$$
$$\left. + \delta_{0m}\delta_{\alpha c}[x^t_{0,L_1+1}(i^w_+) + x^t_{1,L_1+1}(i^w_+)\tau_{L_1}] \right]$$
$$+ T(\mu^a_{i^a_-}, m_{\mathrm{rel}}) \left[\sum_{j=1}^{8N_1} \tilde{C}^m_{\alpha jL_1} g^m_{\alpha jL_1}(i^a_-) e^{-k^m_{\alpha jL_1}\tau_{L_1}} + z^m_{\alpha L_1,0}(i^a_-) e^{-\tau_{L_1}/\mu_0} \right.$$
$$\left. + z^m_{\alpha L_1,1}(i^a_-) e^{+\tau_{L_1}/\mu_0} + \delta_{0m}\delta_{\alpha c}[x^t_{0,L_1}(i^a_-) + x^t_{1,L_1}(i^a_-)\tau_{L_1}] \right]. \quad (1334)$$

Similarly, inserting Eqs. (585) and (587) into Eq. (1333), we obtain

$$\sum_{j=1}^{8N_1} \tilde{C}^m_{\alpha jL_1} g^m_{\alpha jL_1}(i^a_+) e^{-k^m_{\alpha jL_1}\tau_{L_1}} + z^m_{\alpha jL_1,0}(i^a_+) e^{-\tau_{L_1}/\mu_0}$$
$$+ z^m_{\alpha L_1,1}(i^a_+) e^{+\tau_{L_1}/\mu_0} + \delta_{0m}\delta_{\alpha c}[x^t_{0,L_1}(i^a_+) + x^t_{1,L_1}(i^a_+)\tau_{L_1}]$$
$$= R(\mu^a_{i^a_-}, m_{\mathrm{rel}}) \cdot \left[\sum_{j=1}^{8N_1} \tilde{C}^m_{\alpha jL_1} g^m_{\alpha jL_1}(i^a_-) e^{-k^m_{\alpha jL_1}\tau_{L_1}} + z^m_{\alpha L_1,0}(i^a_-) e^{-\tau_{L_1}/\mu_0} \right.$$
$$\left. + z^m_{\alpha L_1,1}(i^a_-) e^{+\tau_{L_1}/\mu_0} + \delta_{0m}\delta_{\alpha c}[x^t_{0,L_1}(i^a_-) + x^t_{1,L_1}(i^a_-)\tau_{L_1}] \right]$$
$$+ T(\mu^w_{i^w_+}, m^{-1}_{\mathrm{rel}}) \cdot \left[\sum_{j=1}^{8N_2} \tilde{C}^m_{\alpha jL_1+1} g^m_{\alpha jL_1+1}(i^s_+) e^{-k^m_{\alpha jL_1+1}\tau_{L_1}} \right.$$
$$\left. + z^m_{\alpha L_1+1,2}(i^s_+) e^{-\tau_{L_1}/\mu_0^w} + \delta_{0m}\delta_{\alpha c}[x^t_{0,L_1+1}(i^w_+) + x^t_{1,L_1+1}(i^w_+)\tau_{L_1}] \right]. \quad (1335)$$

It is useful to simplify the notation in a way similar to when we considered the interface conditions in the upper slab. Thus, we introduce the following simplifications:

$$S_{j1}(i) \equiv S_{jL_1+1}(i) = g^m_{\alpha jL_1+1}(i) e^{-k^m_{\alpha jL_1+1}\tau_{L_1}} \quad (1336)$$
$$R^a(i^a_-) \equiv R(\mu^a_{i^a_-}, m_{\mathrm{rel}}) \quad (1337)$$
$$R^w(i^w_+) \equiv R(\mu^w_{i^w_+}, m^{-1}_{\mathrm{rel}}) \quad (1338)$$
$$T^a(i^a_-) \equiv T(\mu^a_{i^a_-}, m_{\mathrm{rel}}) \quad (1339)$$
$$T^w(i^w_+) \equiv T(\mu^w_{i^w_+}, m^{-1}_{\mathrm{rel}}). \quad (1340)$$

Equation (1334) may now be rewritten as

$$-\sum_{j=1}^{8N_1} T^a(i_-^a) C_{jL_1} A_{jL_1}(i_-^a) + \sum_{j=1}^{8N_2} \left[C_{jL_1+1} S_{j1}(i_-^w) - R^w(i_+^w) C_{jL_1+1} S_{j1}(i_+^w) \right]$$
$$= T^a(i_-^a) b_{L_1}(i_-^a) + R^w(i_+^w) b_{L_1+1}(i_+^w) - b_{L_1+1}(i_-^w) \tag{1341}$$

and Eq. (1335) as

$$\sum_{j=1}^{8N_1} \left[C_{jL_1} A_{jL_1}(i_+^a) - R^a(i_-^a) C_{jL_1} A_{jL_1}(i_-^a) \right] - \sum_{j=1}^{8N_2} T^w(i_+^w) C_{jL_1+1} S_{j1}(i_+^w)$$
$$= R^a(i_-^a) b_{L_1}(i_-^a) + T^w(i_+^w) b_{L_1+1}(i_+^w) - b_{L_1}(i_+^a) \tag{1342}$$

where $b_{L_1}(i_+^a)$ and $b_{L_1+1}(i_+^w)$ are defined by Eqs. (1323) and (1324). Recall that both i_-^w and i_+^w have $4N_1$ elements, which are distributed across the $[1, 4N_2]$ and $[4N_2 + 1, 8N_2]$ intervals, respectively. To bring the system of equations for the noncritical region in matrix form, we define the following matrices and vectors:

$\mathbf{A}^-$ with elements $A^-(i,j) = A_{jL_1}(i = i_-^a),\ i = 1, 4N_1,\ j = 1, 8N_1$
$\mathbf{A}^+$ with elements $A^+(i,j) = A_{jL_1}(i = i_+^a),\ i = 1, 4N_1,\ j = 1, 8N_1$
$\mathbf{t}_a$ with elements $t_a(i) = T^a(i = i_-^a),\ i = 1, 4N_1$
$\mathbf{A}_t^-$ with elements $A_t^-(i,j) = t_a(i) A(i,j),\ i = 1, 4N_1,\ j = 1, 8N_1$
$\mathbf{c}_{L_1}^-$ with elements $C_{jL_1},\ j = 1, 8N_1$
$\mathbf{c}_{L_1}^+$ with elements $C_{jL_1+1},\ j = 1, 8N_2$
$\mathbf{S}^-$ with elements $S^-(i,j) = S_{j1}(i = i_-^w),\ i = 1, 4N_1,\ j = 1, 8N_2$
$\mathbf{S}^+$ with elements $S^+(i,j) = S_{j1}(i = i_+^w),\ i = 1, 4N_1,\ j = 1, 8N_2$
$\mathbf{r}_w$ with elements $r_w(i) = R^w(i = i_+^a),\ i = 1, 4N_1$
$\mathbf{S}_r^+$ with elements $S_r^+(i,j) = r_w(i) S^+(i,j),\ i = 1, 4N_1,\ j = 1, 8N_2$
$\mathbf{t}_w$ with elements $t_w(i) = T^w(i = i_+^a),\ i = 1, 4N_1$
$\mathbf{S}_t^+$ with elements $S_t^+(i,j) = t_w(i) S^+(i,j),\ i = 1, 4N_1,\ j = 1, 8N_2$
$\mathbf{b}_{L_1}$ with elements $b_{L_1}(i) = b_{L_1}(i = i_-^a),\ i = 1, 4N_1$
$\mathbf{b}_{L_1+1}$ with elements $b_{L_1+1}(i) = b_{L_1+1}(i = i_+^w),\ i = 1, 4N_1$
$\mathbf{b}_{L_1}^t$ with elements $b_{L_1}^t(i) = T^a(i = i_-^a) b_{L_1}(i = i_-^a),\ i = 1, 4N_1$
$\mathbf{b}_{L_1}^r$ with elements $b_{L_1}^r(i) = R^a(i = i_-^a) b_{L_1}(i = i_-^a),\ i = 1, 4N_1$
$\mathbf{b}_{L_1+1}^t$ with elements $b_{L_1+1}^t(i) = T^w(i = i_+^w) b_{L_1+1}(i = i_+^w),\ i = 1, 4N_1$
$\mathbf{b}_{L_1+1}^r$ with elements $b_{L_1+1}^r(i) = R^a(i = i_+^w) b_{L_1+1}(i = i_+^w),\ i = 1, 4N_1$.

Using these definitions, we may write Eqs. (1341) and (1342) for the interface conditions of the noncritical region in matrix form as

$$\begin{pmatrix} -\mathbf{A}_t^- & \mathbf{S}^- - \mathbf{S}_r^+ \\ \mathbf{A}^+ - \mathbf{A}_r^- & -\mathbf{S}_t^+ \end{pmatrix} \times \begin{pmatrix} \mathbf{c}_{L_1}^- \\ \mathbf{c}_{L_1}^+ \end{pmatrix} = \begin{pmatrix} \mathbf{b}_{L_1}^t - \mathbf{b}_{L_1+1} + \mathbf{b}_{L_1+1}^r \\ \mathbf{b}_{L_1}^r - \mathbf{b}_{L_1} + \mathbf{b}_{L_1+1}^t \end{pmatrix}. \tag{1343}$$

Critical Region, $\mu_i^w \leq \mu_{\rm crit}$

In the critical region, light incident upon the interface in the lower slab will be totally reflected. Mathematically, this requirement is expressed as

$$\tilde{i}^m_{\alpha\ell}(\tau_{L_1}, i^c_-) = \tilde{i}^m_{\alpha\ell}(\tau_{L_1}, i^c_+), \qquad i^c_\pm = [1, 4N_c], N_c = (N_2 - N_1) \tag{1344}$$

where the quantities $i^c_\pm$ were defined above Eq. (1332). Inserting the general solution for the lower slab [Eq. (587)] into Eq. (1344), we obtain

$$\begin{aligned}
&\sum_{j=1}^{8N_2} \tilde{C}^m_{\alpha jL_1+1} g^m_{\alpha jL_1+1}(i^c_-) e^{-k^m_{\alpha jL_1+1}\tau_{L_1}} + z^m_{\alpha L_1+1,2}(i^c_-) e^{-\tau_{L_1}/\mu_0^w} \\
&+ \delta_{0m}\delta_{\alpha c}[x^t_{0,L_1+1}(i^c_-) + x^t_{1,L_1+1}(i^c_-)\tau_{L_1}], \\
&= \sum_{j=1}^{8N_2} \tilde{C}^m_{\alpha jL_1+1} g^m_{\alpha jL_1+1}(i^c_+) e^{-k^m_{\alpha jL_1+1}\tau_{L_1}} + z^m_{\alpha L_1+1,2}(i^c_+) e^{-\tau_{L_1}/\mu_0^w} \\
&+ \delta_{0m}\delta_{\alpha c}[x^t_{0,L_1+1}(i^c_+) + x^t_{1,L_1+1}(i^c_+)\tau_{L_1}]
\end{aligned} \tag{1345}$$

which may be rewritten as

$$\begin{aligned}
&\sum_{j=1}^{8N_2} C_{j,L_1+1}[S_{j1}(i = i^c_-) - S_{j1}(i = i^c_+)] \\
&= b_{L_1+1}(i = i^c_+) - b_{L_1+1}(i = i^c_-), \qquad i = 1, \ldots, 4N_c
\end{aligned} \tag{1346}$$

where we have used the notation and definitions introduced above. Defining the $4N_c \times 8N_2$ matrices

$$\begin{aligned}
&\mathbf{S}_c^- \text{ with elements } S_c^-(i,j) = S_{j1}(i = i^c_-),\ i = 1, 4N_c,\ j = 1, 8N_2 \\
&\mathbf{S}_c^+ \text{ with elements } S_c^+(i,j) = S_{j1}(i = i^c_+),\ i = 1, 4N_c,\ j = 1, 8N_2
\end{aligned}$$

the $4N_c \times 1$ vectors

$$\begin{aligned}
&\mathbf{b}^+_{L_1+1,c} \text{ with elements } b_{L_1+1}(i) = b_{L_1+1}(i = i^c_+),\ i = 1, 4N_c \\
&\mathbf{b}^-_{L_1+1,c} \text{ with elements } b_{L_1+1}(i) = b_{L_1+1}(i = i^c_-),\ i = 1, 4N_c
\end{aligned}$$

and the $8N_2 \times 1$ vector

$$\mathbf{c}^+_{L_1} \text{ with elements } c^+_{L_1}(i),\ i = 1, 8N_2$$

we may rewrite Eq. (1346) in matrix form as

$$(\mathbf{S}_c^- - \mathbf{S}_c^+)\mathbf{c}^+_{L_1} = \mathbf{b}^+_{L_1+1,c} - \mathbf{b}^-_{L_1+1,c}. \tag{1347}$$

D.4
Layer Interface Conditions in the Lower Slab

The situation is completely analogous to that in the upper slab. Thus, we obtain for each layer interface in the lower slab (water)

$$(\mathbf{S}_\ell - \mathbf{S}_{\ell+1})\mathbf{c}_\ell = \mathbf{b}_{\ell+1} - \mathbf{b}_\ell, \qquad \ell = L_1 + 1, \ldots, L = L_1 + L_2, \tag{1348}$$

where the matrices $\mathbf{S}_\ell$ and $\mathbf{S}_{\ell+1}$ are defined as in Eqs. (1325) and (1326) with the letter A replaced by the letter S. Defining $\mathbf{S}'_\ell \equiv \mathbf{S}_\ell - \mathbf{S}_{\ell+1}$, and applying Eq. (1348) to every layer interface, we obtain the following system of equations for the entire lower slab:

$$\mathbf{S}'_{L_2}\mathbf{c}_{L_2} = \mathbf{b}_{L_2}, \tag{1349}$$

where

$$\mathbf{S}'_{L_2} = \begin{pmatrix} \mathbf{S}'_1 & 0 & \cdots & 0 \\ 0 & \mathbf{S}'_2 & 0 & \vdots \\ \vdots & & & \vdots \\ 0 & \cdots & \mathbf{S}'_{L_2-2} & 0 \\ 0 & \cdots & 0 & \mathbf{S}'_{L_2-1} \end{pmatrix}_{8N_2\cdot L_2 \times 8N_2\cdot(L_2-1)} \tag{1350}$$

and

$$\mathbf{c}_{L_2} = \begin{pmatrix} \mathbf{c}_{L_1+1} \\ \vdots \\ \mathbf{c}_{L_1+L_2} \end{pmatrix}_{8N_2\cdot L_2\times 1}, \qquad \mathbf{b}_{L_2} = \begin{pmatrix} \mathbf{b}_{L_1+2} - \mathbf{b}_{L_1+1} \\ \vdots \\ \mathbf{b}_{L_1+L_2-2} - \mathbf{b}_{L_1+L_2-1} \end{pmatrix}_{8N_2\cdot(L_2-1)\times 1} \tag{1351}$$

The elements of the matrix $\mathbf{S}'_{L_2}$ are given by Eqs. (1321) and (1322) and of the vectors $\mathbf{b}_\ell$ and $\mathbf{b}_{\ell+1}$ by Eqs. (1323) and (1324).

D.5
Bottom Boundary of Lower Slab

At the lower boundary of the lower slab (bottom of the water in a coupled atmosphere–water system), where $\tau = \tau_L$, the Stokes vector must satisfy the following boundary condition:

$$\begin{aligned} \tilde{\mathbf{I}}(\tau_L, \mu, \phi) &= \mathbf{E}_t(\mu)\mathbf{S}_t(\tau_L) + \frac{1}{\pi}\int_0^{2\pi} d\phi' \int_0^1 d\mu' \mu' \mathbf{B}(-\mu', \phi'; \mu, \phi)\tilde{\mathbf{I}}(\tau_L, -\mu', \phi') \\ &\quad + \mathbf{B}(-\mu_0^w, \phi^w; \mu, \phi)\frac{\mu_0^w}{\pi}\mathbf{S}_b^w e^{-\tau_{L_2}/\mu_0^w} \end{aligned} \tag{1352}$$

$$L = L_1 + L_2 \qquad \text{(the total number of layers in the system).}$$

The first term on the right-hand side is due to thermal emission, the second term represents bidirectional reflectance of the diffuse incident radiation field, the third term is due to bidirectional reflectance of the attenuated direct beam reaching the bottom boundary, and

$$\mathbf{S}_b^w = \frac{\mu_0}{\mu_0^w}\mathbf{T}(-\mu_0, m_{\rm rel})\mathbf{S}_b e^{-\tau_{L_1}/\mu_0} \tag{1353}$$

is the direct beam light source just below the interface between the two slabs. The matrix $\mathbf{E}_t$ represents the unpolarized radiation emitted by the lower boundary

$$\mathbf{E}_t(\mu) = \begin{pmatrix} e_{11} & 0 & 0 & 0 \\ 0 & e_{22} & 0 & 0 \\ 0 & 0 & 0 & 0 \\ 0 & 0 & 0 & 0 \end{pmatrix}. \tag{1354}$$

The matrix $\mathbf{B}$ is the bidirectional reflectance of polarized radiation that specifies how the incident radiation field is reflected at the lower boundary.

To isolate the azimuthal dependence in Eq. (1352), we proceed by an expansion in a Fourier series in a manner similar to the treatment of the phase matrix, Stokes vector, and source term in Chapter 4. By performing an integration over ϕ', comparing Fourier modes, and introducing Gaussian quadrature, one arrives at the following equation for each Fourier component m of the cosine modes:

$$\begin{aligned} \tilde{\mathbf{I}}_c^m(\tau_L, +\mu_i^w) = {} & \delta_{0m}\mathbf{E}_t(\mu_i^w)\mathbf{S}_t(\tau_L) + \sum_{j=1}^{N_2} w_j^w \mu_j^w \big[(1+\delta_{0m})\mathbf{B}_c^m(-\mu_j^w, \mu_i^w,)\tilde{\mathbf{I}}_c^m(\tau_L, -\mu_j^w) \\ & - \mathbf{B}_s^m(-\mu_j^w, \mu_i^w)\tilde{\mathbf{I}}_s^m(\tau_L, -\mu_j^w)\big] + \frac{\mu_0^w}{\pi} e^{-\tau_{L_2}/\mu_0^w}\mathbf{B}_c^m(-\mu_0^w, \mu_i^w)\mathbf{S}_b^w \\ & i = 1, \ldots, N_2 \end{aligned} \tag{1355}$$

and similarly for the sine modes

$$\begin{aligned} \tilde{\mathbf{I}}_s^m(\tau_L, +\mu_i^w) = {} & \sum_{j=1}^{N_2} w_j^w \mu_j^w \big[\mathbf{B}_c^m(-\mu_j^w, \mu_i^w)\tilde{\mathbf{I}}_s^m(\tau_L, -\mu_j^w) \\ & + \mathbf{B}_s^m(-\mu_j^w, \mu_i^w)\tilde{\mathbf{I}}_c^m(\tau_L, -\mu_j^w)\big] + \frac{\mu_0^w}{\pi} e^{-\tau_{L_2}/\mu_0^w}\mathbf{B}_s^m(-\mu_0^w, \mu_i^w)\mathbf{S}_b^w \\ & i = 1, \ldots, N_2. \end{aligned} \tag{1356}$$

The next step is to transform these expressions into the $8N$ vector form, and for that purpose the following terms are needed ($\alpha = c, s$):

$\tilde{i}_{tc}^m(\tau_L, i)$ – thermal emission term, only valid for cosine modes ($\alpha = c$)
$\tilde{i}_{b\alpha}^m(\tau_L, i)$ – direct beam reflected at the surface
$\tilde{i}_{d\alpha}^m(\tau_L, i)$ – diffuse radiation reflected at the surface.

D.5.1
Bottom Thermal Emission Term

In the $8N$ notation, the thermal emission term in Eq. (1355) takes the form

$$\tilde{i}_{tc}^{m}(\tau_L, i) = \delta_{0m} \frac{B(T(\tau_L))}{2} \cdot \begin{cases} e_{11}(\mu_{i-4N_2}) & i = 4N_2 + 1, \ldots, 5N_2 \\ e_{22}(\mu_{i-5N_2}) & i = 5N_2 + 1, \ldots, 6N_2 \\ 0 & i = 6N_2 + 1, \ldots, 7N_2 \\ 0 & i = 7N_2 + 1, \ldots, 8N_2. \end{cases} \tag{1357}$$

By definition, there will be thermal emission only for the cosine mode $m = 0$, which explains the presence of δ_{0m} in Eq. (1357). $B(T(\tau_L))$ is the Planck function at temperature T.

D.5.2
Direct Beam Term

We assume that the bidirectional reflectance matrix has the same symmetry with respect to $(\phi' - \phi)$ as the phase matrix. Then

$$\mathbf{B}_c^m = \begin{pmatrix} B_{11c}^m & B_{12c}^m & 0 & 0 \\ B_{21c}^m & B_{22c}^m & 0 & 0 \\ 0 & 0 & B_{33c}^m & B_{34c}^m \\ 0 & 0 & B_{43c}^m & B_{44c}^m \end{pmatrix} \equiv \begin{pmatrix} \mathbf{B}_{1c}^m & \mathbf{0} \\ \mathbf{0} & \mathbf{B}_{2c}^m \end{pmatrix}$$

$$\mathbf{B}_s^m = \begin{pmatrix} 0 & 0 & B_{13s}^m & B_{14s}^m \\ 0 & 0 & B_{23s}^m & B_{24s}^m \\ B_{31s}^m & B_{32s}^m & 0 & 0 \\ B_{41s}^m & B_{42s}^m & 0 & 0 \end{pmatrix} \equiv \begin{pmatrix} \mathbf{0} & \mathbf{B}_{1s}^m \\ \mathbf{B}_{2s}^m & \mathbf{0}. \end{pmatrix}$$

Recall that we combined the $I_{\|}$ and $I_{\perp}$ cosine mode components with the U and V sine mode components, and vice versa, to form the $8N$ system. We now do the same with the the cosine [Eq. (1355)] and sine [Eq. (1356)] mode expressions for the direct beam reflected at the bottom boundary:

$$\tilde{i}_{bc}^{m}(\tau_L, i) = \frac{\mu_0^w}{\pi} e^{-(\tau_{L_1}/\mu_0 + \tau_{L_2}/\mu_0^w)}$$
$$\begin{cases} B_{11c}^m(\mu_{i-4N_2}, -\mu_0^w) T_{11} S_{\|b} + B_{12c}^m(\mu_{i-4N_2}, -\mu_0^w) T_{22} S_{\perp b} & i = 4N_2 + 1, \ldots, 5N_2 \\ B_{21c}^m(\mu_{i-5N_2}, -\mu_0^w) T_{11} S_{\|b} + B_{22c}^m(\mu_{i-5N_2}, -\mu_0^w) T_{22} S_{\perp b} & i = 5N_2 + 1, \ldots, 6N_2 \\ B_{31s}^m(\mu_{i-6N_2}, -\mu_0^w) T_{11} S_{\|b} + B_{32s}^m(\mu_{i-6N_2}, -\mu_0^w) T_{22} S_{\perp b} & i = 6N_2 + 1, \ldots, 7N_2 \\ B_{41s}^m(\mu_{i-7N_2}, -\mu_0^w) T_{11} S_{\|b} + B_{42s}^m(\mu_{i-7N_2}, -\mu_0^w) T_{22} S_{\perp b} & i = 7N_2 + 1, \ldots, 8N_2 \end{cases} \tag{1358}$$

$$\tilde{i}_{bs}^{m}(\tau_L, i) = \frac{\mu_0^w}{\pi} e^{-(\tau_{L_1}/\mu_0 + \tau_{L_2}/\mu_0^w)}$$
$$\begin{cases} B_{13s}^m(\mu_{i-4N_2}, -\mu_0^w) T_{33} S_{Ub} + B_{14s}^m(\mu_{i-4N_2}, -\mu_0^w) T_{44} S_{Vb} & i = 4N_2+1, \ldots, 5N_2 \\ B_{23s}^m(\mu_{i-5N_2}, -\mu_0^w) T_{33} S_{Ub} + B_{24s}^m(\mu_{i-5N_2}, -\mu_0^w) T_{44} S_{Vb} & i = 5N_2+1, \ldots, 6N_2 \\ B_{33c}^m(\mu_{i-6N_2}, -\mu_0^w) T_{33} S_{Ub} + B_{34c}^m(\mu_{i-6N_2}, -\mu_0^w) T_{44} S_{Vb} & i = 6N_2+1, \ldots, 7N_2 \\ B_{43c}^m(\mu_{i-7N_2}, -\mu_0^w) T_{33} S_{Ub} + B_{44c}^m(\mu_{i-7N_2}, -\mu_0^w) T_{44} S_{Vb} & i = 7N_2+1, \ldots, 8N_2. \end{cases} \tag{1359}$$

D.5.3 Bottom Diffuse Radiation

We define

$$\mathbf{R}_c^m \equiv r_{ij}^{mc} \equiv w_j^w \mu_{+j}^w \begin{pmatrix} (1+\delta_{0m})\mathbf{B}_{1c}^m & -\mathbf{B}_{1s}^m \\ \mathbf{B}_{2s}^m & \mathbf{B}_{2c}^m \end{pmatrix}_{4N_2 \times 4N_2} \Bigg|_{+\mu_i^w, -\mu_j^w;\, i,j=1,\ldots,N_2} \tag{1360}$$

and

$$\mathbf{R}_s^m \equiv r_{ij}^{ms} \equiv w_j^w \mu_{+j}^w \begin{pmatrix} \mathbf{B}_{1c}^m & \mathbf{B}_{1s}^m \\ -\mathbf{B}_{2s}^m & (1+\delta_{0m})\mathbf{B}_{2c}^m \end{pmatrix}_{4N_2 \times 4N_2} \Bigg|_{+\mu_i^w, -\mu_j^w;\, i,j=1,\ldots,N_2} . \tag{1361}$$

The combination of the $I_{\parallel}$ and $I_{\perp}$ components of the diffuse radiation in Eq. (1355) with the U and V components of Eq. (1356) gives

$$\tilde{i}_{dc}^{m}(\tau_L, i) = \sum_{n=1}^{4N_2} r_{i-4N_2,n}^{mc} \tilde{i}_c^m(\tau_L, n) \qquad i = 4N_2+1, \ldots, 8N_2. \tag{1362}$$

Substitution of the general solution in the lower slab [Eq. (587)] yields

$$\tilde{i}_{dc}^{m}(\tau_L, i) = \sum_{j=1}^{8N_2} \tilde{C}_{cjL}^m \left[\sum_{n=1}^{4N_2} r_{i-4N_2,n}^{mc} g_{cjL}^m(n) \right] e^{-k_{cjL}^m \tau_L}$$
$$+ \sum_{n=1}^{4N_2} r_{i-4N_2,n}^{mc} \left[z_{cL,2}^m(n) e^{-\tau_L/\mu_0^w} + \delta_{0m}[x_{0,L}(n) + x_{1,L}(n)\tau_L] \right] \tag{1363}$$
$$i = 4N_2+1, \ldots, 8N_2.$$

For the cosine modes, we get the intensity components in the $8N$ notation by combining the three terms, namely thermal emission [Eq. (1357)], reflection of diffuse radiation [Eq. (1363)], and reflection of direct beam radiation [Eq. (1358)]:

$$\tilde{i}_c^{m+}(\tau_L, i) \equiv \tilde{i}_{tc}^m(\tau_L, i) + \tilde{i}_{bc}^m(\tau_L, i) + \tilde{i}_{dc}^m(\tau_L, i) \tag{1364}$$
$$i = 4N_2+1, \ldots, 8N_2.$$

For the sine modes, the expression for reflection of the diffuse radiation is similar to that of the cosine mode except for the missing thermal source term

$$i_{ds}^{m}(\tau_L, i) = \sum_{j=1}^{8N_2} \tilde{C}_{sjL}^{m} \left[\sum_{n=1}^{4N_2} r_{i-4N_2,n}^{ms} g_{sjL}^{m}(n) \right] e^{-k_{sjL}^{m} \tau_L}$$
$$+ \sum_{n=1}^{4N_2} r_{i-4N_2,n}^{ms} z_{sL,2}^{m}(n) e^{-\tau_L/\mu_0^w} \qquad (1365)$$
$$i = 4N_2 + 1, \ldots, 8N_2.$$

Since there is no thermal emission for the sine modes, the lower boundary sine mode components are obtained by adding the reflection of the diffuse radiation [Eq. (1365)] and the direct beam [Eq. (1359)]

$$i_s^{m+}(\tau_L, i) \equiv i_{bs}^{m}(\tau_L, i) + i_{ds}^{m}(\tau_L, i) \qquad i = 4N_2 + 1, \ldots, 8N_2. \qquad (1366)$$

D.5.4
Bottom Boundary Condition

We have found expressions for the cosine [Eq. (1364)] as well as the sine [Eq. (1366)] modes of the upward radiation field at the lower boundary. Thus, for each mode $\alpha = c, s$, we have the following boundary condition for the upward radiation field:

$$i_{\alpha L}^{m}(\tau_L, i) = i_{\alpha}^{m+}(\tau_L, i) \qquad i = 4N_2 + 1, \ldots, 8N_2. \qquad (1367)$$

Substituting the general solution in the lower slab [Eq. (587)] into Eq. (1367), we obtain the following equation:

$$\sum_{j=1}^{8N_2} \tilde{C}_{\alpha jL}^{m} g_{\alpha jL}^{m}(i) e^{-k_{\alpha jL}^{m} \tau_L} + z_{\alpha L,2}^{m}(i) e^{-\tau_L/\mu_0^w} + \delta_{0m}\delta_{\alpha c}[x_{0,L}^{t}(i) + x_{1,L}^{t}(i)\tau_L]$$
$$= \sum_{j=1}^{8N_2} \tilde{C}_{\alpha jL}^{m} \left[\sum_{n=1}^{4N_2} r_{i-4N_2,n}^{m\alpha} g_{\alpha jL}^{m}(n) \right] e^{-k_{\alpha jL}^{m} \tau_L}$$
$$+ \sum_{n=1}^{4N_2} r_{i-4N_2,n}^{m\alpha} \left[z_{\alpha L,2}^{m}(n) e^{-\tau_L/\mu_0^w} + i_{b\alpha}^{m}(\tau_L, i) + \delta_{0m}\delta_{\alpha c}[i_{tc}^{m}(\tau_L, i) + x_{0,L}^{t}(n) + x_{1,L}^{t}(n)\tau_L] \right]$$
$$i = 4N_2 + 1, \ldots, 8N_2$$

which may be reorganized so that the unknown coefficients appear on the left-hand side:

$$\sum_{j=1}^{8N_2} \tilde{C}_{\alpha jL}^{m} \left[g_{\alpha jL}^{m}(i) - \sum_{n=1}^{4N_2} r_{i-4N_2,n}^{m\alpha} g_{\alpha jL}^{m}(n) \right] e^{-k_{\alpha jL}^{m} \tau_L}$$
$$= i_{b\alpha}^{m}(\tau_L, i) + \sum_{n=1}^{4N_2} r_{i-4N_2,n}^{m\alpha} \left[z_{\alpha L,2}^{m}(n) e^{-\tau_L/\mu_0^w} + \delta_{0m}\delta_{\alpha c}[x_{0,L}^{t}(n) + x_{1,L}^{t}(n)\tau_L] \right]$$
$$- z_{\alpha L,2}^{m}(k) e^{-\tau_L/\mu_0^w} + \delta_{0m}\delta_{\alpha c}[i_{tc}^{m}(\tau_L, i) + x_{0,L}^{t}(i) + x_{1,L}^{t}(i)\tau_L] \qquad (1368)$$
$$i = 4N_2 + 1, \ldots, 8N_2.$$

Introducing the simplifications

$$S_j^b(i) \equiv \left[g_{\alpha jL}^m(i) - \sum_{n=1}^{4N_2} r_{i-4N_2,n}^{m\alpha} g_{\alpha jL}^m(n) \right] e^{-k_{\alpha jL}^m \tau_L}$$

$$b^b(i) = \text{ the entire right-hand side of Eq. (1368)}$$

we may rewrite Eq. (1368) in matrix form as

$$\mathbf{S}_b \mathbf{c}_L = \mathbf{b}_b, \tag{1369}$$

where

$$\mathbf{S}_b = \begin{pmatrix} S_1^b(4N_2+1) & \cdots & S_{8N_2}^b(4N_2+1) \\ S_1^b(4N_2+2) & \cdots & S_{8N_2}^b(4N_2+2) \\ \vdots & & \vdots \\ S_1^b(8N_2) & \cdots & S_{8N_2}^b(8N_2) \end{pmatrix}_{4N_2 \times 8N_2} \tag{1370}$$

$$\mathbf{c}_L = \begin{pmatrix} C_1^L \\ \vdots \\ C_{8N_2}^L \end{pmatrix}_{8N_2 \times 1}, \qquad \mathbf{b}_b = \begin{pmatrix} b^b(4N_2+1) \\ \vdots \\ b^b(8N_2) \end{pmatrix}_{4N_2 \times 1}. \tag{1371}$$

References

1. Kepler, J. (2000) *1604. ad vitellionem paralipomena, quibus astronomiae pars optica traditur*, Apud Claudium Marnium & Hæredes Ioannis Aubrii, Frankfurt. See Donahue.
2. Bouguer, P. (1729) *Essai d'Optique, sur la gradation de la lumiere*, Claude Jombert.
3. Lambert, J.H. (1760) *Photometria, sive de Mensura et Gradibus Luminis, Colorum et Umbrae*, Augsburg.
4. Mishchenko, M.I. (2013) 125 years of radiative transfer: enduring triumphs and persisting misconceptions, in *Radiation Processes in the Atmosphere and Ocean (IRS2012): Proceedings of the International Radiation Symposium (IRC/IAMAS)*, vol. 1531, AIP Publishing, pp. 11–18.
5. Schuster, A. (1905) Radiation through a foggy atmosphere. *Astrophys. J.*, **21**, 1–22.
6. Lommel, E. (1887) Die photometrie der diffusen zurückwerfung. *Sitzber. Acad. Wissensch. München*, **17**, 95–124.
7. Chwolson, O.D. (1889) Grundzüge einer mathematischen theorie der inneren diffusion des lichtes. *Bull. l'Acad. Imperiale Sci. St. Petersbourg*, **33**, 221–256.
8. Planck, M. (1914) *The Theory of Heat Radiation*, P. Blakiston's Son & Co, Philadelphia, PA.
9. Milne, E. (1930) Thermodynamics of the stars. *Handbuch der Astrophys.*, **3**, 65–255.
10. Hopf, E. (1934) *Mathematical Problems of Radiative Equilibrium*, Cambridge University Press.
11. Chandrasekhar, S. (1960) *Radiative Transfer*, Dover Publications, New York.
12. Marshak, A. and Davis, A. (2005) *3D Radiative Transfer in Cloudy Atmospheres*, Springer-Verlag.
13. Lenoble, J. (1993) *Atmospheric Radiative Transfer*, A. Deepak Publishing, Hampton, VA.
14. Yanovitskij, E.G. (1997) *Light Scattering in Inhomogeneous Atmospheres*, Springer-Verlag.
15. Sobolev, V.V. (1975) *Light Scattering in Planetary Atmospheres*, Pergamon Press.
16. Hovenier, J.W., der Mee, C.D.V., and Domke, H. (2004) *Transfer of Polarized Light in Planetary Atmospheres*, Kluwer Academic Publishers, Dordrecht.
17. Goody, R.M. and Yung, Y.L. (1989) *Atmospheric Radiation: Theoretical Basis*, Oxford University Press.
18. Thomas, G. and Stamnes, K. (1999) *Radiative Transfer in the Atmosphere and Ocean*, Cambridge University Press, New York.
19. Goody, R.M. and Yung, Y.L. (2002) *An Introduction to Atmospheric Radiation*, Academic Press.
20. Zdunkowski, W., Trautmann, T., and Bott, A. (2007) *Radiation in the Atmosphere: A Course in Theoretical Meteorology*, Cambridge University Press.

Radiative Transfer in Coupled Environmental Systems: An Introduction to Forward and Inverse Modeling, First Edition. Knut Stamnes and Jakob J. Stamnes.

21. Wendisch, M. and Yang, P. (2012) *Theory of Atmospheric Radiative Transfer*, John Wiley & Sons, Ltd.
22. Stokes, G. (1852) On the composition and resolution of streams of polarized light from different sources. *Trans. Camb. Philos. Soc.*, **9**, 399–416.
23. Chandrasekhar, S. (1946) On the radiative equilibrium of a stellar atmosphere, XI. *Astrophys. J.*, **104**, 110–132.
24. Mandel, L. and Wolf, E. (1995) *Optical Coherence and Quantum Optics*, Cambridge University Press.
25. Mishchenko, M.I. (2008) Multiple scattering, radiative transfer, and weak localization in discrete random media: unified microphysical approach. *Rev. Geophys.*, **46** (2), RG2003.
26. Mobley, C.D., Gentili, B., Gordon, H.R., Jin, Z., Kattawar, G.W., Morel, A., Reinersman, P., Stamnes, K., and Stavn, R.H. (1993) Comparison of numerical models for computing underwater light fields. *Appl. Opt.*, **32**, 7484–7504.
27. Gjerstad, K.I., Stamnes, J.J., Hamre, B., Lotsberg, J.K., Yan, B., and Stamnes, K. (2003) Monte Carlo and discrete-ordinate simulations of irradiances in the coupled atmosphere-ocean system. *Appl. Opt.*, **42**, 2609–2622.
28. Hamre, B., Winther, J.G., Gerland, S., Stamnes, J.J., and Stamnes, K. (2004) Modeled and measured optical transmittance of snow covered first-year sea ice in Kongsfjorden, Svalbard. *J. Geophys. Res.*, **109**, doi: 10.1029/2003JC001926.
29. Jiang, S., Stamnes, K., Li, W., and Hamre, B. (2005) Enhanced solar irradiance across the atmosphere-sea ice interface: a quantitative numerical study. *Appl. Opt.*, **46**, 2613–2625.
30. Hestenes, K., Nielsen, K.P., Zhao, L., Stamnes, J.J., and Stamnes, K. (2007) Monte Carlo and discrete-ordinate simulations of spectral radiances in the coupled air-tissue system. *Appl. Opt.*, **46**, 2333–2350.
31. Chen, B., Stamnes, K., and Stamnes, J.J. (2001) Validity of diffusion approximation in bio-optical imaging. *Appl. Opt.*, **40**, 6356–6366.
32. Swanson, D.L., Laman, S.D., Biryulina, M., Nielsen, K.P., Ryzhikov, G., Stamnes, J.J., Hamre, B., Zhao, L., Sommersten, E., Castellana, F.S., and Stamnes, K. (2010) Optical transfer diagnosis of pigmented lesions. *Dermatol. Surg.*, **36** (12), 1979–1986.
33. Nielsen, K., Zhao, L., Ryzhikov, G.A., Biryulina, M.S., Sommersten, E.R., Stamnes, J.J., Stamnes, K., and Moan, J. (2008) Retrieval of the physiological state of human skin from UV-Vis reflectance spectra: a feasibility study. *J. Photochem. Photobiol., B*, **93**, 23–31.
34. Spurr, R., Stamnes, K., Eide, H., Li, W., Zhang, K., and Stamnes, J.J. (2007) Simultaneous retrieval of aerosol and ocean properties: a classic inverse modeling approach: I. Analytic Jacobians from the linearized CAO-DISORT model. *J. Quant. Spectrosc. Radiat. Transfer*, **104**, 428–449.
35. Gordon, H.R. (1997) Atmospheric correction of ocean color imagery in the earth observation system era. *J. Geophys. Res.*, **102**, 17 081–17 106.
36. Mishchenko, M.I. and Travis, L.D. (1997) Satellite retrieval of aerosol properties over the ocean using polarization as well as intensity of reflected sunlight. *J. Geophys. Res.*, **102**, 16 989–17 013.
37. Rodgers, C.D. (2000) *Inverse Methods for Atmospheric Sounding: Theory and Practice*, World Scientific, London.
38. Li, W., Stamnes, K., Spurr, R., and Stamnes, J.J. (2008) Simultaneous retrieval of aerosols and ocean properties: a classic inverse modeling approach. II. SeaWiFS case study for the Santa Barbara channel. *Int. J. Remote Sens.*, **29**, 5689–5698.
39. Jin, Z. and Stamnes, K. (1994) Radiative transfer in non-uniformly refracting media: atmosphere-ocean system. *Appl. Opt.*, **33**, 431–442.
40. Kattawar, G. and Adams, C. (1989) Stokes vector calculations of the submarine light field in an atmosphere-ocean with scattering according to the Rayleigh phase matrix: effect of interface refractive index on radiance and polarization. *Limnol. Oceanogr.*, **34**, 1453–1472.

41. Mishchenko, M.I., Lacis, A., and Travis, L.D. (1994) Errors due to the neglect of polarization in radiance calculations for Rayleigh-scattering atmospheres. *J. Quant. Spectrosc. Radiat. Transfer*, **51**, 491–510.
42. Lacis, A.A., Chowdhary, J., Mishchenko, M.I., and Cairns, B. (1998) Modeling errors in diffuse-sky radiation: vector vs. scalar treatment. *Geophys. Res. Lett.*, **25**, 135–138.
43. Zhai, P.W., Hu, Y., Chowdhary, J., Trepte, C.R., Lucker, P.L., and Josset, D.B. (2010) A vector radiative transfer model for coupled atmosphere and ocean systems with a rough interface. *J. Quant. Spectrosc. Radiat. Transfer*, **111** (7), 1025–1040.
44. Chowdhary, J. (1999) Multiple scattering of polarized light in atmosphere-ocean systems: application to sensitivity analyses of aerosol polarimetry. PhD thesis. Columbia University.
45. Chowdhary, J., Cairns, B., and Travis, L.D. (2002) Case studies of aerosol retrievals over the ocean from multiangle, multispectral photopolarimetric remote sensing data. *J. Atmos. Sci.*, **59** (3), 383–397.
46. Chowdhary, J., Cairns, B., Mishchenko, M.I., Hobbs, P.V., Cota, G.F., Redemann, J., Rutledge, K., Holben, B.N., and Russell, E. (2005) Retrieval of aerosol scattering and absorption properties from photopolarimetric observations over the ocean during the clams experiment. *J. Atmos. Sci.*, **62** (4), 1093–1117.
47. Chowdhary, J., Cairns, B., Waquet, F., Knobelspiesse, K., Ottaviani, M., Redemann, J., Travis, L., and Mishchenko, M. (2012) Sensitivity of multiangle, multispectral polarimetric remote sensing over open oceans to water-leaving radiance: analyses of RSP data acquired during the MILAGRO campaign. *Remote Sens. Environ.*, **118**, 284–308.
48. Chami, M., Santer, R., and Dilligeard, E. (2001) Radiative transfer model for the computation of radiance and polarization in an ocean-atmosphere system: polarization properties of suspended matter for remote sensing. *Appl. Opt.*, **40** (15), 2398–2416.
49. Min, Q. and Duan, M. (2004) A successive order of scattering model for solving vector radiative transfer in the atmosphere. *J. Quant. Spectrosc. Radiat. Transfer*, **87** (3), 243–259.
50. Fischer, J. and Grassl, H. (1984) Radiative transfer in an atmosphere-ocean system: an azimuthally dependent matrix-operator approach. *Appl. Opt.*, **23** (7), 1032–1039.
51. Ota, Y., Higurashi, A., Nakajima, T., and Yokota, T. (2010) Matrix formulations of radiative transfer including the polarization effect in a coupled atmosphere-ocean system. *J. Quant. Spectrosc. Radiat. Transfer*, **111**, 878–894.
52. Lotsberg, J. and Stamnes, J. (2010) Impact of particulate oceanic composition on the radiance and polarization of underwater and backscattered light. *Opt. Express*, **18** (10), 10 432–10 445.
53. Dahlback, A. and Stamnes, K. (1991) A new spherical model for computing the radiation field available for photolysis and heating at twilight. *Planet. Space Sci.*, **39**, 671–683.
54. Spurr, R.J.D. (2008) LIDORT and VLIDORT: linearized pseudo-spherical scalar and vector discrete ordinate radiative transfer models for use in remote sensing retrieval algorithms, in *Light Scattering Reviews*, vol. 3 (ed. A. Kokhanovsky), Springer-Verlag, Berlin.
55. Rozanov, V., Rozanov, A., Kokhanovsky, A., and Burrows, J. (2014) Radiative transfer through terrestrial atmosphere and ocean: software package SCIATRAN. *J. Quant. Spectrosc. Radiat. Transfer*, **133**, 13–71.
56. Chandrasekhar, S. (1958) On the diffuse reflection of a pencil of radiation by a plane-parallel atmosphere. *Proc. Natl. Acad. Sci. U.S.A.*, **44** (9), 933.
57. Shiina, T., Yoshida, K., Ito, M., and Okamura, Y. (2007) Long-range propagation of annular beam for lidar application. *Opt. Commun.*, **279** (1), 159–167.

58. Habel, R., Christensen, P.H., and Jarosz, W. (2013) in *Photon Beam Diffusion: A Hybrid Monte Carlo Method for Subsurface Scattering*, Eurographics Symposium on Rendering, vol. 32. The Eurographics Association and Blackwell Publishing Ltd.
59. Barichello, L. and Siewert, C. (2000) The searchlight problem for radiative transfer in a finite slab. *J. Comput. Phys.*, **157** (2), 707–726.
60. Kim, A.D. and Moscoso, M. (2003) Radiative transfer computations for optical beams. *J. Comput. Phys.*, **185** (1), 50–60.
61. Mitra, K. and Churnside, J.H. (1999) Transient radiative transfer equation applied to oceanographic lidar. *Appl. Opt.*, **38** (6), 889–895.
62. Stamnes, K., Lie-Svendsen, O., and Rees, M.H. (1991) The linear Boltzmann equation in slab geometry: development and verification of a reliable and efficient solution. *Planet. Space Sci.*, **39**, 1435–1463.
63. De Beek, R., Vountas, M., Rozanov, V., Richter, A., and Burrows, J. (2001) The ring effect in the cloudy atmosphere. *Geophys. Res. Lett.*, **28** (4), 721–724.
64. Landgraf, J., Hasekamp, O., Van Deelen, R., and Aben, I. (2004) Rotational Raman scattering of polarized light in the earth atmosphere: a vector radiative transfer model using the radiative transfer perturbation theory approach. *J. Quant. Spectrosc. Radiat. Transfer*, **87** (3), 399–433.
65. Spurr, R., de Haan, J., van Oss, R., and Vasilkov, A. (2008) Discrete-ordinate radiative transfer in a stratified medium with first-order rotational raman scattering. *J. Quant. Spectrosc. Radiat. Transfer*, **109** (3), 404–425.
66. Ge, Y., Gordon, H., and Voss, K. (1993) Simulation of inelastic scattering contributions to the irradiance field in the oceanic variation in fraunhofer line depths. *Appl. Opt.*, **32**, 4028–4036.
67. Kattawar, G. and Xu, X. (1992) Filling-in of fraunhofer lines in the ocean by raman scattering. *Appl. Opt.*, **31**, 1055–1065.
68. Bohren, C.F. and Huffman, D.R. (1998) *Absorption and Scattering of Light by Small Particles*, John Wiley & Sons, Inc., New York.
69. Mobley, C.D. (1994) *Light and Water*, Academic Press, New York.
70. Rayleigh, L. (1920) A re-examination of the light scattered by gases in respect of polarization. I. Experiments on the common gases. *Proc. R. Soc. London, Ser. A*, **98**, 435–450.
71. Rayleigh, L. (1920) A re-examination of the light scattered by gases in respect of polarization. II. Experiments on helium and argon. *Proc. R. Soc. London, Ser. A*, **98**, 57–64.
72. Morel, A. (1974) Optical properties of pure water and pure seawater, in *Optical Aspects of Oceanography* (eds N.G. Jerlov and E.S. Nielsen), Academic Press, San Diego, CA, pp. 1–24.
73. Morel, A. and Gentili, B. (1991) Diffuse reflectance of oceanic waters: its dependence on sun angle as influenced by the molecular scattering contribution. *Appl. Opt.*, **30**, 4427–4437.
74. Rayleigh, L. (1871) On the light from the sky, its polarization and colour. *Philos. Mag.*, **41**, 107–120, 274–279, 447–454.
75. Henyey, L.C. and Greenstein, J.L. (1941) Diffuse radiation in the galaxy. *Astrophys. J.*, **93**, 70–83.
76. Diehl, P. and Haardt, H. (1980) Measurement of the spectral attenuation to support biological research in a "plankton tube" experiment. *Oceanol. Acta*, **3**, 89–96.
77. McCave, I.N. (1983) Particulate size spectra, behavior, and origin of nephloid layers over the nova scotia continental rise. *J. Geophys. Res.*, **88**, 7647–7660.
78. Fournier, G.R. and Forand, J.L. (1994) Analytic phase function for ocean water. SPIE Proceedings on Ocean Optics XII, vol. 2558, pp. 194–201.
79. Mobley, C.P., Sundman, L.K., and Boss, E. (2002) Phase function effects on oceanic light fields. *Appl. Opt.*, **41**, 1035–1050.
80. Petzold, T.L. (1972) Volume Scattering Functions for Selected Ocean Waters. SIO Ref. 72–78. Scripps Institution of Oceanography Visibility Laboratory, pp. 72–78.

81. Hovenier, J.W. and van der Mee, C.V.M. (1983) Fundamental relationships relevant to the transfer of polarized light in a scattering atmosphere. *Astron. Astrophys.*, **128**, 1–16.
82. De Haan, J., Bosma, P., and Hovenier, J. (1987) The adding method for multiple scattering calculations of polarized light. *Astron. Astrophys.*, **183**, 371–391.
83. Siewert, C. (2000) A discrete-ordinates solution for radiative-transfer models that include polarization effects. *J. Quant. Spectrosc. Radiat. Transfer*, **64** (3), 227–254.
84. Cohen, D., Stamnes, S., Tanikawa, T., Sommersten, E.R., Stamnes, J.J., Lotsberg, J.K., and Stamnes, K. (2013) Comparison of discrete ordinate and Monte Carlo simulations of polarized radiative transfer in two coupled slabs with different refractive indices. *Opt. Express*, **21**, 9592–9614.
85. Siewert, C.E. (1981) On the equation of transfer relevant to the scattering of polarized light. *Astrophys. J.*, **245**, 1080–1086.
86. Siewert, C.E. (1982) On the phase matrix basic to the scattering of polarized light. *Astron. Astrophys.*, **109**, 195–200.
87. Sommersten, E.R., Lotsberg, J.K., Stamnes, K., and Stamnes, J.J. (2010) Discrete ordinate and Monte Carlo simulations for polarized radiative transfer in a coupled system consisting of two media with different refractive indices. *J. Quant. Spectrosc. Radiat. Transfer*, **111**, 616–633.
88. Mishchenko, M.I. (1991) Light scattering by randomly oriented rotationally symmetric particles. *J. Opt. Soc. Am. A*, **8**, 871–882.
89. Wiscombe, W.J. (1977) The delta-m method: rapid yet accurate radiative flux calculations for strongly asymmetric phase functions. *J. Atmos. Sci.*, **34**, 1408–1422.
90. Spurr, R.J.D. (2006) Vlidort: a linearized pseudo-spherical vector discrete ordinate radiative transfer code for forward model and retrieval studies in multilayer multiple scattering media. *J. Quant. Spectrosc. Radiat. Transfer*, **102**, 316–342.
91. Hu, Y.X., Wielicki, B., Lin, B., Gibson, G., Tsay, S.C., Stamnes, K., and Wong, T. (2000) Delta-fit: a fast and accurate treatment of particle scattering phase functions with weighted singular-value decomposition least squares fitting. *J. Quant. Spectrosc. Radiat. Transfer*, **65**, 681–690.
92. Zhai, P.W., Hu, Y., Trepte, C.R., and Lucker, P.L. (2009) A vector radiative transfer model for coupled atmosphere and ocean systems based on successive order of scattering method. *Opt. Express*, **17** (4), 2057–2079.
93. Hansen, J.E. and Hovenier, J. (1974) Interpretation of the polarization of venus. *J. Atmos. Sci.*, **31** (4), 1137–1160.
94. Deirmendjian, D. (1969) *Electromagnetic Scattering on Spherical Polydispersions*, Elsevier.
95. Hansen, J.E. and Travis, L.D. (1974) Light scattering in planetary atmospheres. *Space Sci. Rev.*, **16**, 527–610.
96. Mie, G. (1908) Beiträge zur optik trüber medien, speziell kolloidaler metallösungen. *Ann. Phys.*, **330** (3), 377–445.
97. Grenfell, T.S., Warren, S.G., and Mullen, P.C. (1994) Optical properties of deep glacial ice at the south pole. *J. Geophys. Res.*, **99**, 18 669–18 684.
98. Born, M. and Wolf, E. (1980) *Principles of Optics*, Cambridge University Press.
99. Anderson, G.P., Clough, S.A., Kneizys, F.X., Chetwynd, J.H., and Shettle, E.P. (1986) *AFGL Atmospheric Constituent Profiles (0-120 km), AFGL-TR-86-0110 (OPI)*, Hanscom AFB, MA 01736.
100. Berk, A., Anderson, G.P., Acharya, P.K., Bernstein, L.S., Muratov, L., Lee, J., Fox, M., Adler-Golden, S.M., Chetwynd, J.H., Hoke, M.L. *et al.* (2005) Modtran 5: a reformulated atmospheric band model with auxiliary species and practical multiple scattering options: update, in *Defense and Security*, International Society for Optics and Photonics, pp. 662–667.

101. Lacis, A.A. and Oinas, V. (1991) A description of the correlated k distribution method for modeling nongray gaseous absorption, thermal emission, and multiple scattering in vertically inhomogeneous atmospheres. *J. Geophys. Res. Atmos. (1984–2012)*, **96** (D5), 9027–9063.
102. Wiscombe, W. and Evans, J. (1977) Exponential-sum fitting of radiative transmission functions. *J. Comput. Phys.*, **24** (4), 416–444.
103. West, R., Crisp, D., and Chen, L. (1990) Mapping transformations for broadband atmospheric radiation calculations. *J. Quant. Spectrosc. Radiat. Transfer*, **43** (3), 191–199.
104. Meadows, V.S. and Crisp, D. (1996) Ground-based near-infrared observations of the venus nightside: the thermal structure and water abundance near the surface. *J. Geophys. Res. Planets (1991–2012)*, **101** (E2), 4595–4622.
105. Natraj, V., Jiang, X., Shia, R.L., Huang, X., Margolis, J.S., and Yung, Y.L. (2005) Application of principal component analysis to high spectral resolution radiative transfer: a case study of the o2a band. *J. Quant. Spectrosc. Radiat. Transfer*, **95** (4), 539–556.
106. Ahmad, Z., Franz, B.A., McClain, C.R., Kwiatkowska, E.J., Werdell, J., Shettle, E.P., and Holben, B.N. (2010) New aerosol models for the retrieval of aerosol optical thickness and normalized water-leaving radiances from the SeaWiFS and modis sensors over coastal regions and open oceans. *Appl. Opt.*, **49**, 5545–5560.
107. Hess, M., Koepke, P., and Schult, I. (1998) Optical properties of aerosols and clouds: the software package OPAC. *Bull. Am. Meteorol. Soc.*, **79**, 831–844.
108. Gordon, H.R. and Wang, M. (1994) Retrieval of water-leaving radiance and aerosol optical thickness over the oceans with SeaWiFS: a preliminary algorithm. *Appl. Opt.*, **33**, 443–445.
109. Davies, C.N. (1974) Size distribution of atmospheric particles. *J. Aerosol Sci.*, **5**, 293–300.
110. Holben, B.N., Eck, T.F., Slutsker, I., Tanre, D., Buis, J.P., Setzer, A., Vermote, E., Reagan, J.A., Kaufman, Y., Nakajima, T., Lavenu, F., Jankowiak, I., and Smirnov, A. (1998) Aeronet – a federated instrument network and data archive for aerosol characterization. *Remote Sens. Environ.*, **66**, 1–16.
111. Holben, B.N., Tanre, D., Smirnov, A., Eck, T.F., Slutsker, I., Abuhassan, N., Newcomb, W.W., Schafer, J., Chatenet, B., Lavenue, F., Kaufman, Y.J., Castle, J.V., Setzer, A., Markham, B., Clark, D., Frouin, R., Halthore, R., Karnieli, A., O'Neill, N.T., Pietras, C., Pinker, R.T., Voss, K., and Zibordi, G. (2001) An emerging ground-based aerosol climatology: aerosol optical depth from aeronet. *J. Geophys. Res.*, **106**, 12 067–12 097.
112. Hänel, G. (1976) The properties of atmospheric aerosol particles as functions of the relative humidity at thermodynamic equilibrium with the surrounding moist air, in *Advances in Geophysics*, Vol. 19 (eds H.E. Landsberg and J.V. Miehem), Academic Press.
113. Shettle, E.P. and Fenn, R.W. (1979) Models for the Aerosols of the Lower Atmosphere and the Effects of Humidity Variations on their Optical Properties. AFGL-TR-79-0214. Air Force Geophysics Laboratory, Hanscomb AFB, Mass.
114. Yan, B., Stamnes, K., Li, W., Chen, B., Stamnes, J.J., and Tsay, S.C. (2002) Pitfalls in atmospheric correction of ocean color imagery: how should aerosol optical properties be computed? *Appl. Opt.*, **41**, 412–423.
115. Segelstein, D.J. (1981) The complex refractive index of water. MS thesis. Department of Physics. University of Missouri-Kansas City.
116. Smith, R.C. and Baker, K.S. (1981) Optical properties of the clearest natural waters (200–800nm). *Appl. Opt.*, **36**, 177–184.
117. Sogandares, F.M. and Fry, E.S. (1997) Absorption spectrum (340–640 nm) off pure water. I. Photothermal measurements. *Appl. Opt.*, **36**, 8699–8709.

118. Pope, R.M. and Fry, E.S. (1997) Absorption spectrum (380–700 nm) of pure water, II integrating cavity measurements. *Appl. Opt.*, **36**, 8710–8723.
119. Kou, L., Labrie, D., and Chylek, P. (1993) Refractive indices of water and ice in the 0.65 μm to 2.5 μm spectral range. *Appl. Opt.*, **32**, 3531–3540.
120. Hu, Y.X. and Stamnes, K. (1993) An accurate parameterization of the radiative properties of water clouds suitable for use in climate models. *J. Climate*, **6**, 728–742.
121. Warren, S.G. and Brandt, R.E. (2008) Optical constants of ice from the ultraviolet to the microwave: a revised compilation. *J. Geophys. Res.*, **113**, D14 220, doi: 10.1029/2007JD009 744.
122. Jin, Z., Stamnes, K., Weeks, W.F., and Tsay, S.C. (1994) The effect of sea ice on the solar energy budget in the atmosphere-sea ice-ocean system: a model study. *J. Geophys. Res.*, **99**, 25 281–25 294.
123. Stamnes, K., Hamre, B., Stamnes, J.J., Ryzhikov, G., Birylina, M., Mahoney, R., Hauss, B., and Sei, A. (2011) Modeling of radiation transport in coupled atmosphere-snow-ice-ocean systems. *J. Quant. Spectrosc. Radiat. Transfer*, **112**, 714–726.
124. Ackermann, M. *et al.* (2006) Optical properties of deep glacial ice at the south pole. *J. Geophys. Res.*, **111**, D13 203, doi: 10.1029/2005JD006 687.
125. Fialho, P., Hansen, A.D.A., and Honrath, R.E. (2005) Absorption coefficients by aerosols in remote areas: a new approach to decouple dust and black carbon absorption coefficients using seven-wavelength aethalometer data. *J. Aerosol Sci.*, **36**, 267–282.
126. Twardowski, M.S., Boss, E., Sullivan, J.M., and Donaghay, P.L. (2004) Modeling the spectral shape of absorption by chromophoric dissolved organic matter. *Mar. Chem.*, **89**, 69–88.
127. Uusikivi, J. *et al.* (2010) Contribution of mycosporine-like amino acids and colored dissolved and particulate matter to sea ice optical properties and ultraviolet attenuation. *Limnol. Oceanogr.*, **55**, 703–713.
128. Ruddick, K., Bouchra, N. *et al.* (2013) *Coastcoulor Round Robin – Final Report, ftp://ccrropen@ftp.coestcolour.org/RoundRobin/CCRRreport.pdf with annex: ftp://ccrropen@ftp.coastcolour.org/RoundRobin/CCRR_report_OCSMART.pdf.*
129. Babin, M., Stramski, A.D., Ferrari, G.M., Claustre, H., Bricaud, A., Obelesky, G., and Hoepffner, N. (2003) Variations in the light absorption coefficients of phytoplankton, nonalgal particles and dissolved organic matter in coastal waters around europe. *J. Geophys. Res.*, **108**, 3211, doi: 10.1029/2001JC000 882.
130. Babin, M., Morel, A., Fournier-Sicre, V., Fell, F., and Stramski, D. (2003) Light scattering properties of marine particles in coastal and open ocean waters as related to the particle mass concentration. *Limnol. Oceanogr.*, **28**, 843–859.
131. Bricaud, A., Morel, A., Babin, M., Allali, K., and Claustre, H. (1998) Mie-scattering calculation. *J. Geophys. Res.*, **103**, 31 033–31 044.
132. Loisel, H. and Morel, A. (1998) Light scattering and chlorophyll concentration in case 1 waters: a re-examination. *Limnol. Oceanogr.*, **43**, 847–857.
133. Morel, A., Antoine, D., and Gentili, B. (2002) Bidirectional reflectance of oceanic waters: accounting for raman emission and varying particle scattering phase function. *Appl. Opt.*, **41**, 6289–6306.
134. Stramski, D., Bricaud, A., and Morel, A. (2001) Modeling the inherent optical properties of the ocean based on the detailed composition of the planktonic community. *Appl. Opt.*, **40**, 2929–2945.
135. Kostadinov, T.S., Siegel, D.A., and Maritorena, S. (2009) Retrieval of the particle size distribution from satellite ocean color observations. *J. Geophys. Res.*, **114**, C09 015, doi: 10.1029/2009JC005 303.
136. Kostadinov, T.S., Siegel, D.A., and Maritorena, S. (2010) Global variability of phytoplankton functional types from space: assessment via the particle

size distribution. *Biogeosciences*, **7**, 3239–3257.

137. Kostadinov, T.S., Siegel, D.A., Maritorena, S., and Guillocheau, N. (2012) Optical assessment of particle size and composition in the santa barbara channel. *Appl. Opt.*, **114**, 3171–3189.
138. Sieburth, J.M., Smetacek, V., and Lenz, J. (1978) Pelagic ecosystem structure: heterotrophic compartments of the plankton and their relationship to plankton size fractions. *Limnol. Oceanogr.*, **23**, 1256–1263.
139. Ahn, Y.H., Bricaud, A., and Morel, A. (1992) Light backscattering efficiency and related properties of some phytoplankters. *Deep Sea Res. Part A*, **39** (11), 1835–1855.
140. Garver, S.A. and Siegel, D. (1997) Inherent optical property inversion of ocean color spectra and its biogeochemical interpretation 1. Time series from the Sargasso Sea. *J. Geophys. Res.*, **102**, 18 607–18 625.
141. Maritorena, S., Siegel, D.A., and Peterson, A.R. (2002) Optimization of a semi-analytical ocean color model for global-scale applications. *Appl. Opt.*, **41**, 2705–2714.
142. Zhang, X., Twardowski, M., and Lewis, M. (2011) Retrieving composition and sizes of oceanic particle subpopulations from the volume scattering function. *Appl. Opt.*, **50**, 1240–1259.
143. Zhang, X., Gray, D.J., Huot, Y., You, Y., and Bi, L. (2012) Comparisons of optically derived particle size distributions: scattering over the full angular range versus diffraction at near forward angles. *Appl. Opt.*, **51**, 5085–5099.
144. Reif, F. (1965) *Fundamentals of Statistical and Thermal Physics*, McGraw-Hill, Boston, MA.
145. Cox, C. and Munk, W. (1954) Measurement of the roughness of the sea surface from photographs of the sun's glitter. *J. Opt. Soc. Am.*, **44**, 838–850.
146. Hapke, B. (2012) *Theory of Reflectance and Emittance Spectroscopy*, Cambridge University Press.
147. Rahman, H., Pinty, B., and Verstraete, M.M. (1993) Coupled surface-atmosphere reflectance (CSAR) model: 2. Semiempirical surface model usable with noaa advanced very high resolution radiometer data. *J. Geophys. Res. Atmos. (1984–2012)*, **98** (D11), 20 791–20 801.
148. Govaerts, Y., Wagner, S., Lattanzio, A., and Watts, P. (2010) Joint retrieval of surface reflectance and aerosol optical depth from msg/seviri observations with an optimal estimation approach: 1. Theory. *J. Geophys. Res. Atmos. (1984–2012)*, **115** (D2), D02203.
149. Dubovik, O., Herman, M., Holdak, A., Lapyonok, T., Tanré, D., Deuzé, J., Ducos, F., Sinyuk, A., and Lopatin, A. (2011) Statistically optimized inversion algorithm for enhanced retrieval of aerosol properties from spectral multi-angle polarimetric satellite observations. *Meas. Tech*, **4**, 975–1018.
150. Ross, J. (1981) *The Radiation Regime and Architecture of Plant Stands*, Vol. 3, Springer-Verlag.
151. Li, X. and Strahler, A.H. (1992) Geometric-optical bidirectional reflectance modeling of the discrete crown vegetation canopy: effect of crown shape and mutual shadowing. *IEEE Trans. Geosci. Remote Sens.*, **30** (2), 276–292.
152. Wanner, W., Li, X., and Strahler, A. (1995) On the derivation of kernels for kernel-driven models of bidirectional reflectance. *J. Geophys. Res. Atmos. (1984–2012)*, **100** (D10), 21 077–21 089.
153. Maignan, F., Breon, F.M., and Lacaze, R. (2004) Bidirectional reflectance of earth targets: evaluation of analytical models using a large set of spaceborne measurements with emphasis on the hot spot. *Remote Sens. Environ.*, **90** (2), 210–220.
154. Justice, C.O., Vermote, E., Townshend, J.R., Defries, R., Roy, D.P., Hall, D.K., Salomonson, V.V., Privette, J.L., Riggs, G., Strahler, A. *et al.* (1998) The moderate resolution imaging spectroradiometer (MODIS): land remote sensing for global change research. *IEEE Trans. Geosci. Remote Sens.*, **36** (4), 1228–1249.
155. Litvinov, P., Hasekamp, O., Cairns, B., and Mishchenko, M. (2010) Reflection

models for soil and vegetation surfaces from multiple-viewing angle photopolarimetric measurements. *J. Quant. Spectrosc. Radiat. Transfer*, **111** (4), 529–539.

156. Litvinov, P., Hasekamp, O., and Cairns, B. (2011) Models for surface reflection of radiance and polarized radiance: comparison with airborne multi-angle photopolarimetric measurements and implications for modeling top-of-atmosphere measurements. *Remote Sens. Environ.*, **115** (2), 781–792.

157. Cairns, B., Waquet, F., Knobelspiesse, K., Chowdhary, J., and Deuzé, J.L. (2009) Polarimetric remote sensing of aerosols over land surfaces, in *Satellite Aerosol Remote Sensing Over Land*, Springer-Verlag, pp. 295–325.

158. Maignan, F., Bréon, F.M., Fédèle, E., and Bouvier, M. (2009) Polarized reflectances of natural surfaces: spaceborne measurements and analytical modeling. *Remote Sens. Environ.*, **113** (12), 2642–2650.

159. Mishchenko, M.I. (2002) Vector radiative transfer equation fro arbitrarily shaped and arbitrarily oriented particles: a microphysical derivation from statistical electromagnetics. *Appl. Opt.*, **41**, 7114–7134.

160. Mishchenko, M.I. (2003) Microphysical approach to polarized radiative transfer: extension to the case of an external observation point. *Appl. Opt.*, **42** (24), 4963–4967.

161. Stamnes, K., Tsay, S.C., Wiscombe, W.J., and Jayaweera, K. (1988) Numerically stable algorithm for discrete-ordinate-method radiative transfer in multiple scattering and emitting layered media. *Appl. Opt.*, **27**, 2502–2509.

162. Stamnes, K., Tsay, S.C., Wiscombe, W.J., and Laszlo, I. (2000) *DISORT, A General-Purpose Fortran Program for Discrete-Ordinate-Method Radiative Transfer in Scattering and Emitting Layered Media: Documentation of Methodology*. NASA report, *ftp://climate.gsfc.nasa.gov/pub/wiscombe/Multiple Scatt/* (accessed 23 April 2015).

163. Yan, B. and Stamnes, K. (2003) Fast yet accurate computation of the complete radiance distribution in the coupled atmosphere-ocean system. *J. Quant. Spectrosc. Radiat. Transfer*, **76**, 207–223.

164. Kattawar, G. and Adams, C. (1990) Errors in radiance calculations induced by using scalar rather than Stokes vector theory in a realistic atmosphere-ocean system, in *Ocean Optics X* (ed. R.W. Spinrad), Proceedings of the Society of Photo-Optical Instrumentation Engineers.

165. Greenwald, T., Bennartz, R., O'Dell, C., and Heidinger, A. (2005) Fast computation of microwave radiances for data assimilation using the successive order of scattering method. *J. Appl. Meteorol.*, **44** (6), 960–966.

166. Heidinger, A.K., O'Dell, C., Bennartz, R., and Greenwald, T. (2006) The successive-order-of-interaction radiative transfer model. Part I: model development. *J. Appl. Meteorol. Climatol.*, **45** (10), 1388–1402.

167. O'Dell, C.W., Heidinger, A.K., Greenwald, T., Bauer, P., and Bennartz, R. (2006) The successive-order-of-interaction radiative transfer model. Part II: model performance and applications. *J. Appl. Meteorol. Climatol.*, **45** (10), 1403–1413.

168. Kotchenova, S.Y., Vermote, E.F., Matarrese, R., Klemm, F.J. Jr. *et al.* (2006) Validation of a vector version of the 6s radiative transfer code for atmospheric correction of satellite data. Part I: path radiance. *Appl. Opt.*, **45** (26), 6762–6774.

169. Kotchenova, S.Y. and Vermote, E.F. (2007) Validation of a vector version of the 6s radiative transfer code for atmospheric correction of satellite data. Part II. Homogeneous Lambertian and anisotropic surfaces. *Appl. Opt.*, **46** (20), 4455–4464.

170. Lenoble, J., Herman, M., Deuzé, J., Lafrance, B., Santer, R., and Tanré, D. (2007) A successive order of scattering code for solving the vector equation of transfer in the earth's atmosphere with aerosols. *J. Quant. Spectrosc. Radiat. Transfer*, **107** (3), 479–507.

171. Liou, K.N. (1973) A numerical experiment on Chandrasekhar's discrete-ordinate method for radiative transfer. *J. Atmos. Sci.*, **30**, 1303–1326.

172. Stamnes, K. and Swanson, R.A. (1981) A new look at the discrete ordinate method for radiative transfer calculations in anisotropically scattering atmospheres. *J. Atmos. Sci.*, **38**, 387–399.

173. Stamnes, K. and Dale, H. (1981) A new look at the discrete ordinate method for radiative transfer calculations in anisotropically scattering atmospheres. II. Intensity computations. *J. Atmos. Sci.*, **38**, 2696–2706.

174. Stamnes, K. and Conklin, P. (1984) A new multi-layer discrete ordinate approach to radiative transfer in vertically inhomogeneous atmospheres. *J. Quant. Spectrosc. Radiat. Transfer*, **31**, 273–282.

175. Lin, Z., Stamnes, S., Jin, Z., Laszlo, I., Tsay, S.C., Wiscombe, W.J., and Stamnes, K. (2015) Improved discrete ordinate solutions in the presence of an anisotropically reflecting lower boundary: upgrades of the DISORT computational tool. *J. Quant. Spectrosc. Radiat. Transfer*, **157**, 119–134.

176. Weng, F. (1992) A multi-layer discrete-ordinate method for vector radiative transfer in a vertically-inhomogenous, emitting and scattering atmosphere - I. Theory. *J. Quant. Spectrosc. Radiat. Transfer*, **47**, 19–33.

177. Weng, F. (1992) A multi-layer discrete-ordinate method for vector radiative transfer in a vertically-inhomogenous, emitting and scattering atmosphere - II. Applications. *J. Quant. Spectrosc. Radiat. Transfer*, **47**, 35–42.

178. Schulz, F.M., Stamnes, K., and Weng, F. (1999) VDISORT: an improved and generalized discrete ordinate radiative transfer model for polarized (vector) radiative transfer computations. *J. Quant. Spectrosc. Radiat. Transfer*, **61**, 105–122.

179. Schulz, F.M. and Stamnes, K. (2000) Angular distribution of the stokes vector in a plane parallel, vertically inhomogeneous medium in the vector discrete ordinate radiative transfer (VDISORT) model. *J. Quant. Spectrosc. Radiat. Transfer*, **65**, 609–620.

180. Sommersten, E.R. (2005) CAOVDISORT: a discrete ordinate polarized radiative transfer model for a coupled system with two media with different indices of refraction. MS thesis. Department of Physics. University of Bergen, Norway.

181. Kokhanovsky, A.A., Cornet, C., Duan, M., Emde, C., Katsev, I.L., C-Labonnote, L., Min, Q., Nakajima, T., Ota, Y., Prikhach, A. *et al.* (2010) Benchmark results in vector radiative transfer. *J. Quant. Spectrosc. Radiat. Transfer*, **111**, 1931–1946.

182. Weng, F. (1992) A multi-layer discrete ordinate method for vector radiative transfer in a vertically-inhomogeneous, emitting and scattering atmosphere – I. Theory. *J. Quant. Spectrosc. Radiat. Transfer*, **47**, 19–33.

183. Stokes, G. (1862) On the intensity of the light reflected from or transmitted through a pile of plates. *Proc. R. Soc. London*, **11**, 545–556.

184. Twomey, S., Jacobowitz, H., and Howell, J. (1966) Matrix methods for multiple scattering problems. *J. Atmos. Sci.*, **23**, 289–296.

185. van de Hulst, H.C. and Grossman, K. (1968) Multiple light scattering in planetary atmospheres, in *The Atmospheres of Venus and Mars* (eds J.C. Brandt and M.V. McElroy), Gordon and Breach.

186. Twomey, S., Jacobowitz, H., and Howell, J. (1969) Discrete space theory of radiative transfer and its application to problems in planetary atmospheres. *J. Atmos. Sci.*, **26**, 963–972.

187. Wiscombe, W.J. (1983) Atmospheric radiation: 1975–1983. *Rev. Geophys.*, **21**, 997–1021.

188. van de Hulst, H.C. and Grossman, K. (1965) *Radiative Transfer on Discrete Spaces*, Pergamon Press, New York.

189. Stephens, G.L. (1980) Radiative transfer on a linear lattice: application to anisotropic ice crystals clouds. *J. Atmos. Sci.*, **37**, 2095–2104.

190. Plass, G.N., Kattawar, G.W., and Catchings, F.E. (1973) Matrix operator theory of radiative transfer, I,

Rayleigh scattering. *Appl. Opt.*, **12**, 314–329.
191. Tanaka, M. and Nakajima, T. (1986) Matrix formulations for the transfer of solar radiation in a plane-parallel scattering atmosphere. *J. Quant. Spectrosc. Radiat. Transfer*, **28**, 13–21.
192. Lenoble, J. (ed.) (1985) *Radiative Transfer in Scattering and Absorbing Atmospheres: Standard Computational Procedures*, A. Deepak Publishing, Hampton, VA.
193. Waterman, P.C. (1981) Matrix-exponential description of radiative transfer. *J. Opt. Soc. Am.*, **71**, 410–422.
194. Stamnes, K. (1986) The theory of multiple scattering in plane parallel atmospheres. *Rev. Geophys.*, **71**, 299–310.
195. Plass, G.N. and Kattawar, G.W. (1968) Monte Carlo calculations of light scattering from clouds. *Appl. Opt.*, 7 (3), 415–419.
196. Kattawar, G.W. and Plass, G.N. (1968) Radiance and polarization of multiple scattered light from haze and clouds. *Appl. Opt.*, 7 (8), 1519–1527.
197. Marchuk, G., Mikhailov, G., Nazaraliev, M., Dacbinjan, R., Kargin, B., and Elepov, B. (1981) Monte Carlo methods in atmospheric optics. *Appl. Opt.*, **20** (6), 958–1074.
198. O'Brien, D.M. (1992) Accelerated quasi Monte Carlo integration of the radiative transfer equation. *J. Quant. Spectrosc. Radiat. Transfer*, **48**, 41–59.
199. Roberti, L. and Kummerow, C. (1999) Monte Carlo calculations of polarized microwave radiation emerging from cloud structures. *J. Geophys. Res. Atmos. (1984–2012)*, **104** (D2), 2093–2104.
200. Tynes, H.H., Kattawar, G.W., Zege, E.P., Katsev, I.L., Prikhach, A.S., and Chaikovskaya, L.I. (2001) Monte Carlo and multicomponent approximation methods for vector radiative transfer by use of effective mueller matrix calculations. *Appl. Opt.*, **40**, 400–412.
201. Ishimoto, H. and Masuda, K. (2001) A Monte Carlo approach for the calculation of polarized light. *J. Quant. Spectrosc. Radiat. Transfer*, **72**, 467–483.
202. Sommersten, E.R., Lotsberg, J.K., Stamnes, K., and Stamnes, J.J. (2010) Discrete ordinate and Monte Carlo simulations for polarized radiative transfer in a coupled system consisting of two media with different refractive indices. *J. Quant. Spectrosc. Radiat. Transfer*, **111** (4), 616–633.
203. Kattawar, G.W. (1991) *Selected Papers on Multiple Scattering in Plane Parallel Atmospheres and Oceans: Methods*, Society of Photo Optical.
204. Spurr, R.J.D., Kurosu, T.P., and Chance, K.V. (2001) A linearized discrete ordinate radiative transfer model for atmospheric remote sensing retrieval. *J. Quant. Spectrosc. Radiat. Transfer*, **68**, 689–735.
205. Broomhead, D.S. and Lowe, D. (1988) Radial Basis Functions, Multi-Variable Functional Interpolation and Adaptive Networks. Technical report 4148, RSRE.
206. Orr, M.J.L. (1996) Introduction to radial basis function networks. *Neural Comput.*, **7**, 606–623.
207. Bors, A.G. (2001) Introduction of the radial basis function (RBF) networks. Online Symposium for Electronics Engineers, DSP Algorithms: Multimedia, vol. 1, pp. 1–7.
208. Stamnes, K., Tsay, S.C., and Nakajima, T. (1988) Computation of eigenvalues and eigenvectors for discrete ordinate and matrix operator method radiative transfer. *J. Quant. Spectrosc. Radiat. Transfer*, **39**, 415–419.
209. Strang, G. (2005) *Linear Algebra and its Applications*, Thomson–Brooks/Cole, Belmont, CA.
210. Sivia, D.S. and Skilling, J. (2006) *Data Analysis: A Bayesian Tutorial*, Oxford University Press, Oxford and New York.
211. Aster, R.C., Borchers, B., and Thurber, C.H. (2005) *Parameter Estimation and Inverse Problems*, Elsevier Press, Amsterdam and New York.
212. Jeffreys, H. (1998) *The Theory of Probability*, Oxford University Press.

213. Cox, R.T. (1946) Probability, frequency and reasonable expectation. *Am. J. Phys.*, **14** (1), 1–13.
214. Keynes, J. (1921) *A Treatise on Probability*, Vol. 8 of Collected Writings (1973 ed.). Cambridge University Press (April 27, 1990).
215. Jaynes, E. (1978) Where do we stand on maximum entropy, in *The Maximum Entropy Formalism* (eds R.D. Levine and M. Tribus), MIT Press, pp. 15–118.
216. Jaynes, E.T. (1957) Information theory and statistical mechanics. II. *Phys. Rev.*, **108** (2), 171.
217. Jaynes, E.T. (1957) Information theory and statistical mechanics. *Phys. Rev.*, **106** (4), 620.
218. Skilling, J. (1988) The axioms of maximum entropy, in *Maximum-Entropy and Bayesian Methods in Science and Engineering*, Springer-Verlag, pp. 173–187.
219. Strang, G. (2003) *Introduction to linear algebra*, Cambridge Publication.
220. Twomey, S. (2002) *Introduction to the Mathematics of Inversion in Remote Sensing and Indirect Measurements*, Courier Dover Publications.
221. Tikhonov, A. and Arsenin, V.Y. (1977) *Methods for Solving Ill-Posed Problems*, John Wiley & Sons, Inc.
222. Press, W.H., Teukolksy, S., Vetterling, W.T., and Flannery, B. (1995) *Numerical Recipes: The Art of Scientific Computing*, Cambridge University Press, New York.
223. Kylling, A., Stamnes, K., and Tsay, S.C. (1995) A reliable and efficient two-stream algorithm for spherical radiative transfer: documentation of accuracy in realistic layered media. *J. Atmos. Chem.*, **21** (2), 115–150.
224. Anonymous, A. (1994) Environmental Health Criteria 160 – Ultraviolet Radiation. Technical Report, EHC 160, 1994, 2nd edition. World Health Organization (WHO).
225. Dahlback, A., Eide, H.A., Høiskar, B.A.K., Olsen, R.O., Schmidlin, F.J., Tsay, S.C., and Stamnes, K. (2005) Comparison of data for ozone amounts and ultraviolet doses obtained from simultaneous measurements with various standard ultraviolet instruments. *Opt. Eng.*, **44**, 041 010.
226. Dahlback, A., Gelsor, N., Stamnes, J.J., and Gjessing, Y. (2007) UV measurements in the 3000–5000 m altitude region in tibet. *J. Geophys. Res.*, **112**, 308.
227. Kazantzidis, A., Bais, A.F., Zempila, M.M., Meleti, C., Eleftheratos, K., and Zerefos, C.S. (2009) Evaluation of ozone column measurements over greece with NILU-UV multi-channel radiometers. *Int. Remote Sens.*, **30**, 4273–4281.
228. Norsang, G., Kocbach, L., Tsoja, W., Stamnes, J., Dahlbach, A., and Nema, P. (2009) Ground-based measurements and modelling of solar UV-B radiation in Lhasa, Tibet. *J. Atmos. Environ.*, **43**, 1498–1502.
229. Norsang, G., Kocbach, L., Stamnes, J.J., Tsoja, W., and Pingcuo, N. (2011) Spatial distribution and temporal variation of solar UV radiation over the Tibetan Plateau. *Appl. Phys. Res.*, **3**, 37–46.
230. Norsang, G., Chen, Y., Pingcuo, N., Dahlback, A., Frette, O., Kjeldstad, B., Hamre, B., Stamnes, K., and Stamnes, J.J. (2014) Comparison of ground-based measurements of solar UV radiation at four sites on the Tibetan Plateau. *Appl. Opt.*, **53**, 736–747.
231. Dahlback, A. (1996) Measurements of biologically effective UV doses, total ozone abundances, and cloud effects with multichannel, moderate bandwidth filter instrument. *Appl. Opt.*, **35**, 6514–6521.
232. Stamnes, K., Slusser, J., and Bowen, M. (1991) Measurements of biologically effective UV doses, total ozone abundances, and cloud effects with multichannel, moderate bandwidth filter instrument. *Appl. Opt.*, **30**, 4418–4426.
233. Fan, L., Li, W., Dahlback, A., Stamnes, J.J., Englehardt, S., Stamnes, S., and Stamnes, K. (2014) Comparisons of three NILU-UV instruments deployed at the same site in the new york area. *Appl. Opt.*, **53**, 3598–3606.
234. Høiskar, B.A.K., Haugen, R., Danielsen, T., Kylling, A., Edvardsen, K.,

Dahlback, A., Johnsen, B., Blumthaler, M., and Schreder, J. (2003) Multichannel moderate-bandwidth filter instrument for measurement of the ozone-column amount, cloud transmittance, and ultraviolet dose rates. *Appl. Opt.*, **42**, 3472–3479.

235. Fan, L., Li, W., Dahlback, A., Stamnes, J.J., Stamnes, S., and Stamnes, K. (2014) A new neural-network based method to infer total ozone column amounts and cloud effects from multi-channel, moderate bandwidth filter instruments. *Opt. Express*, **22**, 19 595–19 609.

236. Fre, J. and Dozier, J. (1986) The image processing workbench – portable software for remote sensing instruction and research. Proceedings of the 1986 International Geoscience and Remote Sensing Symposium, ESA SP-254, European Space Agency, Paris, pp. 271–276.

237. Baldridge, A.M., Hook, S.J., Grove, C.I., and Rivera, G. (2009) The aster spectral library version 2.0. *Remote Sens. Environ.*, **113**, 711–715.

238. Grenfell, T.C. (2005) A radiative transfer model for sea ice with vertical structure variations. *J. Geophys. Res.*, **96**, 16 991–17 001.

239. Brandt, R.E., Warren, S.G., Worby, A.P., and Grenfell, T.C. (2005) Surface albedo of the antarctic sea-ice zone. *J. Climate*, **18**, 3606–3622.

240. Briegleb, B.P. and Light, B. (2007) A Delta-Eddington Multiple Scattering Parameterization of Solar Radiation in the Sea Ice Component of the Community Climate System Model. NCAR Technical Note NCAR/TN-472+STR. National Center for Atmospheric Research.

241. Aoki, T., Aoki, T., Fukabori, M., Hachikubo, A., Tachibana, Y., and Nishio, F. (2000) Effects of snow physical parameters on spectral albedo and bidirectional reflectance of snow surface. *J. Geophys. Res.*, **105**, 10 219–10 236.

242. Tanikawa, T., Aoki, T., Hori, M., Hachikubo, A., and Aniya, M. (2002) Snow bidirectional reflectance model using non-spherical snow particles and its validation with field measurements. *EARSel eProc.*, **5**, 137–145.

243. Kokhanovsky, A.A., Aoki, T., Hachikubo, A., Hori, M., and Zege, E.P. (2005) Effects of snow physical parameters on spectral albedo and bidirectional reflectance of snow surface. *IEEE Trans. Geosci. Remote Sens.*, **43**, 1529–1535.

244. Fily, M., Bourdelles, B., Dedieu, J.P., and Sergent, C. (1997) Comparison of in situ and Landsat Thematic Mapper derived snow grain characteristics in the Alps. *Remote Sens. Environ.*, **17**, 86–92.

245. Bourdelles, B. and Fily, M. (1993) Snow grain-size determination from Landsat imagery over Terre Adelie, Antarctica. *Ann. Glaciol.*, **17**, 86–92.

246. Nolin, A.W. and Dozier, J. (1993) Estimating snow grain size using AVIRIS data. *Remote Sens. Environ.*, **44**, 231–238.

247. Painter, T.H., Roberts, D.A., Green, R.O., and Dozier, J. (1998) The effect of grain size on spectral mixture analysis of snow-covered area from AVIRIS data. *Remote Sens. Environ.*, **65**, 320–332.

248. Green, R.O., Dozier, J., Roberts, D.A., and Painter, T.H. (2002) Spectral snow reflectance models for grain size and liquid water fraction in melting snow for the solar reflected spectrum. *Ann. Glaciol.*, **34**, 71–73.

249. Hall, D.K., Riggs, G.A., and Salomonson, V.V. (1995) Development of methods for mapping global snow cover using moderate resolution imaging spectroradiometer data. *Remote Sens. Environ.*, **54**, 127–140.

250. Warren, S.G. (1982) Optical properties of snow. *Rev. Geophys. Space Phys.*, **20**, 67–89.

251. Stamnes, K., Li, W., Eide, H., Aoki, T., Hori, M., and Storvold, R. (2007) ADEOS-II/GLI snow/ice products – Part I. Scientific basis. *Remote Sens. Environ.*, **111**, 258–273.

252. Aoki, T., Hori, M., Motoyohi, H., Tanikawa, T., Hachikubo, A., Sugiura, K., Yasunari, T., Storvold, R., Eide, H.A., Stamnes, K., Li, W., Nieke, J., Nakajima, Y., and Takahashi, F. (2007)

ADEOS-II/GLI snow/ice products – Part II. Validation results. *Remote Sens. Environ.*, **111**, 320–336.
253. Hori, M., Aoki, T., Stamnes, K., and Li, W. (2007) ADEOS-II/GLI snow/ice products – Part III. Retrieved results. *Remote Sens. Environ.*, **111**, 274–319.
254. Nakajima, T.Y., Nakajima, T., Nakajima, M., Fukushima, H., Kuji, M., and Uchiyama, A. (1988) Optimization of the advanced earth observing satellite II global imager channels by use of radiative transfer calculations. *Appl. Opt.*, **37**, 3149–3163.
255. Nieke, J., Aoki, T., Tanikawa, T., Motoyoshi, H., and Hori, M. (2004) A satellite cross-calibration experiment. *IEEE Geosci. Remote Sens. Lett.*, **1**, 215–219.
256. Stamnes, K., Li, W., Eide, H., and Stamnes, J.J. (2005) Challenges in atmospheric correction of satellite imagery. *Opt. Eng.*, **44**, 041 003–1–041 003–9.
257. Llewellyn-Jones, D.T., Minnett, P.J., Saunders, R.W., and Zavody, A.M. (1984) Satellite multichannel infrared measurements of sea surface temperature of the Northeast Atlantic Ocean using AVHRR/2. *Q. J. R. Meteorol. Soc.*, **110**, 613–631.
258. Barton, I.J. (1990) Transmission model and ground-truth investigation of satellite-derived sea surface temperatures. *J. Clim. Appl. Meteorol.*, **24**, 508–516.
259. Minnett, P.J. (1990) The regional optimization of infrared measurements of sea surface temperature from space. *J. Geophys. Res.*, **95**, 13 497–13 510.
260. Key, J. and Haefliger, M. (1992) Arctic ice surface temperature retrieval from AVHRR thermal channels. *J. Geophys. Res.*, **97**, 5885–5893.
261. Key, J.R., Collins, J.B., Fowler, C., and Stone, R.S. (1997) High-latitude surface temperature estimates from thermal satellite data. *Remote Sens. Environ.*, **67**, 302–309.
262. Wan, Z. and Dozier, J. (1996) A generalized split-window algorithm for retrieving land-surface temperature from space. *IEEE Trans. Geosci. Remote Sens.*, **34**, 892–905.
263. Hall, D.K., Key, J.R., Casey, K.A., Riggs, G.A., and Cavalieri, D.J. (2004) Sea ice surface temperature products from MODIS. *IEEE Trans. Geosci. Remote Sens.*, **42**, 1076–1087.
264. Brun, E., David, P., Sudul, M., and Brugnot, G. (1992) A numerical model to simulate snow-cover stratigraphy for operational avalanche forecasting. *J. Glaciol.*, **38**, 13–22.
265. Gordon, H., Brown, O., and Jacobs, M. (1975) Computed relationships between the inherent and apparent optical properties of a flat homogeneous ocean. *Appl. Opt.*, **14**, 417–427.
266. Gordon, H. (1987) A bio-optical model describing the distribution of irradiance at the sea surface resulting from a point source embedded in the ocean. *Appl. Opt.*, **26**, 4133–4148.
267. Gordon, H. (1989) Can the Lambert-Beer law be applied to the diffuse attenuation coefficient of ocean water? *Limnol. Oceanogr.*, **34**, 1389–1409.
268. Gordon, H. (1989) Dependence of the diffuse reflectance of natural waters on the sun angle. *Limnol. Oceanogr.*, **34**, 1484–1489.
269. Gordon, H. (1992) Diffuse reflectance of the ocean: influence of nonuniform phytoplankton pigment profile. *Appl. Opt.*, **31**, 2116–2129.
270. Adams, C. and Kattawar, G. (1993) Effect of volume scattering function on the errors induced when polarization is neglected in radiance calculations in an atmosphere-ocean system. *Appl. Opt.*, **32**, 4610–4617.
271. Kattawar, G. and Adams, C. (1992) Errors induced when polarization is neglected, in *Optics for the Air-Sea Interface*, vol. 1749 (ed. L. Epstep), Proceedings of the Society of Photo-Optical Instrumentation Engineers, pp. 2–22.
272. Plass, G. and Kattawar, G. (1969) Radiative transfer in an atmosphere-ocean system. *Appl. Opt.*, **8**, 455–466.
273. Plass, G. and Kattawar, G. (1972) Monte-Carlo calculations of radiative transfer in the earth's atmosphere ocean system: I. Flux in the atmosphere and ocean. *Phys. Oceanogr.*, **2**, 139–145.

274. Gordon, H. and Brown, O. (1973) Irradiance reflectivity of a flat ocean as a function of its optical properties. *Appl. Opt.*, **12**, 1549–1551.
275. Plass, G.N., Kattawar, G.W., and Guinn, J.A. Jr. (1975) Radiative transfer in the earth's atmosphere and ocean: influence of ocean waves. *Appl. Opt.*, **14**, 1924–1936.
276. Morel, A. and Gentili, B. (1993) Diffuse reflectance of oceanic waters. II. Bidirectional aspect. *Appl. Opt.*, **32**, 2803–2804.
277. Kirk, J. (1981) Monte Carlo Procedure for Simulating the Penetration of Light into Natural Waters. Technical Report, Division of Plant Industry Technical Paper 36, Commonwealth Scientific and Industrial Research Organization, Canberra, Australia.
278. Blattner, W., Horak, H., Collins, D., and Wells, M. (1974) Monte Carlo studies of the sky radiation at twilight. *Appl. Opt.*, **13**, 534–547.
279. Stavn, R. and Weidemann, A. (1988) Optical modeling of clear ocean light fields: Raman scattering effects. *Appl. Opt.*, **27**, 4002–4011.
280. Stavn, R. and Weidemann, A. (1992) Raman scattering in ocean optics: quantitative assessment of internal radiant emission. *Appl. Opt.*, **31**, 1294–1303.
281. Mobley, C. (1989) A numerical model for the computation of radiance distributions in natural waters with wind-roughened surfaces. *Limnol. Oceanogr.*, **34**, 1473–1483.
282. Beckmann, P. and Spizzichino, A. (1963) *The Scattering of Electromagnetic Waves from Rough Surfaces*, MacMillan, New York.
283. Jin, Z., Charlock, T.P., Rutledge, K., Stamnes, K., and Wang, Y. (2006) An analytical solution of radiative transfer in the coupled atmosphere-ocean system with rough surface. *Appl. Opt.*, **45**, 7443–7455.
284. Zhai, P.W., Hu, Y., Chowdhary, J., Trepte, C.R., Lucker, P.L., and Josset, D.B. (2007) A vector radiative transfer model for coupled atmosphere and ocean systems with a rough interface. *J. Quant. Spectrosc. Radiat. Transfer*, **111**, 1025–1040.
285. Ottaviani, M., Spurr, R., Stamnes, K., Li, W., Su, W., and Wiscombe, W.J. (2008) Improving the description of sunglint for accurate prediction of remotely-sensed radiances. *J. Quant. Spectrosc. Radiat. Transfer*, **109**, 2364–2375.
286. Wang, M. and Bailey, S. (2001) Correction of sun glint contamination on the SeaWiFS ocean and atmosphere products. *Appl. Opt.*, **40**, 4790–4798.
287. Carder, K., Chen, F., Cannizarro, J., Campbell, J., and Mitchell, B. (2004) Performance of MODIS semi analytical ocean colour algorithm for chlorophyll-a. *Adv. Space Res.*, **33**, 1152–1159.
288. Siegel, D.A., Wang, M., Maritorena, S., and Robinson, W. (2000) Atmospheric correction of satellite ocean colour imagery: the black pixel assumption. *Appl. Opt.*, **39**, 3582–3591.
289. Stamnes, K., Li, W., Yan, B., Eide, H., Barnard, A., Pegau, W.S., and Stamnes, J.J. (2003) Accurate and self-consistent ocean colour algorithm: simultaneous retrieval of aerosol optical properties and chlorophyll concentrations. *Appl. Opt.*, **42**, 939–951.
290. Tsay, S.C. and Stephens, G.L. (1990) *A Physical/Optical Model for Atmospheric Aerosols with Application to Visibility Problems*, Department of Atmospheric Sciences, Colorado State University, Fort Collins, CO.
291. Zhang, K., Li, W., Stamnes, K., Eide, H., Spurr, R., and Tsay, S.C. (2007) Assessment of the modis algorithm for the retrieval of aerosol parameters over the ocean. *Appl. Opt.*, **46**, 1525–1534.
292. Stamnes, K., Li, W., Fan, Y., Tanikawa, T., Hamre, B., and Stamnes, J.J. (2010) A fast yet accurate algorithm for retrieval of aerosol and marine parameters in coastal waters. Ocean Optics XX, Anchorage, Alaska.
293. Stamnes, S., Fan, Y., Li, W., Chen, N., Stamnes, J.J., Tanikawa, T., and Stamnes, K. (2014) Use of polarization to retrieve aerosol parameters in coupled atmosphere-water systems. *Optics Express*, submitted.

294. Moncet, J.L., Uymin, G., Lipton, A.E., and Snell, H.E. (2008) Infrared radiance modeling by optimal spectral sampling. *J. Atmos. Sci.*, **65** (12), 3917–3934.

295. Clough, S.A., Iacono, M.J., and Moncet, J.L. (1992) Line-by-line calculations of atmospheric fluxes and cooling rates: application to water vapor. *J. Geophys. Res. Atmos. (1984–2012)*, **97** (D14), 15 761–15 785.

296. Clough, S.A., Shephard, M.W., Mlawer, E.J., Delamere, J.S., Iacono, M.J., Cady-Pereira, K., Boukabara, S., and Brown, P.D. (2005) Atmospheric radiative transfer modeling: a summary of the AER codes, short communication. *J. Quant. Spectrosc. Radiat. Transfer*, **91**, 233–244.

297. Bass, F.G. and Fuks, I.M. (1979) *Wave Scattering from Statistically Rough Surfaces*, Pergamon Press, Oxford.

298. Goodman, J.W. (1968) *Introduction to Fourier Optics*, McGraw-Hill, New York.

299. Sancer, M.I. (1969) Shadow-corrected electromagnetic scattering from a randomly rough surface. *IEEE Trans. Antennas Propag.*, **17**, 577–585.

300. Tsang, L., Kong, J.A., and Shin, R.T. (1985) *Theory of Microwave Remote Sensing*, John Wiley & Sons, Inc., New York.

301. Smith, B.G. (1967) Geometrical shadowing of a random rough surface. *IEEE Trans. Antennas Propag.*, **15**, 668–671.

Index

Radiative Transfer in Coupled Environmental Systems: An Introduction to Forward and Inverse Modeling,
First Edition. Knut Stamnes and Jakob J. Stamnes.